Synthese Library

Studies in Epistemology, Logic, Methodology, and Philosophy of Science

Volume 384

The aim of *Synthese Library* is to provide a forum for the best current work in the methodology and philosophy of science and in epistemology. A wide variety of different approaches have traditionally been represented in the Library, and every effort is made to maintain this variety, not for its own sake, but because we believe that there are many fruitful and illuminating approaches to the philosophy of science and related disciplines.

Special attention is paid to methodological studies which illustrate the interplay of empirical and philosophical viewpoints and to contributions to the formal (logical, set-theoretical, mathematical, information-theoretical, decision-theoretical, etc.) methodology of empirical sciences. Likewise, the applications of logical methods to epistemology as well as philosophically and methodologically relevant studies in logic are strongly encouraged. The emphasis on logic will be tempered by interest in the psychological, historical, and sociological aspects of science.

Besides monographs *Synthese Library* publishes thematically unified anthologies and edited volumes with a well-defined topical focus inside the aim and scope of the book series. The contributions in the volumes are expected to be focused and structurally organized in accordance with the central theme(s), and should be tied together by an extensive editorial introduction or set of introductions if the volume is divided into parts. An extensive bibliography and index are mandatory.

More information about this series at http://www.springer.com/series/6607

Stefania Centrone
Editor

Essays on Husserl's Logic and Philosophy of Mathematics

Springer

Editor
Stefania Centrone
Institut für Philosophie
Carl von Ossietzky Universität Oldenburg
Oldenburg, Germany

Synthese Library
ISBN 978-94-024-1130-0 ISBN 978-94-024-1132-4 (eBook)
DOI 10.1007/978-94-024-1132-4

Library of Congress Control Number: 2017949262

Printed on acid-free paper

This Springer imprint is published by Springer Nature
The registered company is Springer Science+Business Media B.V.
The registered company address is: Van Godewijckstraat 30, 3311 GX Dordrecht, The Netherlands

To the memory of Richard Tieszen

Papier | Fördert gute Waldnutzung
FSC® C083411

Kolibri 360 GmbH
Ferdinand-Jühlke-Straße 7
99095 Erfurt, Deutschland
produktsicherheit@kolibri360.de

Preface

In my *Logic and Philosophy of Mathematics in the Early Husserl* (Synthese Library, 2010), I set out "to restore the level of the real discussion between Husserl and his important early interlocutors, some of whom made definitive contributions to the development of formal logic as an autonomous discipline in the last two centuries." To this end, I considered Husserl's relationship to the algebraists of logic, in particular George Boole and Ernst Schröder, as well as many connections between his work on logic and philosophy of mathematics and the work of Georg Cantor, Bernard Bolzano, Gottlob Frege, and David Hilbert. However, a sense of dissatisfaction remained upon my completion of this project. Clarifying the relationships between Husserl and his most important interlocutors in the fields of logic and philosophy of mathematics was an impossible task for a single scholar; moreover, Husserl's reception and influence remained almost unexplored. During the years, also thanks to the annual meetings of the Husserl Circle, which were becoming more and more international due to the indefatigable activity of its secretary, Burt Hopkins, I had the opportunity to meet and attend lectures of many scholars and colleagues who seemed to look at Husserl's work from a perspective similar to my own, that is, who seemed to believe that a logico-historical and a phenomenological reading glass do not necessarily exclude each other when it is a matter of making sense of Husserl's writings. Meanwhile, the appearance of *Phenomenology and Mathematics* (Phenomenologica, 2010) by Mirja Hartimo, which "gathers the contributions of the main scholars of the field," most of whom also are contributors to the present volume, seemed to me an important further step in this direction, and, moreover, it gave me the courage to address the question of the relationships of Husserl's key ideas in logic and philosophy of mathematics to other figures. This volume presents contributions from the analytical and phenomenological perspective. It does not aim to give a comprehensive final judgment on Husserl's work but rather to open new and perhaps novel interpretative perspectives, from which to look at that work.

Very special thanks go to all authors in this volume. The encouragement at a decisive moment and the friendly advice I received from Professor Otávio Bueno, the editor in chief of *Synthese*, and from Christi Jongepier-Lue, the assistant publishing editor of Springer Science and Business Media, were truly invaluable. The criticisms and suggestions made by an anonymous referee for Synthese Library who read the penultimate draft of this book were very helpful. To her/him also is my warmest thanks.

Oldenburg, Germany Stefania Centrone

Contents

Contributors

Christian Beyer is professor of theoretical philosophy at the Georg-August-Universität Göttingen since 2007. He studied philosophy, linguistics, and the history of science in Bielefeld and Hamburg, where he also took his PhD (1999). He was a visiting scholar at Stanford University (1994–1995) and served as temporary lecturer in philosophy at the University of Sheffield (2000) and as *Wissenschaftlicher Mitarbeiter* at the University of Erfurt (2000–2005), where he earned his *habilitation*. In 2005, he was awarded a Heisenberg Scholarship of the German Research Foundation, and in 2016, he was fellow at the Centre for Advanced Study of the Norwegian Academy of Science and Letters. He is the author of *Von Bolzano zu Husserl* (Kluwer 1996), *Intentionalität und Referenz* (Mentis 2000), and *Subjektivität, Intersubjektivität, Personalität* (de Gruyter 2006) and co-editor of *Philosophical Knowledge* (Rodopi 2007), *Edmund Husserl 1859–2009* (de Gruyter 2011), and *Content, Context, and Conversation* (Springer 2014).

Ettore Casari is emeritus professor of logic at the Scuola Normale Superiore in Pisa and a member, among others, of the Accademia Nazionale dei Lincei and the Academia Europaea. He is also member of the International Scientific Board of *Rivista di Filosofia*. Ettore Casari is author of many volumes, the most recent one being *Bolzano's Logical System* (Oxford University Press 2016), and numerous articles in the fields of logic, history of logic, and philosophy of mathematics.

Stefania Centrone is currently *privat Dozent* at the University of Hamburg and research assistant at the Carl von Ossietzky University of Oldenburg. In 2012, she was awarded a DFG-Eigene Stelle for the project "Bolzanos und Husserls Weiterentwicklung von Leibnizens Ideen zur Mathesis Universalis" at the CvO University of Oldenburg. In 2016 she was deputy professor of theoretical philosophy at the Georg-August-Universität Göttingen. She is author, among others, of the volumes *Logic and Philosophy of Mathematics in the Early Husserl* (Springer 2010) and *Studien zu Bolzano* (Academia Verlag 2015).

Jairo José Da Silva is full professor of mathematics (retired) at São Paulo State University (UNESP) and a member of the Center of Logic, Epistemology and History of Science (CLE) of the University of Campinas (UNICAMP), both in Brazil. He is also member of the Center of Philosophy, Lisbon University, Portugal, and chief editor of *Cadernos de História e Filosofia da Ciência/Cahiers of History and Philosophy of Science* (CLE-UNICAMP). He authored the books *Sobre o Predicativismo em Hermann Weyl* (CLE 1989) and *Filosofias da Matemática* (Edunesp 2007), both in Portuguese, and co-authored, with Claire Ortiz Hill, the book *The Road Not Taken: On Husserl's Philosophy of Logic and Mathematics* (College Publications 2013). He has also written a number of essays on the philosophy of logic and mathematics, some of them focusing on the thought of Husserl, Poincaré, Weyl, and Wittgenstein.

Mirja Hartimo is currently a postdoctoral researcher working at the Centre for Advanced Study at the Norwegian Academy of Science and Letters, employed by the Norwegian University of Life Sciences. She has graduated with a doctorate in philosophy from Boston University in 2005. Since then, she has worked at the University of Helsinki, University of Jyväskylä, and University of Tampere in Finland until 2015 when she obtained her present position in Norway.

Claire Ortiz Hill holds a BA and a master's degree from the University of California, Riverside, and a maîtrise and doctorat from the University of Paris, Sorbonne. She is the author of *Word and Object in Husserl, Frege and Russell: The Roots of Twentieth-Century Philosophy* (Ohio University Press 1991); *Rethinking Identity and Metaphysics: On the Foundations of Analytic Philosophy* (Yale University Press 1997); *The Roots and Flowers of Evil in Baudelaire, Nietzsche, and Hitler* (Open Court 2006); with Guillermo Rosado Haddock, *Husserl or Frege? Meaning, Objectivity and Mathematics* (Open Court 2000); with Jairo José da Silva, *The Road Not Taken: On Husserl's Philosophy of Logic and Mathematics* (College Publications 2013); and numerous articles and reviews. She has translated Husserl's *Introduction to Logic and Theory of Knowledge* and *Logic and General Theory of Science*. She is a Spanish New Mexican and an independent scholar and has been a religious hermit with the Archdiocese of Paris, France, since the early 1980s.

Burt C. Hopkins, University of Lille, UMR-CNRS 8163 STL. His research continues the tradition of transcendental phenomenology and is focused on the critique of symbolic reason. His books include *Intentionality in Husserl and Heidegger: The Problem of the Original Method and Phenomenon of Phenomenology* (Springer 1993), *The Philosophy of Husserl* (McGill Queen's University Press 2010), and *The Origin of the Logic of Symbolic Mathematics: Edmund Husserl and Jacob Klein* (Indiana University Press 2011).

Carlo Ierna is currently working as lecturer in history of philosophy at the University of Groningen. After working at the Husserl Archives Leuven since 2004, in 2012 Dr. Ierna was awarded a prestigious Dutch NWO VENI grant for his research project on the ideal of "Philosophy as Science" in the School of Brentano. In 2014, he was visiting fellow in philosophy at Harvard where he discovered an unknown letter from Husserl to Brentano. Dr. Ierna is completing his book on early Husserl, which will also contain an edition of Brentano's 1884/1885 lectures on elementary logic.

Ansten Klev is a postdoctoral fellow at the Czech Academy of Sciences. He works mainly on the philosophy and history of logic and of mathematics.

Pierluigi Minari is professor of logic at the University of Firenze. His main research interests include non-classical logics, proof theory, theories of operations (combinatory logic and lambda calculus), and the history of logic as well. He is the author of *Semantic Investigations on Intermediate Propositional Logics* (Bibliopolis 1989) and *Introduction to Logic* (with A. Cantini, Mondadori Education 2009) and of many research papers in international logic journals, among which are the *Journal of Symbolic Logic*, *Studia Logica*, and *Archive for Mathematical Logic*.

Guillermo E. Rosado Haddock is a retired (December 2010) full professor of philosophy at the University of Puerto Rico at Río Piedras – where he had obtained his BA (1966) and MA (1968). He obtained his doctorate in 1973 at the University of Bonn, under the supervision first of Gottfried Martin, who died in October 1972, and then of Gisbert Hasenjaeger. Rosado Haddock is the co-author with Claire Ortiz Hill of the book of essays *Husserl or Frege? Meaning, Objectivity and Mathematics* (Open Court 2000) and the author of *A Critical Introduction to the Philosophy of Gottlob Frege* (Ashgate 2006), *The Young Carnap's Unknown Master* (Ashgate 2008), and the book of essays *Against the Current* (Ontos Verlag 2012), as well as multiple essays and critical studies appearing in journals or publishers in a dozen different countries, mostly dealing with the philosophy of logic and the philosophy of mathematics, be it in Husserl, in Frege, or not restricted to any particular philosopher. Rosado Haddock is presently editing a book of various authors on *Husserl and Analytic Philosophy* and is also preparing a new collection of some of his essays and critical studies.

Richard Tieszen was professor of philosophy at San José State University, located in California's Silicon Valley. He is the author of *After Gödel: Platonism and Rationalism in Mathematics and Logic* (Oxford University Press 2011); *Phenomenology, Logic, and the Philosophy of Mathematics*; and *Mathematical Intuition: Phenomenology and Mathematical Knowledge* (Cambridge University Press 2005), co-editor of a book on the philosophy of mathematics and another on comparative philosophy, and guest editor of several special issues of *Philosophia Mathematica*. He has published numerous papers and reviews on Husserl, Gödel, and other figures and issues in the philosophy of mathematics, logic, and phenomenology. He has been a visiting professor at Utrecht University in the Netherlands and at Stanford University. Richard Tieszen passed away shortly after completing his contribution to this volume. The volume is dedicated to his memory.

Mark van Atten studied artificial intelligence at Utrecht University as well as philosophy at Utrecht and Harvard and currently is directeur de recherche at CNRS in Paris. His main areas of research are philosophy of mathematics and idealistic philosophy, in particular Brouwer, Gödel, and Husserl. His books include *Brouwer Meets Husserl* (Springer 2007) and *Essays on Gödel's Reception of Leibniz, Husserl, and Brouwer* (Springer 2015).

Jan von Plato is currently a professor of philosophy at the University of Helsinki, Finland. He studied mathematics and philosophy there with a doctoral thesis on questions in the foundations of probability theory in 1980. His first book *Creating Modern Probability* (Cambridge University Press 1994) summarizes these early interests. His later work is mainly on logical themes, especially intuitionism and proof theory and their development. Works dedicated to these topics include the books *Saved from the Cellar: Gerhard Gentzen's Shorthand Notes on Logic and Foundations of Mathematics* (2016) and *The Great Formal Machinery Works: Theories of Deduction and Computation at the Origins of the Digital Age* (2017).

Judson Webb is professor of philosophy at Boston University. He is the author of *Mentalism, Mechanism, and Metamathematics* (Reidel 1980). He has written a number of articles on Kant, Hilbert, and Gödel. His most recent publications are "Hintikka on Aristotelian Constructions," "Kantian Intuitions," and "Peircean Theorems," in *The Philosophy of Jaakko Hintikka* (Open Court 2006).

Introduction

This volume sets out to locate precisely Husserl's work in the field of logic and the philosophy of mathematics – a goal surely worth to be pursued, especially in the light of the developments in these fields during the past century. The aim is to provide an in-depth reconstruction and analysis of the discussion between Husserl and his most important interlocutors and to clarify pivotal ideas of Husserl's by considering their reception and elaboration by some of his disciples, such as Oskar Becker and Jacob Klein, as well as their influence on some of the most significant logicians of the past century, such as Rudolf Carnap and Kurt Gödel. Most of the papers focus on Husserl and another scholar – e.g., Leibniz, Bolzano, Kant, Brouwer, and Frege – and trace out and contextualize lines of influence, points of contact, and points of disagreement.

In the following, I will outline the main issues and historical movements and will conclude with a brief overview of each chapter in this volume.

Frege is reported to have said that there is gold in his *Nachlass*. Husserl seems to think that much the same is true of Gottfried Wilhelm Leibniz and really gets many of his own key ideas from him. Chapter 1 of the volume aims at showing how Husserl's idea of formal mathematics as theory of abstract structures has an important source of inspiration in Leibniz. From him, as well as from the reception of Leibniz by the Bohemian mathematician and philosopher Bernard Bolzano, Husserl gets the idea that abstract mathematics is, at root, a priori ontology, "the totality of the laws of possible being," as Ettore Casari once put it.[1] Indeed, already in his youthful *Contributions to a Better-Grounded Presentation of Mathematics* (1810), Bolzano defines mathematics as the "science which deals with the general laws (forms) to which things must conform in their existence"[2] and understands by "things" not only those that actually exist "but also those which simply exist in

[1]Casari 2004.

[2]Bolzano 1810, §8.

our *imagination* [. . .],"[3] thereby meaning "things that are possible," things whose concept is not contradictory: "How must things be made in order that they should be possible?"[4]

Against the framework of these conceptual questions about the *mathesis*, Chap. 1 points out how Husserl's account of *symbolic thinking*, which has such important implications in Husserl's conception of algorithms as well as in his later account of knowledge in the *Logical Investigations*, actually traces back to Leibniz and not to Brentano, in spite of what Husserl himself declares in his first published work, *The Philosophy of Arithmetic* (1891).[5] Even so, among the main topics dealt with in the volume is the relation between Husserl and his teacher Franz Brentano (Chap. 7) as well as a very close reading of Husserl's claims, distinctions, and arguments with respect to Kant (Chap. 2).

The issue of abstract mathematics as *mathesis universalis* is resumed in Chap. 16 (*Husserl and Gödel*) along with that of Leibniz's influence on Husserl. Here we apprehend how the "realization of the Leibnizian idea of a universal ontology as the systematic unity of all conceivable a priori sciences" is thought of to be realizable on the basis of the transcendental methodological method, i.e., according to the late Husserl (in a draft of the *Encyclopedia Britannica* entry), "on a new foundation which overcomes 'dogmatism'." The phenomenological method is shown here "as a way to develop and defend a new kind of rationalism that avoids the excesses of older forms of rationalism but also avoids any kind of mysticism." Furthermore, Chap. 16 points out that the realization of Leibniz's dream of a *universal characteristic*, if the latter is interpreted as a formal system, is not possible in the light of Gödel's incompleteness theorems. The chapter also highlights how Gödel himself does not interpret his own incompleteness theorems as negative results that exclude that every clearly posed mathematical yes-or-no question is solvable by reason and claims that Gödel is here relying on Husserl to show, this time at a variance with Leibniz, that it is possible to solve gradually any intelligible question by reflecting on the concepts and on the way we use them.

Let us now come back to early Husserl and discuss another key point. One of the distinctive traits of early Husserl's work is the simultaneous presence in his logical and mathematical reflections of two different directions of research, (1) the project of a substantial mathematization of logic and (2) a conception of logic as the study of objective relations occurring among certain abstract logical entities, such as concepts and propositions.

As far as point (1) is concerned, we find Husserl's interest in specifically logico-formal issues: he succeeds in grasping with great clarity and insight the implications of the formal-abstract trend in mathematics and, in particular, of its tendency toward algebrization, which he is able to transfer to and elaborate at the logico-theoretical level. Viewed from this perspective, the volume proceeds

---

[3]Loc. cit.

[4]Casari 2004, 161.

[5]Husserl 1891, 215 fn. (Engl. transl., 2003, 205).

by outlining some important similarities and differences between early Husserl's work and that of the algebraists of logic, in particular George Boole (Chap. 5) and Ernst Schröder (Chap. 6), as well as between early Husserl's work and that of the German mathematician, linguist, and physicist Hermann Günther Grassmann (Chap. 4), thus setting out to understand more closely some contributions that may have anticipated Husserl's ideas or influenced them. One of the main issues concerns the abstraction and generality of algebra, which culminate in various attempts by Husserl to characterize universal algebra.[6] In the studies for the second volume of the *Philosophy of Arithmetic*, which actually never saw the light of the day, Husserl characterizes universal algebra as a system of operations on a certain set. The calculus is made up of certain general symbols of operations, which are defined by certain general definitions and obey to certain very general laws.[7] Husserl's account of universal algebra and some of his previous investigations on the concept of system of numeration in a given base[8] very closely resemble the work of the algebraists of logic. Indeed, already in the work of the English algebraists in Cambridge (C. Babbage, G. Peacock, J. W. Herschel) in the period 1830–1840, a distinction was made between (i) abstract algorithm of computation, by means of which conclusions are drawn in a deductive-algorithmic way, and (ii) the possible systems of entities that can satisfy such formal conditions. With the same abstract algorithm of computation, one can provide a unitary treatment for systems of heterogeneous entities that manifest a similar structural behavior. Later on, with the contributions of scholars such as W. R. Hamilton, H. Grassmann, and A. Cayley, there was a progressive distancing from the idea of algebra as "symbolic algebra of magnitudes," which culminated in the explicit disengagement of algebraic research from the quantitative dimension, in particular with Boole and his creation of the algebra of logic. Algebra no longer only treated numbers or magnitudes but also propositions, concepts, and, in general, *qualitative data*. The laws under which they fall are independent from any specific interpretation of the symbolism, and the structural properties of the operations that are reflected in such laws are unleashed from numerical elements and assume the character of abstract algorithmic procedures for "calculations" performed with symbolic expressions. The explicit separation between laws of calculus – purely *formal* laws – and their interpretations is more or less the distinctive trait of modern abstract algebra and mathematics. Chapters 5 and 6 present Husserl's investigations about abstract algorithms of computation along with his reflections on Boole's and Schröder's calculi, while Chap. 4 enlightens Grassmann's *formalization of computation* and traces back to Grassmann another key idea in Husserl's reflection, that of recursive process. Surprisingly enough, Husserl, as far as we know, turns out to be the

---

[6]See *Das Imaginäre in der Mathematik* (December/January 1901/1902), in HGW XII, 430–51 (Engl. transl., 2003, 409–432), and the critical edition Schuhmann & Schuhmann 2001.

[7]*Das Imaginäre in der Mathematik*, cit., 433 (Engl. transl., 2003, 412). Hereto cp. Centrone 2010, 159–61.

[8]Husserl 1891, 243–4 (Engl. transl., 2003, 230–1).

first scholar who, having insisted on the algorithmic meaning of arithmetical operations, explicitly specifies a number of general procedures by means of which new arithmetical (computable) operations are generated from given ones and at the same time attempts to investigate the question concerning the characterization of the class of *computable* arithmetical functions *as a whole*.[9]

As regards point (2) above, a conception of logic is present in early Husserl's work as the study of objective relations occurring among certain abstract semantic entities. A source of inspiration for this claim is the theory of notions (*Vorstellungen an sich*) and propositions (*Sätze an sich*) in Bolzano's *Wissenschaftslehre*, and one of its more remote ancestors is the Stoic doctrine of *sayables* (*lektá*). Chapter 3, *Husserl and Bolzano*, dwells on Bolzano's logical universe, which might turn out to be unfamiliar to the English-speaking scholar, since a complete translation of Bolzano's masterwork, *The Theory of Science*, has been published only in 2014. It is a fact of intellectual history that Bolzano's theory of notions and propositions and Frege's theory of the *senses* of our non-propositional and propositional mental acts or states and linguistic utterances appear to be very close. Both Frege and Bolzano believed in an objective realm of abstract logical entities, distinct from mental states and processes and from the objects people think and talk about. Like Bolzano's notions and propositions, Frege's *senses can* be the contents of our mental acts or states and/or linguistic utterances, can subsist independently of us, and are "capable of being the common property of many."[10] Both Frege and Bolzano take propositions to have a complex internal structure, and it is of the essence of their non-propositional parts to be parts of propositions. Both Bolzano's and Frege's propositions are bearers of *unrelativized* truth or falsity. Both endorse the principle of *bivalence*, though Frege only for propositions expressible in his conceptual notation (*Begriffsschrift*) and not for those expressed in natural languages. Finally, both philosophers take the concept of propositional truth to be *epistemically unconstrained*. Dagfinn Føllesdal conjectured in 1958 that Frege was an important factor in Husserl's conversion from the psychologism of the *Philosophy of Arithmetic* to the anti-psychologism of the *Prolegomena*.[11] This claim has been contested by J. N. Mohanty (1982) and is strongly contested in Chap. 10 of this volume by G. Rosado Haddock. Yet it is a fact that Husserl acknowledges his indebtedness to both Frege and Bolzano. In his *Prolegomena to Pure Logic*, he finds important to stress that his investigations were not "in any sense mere commentaries upon, or critically improved expositions of, Bolzano's thought patterns" but that they "[had] been crucially stimulated by Bolzano ...."[12]

[9]See Husserl 1891, Ch. XIII. This point has been first made by E. Casari (Casari 1991, 46). For a mathematical reconstruction of Husserl's intuitions, see Centrone 2006 (and 2010, 54–61), where I have defended the thesis that the generation procedures that Husserl studies in the 13th chapter of the *Philosophy of Arithmetic* give indeed rise to a class of numerical functions that is extensionally equivalent to the one known in contemporary logic as the class of partial recursive functions.

[10]Frege, "Über Sinn und Bedeutung" (Frege 1892), fn. 5.

[11]See Føllesdal 1958 and Føllesdal 1982, which is a reply by Føllesdal to one of his critics.

[12]Husserl 1900, 227 (English translation, 1970, 224).

Bolzano had, in Husserl's eyes, the great merit of having characterized pure logic as a discipline that is concerned "with the most general conditions of *truth* itself"[13] and deals with the relations among the *contents* of our thoughts. He praised Bolzano's *Wissenschaftslehre* as "a work which ... far surpasses everything that world-literature has to offer in the way of a systematic contribution to logic."[14] In the same *Prolegomena*, he mentions Frege in a footnote and writes: "I need hardly say that I no longer approve of my own fundamental criticism of Frege's antipsychologistic position set forth in my *Philosophy of Arithmetic*. I may here take the opportunity, in relation to all of the discussion of these Prolegomena, to refer to the Preface of Frege's later work *Die Grundgesetze der Arithmetik*, vol. I (Jena 1893)."[15] Indeed, Frege's *Preface* contains much of what Husserl presents on a wider screen in the *Prolegomena*. We find a further hint in the *Husserl-Chronik* (Schuhmann 1977). It is a remark from H. Spiegelberg's *Scrap-Book* (a manuscript quoted with the permission of the author). There we read[16]:

> Andrew Osborn visited H. 1935 in Black Forest to ask him about Frege's influence on the abandonment of the psychological approach of the "Philosophie der Arithmetik." H. concurred, but also mentioned his chance discovery of Bolzano's work in a second-hand book store.

The already mentioned Chaps. 3 and 9 (*Husserl and Frege*) explore in detail the relationship between Husserl and Bolzano, respectively Husserl and Frege as to logical objectivism and the validity of logical laws, while Chap. 10 rejects the claim of Frege's influence. The final judgment is left to the reader.

It is indeed clear that each chapter of the volume brings out specific themes of Husserl's work. At the same time, it is the peculiarity of a collective volume that it could not adopt a harmonic preconception neither with respect to the author's thought nor with respect to certain interpretative questions. The good thing is that the aspects of internal contradiction that are often present in each thought that aspires to call himself philosophical are not given up in favor of a reading that inclines to obscure them on the account of a comprehensive interpretative stance that *privileges* one point of view. Actually, as Chaps. 3 and 9 on the one side and Chap. 10 on the other side nicely disagree, so Chap. 12 (*Husserl and Brouwer*) and Chap. 16 (*Husserl and Gödel*) disagree as to Gödel's project of founding classic mathematics on transcendental phenomenology. As well, Chap. 7 (*Husserl and Brentano*) disagrees with what Ierna calls "the mainstream account of Edmund Husserl's early works," advocated, among others, by Ettore Casari (above all in Casari 1991) and myself (in Centrone 2010). Such account sees Husserl's *Philosophy of Arithmetic* mainly as a work fitting into the general framework of the so-called *research on the foundations* of mathematics, into which two separate influences flow: a mathematical one coming from Karl Weierstrass, whose project of

---

[13] Bolzano 1837, I, 65.

[14] Husserl 1900, 226 (English translation, 1970, 223).

[15] Loc. cit., 169 fn. (318 fn. 6).

[16] Schuhmann 1977, 463. Cp. Føllesdal 1982, 55.

founding analysis on a restricted number of simple and primitive concepts Husserl would have inherited, and a philosophical one coming from Franz Brentano, whose method for identifying the primitive concepts by describing the psychological laws that regulate their formation Husserl would have adopted.

A short overview of each chapter in the volume follows.

Chapter 1. "Husserl and Leibniz: Notes on the Mathesis Universalis" by Stefania Centrone and Jairo da Silva considers Husserl's intellectual debt to Leibniz with respect to two key ideas, the idea of symbolic thinking and the idea of universal mathematics (*mathesis universalis*). In his first major work, the *Philosophy of Arithmetic* (1891), Husserl maintains that the genesis of arithmetic is to be found in the fact that we are almost always forced to limit ourselves to symbolic number presentations. Arithmetic, as a whole, is nothing but *a collection of artificial means* to alleviate the essential incapacity to have a proper presentation of all numbers. As mentioned above, Husserl claims to have taken over the distinction between *proper* (*eigentliche*) and *symbolic* (*symbolische*) presentations from Brentano, but, at root, it is an adaptation of Leibniz's distinction between *cognitio intuitiva* and *cognitio caeca vel symbolica*. Thus, the authors explore the uses Husserl puts the Leibnizian concept of symbolic cognition and how he develops it to explain and justify the sense and functioning of algorithms. Next, they look at another key idea of Leibniz's appropriated by Husserl, namely, the idea of *mathesis universalis.* They start out from Leibniz's conception of mathematical disciplines as branches of the *mathesis* and of the latter as a general science of forms applicable not only to magnitudes but to any object whatever *that exists in our imagination*, i.e., to all objects that are, in principle, *possible*. Thereafter they outline the development of this idea in Husserl, going through early Bolzano's conception of mathematics as the "totality of the laws of possible being."

Chapter 2. "Husserl and His Alter Ego Kant" by Judson Webb begins with the presentation of background material in the works of Kant and Lambert that are relevant to Husserl's lifelong concern with Kant. This intensifies when, in the midst of his own transcendental turn, he recognizes that by virtue of his Copernican turn, Kant became the first to detect the secret longing of modern philosophy for a phenomenological clarification at the sense of being. Kant's transcendental idealism and deductions presuppose a pure ego that he does not adequately analyze but which Husserl finds must survive the phenomenological reduction as pure subjectivity. Husserl's verdict is that Kant does not achieve such a genuine reduction, without which he is unable to eliminate things in themselves from his epistemology or account for intersubjectivity. Husserl improves upon Kant's inadequate accommodation of meaning to intuition in mathematics with his theory of categorical intuition, as well as a method of clarification of concepts that occur in seemingly unintuitive impossibility proofs.

Chapter 3. "Husserl and Bolzano" by Ettore Casari is a contribution to our understanding of Husserl in relation both to earlier and more mature Bolzano as well as to Lotze. The chapter focuses firstly on the idea that Bolzano takes up from the Leibnizian-Wolffian tradition that there is a certain objective connection among truths, independent of the cognitive activity of the subject: certain truths are the "grounds (*Gründe*)" of others and the latter are "consequences (*Folgen*)" of the former. This idea is taken up more than once by Husserl, for instance, in his *Prolegomena to Pure Logic*, when he talks of *Begründung*, and it also has an effect on his conception of proofs and theories. The chapter presents, then, Bolzano's logical universe, in particular, the three main conceptual areas of Bolzano's logical system, namely, *lektology*, that is, the theory of the possible contents of our mental act or states and linguistic utterances; *consecutivity*, that is, his theory of the ground-consequence relation; and his *theory of knowledge*, the theory of the relations between mental acts or states and their abstract logical contents, such as concepts and propositions. Casari then explains how Bolzano's conceptual system was made accessible to Husserl by his study of Lotze and shows how much Bolzano is present in Husserl's *Logikvorlesung 1996*.

Chapter 4. "Husserl and Grassmann" by Jan von Plato distinguishes two late-nineteenth-century approaches to the foundations of arithmetic: one as in Frege, who tries to answer the ontological question of what numbers are, and another as in Grassmann, with numbers as undefined basic concepts but instead definitions of the arithmetic operations that allow their properties to be proved by induction. Husserl's *Philosophie der Arithmetik* stands at the crossroads of these two traditions. Reading it in this light, the lack of a development of formal arithmetic is striking. Husserl tries instead to decide on the basis of general criteria what is reasonable and defensible and what is not. He sees that Grassmann's approach leads to an extremely useful "art of computing and arithmetic" but also criticizes it as a "pure mechanics of computation" that would not be philosophically satisfying.

Chapter 5. "Husserl and Boole" by Stefania Centrone and Pierluigi Minari confronts Husserl's representation of the problem-solving processes with the analysis of "symbolic reasoning" proposed by George Boole in the *Laws of Thought* (1852). This chapter focuses, in particular, on Husserl's lecture "Über die neueren Forschungen zur deduktiven Logik" of 1895, in which an entire section is devoted to Boole. As corroborated by the previous and by the next chapter, Husserl seems to offer a similar criticism of Grassmann's approach as well as of Boole's and Schröder's. On the one hand, according to Husserl, Boole and Schröder are masters in the development of the *logical calculus*, and on the other hand, they are "bad philosophers of mathematics"; they don't grasp the real sense and meaning of the calculus they are setting up. The chapter also focuses on the different reception by Husserl and Boole of the Leibnizian idea of "parallelism" between algorithm and field of experience to be interpreted by the algorithm. While Boole explicitly admits the possibility that not all the steps of the problem-solving processes are *interpretable*, Husserl's position appears to be the opposite: the parallelism between the symbolic-algorithmic level of the problem-solving process and the conceptual one should be preserved along the entire process.

Chapter 6. "Husserl and Schröder" by Stefania Centrone and Pierluigi Minari pursues the goal of clarifying to what extent the work of the German mathematician Ernst Schröder (1841–1902) on the algebra of logic is taken into consideration and criticized in the work of early Husserl, focusing on Husserl's 1891 *Review* of the first volume of Schröder's monumental *Vorlesungen über die Algebra der Logik* (*Exakte*

Logik) and on Husserl's text *Der Folgerungskalkül und die Inhaltslogik* written in the same year of Husserl's *Review*. Husserl levels the same kind of criticism against Schöder that, as the two previous chapters show, he had leveled against Grassmann and Boole: while praising Schröder's calculus, he strongly criticizes Schröder's attempt at a philosophical clarification and justification of it. The chapter also considers Frege's famous letter to Husserl dated 24.5.1891 and briefly compares Frege's and Husserl's objections to Schröder.

Chapter 7. "The Brentanist Philosophy of Mathematics in Edmund Husserl's Early Works" by Carlo Ierna begins by challenging the mainstream account of Edmund Husserl's early works as a combination of two separate influences: a mathematical one in Berlin from i.a. Karl Weierstrass and a philosophical one in Vienna and Halle from Franz Brentano and Carl Stumpf. Instead, this chapter builds a case for a preexisting framework of a Brentanist philosophy of mathematics. Thus, rather than an original application of Brentano's psychology to the foundations of mathematics, Husserl's early writings represent a further elaboration on topics that had already been discussed previously in the School of Brentano. Starting from Brentano's own discussion of mathematical topics in his lectures and considering the various works on the philosophy of mathematics by his students, including Stumpf and Ehrenfels, Ierna then analyzes Husserl's position in the *Philosophy of Arithmetic* against this background. In the last section of the chapter, Ierna then traces the continuities and discontinuities of Husserl's position during the period leading up to the *Logical Investigations*. The result is that through the reconstruction of the Brentanist framework, Husserl's early works gain stronger connections both to their background in the School of Brentano as well as to the development of his early phenomenology.

Chapter 8. "Husserl and Cantor" by Claire Ortiz Hill seeks to shed light on the complex period in the development of Husserl's thought from 1886 to 1900, during which he maintained close personal and collegial ties with the creator of set theory, Georg Cantor. She looks how their ideas about psychologism, Platonism, set abstraction, metaphysics, arithmetization, manifolds, actual consciousness, and pure logic overlapped and crisscrossed during those years. She considers that although Husserl's ideas changed dramatically and definitively during that time, and while it might be tempting to think that those changes were conditioned, if not actually induced by Cantor's bold experiments in mathematics, metaphysics, and epistemology, Husserl himself described the time as one of intellectual crisis, of lonely hard work during which he saw all around him only ambiguously defined problems and profoundly unclear theories. He said that, sick of the confusion and afraid of sinking into an ocean of endless criticism, for the sake of philosophical self-preservation, he had felt compelled to draw to strike out on his own. It was, he said, Lotze's more sophisticated conceptions about Platonic ideas that were responsible for his radical rejection of psychologism, espousal of Platonism, and newfound comprehension of Bolzano's more sophisticated work on pure logic that led him to adopt metaphysical and epistemological views that he had been taught to consider odious and despicable.

Chapter 9. "Husserl and Frege on Sense" by Christian Beyer begins by describing some aspects of the first (1891) phase of a debate between Husserl and Frege, in their correspondence, regarding what Frege calls "sense." Following this, Frege's main arguments for the distinction between sense and reference and some of his most important theses regarding sense are presented, in order to set the stage for both the presentation of the second (1906) and last phase of their debate and for some further comparisons between Frege and Husserl in the course of a presentation of some important features of Husserl's conception of sense. It is argued that while the similarities between their views speak in favor of the so-called Fregean interpretation of Husserl's notion of noematic sense, there are also important differences. With regard to the latter, it is argued that Husserl's view yields a more general criterion of propositional difference and also provides a more detailed conception of the use of indexicals and (other) non-descriptive singular terms and of (what determines) their reference. In this context, Husserl's conceptions of constitution and genetic constitution analysis, respectively, are invoked and interpreted in terms of the epistemic notion of processing mental files or individual "concepts" (as Husserl calls them in *Erfahrung und Urteil*).

Chapter 10. "Husserl and Riemann" by Guillermo E. Rosado Haddock traces the origins of Husserl's conception of mathematics as a *Mannigfaltigkeitslehre* to Riemann. Certainly, it has been difficult for scholars to trace the possible influences on Husserl's mature views on logic, mathematics, and their relation in his *opus magnum*, *Logische Untersuchungen*, some erroneously believing that Frege's review of 1894 of Husserl's youth work, *Philosophie de Arithmetik*, represented a decisive influence, as presumably was Frege's distinction between sense and referent. Both assertions have been proven to be false. Frege simply played for Husserl a similar role as that of Hume for Kant. It has also been thought that Cantor exerted a decisive influence on his younger friend Husserl, and it is not excluded that he had some influence, though not so much on Husserl's conception of *Mannigfaltigkeit*, a term used by Cantor more or less as a synonym for what we now call "set." The fact of the matter is that the student and assistant of the great Karl Weierstrass and friend of Cantor used the word *Mannigfaltigkeit* in the sense that Bernhard Riemann had used it, and his conception of mathematics as a *Mannigfaltigkeitslehre* can very well be conceived as a generalization of Riemann's usage in his duly famous "Über die Hypothesen, welche der Geometrie zu Grunde liegen." Moreover, it is shown that Riemann influenced Husserl in another respect, namely, in Husserl's conception of physical space as a manifold whose structure is to be determined empirically, not by a priori considerations. By the way, Husserl is a counterexample to the simplistic distinction made by historians of mathematics between mathematical schools: his formation was in the Berlin school but was more decisively influenced by a prominent member of the Göttingen school.

Chapter 11. "Husserl and Hilbert" by Mirja Hartimo presents Husserl's approach to mathematics as complementary to but also critical of Hilbert's approach. The paper examines first Husserl's and Hilbert's approaches around the turn of the century in terms of "the division of labor between mathematicians and philosophers" advocated by Husserl in 1900. The division of labor reflects Husserl's, as well as

Hilbert's, non-revisionist approach toward mathematics. According to it, mathematics comes first, whereas the philosophers' task is to examine its conceptual essence. In his later writings, Husserl was more critical toward Hilbert's approach. The paper argues that for both Husserl and Hilbert mathematics was a study of structures. While Husserl's view of formal mathematics in the *Formal and Transcendental Logic* is mainly in line with that of Hilbert's, he was critical of Hilbert's attempt at founding it formalistically on sensuous signs. In contrast to Hilbert, Husserl wanted to show it is reducible to judgments about individuals.

Chapter 12. "Construction and Constitution in Mathematics" by Mark van Atten argues that Brouwer's notion of the construction of purely mathematical objects and Husserl's notion of their constitution by the transcendental subject coincide. Various objections to Brouwer's intuitionism that have been raised in recent phenomenological literature (by Hill, Rosado Haddock, and Tieszen) are addressed. On this basis, an objection to Gödel's project of founding classical mathematics on transcendental phenomenology is presented. The problem for that project lies not so much in Husserl's insistence on the spontaneous character of the constitution of mathematical objects, or in his refusal to allow an appeal to higher minds, as in the combination of these two attitudes.

Chapter 13. "Husserl and Weyl" by Jairo da Silva focuses on the influence Husserl's teachings might have had on the philosophical and scientific ideas of the great mathematician, physicist, and philosopher of science, Hermann Weyl. Husserl is explicitly mentioned in two important works of Weyl, *Das Kontinuum* and *Raum, Zeit, Materie*, both published at roughly the same time (1918 and 1919, respectively). The author's goal, however, is not simply to measure the extent of Husserl's influence on Weyl, although this is an ever-present concern, but to clarify the views of one by contrasting them with those of the other. To accomplish the task, da Silva carries out a comparative study of the views of both thinkers on issues such as mathematical existence, mathematical intuition, the validity of classical logic, the concept of logical definiteness, the nature of symbolic mathematics, the role of mathematics in empirical science, scientific theories vis-à-vis perception, space representation, the philosophy of geometry, and intentional constitution in general. Da Silva concludes that despite some points of divergence, Husserl is an influence to be reckoned with, although sometimes in an elusive way. As da Silva shows, this influence is not restricted to the field of philosophy, extending to Weyl's scientific ideas as well.

Chapter 14. "Paradox, Harmony, and Crisis in Phenomenology" by Judson Webb tries to analyze to what degree the discovery of logical paradoxes factors into Husserl's view of the crisis of modern science, as well as the extent to which they may have weakened his confidence that his transcendental phenomenology could account for the wonderful affinity between mathematical thought and things that the natural sciences evince – in short, that pure phenomenological consciousness should actually constitute nature without simply assuming a doctrine of pre-established harmony. Husserl develops to this end a supramathematics of all possible deductive theory forms for whose arithmetical theories he offers completeness proofs. Close attention to these proofs seems to show that he was vaguely aware of the difficulties that lead to incompleteness theorem, but he did not pursue his examination of the logical paradoxes to a point that might have highlighted these difficulties. Finally, just as the work of Weyl in general relativity invoking phenomenological themes encouraged Husserl to believe that his phenomenology could bring clarity a priori to physical science, the rise of quantum mechanics convinced Weyl that intuitionism and phenomenology could not account for it after all. For his students Becker and Mahnke, this created "a crisis of phenomenological method" itself, but Husserl concluded that it did so only for Becker's mantic phenomenology, not his own.

Chapter 15. "Husserl and Carnap on Regions and Formal Categories" by Ansten Klev is concerned with what may be called Husserl's doctrine of categories. This chapter offers a detailed explanation of Husserl's technical definition of a region in the *Ideas* (1913). The explanation displays close similarities between Husserl's doctrine of essences and the doctrine of concepts in traditional logic. Moreover, both the notion of region and the notion of formal category are elucidated by means of concepts from Husserl's philosophy more generally. The second part of the chapter considers how the notions of region and formal category can be made sense of within Rudolf Carnap's so-called constitution system in his *The Logical Structure of the World* (1928). This constitution system is divided into different sections corresponding to different kinds of objects and, therefore, it is noted, also to regions in Husserl's sense. Formal categories in this context will be the elements of the logical framework, namely, simple type theory, used in the construction of the constitution system.

Chapter 16. "Husserl and Gödel" by Richard Tieszen is concerned with Kurt Gödel's study of the philosophy of Husserl, going back to 1959. Gödel turned to Husserl to find a philosophical foundation for logic and mathematics, along with a methodology that would deepen and expand his efforts to develop a rigorous philosophy that would include an account of rational intuition, meaning clarification, an analysis of mind that would not be mechanistic, a kind of Platonism, and a monadology that would improve upon Leibniz's "preliminary" version. The logician Hao Wang tells us that Gödel was interested primarily in the work that Husserl did after 1906, which is when Husserl took his turn into transcendental eidetic phenomenology and in which transcendental egos were soon to be described by Husserl as "monads." In the chapter, Richard Tieszen discusses what is known about Gödel's study of Husserl, based on Gödel's 1961 lecture manuscript on Husserl, some items in the Gödel *Nachlass*, Gödel's comments about Husserl recorded by Wang, and Wang's own notes of his discussions with Gödel about Husserl. It is clear that, starting in 1959, Gödel was influenced by Husserl in a number of ways. The famous philosophical supplement to the 1964 version of "What Is Cantor's Continuum Problem?," for example, was influenced by his reading of Husserl, although hardly any of the literature on these passages has displayed awareness of this fact, due to lack of knowledge about Husserl's work among most "analytic" philosophers. Several other notes and letters that Gödel wrote after 1959, apart from the 1961 paper, also show Husserl's influence. There is some irony in the fact that one of the greatest logicians of all time would have turned to Husserl's transcendental phenomenology of logic when such a view would have been reviled as meaningless or irrelevant in certain traditions in philosophy with which Gödel was very familiar.

Chapter 17. "Husserl and Jacob Klein" by Burt Hopkins explores the relationship between Husserl's and Klein's accounts of the foundations of modern symbolic mathematics. It argues that the problem of this foundation is twofold. On the one hand, it concerns the epistemological account of the origin of the concept of unity that characterizes the many-as-one. On the other hand, it concerns the gap between proper number concepts and symbolic ones that are actually used in calculations. Hopkins argues that Husserl's self-acknowledged failure to provide either of these foundational accounts in the *Philosophy of Arithmetic* is understandable when viewed from the perspective of Klein's philosophical-mathematical account of the historical origin of modern algebra in François Viète's *The Analytic Art*. Klein's account shows that, on the one hand, the problem of many-as-one was the very problem faced by ancient philosophies of mathematics and cannot be resolved by filtering it back thorough current logic technicalities. On the other hand, Klein shows that Husserl's attempt to ground symbolic mathematics on the concept of positive whole number was doomed to fail, because of an untenable (historical) identification of proper number concept and symbolic one, i.e., the one behind the origin of modern algebra. Moreover, Klein shows that this identification is something with which we live to this day.

Institut für Philosophie
Carl von Ossietzky Universität Oldenburg
Oldenburg, Germany

Stefania Centrone

Chapter 1
Husserl and Leibniz: Notes on the *Mathesis Universalis*

Stefania Centrone and Jairo José Da Silva

> *[L]a bonne caractéristique est.*
> *une des plus grandes aides*
> *de l'esprit humain ([A] good symbolism*
> *is one the greatest aids to the human mind*
> Leibniz, *N.E. IV*, Ch.7, § 6)
> [O]*hne die Möglichkeit symbolischer* [. . .] *Vorstellungen*
> *gäbe es kein höheres Geistesleben, geschweige denn eine*
> *Wissenschaft*
> ([W]*ithout the possibility of symbolic representations* [. . .]
> *there would not exist a higher spiritual life, and even less*
> *science*
> Husserl, *Zur Logik der Zeichen, PdA* 349)

Abstract The notion of *mathesis universalis* appears in many of Edmund Husserl's works, where it corresponds essentially to "a universal a priori ontology". This paper has two purposes; one, largely exegetical, of clarifying how Husserl elaborates on Leibniz' concept of *mathesis universalis* and associated notions like *symbolic thinking* and *symbolic knowledge* filtering them through the lesson of the so called "bohemian Leibniz", Bernard Bolzano; another, more properly philosophical, of examining the role that the *universal mathesis* is allowed to play, and the space it occupies in Husserl's intuition-based epistemology.

Keywords Husserl • Leibniz • Bolzano • *Mathesis Universalis* • Symbolic • Sign

---

S. Centrone (✉)
Institut für Philosophie, Carl von Ossietzky Universität Oldenburg, Oldenburg, Germany
e-mail: stefania.centrone@uni-oldenburg.de

J.J. Da Silva
Sao Paulo State University, Sao Paulo, Brazil
e-mail: dasilvajairo1@gmail.com

S. Centrone (ed.), *Essays on Husserl's Logic and Philosophy of Mathematics*,
Synthese Library 384, DOI 10.1007/978-94-024-1132-4_1

1.1 Introduction

Under the heading "*mathesis universalis*" René Descartes and, later, G.W. Leibniz understood a most general science built on the model of mathematics. Though the term, along with that of "*mathesis universa*", was already used during the XVI century,[1] it is with Descartes und Leibniz that it became customarily to designate with it a universal mathematical science that unifies all formal a priori sciences.[2] In his comment to the *IV Rule to the Direction of the Mind*[3] Descartes talks of a general discipline "that should contain the primary rudiments of human reason and extend to the discovery of truths about any object whatsoever",[4] he maintains to see some traces of this true mathematics already in Pappus and Diophantus,[5] identifies its method with *algebra* or, at least, with algebra "divested of the multiplicity of numbers and incomprehensible figures which overwhelm it"[6] and, instead, endowed with that "abundance of clarity and simplicity [...] that the true *mathesis* ought to have"[7] and adds that this most general mathematical science should have as branches not only arithmetic and geometry but also astronomy, music, optics and mechanics. To the question as to the criterion according to which we decide whether a science belongs to this general mathematics or not, Descartes answers that the concern proper of this science "is with questions of order or mesure and it is

[1]Cp. Piccolomini's *Commentarium de certitudine mathematicarum* (1547), ch. 7; Dasypodius' mathematical writings (1564a, b, 1571, 1593); van Roomen's *Apologia pro Archimede* (1597). All these Authors, except for van Roomen, refer to Proclus' *Commentary on the First Book of Euclid's Elements*, especially to the first prologue of the *Commentary*, which deals with the mathematical sciences in general. In Proclo's commentary there are many hints at a common mathematical discipline that shall precede all other mathematical disciplines and has, therefore, a more general character. Crapulli 1969 offers a very interesting investigation about the development of the idea of the *mathesis* in the XVI century.

[2]Before Descartes and Leibniz there were mainly two criteria for gathering together mathematical disciplines, namely, that they all had, in some way, quantity as their object (think of the disciplines traditionally unified in the *Quadrivium*) and an higher degree of certitude in proofs with respect to non-mathematical disciplines.

[3]Descartes *Rule IV*, *AT.X*, 372–379, here 374f. Actually it is a recurrent theme by Descartes that "all sciences are concatenated", "that it is much more easier to apprehend them all together, then to separate one of them from the others" and focus solely on a unique one (*Rule I*, *AT.X*, 361). Hereto also cp. *AT.X*, 255: "[one single science] cannot be brought to perfection, without doing the same with the others". Responsibility for translations from German, French and Latin is ours, even when we refer to, benefit from, or simply echo published translations.

[4]*AT.X*, 376.

[5]Pappus of Alexandria (290–350 c. AD) and Diophantus of Alexandria (201/215(?)-285–299(?) AD) were Alexandrian Greek Mathematicians.

[6]*AT.X*, 378.

[7]loc. cit.

irrelevant whether the measure [or order] in question applies to numbers, shapes, stars, sounds, or any other object whatever".[8] He continues as follows[9]:

> [t]here must be a general science which explains all the points that can be raised concerning order and measure irrespective of the subject-matter, [. . .] this science should be termed *mathesis universalis* – a venerable term with a well-established meaning – for it covers everything that entitles these other sciences to be called branches of mathematics.

For our purposes, it is important to notice that talk about "mathesis universalis" and "symbolic", not only in Leibniz, but in Descartes as well, usually go together and are put to uses that are epistemological in nature. Descartes, in his "Géométrie",[10] developed a very efficient method for tackling geometric problems by *purely algebraic*, that is *symbolic* means. The geometric constructions required to solve a geometric problem are symbolically represented by an algebraic equation, which once solved by purely algebraic means reveals the sequence of steps required for the geometrical construction to be actually performed and the original geometric problem solved by geometric means. This is a powerful illustration of the efficacy of symbolic calculi as tools for *thinking* and *knowing* that Descartes could not fail to appreciate.

However, Descartes was not the first to realize the epistemic and heuristic efficacy of symbolic calculi. The Italian algebraists of the Renaissance (Cardano,[11] del Ferro,[12] Tartaglia,[13] Bombelli[14]) had already succeeded in developing algorithmic methods for dealing with algebraic equations of the third degree, some of them involving signs without meaning being manipulated as *bona fide* number-denoting symbols.[15] Whereas in Descartes' algebraic geometry algebraic symbols always denoted geometric entities,[16] in Bombelli's algebra signs like the square root of -1, which in those days did not denote anything, could nonetheless participate in

[8]loc. cit.

[9]*AT.X*, 378–379.

[10]Descartes 1954.

[11]Gerolamo Cardano (1501–1576) was an Italian mathematician, physicians and astrologers. He occupies an important place in the history of Renaissance philosophy. He wrote more than 200 works in the most disparate fields, but especially, mathematics, medicine, philosophy and astrology. Among his works we remember his *Artis magnae sive de regulis algebraicis liber unus* (1545), commonly known as *Ars magna*, that draws the anger of Nicolò Tartaglia for having published the solution of third degree equations revealed to him by Tartaglia 6 years earlier, though under oath, not to reveal it.

[12]Scipione del Ferro (1465–1596) was an Italian mathematician who first discovered a method to solve the depressed cubic equations.

[13]Niccolo Fontana Tartaglia (1499/1500–1557) was an Italian mathematician famous for having been the first to translate Euclid and Archimedes into Italian as well as for his controversy with Cardano as to the solution of cubic equations.

[14]Raphael Bombelli (1526–1572) was an Italian mathematician. He is Author on a treatise on algebra (1572) and gave important contributions in the understanding of imaginary numbers.

[15]Cardano 2007.

[16]Descartes 1954.

the algebraic calculus on equal terms with denoting, meaningful symbols.[17] The surprising thing, which Leibniz certainly realized, was that by treating empty symbols as denoting symbols these algebraists succeeded in developing a very efficient calculus and put it to good *heuristic* uses.

Leibniz too invented (concomitantly with but independently of Newton) a symbolic method for dealing with geometric problems, the *infinitesimal calculus*, that shares in the spirit of both Descartes' and Bombelli's creations.[18] Leibniz' calculus operates with symbols for infinitesimals, i.e. non-vanishing quantities that, however, are smaller than any given quantity. Leibniz did not believe that infinitesimals actually existed, but likewise $\sqrt{-1}$ in Bombelli's calculus, to treat symbols for infinitesimal quantities on equal terms with symbols for real quantities (and operate with them) proved to be a very useful strategy. Leibniz's calculus is essentially a set of rules for operating with symbols for infinitesimal quantities, which Berkeley, referring to the correspondent quantities in Newton's calculus, called "ghosts of departed quantities".[19] Previously to Leibniz, Cavalieri[20] and Kepler,[21] among others, had already developed methods for dealing with geometrical problems (calculation of volumes of solids, for example) by supposing that bodies are *actually* formed by an infinity of infinitesimal parts.[22] The reliability of these methods, however, depended heavily on the mathematician's ability to sum infinite series. Leibniz calculus, instead, reduced everything to rule following.[23]

For Leibniz, however, mathematics was not an end in itself; in a letter to Countess Elizabeth, for instance, he says: "As for myself, I cherished mathematics only because I found in it the traces of an art of *invention in general*".[24] Leibniz considered his calculus superior to Descartes' analytic geometry, and so a proof of the superiority of his philosophical system over Descartes.[25] The success of the method pointed towards *generalization* and *extension* of its core idea. The *mathesis*

---

[17]See Nahin 1998.

[18]Leibniz 1684.

[19]Berkeley 1948–57.

[20]Bonaventura Cavalieri (1598–1647) was a pupil of Galileo. He developed Galileo's thoughts into a geometrical method and published in 1635 a work on the subject: *Geometria Indivisilibus Continuorum Nova quadam Ratione Promota*.

[21]Johannes Kepler (1571–1630) is one of the most representative figures in the Renaissance. He was strongly influenced by Copernicus and endorsed the Platonic conception that universe is ordered according to a pre-established mathematical plan. He his Author of many works, among them a *Mysterium Cosmographicum* (1596) and his *Astronomia Nova* (1609), in which we find him saying: "The chief aim of all investigations of the external world should be to discover the rational order and harmony which has been imposed on it by God and which He revealed to us in the language of mathematics" quoted after Pearcey & Thaxton 1994, 126.

[22]See Mancosu 1996.

[23]This comes out clearly in Leibniz's *practice* of the calculus he invented, see Leibniz 1684 note 18.

[24]Leibniz 1678, quoted after Ariew & Gaber 1989, 236.

[25]See Belaval 1960.

universalis was, for Leibniz, essentially a mathematically inspired symbolic calculus for reasoning in general.

Like Descartes and Leibniz, Husserl was a mathematician, and he too was sensitive to the extent to which symbolic means of knowing and thinking can be put to good uses in mathematics, logic, empirical sciences and cognition in general. But symbolic thinking and symbolic knowledge are, for him, the very opposite of intuitive thinking and intuitive knowledge; Husserl's epistemology is, as we know, intuition-centered. This raises a problem: if knowledge proper is, as the *Sixth Logical Investigation* tells, the fulfillment of empty intentions with *intuitions*, how can symbolic calculi have any *epistemic* relevance? The answer to this question is, as we shall see, multifold. But before providing a logical-epistemological justification for symbolic means of knowledge, Husserl had to find their natural context, which he claims to be, as we will see, the realms of formal logic, substantially enlarged vis-à-vis traditional conceptions. This enlarged notion of formal logic is what Husserl understands by *mathesis universalis*. Let us now get down to the details.

1.2 Early Husserl on the Shoulders of Leibniz: Symbolic Thinking and Symbolic Knowledge

There is consensus in the literature that Leibniz was an important influence on Husserl. In a letter, dated 5.I.1917, to his student Dietrich Mahnke, Husserl writes:

> I can understand your love for Leibniz very well. In my youth I often read with open eyes in the Erdmann Edition of L[ei]b[niz], and without doubt that affected me strongly, even as my attitude back then was still very different. Thus I became receptive for some important expositions of Lotze, as well as for some of Lambert and Bolzano, and capable of the turnabout that was decisive for me. Today I still feel close to Leibn[iz].

Husserl's library contained the *New Essays* (BQ 250) and Leibniz's *Hauptschriften zur Grundlegung der Philosophie* (BQ 251/2), both translated, from French and Latin respectively, by Ernst Cassirer, as well as Leibniz's *Kleinere Philosophische Schriften*, translated by Robert Habs. As Husserl himself reports in his letter to Mahnke, his library also contained the Erdmann Edition of Leibniz, *God. Guil. Leibnitii Opera Philosophica quae exstant Latina Gallica Germanica Omnia* (two vol., Berlin 1839/40). In particular, the *Hauptschriften zur Grundlegung der Philosophie* contain the *Meditations on Knowledge, Truth and Ideas*,[26] which are very important for a correct understanding of Husserl's use of Leibniz's notion of symbolic. Notice that when Husserl says to Mahnke that he became receptive

[26]Leibniz 1684. Henceforth quoted as *MED*. *MED* is Leibniz's first mature philosophical publication. It appeared in November 1684 in the Leipzig Journal *Acta Eruditorum* as Leibniz's contribution to the famous Arnauld-Malebranche controversy, triggered by the publication of Arnauld's *Des vrais et des fausses idées* in 1683, an attack on Malebranches's philosophy.

for Lotze, Lambert and Bolzano through Leibniz he is implicitly giving a Platonist interpretation of Leibniz.

In addition, recall that Friedrich Trendelenburg published in 1856 a paper entitled *Über Leibnizens Entwurf einer allgemeinen Charakteristik* that was a common source of inspiration for Gottlob Frege as well as for Husserl himself and for the German mathematician Ernst Schröder. Trendelenburg's paper presumably inspired Frege's decision to entitle his first major work *Begriffsschrift, einer der arithmetischen nachgebildeten Formelsprache des reinen Denkens* (1879).[27] It was also the most important source of information concerning Leibniz's writings for Schröder at the time he wrote the Introduction to his *Vorlesungen über die Algebra der Logik*, whose first volume (*Vorlesungen über die Algebra der Logik* (*Exakte Logik*) (1890) Husserl reviewed in 1891[28] and about which Husserl also wrote the essay *Der Folgerungskalkül und die Inhaltslogik*[29] in the same year.[30]

In particular, some distinctions Leibniz introduced in *MED* were taken up by Husserl at different stages of his intellectual development, notably in the early work *Philosophy of Arithmetic*,[31] the middle period *Logical Investigations*,[32] and the later *Formal and Transcendental Logic*.[33] In the first two works, Husserl *uses* Leibniz's notion of *symbolic* to explain (i) the logical nature of number systems and algorithms (*Philosophy of Arithmetic*) and (ii) the nature of rational thinking (*Logical Investigations*). In his later work (*Formal and Transcendental Logic*) he (iii) radically modifies Leibniz's notions of *distinctness and clarity* to ground "the differences in evidence that substantiate the separating of levels" within his threefold stratification of formal logic. In the following, we will take into consideration only claims (i) and (ii).

MED is perhaps the most important place where Leibniz's notion of *symbolic* appears. Leibniz introduces here the conceptual pairs clear/obscure, adequate/inadequate, intuitive/symbolic. Given our goals here, we set aside the enormous exegetical problems raised by *MED*, *in primis*, the fact that Leibniz is not at all clear about the nature of the bearers of clarity, distinctness, etc., i.e., whether they are mental act or states, linguistic utterances, abstract logical contents, dispositions or something else.[34] In the following quotes, we preserve the Latin names of the bearers of clarity, distinctness etc., circumventing the problem just

[27]Henrich Scholz hypothesizes Trendelenburg as source. See Husserl's and Scholz's *Anmerkungen* in Frege's 1964.

[28]Husserl 1891a.

[29]Husserl 1891b.

[30]Hereto cp. Chap. 6 in this volume.

[31]Husserl, *Philosophie der Arithmetik*. Mit ergänzenden Texten (1890–1901), HGW XII, 1–283. Henceforth cited as *PdA*. English translation cited as *PoA*.

[32]Husserl, *Logische Untersuchungen* II/1 (Tübingen: Max Niemeyer, 19937). Henceforth cited as *LU*. English translation cited as *LI*.

[33]*Formal and Transcendental Logic* [ed. 1929] (henceforth cited as *FTL*), 49 ff; Engl. transl. 56 ff.

[34]Hereto see Centrone 2010b.

mentioned of determining precisely what they are. Leibniz begins with the following hierarchy[35]:

> [...] *cognitio* is either obscure or clear, and again, clear *cognitio* is either confused or distinct, and distinct *cognitio* either inadequate or adequate, and adequate *cognitio* is either symbolic or intuitive, and, indeed, if *cognitio* were, at the same time, both adequate and intuitive, it would be absolutely perfect.

He then explains what it means for a notion (the bearer has changed!)[36] to be obscure or clear, etc. He writes[37]:

> A *notio,*[38] which is not sufficient for recognizing the thing represented is *obscure* [...]. [*C*]*ognitio*[39] is *clear* when I am able to recognize the thing represented. *Clear cognitio,* again, is either *confused* or *distinct*. It is confused when I cannot enumerate one by one marks sufficient for differentiating a thing from others [...] [It is distinct when it is] connected with marks (*notas*) and tests (*examina*) sufficient to distinguish a thing from all others similar bodies [...]

Leibniz also claims that to have a distinct notion of a certain thing is to have a nominal definition of the term expressing this notion. He then continues[40]:

> Also, one has distinct knowledge (*cognitio distincta*) of an indefinable *notio*, since it is primitive, or its own mark (*nota suis ipsius*), that is, since it is irresolvable and is understood only through itself and therefore lacks requisites. But in composite notions (*in notionibus compositis*), since again, the individual marks composing them are sometimes understood clearly but confusedly, [...] such knowledge (*cognitio*) [...] may be distinct, yet inadequate.

Assuming that he takes here a *notio* to be a concept, he is saying that when the conceptual determinations belonging to the content of a concept are not themselves completely analyzed, our knowledge of this concept is not *adequate*. He then goes on[41]:

> *When everything that enters into a distinct notion* [notitia [!]] *is*, again, *distinctly known*, or when analysis has been carried to completion, then knowledge (*cognitio*) is *adequate*. *I don't know whether humans can provide a perfect example of this*

---

35 *MED* 585–586.

36 Talk of "*notio*" as well as "nominal definition" below indicates that Leibniz conceives of "*notiones*" as abstract objects. The question is controversial, for this is against his official position presented in the short essay *Quid sit Idea?* (Leibniz 1677a), in which he maintains that an idea is a certain power of the mind. With "idea", he writes there, we mean "something that is in our mind (*aliquid, quod in mente nostra est*)": "The idea is [...] not a certain mental act, but rather a power, and thus we say that we have the idea of an object even if we don't actually think of it but have the ability to do so on any given occasion. (*Idea* [...] *non in quodam cogitandi actu, sed facultate consistit, et ideam rei habere dicimur, etsi de ea non cogitemus, modo data occasione de ea cogitare possimus.*)" (Ak VI.4B, 1.370).

37 *MED* 585.

38 The bearer has changed! It is no longer a *cognitio*!

39 The bearer has changed again!

40 *MED* 585.

41 *MED* 585–586 (our emphasis).

> [...] *However, we don't usually grasp the entire nature of a thing all at once, especially in a more lengthy analysis, but in place of the things themselves we make use of signs, whose explicit explanation we usually omit for the sake of brevity, knowing or believing that we have the ability to produce it at will.* And so when I think about a chiliagon, that is, a polygon with a thousand equal sides, I don't always consider the nature of a side, or of equality, or of thousandfoldness ... *but in my mind I use these words, whose sense appears only obscurely and imperfectly to the mind, in place of ideas I have of these things, since I remember that I know the meaning of those words, and I decide that explanation is not necessary at this time. I usually call such thinking, which is found both in algebra and arithmetic and, indeed, almost everywhere, blind or symbolic. And indeed, when a notio is very complex, we cannot consider all of its component notions at the same time. When we can, or indeed insofar as we can, I call knowledge intuitive. There is no knowledge (cognitio) of a distinct primitive notion except intuitive, just as our thinking about composites is for the most part symbolic.*

The reader familiar with the early Husserl will immediately identify, in the *Philosophy of Arithmetic* and later works,[42] echoes of what Leibniz is saying here, but also in other works.[43] For instance, the idea that *adequate knowledge* is *intuitive knowledge*, that intuitive knowledge is sparse, that in algebra and arithmetic, particularly, we most often use signs instead of the concepts for which the signs stand, assuming that it is not necessary to bring the meaning of the signs to consciousness every time but that we could do so if we so wanted. We find in Husserl, as we do also in Leibniz, the idea that rational thinking makes use almost everywhere of signs in place of proper concepts. Husserl too calls thinking-through-signs "*blind*" or "*symbolic*". Husserl also assumes, as Leibniz does, that our representational capacity is limited, that we cannot represent in a *proper* (*direct*) *way* a too numerous many-as-one, i.e. we cannot have a proper representation of collections or numbers that exceed a very modest threshold. One of the eight theses Husserl chose to defend in a formal disputation in the summer of 1887 stated that "in the authentic sense one can barely count beyond three".[44]

The distinction between proper and symbolic number presentations is essential in the *Philosophy of Arithmetic*.[45] In this as well as in all his early writings Husserl uses the words "concept" and "presentation" in essentially the same way.[46] He claims to have taken the distinction between proper (*eigentliche*) and symbolic (*symbolische*) presentations over from Brentano, partially modifying it.[47] Curiously enough, he

---

42 See e.g. *LU I*, §20; *LI* 304–306 (*Thought without intuition and the 'surrogative function' of signs*).

43 *NE* II, Ch. 9, §§ 8–10; and Ch. 29 that bears the title "*Of Clear and Obscure, Distinct and Confused Ideas*."

44 *PdA* 339; *PoA* 357.

45 See also Husserl, *Zur Logik der Zeichen* (*Semiotik*), in: *PdA* 340–373. Here *PdA* p. 349f.

46 Cp. Willard 1984, 26.

47 "To [Brentano] I owe the deeper understanding of the vast significance of inauthentic presentations for our whole mental life; this is something which, so far as I can see, no one before him had grasped" (*PdA* 193; *PoA* 205). See also Husserl, *Zur Logik der Zeichen* (*Semiotik*): *PdA*, p. 340.

does not mention Leibniz in this regard. Nonetheless, he claims in a way consistent with Leibniz, that symbolic presentation, i.e. our capacity to refer to things that we do not 'intuit directly', plays an essential role on our psychical and intellectual life in general and, in particular, in the constitution of algorithms – indeed, it is precisely this capacity that constitutes the very possibility of arithmetic. Arithmetic is nothing but 'a collection of artificial means' to alleviate the essential incapacity of a proper presentation of all numbers[48]:

> If we had authentic [*eigentliche*] representations of all numbers, as we do of the first ones in the series, then there would be no arithmetic, for it would then be completely superfluous. [...] But in fact we are extremely limited in our representational capacities. That some sort of limits are imposed upon us here lies in the finitude of human nature. Only from an infinite understanding can we expect the authentic representation of all numbers; for, surely, therein would ultimately lie the capability of uniting a true infinitude of elements into an explicit representation.

The presentation of something is proper when it is given to us, "as it were, in persona".[49] We have a proper presentation of, say, a house when we actually see it. Symbolic presentations are presentations through signs that univocally denote their objects. In this case, the signs are presented properly, while the objects are presented only indirectly, through the signs. We have a symbolic presentation of a house when somebody describes it to us univocally e.g., as "the house at the corner of such and such with such and such streets."[50]

A symbolic presentation denotes its object in such a way that it can always be identified. It has the function of "standing for (*surrogieren*)" the object when it is not or cannot be intuited. Such a substitution can be temporary, as is the case of, say, the house, or permanent, if a proper presentation is forever beyond our cognitive reach; in this latter case, as Husserl puts it, the symbolic presentation has the function of a permanent surrogate for its object. Between the proper presentation of an object and its various improper or symbolic presentations there is a relationship of "logical equivalence".[51] The possibility of substituting symbolic representations for proper ones in judgments is founded on this. Husserl stresses that he has brought forward, in a much stronger way than Brentano, the fact that symbolic presentations allow *univocal identification* of the objects for which they stand. This one-to-one-correspondence between objects and signs, or between number-concepts and number-signs, when we restrict ourselves to algorithms, plays a most important role for the *logical-epistemological justification* of algorithms. The strict parallelism between concept and sign is an idea that Leibniz presents in *De Arte Combinatoria* (1666),[52] but also in other works. For reason of space, we limit ourselves to hint at

[48] *PdA* 193; *PoA* 201–202.

[49] *LU* IV, §45, 144; *LI* 785–786.

[50] *PdA* 194; *PoA* 205;

[51] *PdA* 193–194, 237–239; *PoA* 205–206, 247–240. See also Husserl, *Zum Begriff der Operation* (*On the Concept of Operation*), in: *PdA* 408–429. *PoA*, 385–408. Here: *PdA*, p. 418; *PoA*, p. 395.

[52] Leibniz 1666, 5.

Leibniz's *Dialogue on the Connection between Things and Words* (1677) where – the target being Hobbes' position that truth is arbitrary – after some discussion, the participants in the dialogue, who are called A and B, agree that (i) thinking is impossible without signs of some kind and (ii) there subsist a relation between the characters (in natural languages as well as in calculations) that fits the relation in the things[53]:

> There is in the characters, especially when their choice is a good one, a relation or order that corresponds to a relation or order in the things. [...] If characters can be used for ratiocination, there is in them a kind of complex mutual relation or order with fits the things; if not in the single words at least in their combination and inflection although it is even better if found in the single words themselves. [...] Even if the characters are arbitrary their use and relation have something that is not arbitrary, indeed there is some correspondence between characters and things [...] and that correspondence or relation is the ground of truth. For it is the reason why, regardless which characters we choose, the result either remains the same or we get an equivalent [...].

Trendelenburg[54] expressly emphasizes this idea by Leibniz in the article mentioned above and both Husserl and Schröder make an essential use of it in their writings. Schröder sometimes refers to the parallelism as "univocal correspondence".[55] Trendelenburg stresses that, according to Leibniz, the formation of number-signs strictly mirrors that of number-concepts. Husserl elaborates on this idea in the *Philosophy of Arithmetic* and connects it with another idea of Leibniz, namely that if the *productive* functions of the signs. Compelled by the finitude of human nature, we set up number systems, which Husserl conceives as "systems of numeration in a given basis", the basis being constituted by the numbers that are properly given to us. The arithmetic operations, either *elementary* operations such as *addition*, *multiplication*, *subtraction* and *division* or higher-level ones such as *exponentiation* are conceived of as *procedures of numerical construction* that *generate and at the same time designate* numbers systematically. The essence of number systems resides therein that (i) they *construct* all numerical concepts using a few elementary concepts and operations, and (ii) *designate* the concepts so constructed through the constructions themselves. As we will see below, *the strict correspondence between numerical concepts and numerical signs is the basis on which the logical-epistemological justification of the arithmetic calculus rests.*

In his investigations on the nature of algorithms, Husserl goes still further and sharpens ideas that, to some degree, are to be found also in Leibniz. He observes that algorithms present a characteristic feature: they function perfectly also when unleashed from the conceptual correlates of the signs employed. Systems of signs

---

[53]Leibniz 1677b, in: Ak. IV.4.A, 23–24. Also cp. Leibniz 1688?: "Characterem voco, notam visibilem cogitationes repraesentantem. Ars characteristica est ars ita formandi atque ordinandi characteres, ut referant cogitations, seu ut eam inter se habeant relationem, quam cogitationes inter se habent. Expressio est. aggregatum characterum rem quae exprimitur repraesentantium."

[54]Trendelenburg 1856, 39f.

[55]Schröder 1890, 48; 49. Hereto cp. ch. 6, section 4 in this volume.

are *polyvalent*: they allow a uniform treatment of "similar" situations.[56] One and the same sign system can be used to interpret different conceptual systems[57]:

> It is a fact highly significant for the deeper understanding of mathematics that one and the same system of symbols can serve in *several* conceptual systems, which, different as to their content, exhibit analogies solely in their structural form.

But this is not yet the end of the story. The early Husserl exploits further Leibniz's intuitions. In the first draft of his Schröder's *Review*, written in the same year in which the *PoA* was published (1891), he endorses the thesis that the signs of a symbolic calculus have, *qua signs of a calculus*, only the formal meaning given by the rules for manipulating them (independently of having also, perhaps, considered independently of their role in a calculus, the material meaning provided by the things for which they stand, if indeed they stand for anything): "algorithmic signs ... have a meaning which is exhausted by certain operational rules".[58] The thesis that meaning is use was, of course, defended later by Wittgenstein and Gentzen, but with a much broader scope[59]:

> Calculatory signs (*Rechenzeichen*) or arithmetical signs as algorithmic signs have their meaning (*Bedeutung*) exclusively in the rules of connection, separation, replacement, in short in the operational rules which together make the algorithm an algorithm. It is in the capacity of being subject to different rules in such connections, the permission to replace connections of one definite sort by connections of another definite sort, and the prohibition against connections and replacements which contradict the rules that the general meaning lies which gives the signs their general meaning [...] Meaning is here constituted by completely external prescriptions [...].

The last step in early Husserl's elaboration of Leibniz' ideas is his rendition of Leibniz's claim that our thinking is "almost everywhere blind or symbolic". Husserl undertakes this step in the *Logical Investigations*. There he speaks of "semantic acts" as the constitutive acts of rational thought and of science meant as "cumulative knowing". By "*semantic* act" he means *every* mental act or state and/or linguistic utterance that has a meaning. In the *Logical Investigations* Husserl repeatedly claims that rational thought represents an abstract symbolic way of intending the object of reference. It is a matter of fact that we have the capacity to refer to things of which we do not have or cannot have a direct intuition, e.g., imaginary numbers or impossible objects. But also in cases in which direct intuition is possible, we often give preference to thinking through signs. Otherwise said, direct intuition is not generally necessary for the development of discursive-rational activity, except for a very few cases.

[56]See da Silva 201x.

[57]*PdA* 258; *PoA* 273 (italics in the original).

[58]Husserl, *Aufsätze und Rezensionen*, 393.

[59]loc. cit.

1.3 On the Shoulders of Leibniz and Bolzano: Husserl on the *Mathesis Universalis*

In a fragment entitled *Elementa Nova Matheseos Universalis* (Leibniz 1683?; cp. Couturat 1903, 348–351) Leibniz writes: "the *Mathesis Universalis* shall deliver the method, through which things that are conceivable can be exactly determined (*Mathesis Universalis tradere debet Methodum aliquid exacte determinandi per ea quae sub imaginationem cadunt, sive ut dicam Logicam imaginationis*)". In another fragment: "the *Mathesis* is the science of things that are conceivable" (Couturat 1901, 291).

Leibniz conceives all mathematical disciplines as branches of the *mathesis* and conceives the *mathesis* as a general science of forms applicable not only to magnitudes, but to any object whatever *that exists in our imagination*; i.e., to any objects that are *possible*, at least in principle. Notice that Leibniz displays an Aristotelian conception of alethic modalities.[60] He conceives necessity primarily as impossibility of the opposite: A proposition is necessary iff its opposite implies a contradiction.[61] Also, recall that though Aristotle does not explicitly characterize contingency,[62] Leibniz does and takes "contingent" as both "not- necessary"[63] and "actual", while he takes "possible" as tantamount with "non-contradictory". Thus, a complex concept is possible when the conjunction of its conceptual marks does not involve a contradiction. Let us now come back to the *mathesis*.

As a general *science of forms* the *mathesis* investigates possible relations between "objects in general (*objets quelconques*)". It is an abstract theory of combinations and relations among any objects whatever. Thus, the idea of *mathesis universalis* as formal ontology is, in a way, present already in Leibniz. But Leibniz makes even a step further and outlines a *general theory of operations* that has as its proper objects properties of and relations between operations.[64] His combinatorics contains the *general* and *formal* part of the *mathesis*, being at the same time an *art of discovery* or heuristics. We have hinted above at Husserl's elaboration on the Leibnizian *productive function of signs*. Note that in the last chapter of the *Philosophy of*

---

[60]Cp. Aristotle, *Metaphysics*.Δ 5, 1015a 34–36: "We say that that which cannot be otherwise is necessarily as it is. And from this sense of 'necessary' all the others are somehow derived." Cp. Leibniz, *Confessio philosophi* (Fall 1672-Winter 1672/73): "*Impossibile*, quod possibile non est. *Necessarium* cuius oppositum impossibile est., *Contingens* cuius oppositum possibile est" (Ak VI.3127).

[61]Cp. e.g. *Generales Inquisitiones de analysi notionum et veritatum* (1686): "Necessaria autem proposition est [. . .] cujus oppositam assumendo per resolutionem devenitur in contradictionem" (Ak VI.4A, 761).

[62]In *De Interpretation* 13 he defines "possible" ("endekomenon", "dunaton") as "that which is not impossible", In *An. Pr.* 13bis and 22 he takes "endekomenon" to mean "neither possible nor impossible", that is the characterization we usually give nowadays of "contingent".

[63]Cp. Leibniz, *Definitiones: ens, possibile, existens* (1687–96?) (Ak VI.4A, 867). Also cp. *Definitiones: terminus vel aliquid, nihil* (1688–89) (Ak VI.4A, 936).

[64]See Couturat 1901, 283–322.

Arithmetic (*The Logical Sources of Arithmetic*) Husserl sees the fundamental task of arithmetic, conceived as a general theory of operations [!], in (i) discovering all conceivable operations and (ii) finding for each conceivable operation an efficient algorithmic procedure to execute it.[65]

In the *Prolegomena to Pure Logic* Husserl maintains that Leibniz is the philosopher to whom his own conception of logic looks back, and that his own relation to Leibniz is "relatively of the closest".[66] Leibniz's great merit consists in his constant efforts to transform traditional logic "into a discipline of mathematical form and strictness, into a universal mathematics in the highest and most comprehensive sense".[67] His intuitions according to Husserl, "which speed forth so far beyond their time, seem sharply defined and highly admirable to a modern acquainted with 'formal' mathematics and mathematical logic".[68]

Husserl comes repeatedly to the topic of the *mathesis universalis* throughout his intellectual development. In his *Introduction to Logic and Theory of Knowledge* (lectures from 1906/07), for example, he characterizes Leibniz's *mathesis universalis* thus[69]:

> Under the heading of *mathesis universalis*, Leibniz [...] developed an extraordinarily expanded idea of pure mathematics. This most universal mathematics, according to him, no longer merely deals with what is quantitative, with quantity and number, but also with what is non-quantitative in terms of its mere form. All *argumenta in forma* belong to it, including the entire supply of formal theories of traditional Aristotelian-Scholastic logic. Very recently, Lotze said of arithmetic that it was no more than an independently developed branch of pure logic, and upon occasion Riehl agreed with him. Even though these scholars remained isolated in their views and even though they failed to investigate the matter in greater depth, they did, as we can observe, doubtlessly see what was right.

So, according to Husserl, for Leibniz the *mathesis universalis* is generalized formal mathematics, which he conceives as *mathematics of pure forms*. Later in the same work he says[70]:

> the field of reason as pure reason is then completely filled by pure logic as *mathesis universalis*, and the universality of this mathesis embraces apophantic logic, the mathematical discipline in the categorial and original sense and mathematics in the highest sense, the mathematics of form prototypes of possible theories in general and of possible mathematics in general.

So, for Husserl, *mathesis universalis* is nothing but pure logic, which covers the entire field of pure reason. Indeed[71]:

---

[65] *PoA* 277; *PdA* 262. Hereto cp. Centrone 2010a, 45ff.

[66] *PR* 219–222 (§ 60 *Anknüpfungen an Leibniz/ Links with Leibniz*), *Pre* 138–141. Henceforth: *PR* = Husserl, *Logische Untersuchungen* I, *Prolegomena zur reinen Logik*, Tübingen 1993; *PRe* = English translation thereof, in: *Logical Investigations*, London 1970, Vol. I, 51–247.

[67] loc. cit.

[68] loc. cit.

[69] *LV '06/07*, 54–55.

[70] *LV '06/07*, 91.

[71] *LV '06/07*, 104.

> Logic would, therefore, be the complex of disciplines of mathesis universalis since, in terms of form, it pertains to the possibility of knowledge of reality, and logic would be universal real ontology since, in terms of content, it pertains to the possibility of knowledge of reality.

Or still[72]:

> The concept of pure logic as of the science of the ideal constituents and laws of theory in general or, as we can also say, of the science of truth and objectivity in general. To the extent that it is understood as it must be understood, pure logic is identical with the *mathesis universalis*.

In *Ideas I,*[73] Husserl distinguishes two groups of *eidetic* sciences, formal and material; the first group, which he identifies with *formal mathesis universalis*, includes formal logic (i.e. formal apophantic) and formal ontology. Thus, the *mathesis universalis* includes the whole of formal logic, comprising both formal analytics and formal ontology.

In *Experience and Judgement*, Husserl still says that[74]:

> Taken in its original sense, formal logic is *not only* [*apophantic logic*]; for, in a fully developed formal logic, i.e. in a logic which, as formal *mathesis universalis*, includes formal mathematics, formal apophantics is the counterpart of formal ontology, the theory of something in general and of its derived forms [...].

In *Formal and Transcendental Logic*, Husserl characterizes Leibniz's *mathesis universalis* as syllogistics *plus* formal mathematics[75]:

> The [...] synthesis of traditional syllogistics and formal analysis [is] the Leibnizian idea of a mathesis universalis.

By the characteristics of this work, in which Husserl gradually enlarges the scope of formal logic to eventually include formal ontology alongside with apophantics, that is, the logic of being side by side with the logic of discourse, the concept of *mathesis universalis* is also gradually extended to finally coincide with the totality of formal logic as the a priori science of science. It seems that, at times, Husserl reduces the scope of *mathesis universalis* to formal analytics only[76]:

> Mathesis universalis [includes] mathesis of non-contradiction [and] mathesis of possible truth and then the interpretation of the former as a mathesis of pure senses.

Or still[77]:

> [M]athesis universalis [*is*] a science of apophantic senses [...]

But this is not a final characterization; although *mathesis universalis* is said to *include* both the logic of non-contradiction and truth-logic, it is *not* said that

---

[72] *LV '06/07*, 376.

[73] *Ideen* I, §7, 16ff. (o.p.).

[74] *EU* 1–2; Engl. transl. 11.

[75] *FTL* 65; Engl. transl. 73–74.

[76] *FTL* 122; Engl. transl. 138.

[77] *FTL* 127; Engl. transl. 143.

this is *all* that it includes. Husserl discloses the ontological domain by a change of perspective, from assertions (apophantic level) to the about-what of assertions (ontological level) insofar as the *particular material* content of assertions is considered abstractly in the mode of the something whatsoever. Formal-ontological senses are only apophantic senses from a different perspective; for this reason, by saying that[78]:

> [t]he whole mathesis universalis is accordingly the analytics of possible categorialia, the theory of their essential forms and eidetic laws.

Husserl is not leaving formal ontology, comprising also formal mathematics, outside the scope of the universal *mathesis*. This is clear when he discusses in §23b Leibniz's conception of *mathesis* (*The emerging of the idea of an enlarged analytics, Leibniz's "mathesis universalis," and the methodico-technical unification of traditional syllogistic and formal mathematics*) and in §26d, when he discusses the conception of formal ontology displayed by the Bohemian mathematician and philosopher Bernard Bolzano, as we shall see.

In his last work, The *Crisis of European Sciences and Transcendental Phenomenology*, the inclusion of formal ontology, formal mathematics particularly, in *mathesis universalis* is still more clearly vindicated[79]:

> The more physics advances in the effective mathematization of nature that is primarily given as the intuitive world around us, the more numerous are the physical-mathematical principles at its disposal and, at the same time, the more elaborate is the instrument it possesses for that (the mathesis universalis) [...]

In this quote, *mathesis universalis* is clearly characterized as containing the symbolical-mathematical instruments of investigation of mathematized nature. Husserl uses also the expression "formal mathesis" and indicates that, for him, formal mathesis constitutes a "system of formal laws", which, as he tells us, regulate not only the realm of senses, but also that of *possible being* exclusively as to their *form*.

According to Husserl, after Galileo had set the basis for the new mathematical science of nature, Descartes conceived the idea of a *systematic* universal philosophy, giving it in fact the sense of a mathematical rationalism, philosophy as "universal mathematics". But Descartes could not take this idea to full completion for it required a "*higher systematic elaboration of pure mathematics*" in accordance with this new idea of universality "that appeared in a relatively mature form only with Leibniz as *mathesis universalis*, which is today object of active investigation in the still more mature form of *the mathematics of definite multiplicities*". Clearly, as this last quote shows, formal mathematics is for Husserl too an integral part of *mathesis universalis*.

Already Early Husserl stresses the ontological character of the *mathesis*. It is true that in the last chapter of the *Philosophy of Arithmetic* we find investigations

---

[78] *FTL* 123; Engl. transl. 138.

[79] *Krisis* §9 g.

on *universal arithmetic* that seem to be more on the "apophantic side" and that 10 years later, in a *Double Lecture* presented to the *Mathematische Gesellschaft* in Göttingen in the winter semester of 1901/02,[80] and in the preparatory writings for this lecture,[81] Husserl explicitly says that Leibniz conceives the *mathesis* as a formal *theory of theories*, thus again of the "apophantic side". Indeed he writes[82]:

> Originally limited to the field of numbers and quantities, mathematics has grown far beyond that field. It has increasingly approximated to the goal that *Leibniz* had already clearly conceived, namely, the goal of being a pure theory of theories (*Theorienlehre*), free of all special fields of knowledge and insofar formal. Mathematics in the highest and most inclusive sense is the science of theoretical systems in general, in abstraction from that which is theorized in the given theories of the various sciences.

But Husserl is also always interested in studying the ontological correlate of formal theories. He is always aware of the fact that to formal theories there correspond "possible fields of experience" that are axiomatized by those theories. In this sense one can maintain that already Early Husserl rightly stresses the *ontological* character of the *mathesis*. Abstract mathematics, as *mathesis universalis*, is, viewed from the ontological side, a most general theory of structures each constructed with its own axiom-system.

But did Husserl get the idea of the ontological character of the *mathesis* directly from Leibniz? Given that late Husserl himself mentions Early Bolzano in *FTL*, it is interesting to look closer at Bolzano's conception of the *mathesis* as displayed in his *Contributions to a Better Grounded Presentation of Mathematics* (1810) – which, by the way, is the text on which Husserl focuses his *Logic Lecture 1906/07* and in §26d of *FTL*.

Like Husserl almost 100 years later in the *Double Lecture*, in the *Contributions* Bolzano starts off with a critique of the definition of mathematics as 'the science of quantity' (§1) and comes gradually to the following definition of mathematics: it is "a *science which deals with the general laws (forms) to which things must conform* [*sich richten nach*] *in their existence* [*Dasein*]" (§8). He explains that he understands by "things", not only those that actually exist, "but also those which simply exist in our *imagination* [...] in other words, *everything which can in general be an object of our representational capacity* [*Vorstellungsvermögens*]" (§8). Bolzano means by "things, which simply exist in our *imagination*", exactly like Leibniz more than one century before, "things that are possible" or "things that can be represented, since their concept does not imply a contradiction". From his definition of mathematics as the "totality of the laws of possible being",[83] Bolzano derives a classification of the *mathesis* in several particular disciplines. *The laws to which things must*

---

[80]Husserl, *Das Imaginäre in der Mathematik. I: Zu einem Vortrag in der mathematischen Gesellschaft in Göttingen 1901*; in: *PdA* 430–451; *PoA* 409–452.

[81]See also *Drei Studien zur Definitheit und Erweiterung eines Axiomensystems,* in: *PdA* 452–469, *PoA* 432–438 & 453–464.

[82]*PdA* 430; *PoA* 409–410.

[83]We borrow here terminology from Casari 2004.

conform in their existence can be either so general that they are applicable to *all things completely without exception*, or not. The former laws constitute the *general mathesis*; the latter, the single branches of the *mathesis* (§11). Incidentally, it is worth notice that Bolzano sometimes uses "pure general mathesis" as synonymous of "arithmetic". Thus, arithmetical laws are laws that apply to all things without exception. Bolzano conceives the general *mathesis* and its classification according the following schema:

A
General mathesis
(things in general)
B
Particular mathematical disciplines
(particular things)
I
Aetiology
(things which are not free)
II
(sensible things which are not free)
a
(form of these things *in abstracto*)

α
Theory of time
(time)

β
Theory of space
(space)

b
(sensible things *in concreto*)

α
Temporal aetiology
(sensible things in time)

β
Pure natural science
(sensible things in time and space)

The *general mathesis* includes arithmetic and the theory of combinations. These are laws that are applicable to all things without exception. Bolzano does not mention explicitly the particular mathematical disciplines that make up the different branches of the *mathesis*, but we believe that he is thinking of something like set-theory, theory of manifolds, mereology (as we would say nowadays), and so on. "Particular things", in the schema above, means, as we interpret Bolzano, sets, manifolds, parts of a whole, and so on. In the disciplines belonging to the general *mathesis,* we have just signs, combinations-signs and laws that prescribe how to obtain new signs and signs-combination from given ones. In the various *particular mathesis*, we restrict the field of application, we say, for instance: the signs vary on sets, manifolds ... and the combinations-signs denote combinations that make sense in the field in question. The various disciplines that constitute as many branches of the *particular mathesis* stand with respect to the *general mathesis*, that is, with the disciplines that belong to *general mathesis*, in a species-genus relationship.

Bolzano justifies the partition of B into (I) and (II) of the schema above as follows. Everything, which *exists*, is either *necessary or free* in its existence. Things that are *free* in their existence (as, for example, human decisions) are subject to no conditions and laws and are therefore not an object of mathematics. Things that are *necessary* in their existence are so, either *simply* or *conditionally* (i.e. on the presupposition of something else).

> The necessary in itself is called God and is considered in metaphysics not as a merely possible object but as an actual object. Therefore, there remains only the hypothetically necessary, which we consider as produced through some ground (*Grund*). Now there are certain general conditions according to which everything, which is produced through a ground (in or out of time) must be regulated in its becoming or existence. These conditions taken together and ordered scientifically will therefore constitute the first main part of the particular mathesis, which I call, for want of a better name, the theory of grounds (*Grundlehre*) or aetiology (§ 13).

All that is, is divided in things that have spatial and temporal determinations and such that do not. Thus, time and space, according to Bolzano, are also conditions to which things that are actual must conform in their existence. "Therefore if we develop the properties of time and space in abstracto and order them scientifically, these sciences must also be counted as mathematics .... We therefore have the second and third components of particular mathesis, the theory of time (chronometry), and the theory of space (geometry)" (§14). If, finally, time and space are not considered merely *in abstracto* but as occupied with *actual things* and, precisely, with actual things that are *not* free in their existence but are subject to the laws of causality, then two *new sciences* appear, namely:

(a) *the theory of causes* (*Ursachenlehre*) or *temporal aetiology*, i.e., the general laws to which *unfree things*, which are in *time*, must conform in their existence (and in their changes) and
(b) *pure natural science, that is, the totality of laws that govern unfree things which are both in time and in space at once (§15).*

It is not important for us to establish whether Bolzano's subdivision of the *mathesis* is something we can live with or not, but it is important to stress that, to put it with Judson Webb, Husserl "had really gotten key ideas from [his] teachers and predecessors"[84] and that it is not possible to really understand these ideas without knowing these antecedents. Husserl repeatedly talks in his writings of "something in general (*Etwas überhaupt*)" and of the *mathesis* as the totality of "determination-forms of the something in general". "*Etwas überhaupt*" is another wording for "object in general". "*Mathesis universalis* as the totality of the determination-forms of the something in general" means that it is possible to think of objects whatever, only in certain forms, for instance, as sets or manifolds. The theory of sets, the

[84] Webb 1980, xii.

theory of manifolds ... are most general theories. They are made up of laws to which objects must conform in their existence – as Bolzano would put it – if they are to be thought by us.[85]

Finally, let us turn our attention again to the *LV 06/07* (the first lecture series held by Husserl in Göttingen.) It consists of three sections (Section I: *Die Idee der reinen Logik als einer formalen Wissenschaftstheorie*; Section II: *Noetik, Erkenntnistheorie und Phänomenologie*; Section III: *Die Objektivationsformen*) and several supplementary texts. Chapter II (*Die reine Logik als Theorie der Wissenschaft*) of Section I turns out to be particularly interesting for our topic, for Husserl displays there his own understanding of the *mathesis* as science of possible being. In particular, in §15 (*Die Einordnung der formalen Mathematik in die Wissenschaftstheorie*) Husserl upholds his interpretation of formal mathematics as formal ontology, i.e., of formal theory of being in general[86]:

> [T]his most universal theory of objects, this formal ontology, [encompasses] all of formal mathematics.

Like Bolzano Husserl states that: "formal mathematics [excludes] geometry",[87] since geometry is the theory of things that are in space and not of things in general. The general *mathesis* encompasses[88]:

> [T]he pure theory of cardinal and ordinal numbers, the theory of combinations, and all the disciplines of so-called analysis, theory of numbers, theory of functions, algebra, differential and integral calculus, [the] theory of Euclidean and non-Euclidean multiplicities and all theory of multiplicities generally

Taking Leibniz as forerunner, Husserl remarks[89]:

> [Es ist] doch auffallend, daß ein Leibniz, dessen historische Größe mit der fortschreitenden Kenntnis seines beispiellos umfangreichen Nachlasses so gewaltig gewachsen ist, sich unter dem Titel ‚mathesis universalis" eine außerordentlich erweiterte Idee der reinen Mathematik gebildet hat. Diese allgemeinste Mathematik hat es nach ihm nicht mehr bloß mit dem Quantitativen zu tun, mit Größe und Zahl, sondern auch mit dem Nichtquantitativen seiner bloßen Formen .

Formal mathematics and formal logic are viewed, in this perspective, as the "capital" of pure conceptual truths that are at disposal of each science in general.[90]

---

85 Hereto cp. Casari 2004.

86 *LV 06/07*, 55.

87 loc.cit. (quotation marks added).

88 loc.cit.

89 *LV 06/07*, 55 (quotation marks added).

90 Cp. *LV 06/07*, 59 ff.

1.4 Concluding Remarks

We can summarize our demarche so far thus: Husserl inherits from Leibniz the notion of symbolic and symbolic means for thinking and knowing, together with his conception of sign as an indirect means of referring, that is, a means of univocal identification, and from Bolzano the understanding that Leibniz' *mathesis universalis* consists of the totality of laws of possible being. Taking inspiration from both Bolzano and Leibniz Husserl develops his own conception of *mathesis universalis* as coinciding with the whole of formal logic, encompassing both formal apophantics and formal ontology, thus providing a better delimitation and a clearer characterization of this notion, improving on the inherited tradition.

But some important questions remain which must be addressed, bearing on the role Husserl allowed the universal *mathesis*, that is, formal logic, to have in his epistemology. Two such questions stand out: (1) in view of his characterization of knowledge as *intuitively filled* intentions, in what sense is formal logic, for Husserl, a form of knowledge, if indeed it is a form of knowledge? Given that formal mathematics is, for Husserl, part of formal logic (formal ontology, more precisely), the answer to this question has implications for Husserl's epistemology of formal mathematics. (2) In any case, regardless of whether it is a form of knowledge or not, how can formal logic contribute, if it indeed can, to knowledge *proper*, that is, knowledge in principle capable of intuitive validation? This question has to do, in particular, with the role Husserl allowed formal mathematics to play in the empirical sciences, more notably in physics. In the following, we will summarize the solution to these problems presented in da Silva (2012, 2013).

Husserl does indeed require, in the *Sixth Logical Investigation*, that once the object of an empty intentional representation *can* be intuited then it *must* be so in order for that representation to qualify as knowledge in the proper sense. But what if the object of representation cannot be intuited, as is the case of objects that do *not* exist, but cannot be a priori ruled out as unable to come into existence, that is, mere *possibilia*? Of course, their intuitive presentation cannot be required, and then, if knowledge of *possibilia* is not, as Husserl certainly believed, dismissible out of hand, then it cannot, of course, be intuitive knowledge.

As the previous discussion made clear, formal logic, formal apophantics and formal ontology (comprising formal mathematics), has to do with things – senses or objectualities – that are not necessarily actual. Formal apophantics is concerned with assertions regardless of whether they are actually true or only merely meaningful, and then *capable in principle* of being true; formal ontology, on its turn, is concerned with ontological categories and formal objectual structures regardless of material content.

So, formal logic – Husserl's *mathesis universalis* – is in itself a form of knowledge – *formal knowledge* to give it a name – understood as knowledge of *forms*, *logical* forms. Formal knowledge is propaedeutic to *material* knowledge, which requires the intuitive presentation of the content of knowledge as fulfilling that which is asserted of it. Husserl's *mathesis universalis* is the (meta) science of science with regard to its formal a priori aspects. In the particular case of formal mathematics, its task is to investigate all possible forms or structures that domains of objects, actually existing or not, can *possibly* display – clearly a variant of structuralism.

Our second question can be put thus: can formal knowledge contribute to material knowledge, how? In some aspects, the answer is obvious. Since they hold in general, the laws of the logic of deduction, for example, hold also for *any* system of materially filled assertions whatsoever. The same is true for formal truth-logic (logic of deduction and truth-logic belong, both, according to Husserl, to formal apophantics). Formal-logical laws regulating ontological categories (whose investigation belongs to formal ontology) are also, of course, valid in whatever actually given domain of objects. The same is true for formal mathematics; any formal mathematical theory in abstracto instantiates as proper knowledge of any of its materializations (anything that holds for groups in abstracto, for example, holds for any concretely given group). In short, anything that is true for what *can possibly* exist is per force true of what *actually* exists.

As we can see, for Husserl, *mathesis universalis* is both in itself a body of (formal) knowledge and an instrument of investigation at the service of the particular sciences, either formal or material.

References

Aristotle, *The Complete Works*, vol. I–II, ed. by J. Barnes (Princeton University Press, Princeton, 1984)

Y. Belaval, *Leibniz Critique de Descartes* (Gallimard, Paris, 1960)

G. Berkeley, The analist; or a discourse addressed to an infidel mathematician (1734), in A.A. Luce, T.E. Jessop (eds.), *The Works of George Berkeley Bishop of Cloyne*, vol. 4. (Nelson, London/Edinburgh, 1948–57), pp. 53–102

B. Bolzano, *Beyträge zu einer begründeteren Darstellung der Mathematik. Erste Lieferung* (Caspar Widtmann, Prague, 1810). Reprint: Wissenschaftliche Buchgesellschaft, Darmstadt 1974. English translation in S. Russ (ed.), *The Mathematical Works of Bernard Bolzano* (Oxford University Press, Oxford, 2004), pp. 82–137

G. Cardano, *The Rules of Algebra (Ars Magna)* (Dover, Mineola, 2007)

E. Casari, Logic and the laws of possible being, in *Logic and Metaphysics*, ed. by M. Marsonet, M. Benzi (Name, Genova, 2004), pp. 16–43

S. Centrone, *Logic and Philosophy of Mathematics in the Early Husserl* (Springer, Berlin, 2010a)

S. Centrone, Bolzano und Leibniz über Klarheit und Deutlichkeit, Archiv für Geschichte der Philosophie. **92**(3), 256–289 (2010b)

L. Couturat, *La Logique de Leibniz: d'après des documents inédits* (F. Alcan, Paris, 1901)

L. Couturat, *Opuscules et fragments inédits de Leibniz* (F. Alcan, Paris, 1903)

G. Crapulli, *Mathesis universalis. Genesi di un'idea nel XVI secolo* (Edizioni dell'Ateneo, Roma, 1969)

J.J. Da Silva, Away from the facts, symbolic knowledge in Husserl's philosophy of mathematics, in *Symbolic Knowledge from Leibniz to Husserl, Studies in Logic*, vol. 41, ed. by A.L. Casanave (College Publications, London, 2012), pp. 115–136

J.J. Da Silva, Beyond Leibniz, Husserl's vindication of symbolic mathematics, in *The Road Not Taken, On Husserl's Philosophy of Logic and Mathematics,* ed. by C.O. Hill, J.J. da Silva. (College Publications, London, 2013), pp. 61–80

J.J. Da Silva, Husserl and Hilbert on completeness, still, *Synthese* 201x (online first 16/07/2015)

C. Dasypodius, *Euclidis quindecim elementorum geometriae primum* (C. Mylius, Strasburg, 1564a)

C. Dasypodius, *Euclidis quindecim elementorum geometriae secondum* (C. Mylius, Strasburg, 1564b)

C. Dasypodius, *Euclidis elementorum liber primus* (C. Mylius, Strasburg, 1571)

C. Dasypodius, *Protheoria Mathematica* (Argentorati, Strasburg, 1593)

R. Descartes, *[AT.X] Oeuvres*, vol. X, ed. by C. Adam, P. Tannery (Léopold Cerf, Paris, 1908)

R. Descartes, *The Geometry of René Descartes*, ed. by D.E. Smith, M.L. Lathan (Dover, Mineola, 1954)

G. Frege, *Begriffsschrift und andere Aufsätze. Mit E. Husserls und H. Scholz' Anmerkungen*, ed. by I. Angelelli (Wissenschaftliche Buchgesellschaft, Darmstadt, 1964)

E. Husserl, *[Krisis] Die Krisis der europäischen Wissenschaften und die transzendentale Phänomenologie* (M. Nijhoof, The Hague, 1954)

E. Husserl, Besprechung von: E. Schröder, *Vorlesungen über die Algebra der Logik*, I. Band, Leipzig 1890, Göttingsche gelehrte Anzeigen (1891a), pp. 243–270. English translation in Husserl (1994), pp. 52–91

E. Husserl, Das Folgerungskalkül und die Inhaltslogik, Vierteljahrsschrift für wissenschaftliche Philosophie. **15**, 168–189 (1891b). English translation in Husserl (1994), pp. 92–114

E. Husserl, [*FTL*] *Formale und transzendentale Logik. Versuch einer Kritik der logischen Vernunft*, Jahrbuch für Philosophie und phänomenologische Forschung X (1929). Critical Edition: *Formale und transzendentale Logik. Versuch einer Kritik der logischen Vernunft*, ed. by P. Jannsen, HGW XVII, 1975. English translation: *Formal and Transcendental Logic*, ed. by D. Cairns (Martinus Nijhoff, Den Haag, 1969)

E. Husserl, *Logische Untersuchungen*, Bd I, [*PR*] *Prolegomena zur reinen Logik*, Max Niemeyer, Halle, 1900 (^{2}1913); Bd II, [*LU*] *Untersuchungen zur Phänomenologie und Theorie der Erkenntnis*. Max Niemeyer, Halle a.d.S., 1901 (^{2}1913–1921, ^{7}1993). English translation: *[PRe/LI] Logical Investigations*, ed. by J.N. Findlay (Routledge, New York, 1970)

E. Husserl, [*EU*] *Erfahrung und Urteil* (Academia Verlagsbuchhandlung, Prag 1939). English translation: *Experience and Judgement* (Northwestern University Press, Evanston, 1973)

E. Husserl, [*Ideen*] *Ideen zu einer reinen Phänomenologie und phänomenologischen Philosophie*, Jahrbuch für Philosophie und phänomenologische Forschung I (1913) [as a separate volume: Niemeyer, Halle a.d.S. 1913; ^{2}1922; ^{3}1928]. Critical Edition: *Ideen zu einer reinen Phänomenologie und phänomenologischen Philosophie*, ed. by K. Schuhmann, HGW III/1 (1976)

E. Husserl, *Aufsätze und Rezensionen (1890–1910)*, ed. by B. Rang, HGW XXII (1979)

E. Husserl, *Briefwechsel*, ed. by K. Schuhmann, Husserliana Dokumente vol. III (Kluwer, Dordrecht/Boston/London, 1994a)

E. Husserl, *Early Writings in the Philosophy of Logic and Mathematics* (Kluwer, Dordrecht, 1994b)

E. Husserl, *Philosophie der Arithmetik. Logische und psychologische Untersuchungen*, Bd. I, Pfeffer, Halle a.d.S. 1891. Critical Edition: [*PdA*] *Philosophie der Arithmetik. Mit ergänzenden Texten (1890–1901)*, ed. by L. Eley, HGW XII (1970). English translation: *[PoA] Philosophy of Arithmetic, Psychological and Logical Investigations with Supplementary Texts from 1887–1901*, ed. by D. Willard (Kluwer, Dordrecht, 2003)

E. Husserl, [HGW] *Husserliana, Gesammelte Werke,* I–XXVI: Martinus Nijhoff, Den Haag, 1950–, XXVII–XXXVII: Kluwer, Dordrecht, 1989–; XXXVIII–, Springer, New York, 2005

E. Husserl, [*LV'06–07*] *Einleitung in die Logik und Erkenntnistheorie Vorlesungen 1906/07*, ed. by U. Melle, HGW XXIV, 1985. English translation: *Introduction to Logic and Theory of Knowledge* (Lectures 1906/1907), ed. by C.O. Hill (Springer, Dordrecht, 2008)

G.W. Leibniz, [1666] *Dissertatio de Arte Combinatoria*, Akademieausgabe VI.1.A (Akademie Verlag, Berlin, 1999a), pp. 163–230

G.W. Leibniz, [*MED*] *Meditationes de cognitione, veritate et ideis* (1684), Akademieausgabe IV.4.A (Akademie Verlag, Berlin, 1999b), pp. 585–592

G.W. Leibniz, [1672/73] *Confessio Philosophi*, Akademieausgabe VI.3 (Akademie Verlag, Berlin, 1981), pp. 115–149

G.W. Leibniz, [1677a] *Quid sit Idea?*, Akademieausgabe VI.4.B (Akademie Verlag, Berlin, 1999), pp. 1369–1371

G.W. Leibniz, [1677b] *Dialogus*, Akademieausgabe IV.4.A (Akademie Verlag, Berlin, 1999), pp. 20–25

G.W. Leibniz, [1678] *Leibniz an die Pfalzgräfin Elizabeth* (?), Akademieausgabe II.1.A (Akademie Verlag, Berlin, 1987), pp. 659–666. English translation: *Letter to Countess Elizabeth. On God and Formal Logic*, ed. by R. Ariew, D. Garber (Hackett Publishing Company, Indianapolis/Cambridge, 1989)

G.W. Leibniz, [1683?] *Elementa Nova Matheseos Universalis*, Akademieausgabe IV.4.A (Akademie Verlag, Berlin, 1999), pp. 513–524

G.W. Leibniz, [1686] *Generales inquisitiones de analysi notionum et veritatum*, ed. by F. Schupp (Meiner, Hamburg, 1982)

G.W. Leibniz, [1687a] *Non inelegans specimen demonstrandi in abstractis*, Akademieausgabe IV.4.A (Akademie Verlag, Berlin, 1999), pp. 845–855

G.W. Leibniz, [1687b/96?] *Definitiones: ens, possibile, existens*, Akademieausgabe VI.4.A (Akademie Verlag, Berlin, 1999), pp. 867–870

G.W. Leibniz, [1688a?] *De Characteribus et de Arte Characteristica*, Akademieausgabe IV.4.A (Akademie Verlag, Berlin, 1999), p. 916

G.W. Leibniz, [1688b–89] *Definitiones: terminus vel aliquid, nihil*, Akademieausgabe VI.4.A (Akademie Verlag, Berlin, 1999), pp. 934–936

G.W. Leibniz, [*NE*] *Nouveaux Essais sur l'entendement humain* (1704), Akademieausgabe IV.6 (Akademie Verlag, Berlin, 1962)

G.W. Leibniz, *Nova Methodus* (1684), ed. by D.J. Struik, *A Source Book in Mathematics, 1200–1800* (Princeton University Press, Princeton, 1969)

G.W. Leibniz, *Logical Papers*, ed. by G.H.R. Parkinson (Clarendon Press, Oxford, 1966)

G.W. Leibniz, *Die Grundlagen des logischen Kalkuls*, ed. by F. Schupp, S. Weber (Meiner, Hamburg, 2000)

P. Mancosu, *Philosophy of Mathematics and Mathematical Practice in the Seventeenth Century* (Oxford University Press, Oxford, 1996)

P.J. Nahin, *An Imaginary Tale. The History of* $\sqrt{-1}$ (Princeton University Press, Princeton, 1998)

N.R. Pearcey, C.B. Thaxton, *The Soul of Science* (Crossway Books, Wheathon, 1994)

A. Piccolomini, *Commentarius de certitudine mathematicarum* (Antonio Baldum Asulanum, Roma, 1547)

E. Schröder, *Vorlesungen über die Algebra der Logik (Exakte Logik)*, Bd. I, (Teubner, Leipzig, 1890)

F.A. Trendelenburg, *Über Leibnizens Entwurf einer allgemeinen Charakteristik*, Königl (Akademie der Wissenschaften, Berlin, 1856)

A. Van Romen, *Apologia pro Archimede* (Fleischmann, Würzburg, 1597)

J. Webb, *Mechanism Mentalism and Metamathematics* (Reidel, Dordrecht, 1980)

D. Willard, *Logic and the Objectivity of Knowledge: A Study in Husserl's Philosophy* (Ohio University Press, Athens, 1984)

Chapter 2
Husserl and His Alter Ego Kant

Judson Webb

Abstract Husserl's lifelong interest in Kant eventually becomes a preoccupation in his later years when he finds his phenomenology in competition with Neokantianism for the title of transcendental philosophy. Some issues that Husserl is concerned with in Kant are bound up with the works of Lambert. Kant believed that the role played by principles of sensibility in metaphysics should be determined by a "general phenomenology" on which Lambert had written. Kant initially believed that man is capable only of symbolic cognition, not intellectual intuition. Lambert saw an increasing need in mathematics for symbolic cognition as exemplified in his proofs of the irrationality of π and e. Kant takes from Leibniz and Lambert an unrestricted notion of construction, allowing him to view mathematics as constructing its concepts in intuition, while Lambert's proofs convince him that all mathematical problems are eventually solvable. Husserl criticizes Kant's intuitionism for its inadequate accommodation of meaning to intuition, which he redresses with his theory of categorial intuition. This may improve on Kant's intuition of the axiom of parallels but not so clearly on his spatial intuition. Husserl opposes Kant's view of space as the form of outer intuition with his own view of it as the form of things. Husserl's exploration of the geometry of visual space, which involves his earliest uses of epoché and reduction, converges however with Hilbert's logical analysis of Kantian spatial intuition in leading to Euclidean spatial judgments. Hilbert's analysis leads him to affirm the solvability of all well posed mathematical problems, a thesis complicated by the outbreak of logical paradoxes. Untroubled by such paradoxes, Husserl develops a supramathematics of all possible deductive systems whose completeness implicitly would also provide solutions of all such problems. Husserl's full transcendental turn coincides with his realization that in effecting his Copernican turn, Kant was really the first to detect the "secret longing" of modern philosophy for a phenomenological clarification of being. Husserl now finds that Kant's transcendental deductions presuppose a pure ego not adequately analyzed by Kant that survives the phenomenological reduction. Husserl's idealism leads him to develop his own intuitionism, which adds to Kant's, a "method of clarification" of mathematical concepts intended to clarify difficult impossibility proofs, but neither

---

J. Webb (✉)
Department of Philosophy, Boston University, Boston, MA, USA
e-mail: ahooper4@verizon.net

S. Centrone (ed.), *Essays on Husserl's Logic and Philosophy of Mathematics*, Synthese Library 384, DOI 10.1007/978-94-024-1132-4_2

Husserl nor Kant base arithmetic on time. Husserl's critique of the room left in Kant's idealism, for things in themselves, leads to his own monadological solution of the problem of intersubjectivity. Husserl's final judgment on Kant is that his form of transcendental idealism did not enable him to achieve absolute subjectivity through a genuine transcendental reduction.

Keywords Kant • Lambert • Symbolic cognition • Phenomenology • Epoche • Solvability • Unprovability • Categorial intuition • Clarification • Intuitionism • Husserl

2.1 Introduction

Kant's influence on modern philosophy is immense and understandable. His influence in the philosophy of mathematics is also considerable but is curiously disproportionate with his modest level of mathematical knowledge and his allegiance to Aristotelian logic. He has nevertheless influenced the mathematical philosophy of arguably the four greatest mathematicians of the early twentieth century: Poincaré, Hilbert, Brouwer, and Weyl. It would be interesting to relate some of their differences to the specific sections of the *Critique of Pure Reason* which they have found most attractive, but our interest is with his influence on Husserl, in whose case it was clearly the *Transcendental Analytic* with deductions of the categories that increasingly grip his imagination. Despite a number of criticisms he makes of Kant along the way, Husserl eventually finds that Kant is the first philosopher to be consciously attuned to the "secret longing" of modern philosophy for a phenomenological clarification of being. I shall try to analyze, first, how Kant came to deserve such a compliment, and then, why Husserl came to deny that Kant was ever able to satisfy such a longing.

In Sect. 2.2 I analyze how Kant's defense of his infinite Newtonian cosmology obliged his embrace of the *symbolic cognition* of its concepts of infinity and continuity and his denial that we could have any *intellectual intuition* of them. Kant also explores his own intellectual horizons to see whether difficult questions such as squaring the circle are within them, concerning which he takes from Leibniz a certain tolerance as to what "constructions" might count towards a solution of the problem. From Lambert he learns that the widespread and long held belief, that the problem can be solved by finding a rational value for π, is impossible. Lambert's proof of the irrationality of π eventually joins forces with Kant's belief that all *a priori* mathematical concepts are constructible in intuition, to persuade him that all mathematical problems can be solved. But this only occurs when the doctrine of the *Dissertation*, with its absolute separation of the sensible and intellectual worlds, gives way in the *Critique* to a theory of schematisms that allow for the cooperation and interaction of the faculties of sensibility and the understanding. Here again the crucial influence is from Lambert, who has actually been called Kant's alter ego. However, Lambert did not persuade Kant that mathematics is becoming more dependent on symbolic cognition.

In Sect. 2.3 I analyze Husserl's attempt to justify and improve upon Kant's claims of the importance of intuition in mathematics by formulating his theory of a categorial intuition which he says makes up for Kant's "inadequate accommodation of meaning to intuition." The results on this front are mixed. Husserl's critique of Kant's ideality of space leads him to his own transcendental exploration of the geometry of visual space, with the conclusion that it is Euclidean. Meanwhile, Hilbert's logical analysis of spatial intuition, while bracketing the question of the origin of such intuition and taking Kant's space axiom as its point of departure, converges with empirical research that also depends on Husserl's epoche to the same conclusion. The question is whether and in what sense we have a categorial intuition of this axiom, which postulates the existence of a unique plane through any three non-collinear points of space. Hilbert's logical analysis depends crucially on a number of underivability results, which prompt his proposal that "the modern principle is the proof of unprovability," and which further emboldens him to propose that all well posed mathematical problems are solvable. I analyze Husserl's interesting more reserved attitude towards this mark of modernity, as well as his own more sweeping Leibnizian vision of a supramathematics of all possible complete theory forms which implicitly would in principle decide any such problem.

In Sect. 2.4 I examine Husserl's full engagement with Kant's transcendental idealism and his efforts to improve and extend it to a full phenomenological idealism. His effort to eliminate Kantian things in themselves eventually leads to his embrace of a monadological solution to the problem of intersubjectivity. I try to explain why this problem was more pressing for Husserl than it was for Kant, at least with respect to mathematical knowledge. Aside from intersubjectivity, Husserl squarely faces the problem of clarifying the epistemic stature of those long indirect impossibility proofs that are needed to sustain the thesis of the solvability of all logical and mathematical problems, where it may seem as if a categorial intuition of such impossibility is out of the question. Be that as it may, Husserl's method of clarification leads him to his own brand of phenomenological intuitionism, quite different from Bouwer's in not basing everything on an intuition of time. In fact, I try to show that Kant did not regard an intuition of time as the basis of arithmetic, which he sees as a "purely intellectual synthesis." Finally, I examine the meaning of Husserl's claim that Kant failed to mount a genuine phenomenological reduction.

2.2 Symbolic Cognition, Horizons, Intuition

On October 2, 1770 Kant wrote to Lambert asking for his judgment on the *Dissertation*, especially those sections dealing with the distinction between sensible and intellectual things and the proper use of this distinction in metaphysics. Kant motivated his new idealism of space and time by reminding Lambert that:

> The most universal laws of sensibility play a deceptively large role in metaphysics, where, after all, it is merely concepts and principles of pure reason that are at issue. A quite special, though purely negative science, general phenomenology (*phenomenologia*

> *generalis*), seems to me to be presupposed by metaphysics. In it the principles of sensibility, their validity and their limitations, would be determined, so that these principles could not be confusedly applied to objects of pure reason, as has heretofore almost always happened.[1]

Before formulating his arguments that space and time are pure intuitions, Kant previews his case for their ideality by announcing that:

> There is (for man) no *intuition* of what belongs to the understanding, but only a *symbolic cognition*; and thinking is only possible for us by means of universal concepts in the abstract, not by means of a singular concept in the concrete. For all our intuition is bound to a certain principle of form, and it is only under this form that anything can be *apprehended* by the mind immediately or as singular and not merely conceived discursively by means of general concepts. But this formal principle of our intuition (space and time) is the condition under which something can be the object of our senses. Accordingly, this formal principle . . . is not a means to intellectual intuition.[2]

Our lack of such intuition raised a problem for Kant's new cosmology in which he posited that Newtonian attraction is "the cause of the systematic constitution of the fixed stars in the Milky Way and is precisely that universal relation which unites the parts of our nature in *one* space . . . into all reaches of its infinity."[3] Some critics had argued that the notions of continuous magnitude and infinity were not humanly representable, to which Kant now replies that "From this it is clear how, since unrepresentable and impossible are commonly treated as having the same meaning, the concepts of both the *continuous* and the *infinite* are frequently rejected. For indeed, *according to the laws of intuitive cognition*, any representation of these concepts is absolutely impossible."[4] Kant grants that whatever conflicts with the laws of the understanding and the laws of reason is impossible, but claims that the "lack of accord" between sensibility and understanding "points only to the fact that abstract ideas which the mind entertains when they have been received from the understanding very often cannot be followed up in the concrete and converted into intuitions. But this *subjective* resistance often creates the false impression of an objective inconsistency."[5] Kant argues that the admitted absurdity of an infinite *number* pressed by his critics by no means rules out infinite *multiplicities* larger than any number, which they had overlooked: otherwise "they would have seen clearly that things which do not accord with a fixed law of a certain subject do not, for that reason, pass beyond all understanding. For there could be an understanding, though certainly not a human understanding," that could apprehend an infinite multiplicity "at a single glance" without having to run through it successively.[6] Such an understanding would require an intellectual intuition that Kant denies us: "Divine intuition, however, which is the principle of objects, and not something governed by

[1]Kant 1999, p. 108.

[2]Kant 1992b, p. 389.

[3]Kant 1981, p. 150.

[4]Kant 1992b, p. 378.

[5]loc. cit., p. 379.

[6]loc. cit. p. 379.

a principle, since it is independent, is an archetype and for that reason perfectly intellectual."[7] It is this invocation of an archetype with intellectual intuition that Husserl will brand as the chief of Kant's "mythical constructions."

Lambert's extraordinary reply of October 13, 1770 gave his approval to much of Kant's new doctrine, with the exception of the ideality of time and its overly sharp separation of sensibility and understanding. He argues that the former is incompatible with the reality of *change*: "All changes are bound to time, and are inconceivable without time. *If changes are real, then time is real*, whatever it may be. *If time is unreal, then no change can be real*. I think though, that even an idealist must grant that changes really exist and occur in his representations, for example, their beginning and ending."[8] Implicit in Lambert's argument is that "even an idealist" must admit he observes a succession of changes in his *self*, and hence knows an object suffering changes in time.[9] Kant cannot answer this from the standpoint of the *Dissertation*, where time is said to be "an internal law of the mind" which knows things of the intelligible world as they *are*, not just as they appear. He will have to invoke an "inner *sense*" for time to be the "form" of, so that he can say that even an idealist can apprehend himself only as changes in his representations appear to this "sense," and can only establish those "finite determinations of duration" Lambert said were inseparable from his existence by means of the motions of external bodies.

Lambert's main criticism of the *Dissertation* was directed at Kant's complete separation, "in accordance with the old *phenomenon* and *noumenon* distinction," of the sensible and intellectual worlds, so that "knowledge that comes from the senses thus is and remains sensible, just as knowledge that comes from the understanding remains peculiar to the understanding."[10] Lambert urged that a science with intellectual concepts cannot be *tested* unless they are contaminated with sensible concepts. Euclid's fourth common notion is a good example, contaminating equality with coincidence. Kant will take up this point in the *Critique* by incorporating the apparatus of the Schematism into his epistemology.

As for Kant's case for man's dependence on symbolic cognition in the absence of any intellectual intuitions, Lambert assured him that:

> no one has formed himself a clear representation of all members of an infinite series, and no one is going to do so in the future. But we are able to do arithmetic with such series, to give their sums, and so on, by virtue of the laws of symbolic cognition. We thus extend ourselves far beyond the borders of our "real" thinking. The sign $\sqrt{-1}$ represents an unthinkable non-thing. And yet it can be used very well in finding theorems. What are usually regarded as specimens of pure understanding can be viewed most of the time as specimens of symbolic

---

[7] loc. cit. p. 389.

[8] Kant 1999, p. 116.

[9] See Falkenstein 1991 for thorough analysis of this argument.

[10] loc. cit., p. 114–115. Kant sought to protect intellectual knowledge from "contamination" by sensible concepts.

cognition. This is what I said in §122 of my *Phenomenology*... And I have nothing against your making it quite general.[11]

Lambert had summarized the need for symbolic cognition by observing that "insofar as sensible images do not represent abstract truth in themselves, the pure understanding must represent them without such images... by means of words and suitably chosen symbols. We find in numbers no clarity other than what we represent by numerals. By means of language and other symbols our cognition thereby rises above images and the limits of our imagination, of which algebra is the most perfect example."[12] Kant evinces a similar view of numbers after explaining that pure mathematics deals with space in geometry and time in pure mechanics: "In addition to these concepts, there is a certain concept which itself, indeed, belongs to the understanding but of which the actualization in the concrete requires auxiliary notions of time and space (by successively adding a number of things and setting them simultaneously side by side). This is the concept of number which is treated in arithmetic."[13]

Kant will take over a decade to come to grips with Lambert's critique of the *Dissertation*, which he tells Herz in 1772 will result in a work he may call *The Limits of Sensibility and Reason*.[14] He admitted to Herz that in the *Dissertation* he had not explained how intellectual representation can refer to an object without being affected by it, but he argues that "in mathematics this is possible, because the objects are quantities and can be represented as quantities only because it is possible

[11] loc. cit., p. 118. Lambert found the need for symbolic cognition in four distinct but not unrelated contexts: the theory of infinite series, the logic of contradiction, the use of imaginary numbers, and the problem of parallels.

[12] *Phenomenologie*, §122. The *Phenomenologie* comprises Lambert's effort to develop the theory of sensible appearance and semblance into a "transcendent optics" that incorporates "points of view" and "perspectives." It is the last of the four books into which Lambert 1764 is divided, beginning with *Dïanoïologie*, the theory of thinking, then *Alethiologie*, the theory of truth, and *Semeotik*, the theory of signs and their meaning. Concerning these first two books Lambert explains that "These two sciences would suffice were the human intellect not obliged to bind its knowledge to words and signs and if truth did not often appear to the intellect in a very different semblance." (Lambert 1764, Preface, p. 4) For an insightful discussion of the kinship and differences of Lambert's phenomenology with Husserl's, see Orth 1984.

[13] Kant 1992b, p. 390.

[14] Kant will take 7 years to reply to Lambert's letter which will consist of nothing less than the version of the *Critique of Pure Reason* he announced for publication in 1776, for which he wrote the following dedication to Lambert: "You have honored me with your letters. The endeavor to give at your request an idea of the method of pure philosophy has given rise to a series of reflections on how to develop this idea lying still obscurely within me, and as the possibilities widen within each step forward, my answers have been continually delayed. This work can serve in lieu of an answer, as far as the speculative part is concerned. Since it is due to your instigation and pointers, I would hope that it could always be with you by the endeavor to add to it in your research... It is hardly necessary to add that this work borrows a suggestion from your letter and is not the end." (AA18, p. 64). But Lambert died suddenly in 1777, and Kant took four more years to add to his *Critique* himself before he published it in 1781. We know from Kern 1964, p. 430 that Husserl studied their correspondence, so Kant's dedication would not have surprised him.

for us to produce their mathematical representation (by taking numerical units a given number of times). Hence the concepts of the quantities can be spontaneous and their principles can be determined a priori."[15] Kant asks himself:

> How it happens that objects correspond to that which is merely a product of our isolated minds and how these objects are subjected to the laws that we prescribe to them. Pure mathematics and metaphysics teach that there are such *a priori* cognitions: it is, however, an important investigation to have insight into the ground of their possibility.[16]

Reflection convinces Kant that "if I assume that an *a priori* concept has an object then I must be able to cognize *a priori* everything about that object the condition of which is contained in that concept."[17] The ability to cognize *a priori* everything about the object of mathematical definitions will convince Kant by 1781 that we can find "clear and exact" solutions to any mathematical problem – but only after learning of new discoveries in mathematics.

In the 1770s Kant lectured on "horizons" of our knowledge as expounded in Meier's logic text where a matter was said to be "elevated above the horizon of human cognition when human ignorance of it is absolutely necessary."[18] Kant reflected that "if a subject is placed in a probable manner above the horizon and the pursuit of it cannot be maintained without prejudice, then one must give it up. Squaring the circle."[19] Kant echoed Lambert's concern with fanatic pursuit of this problem when he lectured that "Everyone must think that according to his capabilities, be they great or small, he has a certain determinate horizon suited to himself. E.g. he who knows something of geometry, assuming that it is still not much, must and can well think that finding out how to square the circle is beyond his horizon."[20] He admits that "It is very hard to determine what is really beyond our horizon. If it frequently appears to me as if something were beyond my horizon, so that I hold it to be an impossibility, such a thing can still come within my horizon due to investigation and industry."[21] Kant soon explores how the problem of evaluating π fits his own horizon. Lambert warned that the problem of squaring the circle was a complicated one, requiring one to "make precise what one is actually looking for and why one wants it ... Thus it is still impossible to say whether, among all sectors and segments cut off from a circle, even a single one can be squared without using

[15] Kant 1999, p. 134.

[16] Kant 2005, p. 139.

[17] loc. cit., p. 201.

[18] AA16, p. 174.

[19] loc. cit., p. 175.

[20] Kant 1992c, p. 51. Lambert 1770 doubted that his exposition of the problem "will be read by those who could profit most by it, namely who spend time and effort trying to square the circle. There will always be many such persons who understand little of geometry, and are unable to assess their own powers." (p. 140)

[21] loc. cit., p. 50. Later in the 1790s Kant is convinced that "to determine the rational horizon of human cognition is one of the noblest and hardest occupations of the human spirit." (loc. cit. p. 449)

infinite series or imaginary quantities."[22] To know how one can square the circle, one must see *if infinite series are needed to evaluate* π, such as Leibniz': $\frac{\pi}{4} = 1 - \frac{1}{3} + \frac{1}{5} - \frac{1}{7} + \frac{1}{9} - \frac{1}{11} + \frac{1}{13} \cdots$ If this sum is rational, one can square the circle with a ruler and a compass.

The problem was longstanding ignorance whether π *was* rational. In replying to Locke's suggestion we "perhaps shall never be able to find a circle equal to a square, and certainly know whether there is such a thing," Leibniz explained that "Archimedes has already shown that 'there is such a thing'. It is the square whose side is the mean proportional between the radius and semi-circumference of the circle. He even determined a straight line equal to the circumference of the circle, by means of a tangent to the spiral as others did to the quadratrix . . . There are those who require that the construction be done with nothing but ruler and compass, but there are few geometrical problems in which the construction can be done in that way."[23] Archimedes proved the existence of a square equal to a circle *apagogically*, showing that if I am given a straight line equal to the circumference of a circle, the right triangle with legs equal to this line and the radius of the circle is equal to the circle.

Leibniz wrongly believed he had proved the sum of his series to be irrational, but asks if "a precise abbreviated form of this series can be found. But finite expressions – and especially the irrationals, if one goes out to the supersurds – can vary in so many ways that one cannot enumerate them and easily determine all the possibilities."[24] This shows, he says, that "the human mind raises questions which are so strange, especially when infinity is involved, that it is not surprising that it is hard to get to the bottom of them."[25] Leibniz focused on three issues raised by the problem of squaring the circle: infinity, constructability, and solvability. His acceptance of Clavius' 'construction' of a quadratrix to solve it encouraged Kant's liberal idea of construction.[26]

Kant's first mathematical reflections explore the circumference and area of a circle by the construction of series of circumscribed polygons. That he was trying to estimate π is clear from the problem he set himself: "to find the infinite sequence of triangles whose number redoubles and which subtracted from the square abgd yields a quadrant of the circle, and hence the ratio of the circle to the square of its diameter."[27] Adickes suggested one of his figures was "perhaps an unsuccessful attempt to draw a quadratrix."[28] Kant was probing the problem of squaring the circle to see if it was within his own horizon, but Adickes concluded from the errors in his figures and equations that he had very little "immediate intuitive grasp of

[22] *Dianoilogie*, §495

[23] Leibniz 1981, p. 376. Some geometers already rejected the quadratrix as not being constructible.

[24] loc. cit., p. 377.

[25] loc. cit., p. 377.

[26] Kant never limits the means of constructing mathematical concepts.

[27] AA14, p. 6.

[28] loc. cit., p. 9.

geometrical relations."[29] Kant's philosophy of mathematics invokes such intuition, but it will prove difficult to get such a grasp of Lambert's landmark apagogic proof of the irrationality of π.[30] He proved by contradiction his "surprising theorem" that tan υ must be irrational whenever $\upsilon \neq 0$ is rational; since $\tan \frac{\pi}{4} = 1$, $\frac{\pi}{4}$ must be irrational. Thus, infinite series are required to deal with this problem of geometry.

This result inspired Lambert's novel approach to Newton's three-body problem that yielded "an absolute and general solution by means of infinite series."[31] He argued that it no longer required a solution in integrals given by "finite formulas" but that a proof of the convergence of the infinite series arising from the differential equations of motion would show that the orbits of motions were completely determined by their initial state. Lambert stressed that this problem was no longer one of astronomy or mechanics but of "pure geometry," whose solution yields a rule allowing us "to calculate or construct orbits".[32] This was possibly the stimulus for Kant's reference to an "analogue of construction" in his *Duisburg Nachlass*, in connection with his proof based on the "continuity of alteration," that "whatever happens has a ground, i.e. is determined through something in accordance with a rule."[33] While not convinced by Lambert of the pervasive need of symbolic cognition in mathematics, his work on the three body problem may well have widened the horizon of construction for Kant who, unfettered by criteria for constructions, is clear by 1776 that "Mathematics as synthetic *a priori* cognition bases its possibility on the fact that its concepts can be constructed, for they have to do only with space and time, in which objects of intuition can be given *a priori.*"[34] By 1780 he is ready to "determine the horizon of mathematics."[35] That is, the domain of objects to which mathematics can and cannot be applied.

In the *Critique*, Kant embeds his ideas about mathematics in a transcendental theory of the horizons of the understanding and a dialectical account of the limits of reason itself. His philosophy of mathematics now comes into focus as a foil for the metaphysical excesses of Plato's. True metaphysics is now shown to be "nothing but the *inventory* of all we possess through *pure reason*, considered systematically.

---

[29] Adickes 1924, p. 21.

[30] In his memoir Lambert 1768 expresses tanυ as the quotient of the infinite power series for sinυ and cosυ, and uses a generalization of the Euclidean algorithm and complicated recursions to obtain an infinite continued fraction for tanυ; he then proves that the existence of integers a and b, for which $\tan \upsilon = \frac{a}{b}$ for a rational υ, would imply an infinite decreasing sequence of integers. Lambert concluded that if, as he conjectured, π were transcendental, it would not admit of a "geometrical construction", i.e. with ruler and compass. (p. 322)

[31] Lambert 1769.

[32] loc. cit., p. 357. Lambert observed that such series may not converge for all values of the time.

[33] Kant 2005, p. 176. Kant associates this forerunner of his Second Analogy with the Newtonian principle that "the cause of appearances must be in *commercio* with the world." (p. 177) Kant is known to have followed progress on the three-body problem with interest, especially the most important case of the mutual perturbations of Jupiter and Saturn.

[34] loc. cit., p. 247.

[35] AA24, p. 522.

Nothing can escape us here, because what reason brings forth entirely out of itself cannot be hidden, but is brought to light by reason itself as soon as reason's common principle has been discovered."[36] This transparency of reason allows for much *a priori* knowledge: "Mathematics gives a splendid example of how far we can go with *a priori* cognition independently of experience. Now it is occupied, to be sure, with objects and cognitions only so far as these can be exhibited in intuition. This circumstance, however, is easily overlooked, since the intuition in question can itself be given *a priori*, and thus can hardly be distinguished from a mere pure concept."[37] Overlooking this subtle but crucial distinction, "Plato abandoned the world of the senses because it posed so many hindrance for the understanding, and dared to go beyond it on the wings of the ideas, in the empty space of pure understanding."[38]

Kant is, however, deeply ambivalent about Plato: while hailing the importance of his appeal to ideas of noumenal perfection for morality, he deplores Plato's extension of them "even to mathematics, even though mathematical cognitions have their objects nowhere except in possible experience."[39] The opposition in the Antinomies, between the dogmatic empiricism of the antitheses and the dogmatic intellectualism of the theses, is said by Kant to be respectively that between Epicureanism and Platonism: "Each of the two says more than it knows, but in such a way that the *first* encourages and furthers knowledge, though to the disadvantage of the practical, the *second* provides principles which are indeed excellent for the practical, but in so doing allows reason, in regard to that of which only a speculative knowledge is granted to us, to indulge in ideal explanations of natural appearances, and to neglect the physical investigations of them."[40] The antinomies mark the limits and boundaries of reason that it transgresses only on pain of contradiction.

Room for the free noumenal self Kant requires for morality is made by the transcendental ideality of time, which he "elucidates" by allowing that Lambert's change argument does prove that time is *empirically* real; but he wants to "dispute all claim of time to absolute reality, namely where it would attach to things absolutely as a condition or property even without regard to the form of our sensible intuitions."[41] From his exposition of time as a pure intuition, Kant "concluded" that time could not "attach to things as an objective determination, and thus remain if one abstracted from all subjective conditions of the intuitions of things," for then time could not be "cognized and intuited *a priori* through synthetic propositions."[42] Kant tries to *elucidate this abstraction* by arguing that "if I or another being could intuit myself without this condition of sensibility, then these determinations, which

---

[36]Kant 1997, Axx.

[37]loc. cit., A5/B8.

[38]loc. cit., A5/B9.

[39]loc. cit., A314/B371.

[40]loc. cit., A472/B500.

[41]loc. cit., A35.

[42]loc. cit., A33. Kant refers here to his two "axioms of time" which are taken from the four axioms and three postulates of chronometry formulated in Lambert 1771.

we now represent to ourselves as alternations, would yield us a cognition in which the representation of time and thus also of alternation would not occur at all."[43] But the counterfactual, that I could intuit my inner state as changeless if freed from the bondage of sensibility, would beg the question. Kant requires that "another being" *exists* who would so intuit my inner state, namely, "a primordial being" with an intellectual intuition of myself as it is in itself. Such a thing in itself intuited in such a manner by such a being is for Husserl a "mythical construction," but Kant claims that it is not contradictory which, together with his dialectical resolution of the third antinomy, is enough to justify the consistency of the freedom of the noumenal selves for morality.

The indirect proof of transcendental idealism that Kant mined from the antinomies shows, however, that they are not problems inscrutable to human reason. Some questions about *nature* "must remain uncertain", he grants, but "transcendental philosophy has the special property that there is no question at all dealing with an object given by pure reason that is unsolvable by this very same human reason, for the very same concept that puts us in a position to ask the question must also make us competent to answer it, since the object is not encountered at all outside the concept."[44] Kant claims this also holds "in *pure mathematics,*" where we can

> demand and expect clear and certain solutions to all the question belonging within it, even if up to this time they have not be found... Has it ever been proposed that because of our necessary ignorance of conditions it is uncertain exactly what relation, in rational or irrational numbers, the diameter of a circle bears to its circumference? Since it cannot be given congruently to the former, but has not been found through the latter, it has been judged that at least the impossibility of such a solution can be known with certainty, and Lambert gave a proof of this.[45]

Kant's *thesis* that *all mathematical problems are solvable* stems from his Reflection that "I must be able to cognize *a priori* everything about the object of an *a priori* concept" that I have constructed, such as a circle. But he could hardly have been confident that he could cognize whether π was rational until Lambert proved its irrationality. He knew from trying to estimate π and the history of attempts to square the circle by assuming that π is rational, that Lambert had solved a hard longstanding problem, indeed one showing that, as Leibniz said, "the human mind raises questions which are so strange, especially when infinity is involved, that it is not surprising that it is hard to get to the bottom of them." As Kant's mathematical advisor and commentator Schultz admitted, "it has always been an irksome circumstance for mathematicians to be unable to say *with complete conviction* to so-called circle squarers, who supposed they could give the ratio of the diameter to the circumference of a circle by a rational fraction, that their

---

[43]loc. cit., A37/B54.

[44]loc. cit., A477/B505. Kant admitted some questions about *nature* "cannot be solved at all, e.g. how pieces of matter attract on another" (Kant 2002, p. 138–139), even though Newton's inverse square law governing such incomprehensible attractions is "cognizable a priori".

[45]loc. cit., A480/B508.

undertaking was ludicrous because it was impossible."[46] He stressed that the proof of its irrationality is "extremely difficult," but Kant now believes such proofs will be found "even if up to this time they have not been found".

In the exposition of his proof in *Vorläufige Kenntnisse* that Kant read, Lambert stressed its apagogic nature and the computational side of the problem. His search for a "fast converging formula" for π was so successful that he could have claimed to have *solved* the problem of squaring the circle as understood by expert opinion in 1750.[47] Goldbach wrote to Euler that to test an invention for squaring the circle one must determine "the degree of facility with which this number [π] is expressed by the inventor."[48] One could be said to have squared the circle if he gave a method for approximating π "as easy as one can find $\sqrt{2}$ by extraction of a square root." Euler wanted a procedure yielding a decimal digit for every computational step. Lambert's continued fraction for $\tan\upsilon$ has been shown to produce 765 correct decimal digits for π in 1000 computational steps.[49] By allowing trigonometric functions to take imaginary arguments he also proved the irrationality of *e*, exemplifying what he called in his 1770 letter to Kant symbolic cognition of $\sqrt{-1}$. Did appeal to such a proof as evidence for solvability compromise Kant's claim that "in mathematics it is *a priori* intuition that guides my synthesis, and there all inference can be immediately drawn from pure intuition?"[50]

These words suggest no problem with mathematical proofs, but Kant's third rule for the discipline of pure reason "is that its proofs must never be *apagogic,* but always *ostensive*. The direct or ostensive proof is in *all* kinds of cognition that which is combined with the conviction of truth and simultaneously with insight into its sources; the apagogic proof, on the contrary, can produce certainty, to be sure, but never comprehensibility of the truth in regard to its connection with the grounds of its possibility. Hence the latter are more of an emergency aid than a procedure which satisfies all the aims of reason."[51] *These* words fairly describe Lambert's apagogic proof of the irrationality of π. Kant grants that apagogic proofs "have an advantage in self-evidence over the direct proofs in this: that a contradiction always carries with it more clarity of representation than the best connection and thereby more closely approaches the intuitiveness of a demonstration."[52] The "real cause," says Kant, of the need for apagogic proof in science is this: "If the grounds from which a certain cognition should be derived are too manifold or lie too deeply hidden, then one tries whether they may not be reached through their consequences."[53] In such

[46]Schultz 1803, p. 1x.

[47]This is explained in the informative article by Bullynck 2009.

[48]Quoted from Bullynck, loc. cit., p. 157.

[49]See Bauer 2005, who also finds that Leibniz' series yields four such digits in 1000 steps!

[50]A782/B810.

[51]loc. cit., A789/B819. Last emphasis mine.

[52]loc. cit., A790/B818.

[53]loc. cit., A790/B818. Kant may have been influenced by Lambert's arguments that "apogogical proofs always carry more necessity than direct proofs, since they expose the impossibility or

cases *modus ponens* is unusable, and Kant recommends the use of *modus tollens*, precisely what Lambert used to infer the irrationality of $\frac{\pi}{4}$ from the instance of his theorem that 'if $\frac{\pi}{4}$ is rational, then tan $\frac{\pi}{4}$ is irrational.'[54]

Some argue, that Kant's 'constructivism' justifies the law of excluded middle on which apagogic proofs depend.[55] This law is implicit in his solvability thesis: every mathematical question can be settled, either by a direct proof or by showing the proposition in question to be impossible, as did Lambert. But what did his proof "construct"? On the computational interpretation of squaring the circle Lambert produced a nearly optimal 'construction' of the irrational number π. But π is the ratio of the circumference to the diameter of a circle, and its irrationality means that if one starts from any point on a circle laying off arcs equal to its radius, one generates an infinite sequence of its points without ever arriving at the same point twice. This is conceptually clear but hardly evident in 'pure intuition.' Kant's paradigm case of such intuition occurs in the geometer's proof of the angle sum $\sum = 2R$ for triangles. He begins by constructing a triangle and, after extending one of its sides through its vertex, draws an auxiliary parallel to its base, whereupon he "sees that there arises an external adjacent angle equal to an internal one, etc. In such a way, through a chain of inference that is always guided by intuition, he arrives at a fully illuminating and at the same time general solution to the question."[56] The focal point of what the geometer "sees" here is the relation his auxiliary parallel bears to the rest of his figure, but this is not enough: he must see that is is the *only* such line he could have drawn. It is one thing to have an intuition *of* a single line with some property, another to intuit *that* it is the only such.[57] The uniqueness of parallels is equivalent to Euclid's axiom of parallels whose status and proof, Kant has just learned, have become a contested issue.[58] Scores of alleged proofs of it were being found fallacious. Whatever one felt about the certainty or evidence of the axiom itself, the question of its *provability* had clearly become very uncertain.

---

absurdity of the contrary and thereby make the theorem itself *necessarily true* given that the contrary would under the conditions assumed be impossible or absurd." (*Dianoilogie*, §352).

[54]In a very interesting paper Vuillemin 1961, argues that Lambert's proof was actually adopted by Kant as a model of his transcendental idealism, in particular his indirect proof of it in the Antinomies.

[55]Mancuso 1996 suggested that by grounding reductio proofs on his constructivist assumptions Kant followed "a long tradition critical of their use and determined to limit their use as a last resort." (p. 107) Brouwer gave what he called "an appropriate reworking of Lambert's negative irrationality proof of π" into a positive one (see Mancuso 1998, p. 33). But Kant's claim, which he repeated often in lectures, that "contradiction always carries with it more clarity of representation than the best connection and thereby more closely approaches the intuitiveness of demonstration" leaves doubt he would find Brouwer's proofs more "intuitive" than Lambert's.

[56]A716/B744.

[57]Parsons 1980 observed that even if not an official part of his doctrine, this distinction was implicit in Kant's arguments.

[58]In his extensive reflections on the problem of parallels Kant finally settles for a conceptual proof of an equivalent of Euclid's axiom, that is, one that foregoes any construction. (AA14, p. 23–48)

When he wrote to Kant in 1766, Lambert had just finished his *Theory of Parallel Lines* in which he poses the first coherent formulation of the problem of Euclid's axiom:

> *Whether it can be derived in proper order from the Euclidean postulates together with his other axioms.* Or, *if these are not sufficient, whether there cannot be produced other postulates or axioms which have the same evidence as the Euclidean do, and from which his eleventh axiom can be proved*? In the first part of this question, one can abstract from... the *representation of the matter*. And since Euclid's *postulata* and other axioms have been expressed in words, it can and should be demanded that the proof never appeal to the matter itself, but that the proof should be carried out purely symbolically – when this is possible. In this respect, Euclid's postulates are as it were like so many algebraic equations which one already has in front of oneself and from which one is to compute x, y, z... without looking back to the matter itself.[59]

Lambert tries to prove the axiom indirectly by symbolically deriving contradictions from the hypotheses H_2 and H_3 that $\sum > 2R$ and $\sum < 2R$ respectively. He uses the extension postulate to derive a contradiction from H_2, but realizes that H_2 nevertheless holds on a sphere. His proof that under H_2 the area of triangles is proportional to their excess together with the formula $A = r^2(\alpha + \beta + \delta - \pi)$ for spherical triangles, amazes him: "Here it seems to me remarkable that the second hypothesis is fulfilled when instead of a plane triangle one takes a spherical one, for in this both the sum of the angles is greater than 180° and the area of a triangle is proportional to its excess."[60] But after concerted effort he could derive no contradiction from H_3, and seeing that on substituting $\sqrt{-1}r$ for r the above formula becomes $A = r^2(\pi - \alpha - \beta - \delta)$, a special case of what he proved for plane triangles under the third hypothesis H_3, he explains his difficulty in refuting H_3 by observing that "I should almost conclude from this that the third hypothesis holds on an imaginary sphere. At least there must always *exist something* to show why it cannot be refuted nearly as easily in the plane as in the case of the second hypothesis."[61] Symbolic cognition of $\sqrt{-1}$ helps Lambert explain his inability to refute H_3 "purely symbolically." This is a paradigmatic specimen of the kind of symbolic cognition for which Husserl will seek a logical rationale. Here Lambert anticipates our completeness principle for purely logical refutations, suggesting that if you cannot derive a contradiction from a hypothesis using other axioms "purely symbolically," then there must be a sphere or structure (however imaginary!) in which it is fulfilled – and which explains why you cannot.

---

[59]Ewald 1996, p. 166.

[60]Lambert 1895, p. 199.

[61]loc. cit., p. 203.

2.3 Categorial Intuition, Solvability, Unprovability

1900 was an opportune time for manifestos of modernity. In the *Prolegomena* Husserl enthused that Leibniz' intuitions of an *ars combinatoria and calculus ratiocinator* had become attractive "to a modern acquainted with 'formal' mathematics and mathematical logic."[62] Kant, on the other hand, had bequeathed to us an "utterly defective logic" limited in principle to Aristotelian syllogisms, as well as "confusing mythic concepts of understanding and reason... which he uses in such questionable demarcations, as being, in a proper sense, faculties of the soul."[63] Nevertheless, "the authority of Leibniz must, however, count even less for us than that of Kant or Herbart, since he could not give to his great intentions the weight of completed achievement. He belongs to a past age, beyond which modern science feels that it has traveled a long way."[64] Husserl is very sympathetic to Kant's belief in the possibility of making logic an *a priori* science: others tie it to psychology or empirical science, "whereas on Kant's view, we rather dig down to an internally closed, independent, and... a priori field of theoretical knowledge."[65] Still, says Husserl, "It was ominous that Kant (to whom we nonetheless feel ourselves quite close) should have thought he had done justice to the domain of pure logic in the narrowest sense, by saying that it fell under the principle of contradiction."[66] And while praising Kant for the "great role" he gives to categorial logical functions, Husserl stresses his failure "to achieve our fundamental extension of the concepts of perception and intuition over the categorial realm, and this because he fails to appreciate the deep difference between intuition and signification, their separation and their usual commixture."[67] Kant's inadequate "adaptation of meaning to intuition" allows him to prematurely launch "a critical 'saving' of mathematics, natural sciences, and metaphysics." But although Husserl gives many examples of categorial intuition and its special case of eidetic variation in perception, he gave hardly any in mathematics, where he says Kant's attempt to 'save' it was premature because uninformed by such intuition.

In an admirably clear account of such intuition, Dieter Lohmar gives examples of how its use in geometry and arithmetic might secure *a priori* validity in Husserl's sense. Thus, in order to have *a priori* insight that 'two lines on a plane, not running parallel, intersect at a point,' he explains that "we have to vary all kinds of imagined lines on planes not running parallel. And here we are able (by definition) to find

[62] Husserl 1970a, p. 220. Reviewing how the ideal of grounding the deductive calculus on forms that are most primitive and natural was pursued by Boole, Pierce, and Schröder, Husserl said that this ideal "can be attained; and older German logicians – above all, Leibniz and Lambert – were on the way to attaining it" (Husserl 1994, p. 102).

[63] loc. cit., p. 214.

[64] loc. cit., p. 221.

[65] loc. cit., p. 76.

[66] loc. cit., p. 833.

[67] loc. cit., p. 833.

in every imagined case one direction that shows a progressive approximation of the two lines imagined. Therefore, we can be sure that there will be a point of intersection of the two lines in every possible case. Thus it is valid a priori in the Husserlian sense."[68] This example adds little to the intuition Kant would claim for this intersection; since he accepts the definition of parallel it turns on, it would hardly be a case of his inadequately "adapting meaning to intuition." But what about the intuition Kant claimed of the parallel his geometer drew that guided his proof of $\Sigma = 2R$? We saw that it was not enough just to have an intuition of this parallel: he had also to intuit that it was the *only* such parallel he could draw. Husserl's own analysis of categorial intuition does apparently hone in on this point as follows:

> If we ponder the peculiarity of eidetic abstraction, that it necessarily rests on individual intuition, but does not for that reason mean what is individual in such intuition..., *then the possibility of universal intuitions arises, intuitions which not merely exclude all individuality, but also all sensibility from their intentional purview.* In other words, we distinguish between *sensuous abstraction*, which yields *sensuous concepts*... and *purely categorial abstraction*, which yields *purely categorial concepts. Colour, house, judgement, wish* are purely sensuous concepts; *colouredness, virtue, the axiom of parallels*, etc., have a categorial admixture, while *unity, plurality, relation, concept,* are purely categorial.[69]

Here parallel is a purely sensuous concept, as was circularity for Kant. But the axiom of parallels asserts precisely that there is *one* parallel to a line through a point not on it, thus comprising an "admixture" of this sensuous concept with the category of unity, which is just what Husserl's categorial intuition of this axiom amounts to.[70] But for Kant the axiom that literally depends on the intuition of *space*, considered as the form of outer intuition, is that any three non-collinear points in space lie in a unique plane, for which Husserl suggests no analogous admixture of the purely sensuous concept of plane and the category of unity. In fact, he rejects this Kantian ideality of space.

For Husserl space is the necessary *form of things*: "form of intuition" is a false expression implying "even in Kant, a fatally erroneous position."[71] But the phenomenological constitution of space faces the difficulty that all perception of spatial things is one-sided, so that a proper "three dimensional intuition" presenting all sides at once is impossible: "There is only an improper intuition of space... in relation to the presentational forms and means which we have at our disposal

---

[68]Lohmar 2010, p. 83–84.

[69]Husserl 1970a, p. 818.

[70]Would this accommodate Lobachevsky's categorial intuition of his axiom which postulates a plurality of parallels to a given line through an external point? Is the Euclidean admixture of the concept of parallel with the category of unity more "intuitive" than Lobachevsky's admixture of it with the category of plurality? Husserl says that "if we use the term 'space' of the familiar type of order of the world of phenomena, talk of 'spaces' for which, e.g. the axiom of parallels does not hold, is naturally senseless." (loc. cit., p. 243) But his account of categorial intuition provides no reason at all why the admixture of the concept of parallels with the category of plurality would even be unintuitive, much less "senseless". In fact, Lobachevsky argued that hyperbolic geometry based on his axiom was more coherent than Euclidean geometry.

[71]Husserl 1997, p. 43.

phenomenologically."[72] We can introduce movement into the visual field: "But here we arrive at the difficulty that space is infinite (as Kant says, it is represented as an infinite given magnitude. In any case we do not doubt the evidence of the infinite). But the field is finite."[73] Husserl admits that the "constitution of full givenness" is an ideal limit, whether of the object or of its relation to the Ego: "Our considerations are now allowing us to understand to what extent neo-Kantianism was justified in designating the ultimate determination of the Object an infinite task; we can now see where the phenomenological grounds of that position reside."[74] Husserl approaches this task by constituting expansions of the field, but finds that planes appropriate for projecting perspectival images of spatial things in the oculomotor fields he considers will not in general exist.[75]

Husserl claims that phenomenological reductions are necessary for uncovering the geometrical structure of visual space. He asks whether the visual field has "the character of a plane surface occupying its place in objective space" and observes that when we "see into objective space," changing apprehensions of visual contents can yield "very different intuitions of space. But if we reduce the 'visual field,' if we exclude everything that transcends sensing from the moments of apprehension and from the sense, then something results that is no longer a plane surface, no longer a field in objective space. I am of the opinion that this reduction is possible with evidence and that it is also possible to grasp with evidence the 'broad' relationships in this reduced 'spatiality:' relationships of 'next to one another,' 'in one another,' and so on."[76] Such a reduction is possible for time as well, but "just as one emphatically does not ask about what is 'originally spatial,' about the material out of which the intuition of objective space comes into being *genetically*... so ones does not ask [in the phenomenology of time] about the 'original' from which the intuition of objective 'time' arises."[77] Now as Hilbert explained in his lecture for the *Festschrift*, this was precisely what his "logical analysis of spatial intuition" consciously avoided: any hypothesis about the genetic origin of such intuition. What light then can either phenomenology or logical analysis throw on the geometrical structure of such intuition?

Husserl suggested that "the spatial form of visual space is Euclidean," a form it retains at a high level of constitution. Though he distinguished visual space from "objective perceived space," he held both to be homologous spaces of constant curvature. Meanwhile, psychologists examined the geometric structure of pure

[72] loc. cit., p. 44.

[73] loc. cit., p. 99.

[74] loc. cit., p. 111.

[75] loc. cit., p. 204–205. See Boi 2004 for a thorough analysis of the difficulties Husserl encounters with these expansions of the visual field. A clear and instructive account of Husserl's theory of the basis of geometrical intuition in perceptual space, as well as his synthesis of the latter from various kinesthetic systems, can be found in da Silva 2013.

[76] loc. cit., p. 199.

[77] Husserl 1991, p. 144.

binocular visual space, beginning with Blumenfeld's alley experiments of 1913, to whose subjects he truly had to apply Husserl's prescription for the phenomenology of space, namely, "to exclude all judgmental positings of the experiencing persons and of the experimental things."[78] This was not easy. A subject seated is unable to move his head in a dark room in which only the light of tiny movable lamps is visible. Two such lights are fixed a foot apart, a few feet in front of the subject, who instructs the experimenter to first (i) "Adjust the lights until the two rows of lights appear to be straight and parallel to each other and parallel to the median," and then (ii) "With only the fixed lights P and Q left on, set successive pairs of lights symmetric to the median and at the same apparent separation as the two fixed lights."[79] (i) and (ii) yield respectively so called 'parallel alleys' and 'equidistant alleys,' and except for the occasional subject who says that 'parallel' *means* equidistant, the results are remarkably stable and consistent: the equidistant alleys diverge outside the parallel alleys. This is just the relation between such pairs of straight lines in hyperbolic geometry, prompting the Luneberg hypothesis that pure binocular visual space is hyperbolic. But he presupposed that this space has constant curvature, which long proved elusive to experimental verification. Busemann combined Hilbert's logical analysis of Desargues' theorem with his solution for the plane case of his fifth problem to show, without appealing to differentiability, that Riemannian spaces whose geodesies satisfy Desargues' theorem must have constant curvature.[80] Researchers were able to verify the Desarguesian property of binocular visual space and hence its hyperbolic character under the restricted cue conditions of the alley experiments. Allowing full lighting and motion of subjects, however, leads to Euclidean judgments of distance, providing a naturalization of Husserl's claim. Husserl would naturally only set store in its phenomenological justification, such as Oskar Becker's, but it has not born up under scrutiny.

Let us see how Kant's spatial intuition figures into Hilbert's logical analysis of such intuition, which consciously bracketed the question of its origin. He divided his axioms into five groups "each of which expresses certain basic facts of our intuition which belong together." The most direct expressions of spatial intuition are the space indices axioms, e.g. that any three noncollinear points lie in a plane and that it is unique, which we may call "Kant's axiom" to remind us that it was his paradigmatic example of an axiom that we intuit in space. This axiom immediately implies Desargues' theorem (D), the fundamental theorem of linear perspective. Hilbert showed that (D) cannot be proved from the axioms of plane incidence, order, and parallels, without the axioms of congruence, so that without

[78] Husserl 1997, p. 118.

[79] Zage 1980, p. 289.

[80] Busemann 1955, pp. 65–81. Desargues' theorem states that two triangles are in central perspective from a point in space if and only if they are axially perspective, i.e. the intersections of their pairs of corresponding sides are collinear. This intuitive "Raumsatz," as Hilbert calls it, implies that the two triangles are optically indistinguishable from a point in space, i.e. one is the perspective image of the other, just in case they both vanish when viewed from any point on their axis. The theorem can be proved for planar triangles simply by projection from a point in space.

this latter concept foreign to (D), it does depend logically on the Kantian expression of space intuition. This answers a question Husserl raised about whether there are propositions expressible in the geometry of the plane that can only be proved in space.[81] Hilbert also proved that a plane can be embedded in space if and only if (D) holds in it, allowing him to describe the significance of (D) as "the result of eliminating the space axioms."[82] Next he takes up the question Husserl had identified as most pressing for "modern geometry." As he put it to Stumpf: "Little has been done for the logic of geometry. It is incredible that even logical justification of the old method of Cartesian co-ordinates (and that justification is by no means something obvious) still has not been demonstrated."[83] Husserl envisaged using the algorithms of *arithmetica universalis* suitably symbolized, for "using intuition and actual construction one simply never gets very far."[84]

Hilbert uses axioms of plane incidence and parallels to set up a co-ordinate system and defines by purely geometrical constructions the 'addition' and 'multiplication' of segments and then uses (D) to prove all the 'algebraic laws' for these two operations – commutativity and associativity of the addition, associativity of the multiplication, distributivity of the multiplication over the addition – with the exception of the commutativity of the multiplication, which he discovered to be equivalent to Pascal's theorem (P). He had constructed a *Streckenrechnung* comprising an autonomous analytic geometry making no reference to numbers. If, as he originally believed, (D) logically implied (P), this analytic representation of the plane would have depended only on Kantian space intuition and the order axioms. But Hilbert discovered, by devising a complicated non-Archimedean model, that (P) cannot be derived from (D) unless continuity axioms are used, and hence that either congruence involving the concept of rigid motion or the Archimedean axiom referring to number is needed for analytic geometry. For the motto of the *Festschrift* Hilbert had quoted Kant: "All human knowledge begins with intuitions, proceeds to concepts, and ends with ideas." Our geometrical knowledge of space begins with the *intuitions* of planes, which yields (D) immediately in space; but to prove it in the plane, one needs a *concept* of congruence foreign to (D), while to prove (P) from it alone, it is necessary to invoke a continuity axiom involving an *idea* of infinity.

Kant's words explicitly infused the *Festschrift*, but Lambert's may have influenced it at a logically deeper level. His *Theory of Parallel Lines* was studied intensively in Göttingen when it was republished in 1895. There is reason to believe that it was Lambert's formulation of the problem of parallels, with its comparison of Euclid's axioms to "so many equations that one has to solve for x, y, z, ... without looking back to the matter itself," that led Hilbert to formulate the problem of the consistency of Euclidean geometry with such equanimity. Indeed, this is

[81] Husserl 2003, p. 462.

[82] Hilbert 1971, p. 88.

[83] Husserl 1994, p. 17.

[84] loc. cit., p. 17.

how Hilbert formulates its axioms himself, asking us to "imagine three classes of things called points, lines, and planes denoted by variables a, b, c" and satisfying his axioms. In a review of the *Festschrift,* his assistant J. Sommer explains that "J. H. Lambert compares the axioms to as many equations that can be combined in innumerable ways. Professor Hilbert, to decide the question of consistency, imagines the domain of an enumerable ensemble of numbers, and represents a point by two numbers of the domain, a straight line by the ratios of three numbers."[85] Lambert's *aperçu* that his third hypothesis H_3 was fulfilled in an imaginary sphere showed the value of symbolic cognition. Mach noted that Lambert's sphere with radius $\sqrt{-1}$ was not a visualizable geometric construct but rather an analytic surface with constant negative curvature; nevertheless "it is evident again from this example how experimenting with *symbols* may direct inquiry to the right path, in periods where other points of support are entirely lacking and where every helpful device must be esteemed at its worth."[86]

In the *Prolegomena* Husserl quoted Mach's "excellent remark" to the effect that doing mathematics symbolically without attaining clarity about it "is often to have the disagreeable impression that one's paper and pencil are cleverer than oneself."[87] While apparently unfamiliar with Lambert's work on geometry, Husserl did quote his remark on the value of symbolic reduction for calculation, by way of explaining that:

> In arithmetical thought mere signs genuinely do duty for concepts. 'The reduction of the theory of things to the theory of signs' (to quote Lambert) is what all calculation achieves. Arithmetical signs are 'so selected and perfected, that the theory, combination, transformation, etc. of signs can do what would otherwise have to be done by concepts.'[88]

This formulation sparks Husserl's incisive critical analysis of the idea of "meaningless signs," in which he argues that such signs, like chess pieces, acquire meaning not from their phenomenal or physical properties but from the *rules* which bestow on them a "*game-meaning.*" Husserl argues that "signs taken in a certain *operational or game-sense* do duty for the same sign in full *arithmetical meaningfulness.*"[89] This removes confusions about the nature of 'merely symbolical' mathematical thought. The question remains whether the likes of Lambert's symbolic cognition

[85] J. Sommer 1900, p. 291. This allows Hilbert to satisfy his axioms in a countable algebraic field.

[86] Mach 1960, p. 121. On the question of Lambert's anticipation of Beltrami's pseudosphere, see Schur 1905. As I have indicated elsewhere in this volume, it was Mach's quest for a "universal physical phenomenology" that Husserl credited with pioneering the descriptive phenomenology that Husserl subsequently radicalized. But he was wrong to think that Mach's motivation was to ground mathematical physics in intuition.

[87] Husserl 1970a, b, p. 201.

[88] loc. cit, p. 305. Lambert was talking about algebra, not arithmetic, in his passage from his *Semiotik*. It is a pity Husserl did not consider his fuller discussion of this issue in his *Phenomenologie*.

[89] loc. cit., p. 305.

of an imaginary sphere are generally needed to demonstrate the *unprovability* of propositions from given axioms and whether it qualifies as a specimen of categorial intuition.[90]

Hilbert's analysis of intuition depends crucially on non-derivability proofs. In fact, most of his main results establish the unprovability of this or that geometrical proposition from various groups of axioms, which motivates his "ground rule" for modern mathematics:

> In this investigation the ground rule was to discuss every question that arises in such a way as to find out . . . whether it can be answered in a specified way with some limited means . . . if in the course of mathematical investigations, a problem is encountered, or a theorem is conjectured, the drive for knowledge is then satisfied only if either the complete solution of the problem and the rigorous proof of the theorem are demonstrated successfully, or the basis for the impossibility of success and hence the inevitability of failure are clearly seen. The impossibility of certain solutions and proofs thus plays a prominent role in modern mathematics, and the drive to answer questions of this type is often the cause for the discovery of new and fruitful areas of investigation.[91]

As he put it in his lecture for the *Festschrift*: "the modern principle is the proof of unprovability."[92] The accumulation of such proofs, Hilbert mentioned the impossibility of proving the axiom of parallels and "the impossibility of constructing numbers e and π algebraically," embolden him to claim that all well posed mathematical problems, however hard, are solvable. The same claim was made by Kant for similar reasons.

Husserl was less impressed by such proofs. In a review of Schroder's logic lectures Husserl noted that he could only prove half of the "fundamental theorem" of his calculus, namely, the distributive law, and that he devoted a special section to the unprovability of the other half, where:

> The author succeeds in rigorously demonstrating that this latter subsumption is not deducible by means of the concepts and principles introduced up to this point. He produces, namely, a calculus of "groups" for which the basic formulae all hold true, whereas that subsumption certainly does not. Therefore it cannot be a mere algorithmic consequence of the basic formulae presupposed. All of this I admit but find nothing so impressive about it.[93]

Without meaning "to place a low estimate" on such proofs, Husserl acknowledges that "it is of value to know what is necessarily conditioned upon restricted parts of the foundations of the calculus taken by themselves. In this way one gains the means of transferring at least a relevant partial algorithm over to foreign domains, where the whole algorithm would be no longer transferable."[94] We have seen Husserl's interest in subdomains of complete domains of geometry, where Hilbert

---

[90]Husserl's own loss of confidence in his theory of categorial intuition over the years is indicated by Lohmar 2002.

[91]Hilbert 1971, p. 106.

[92]Hilbert 2004, p. 284.

[93]Husserl 1994, p. 85.

[94]loc. cit. p. 134.

established the minimal purely geometric axioms for constructing their own analytic representation and algebraic algorithm. In his notes for the review Husserl chided Schröder for "something of that zeal for *proving*, which will not be content with the directly Evident, and believes it possible to impart a higher degree of certainty by means of a derivation within a symbolic calculus."[95] Still, Husserl grants that he "attains to the insight that the proof is impossible" for half of his law.

In Husserl's own Logic Lecture of 1896 he marshals an impressive system of logical axioms and rules and compares the prospects of their completeness with the arithmetician's search for axioms: "Has he any *a priori* conceptions assuring him of the completeness of derived arithmetical laws? None at all."[96] As for axioms, he says that "arithmeticians have taken infinite pains to establish the minimum number of arithmetical axioms. Yet it is still in question whether one of the axioms is merely a consequence of the others, without their having noticed it. At least we still lack any systematic proof."[97] But this speaks more to the independence of the axioms than their completeness, and by the time of his Double Lecture in 1901 Husserl has convinced himself that any arithmetic, at any level, is comprised of a set of axioms whose completeness he purports to prove. In the Logic Lecture for 1902/1903 he seems to believe that logic is complete and decidable, since with the algebraic method scientific cognition secures first "an unrestricted control of all deductive possibilities, and then a complete command over all possible problems and methods of solutions, which is the precondition for the most exact and general solutions of problems of the domain in question."[98] Husserl has already indicated in the *Prolegomena* that in the context of his theory of all possible theory forms: "The solution of problems raised within a theoretical discipline, or one of its theories, can at times derive the most effective methodological help from recourse to the categorial type or . . . to the form of the theory, and perhaps also by going over to a more comprehensive form of class of forms and to its laws."[99] Husserl appears then to have begun the new century, persuaded that he was poised to realize Leibniz' dream of a *mathesis universalis* that could solve all problems of the *a priori* sciences.

Nor was he alone. His new colleague Hilbert had just proposed twenty-three problems from far-flung areas of mathematics whose solutions he thought would not only stimulate new methods of research but eventually reveal the inner unity of mathematical science. Such problems would have to be well posed but hard, and Hilbert urged that all such problems were solvable. The reasons he gave for believing this were similar to those Kant gave when he cited Lambert's proof of the irrationality of π: he had solved the old problem of a rational value for π, said Kant, by showing that "at least the impossibility of such a solution can be known with

---

[95] loc. cit., p. 437.

[96] Husserl 2001a, p. 244.

[97] loc. cit., p. 243–244.

[98] Husserl 2001b, p. 37.

[99] Husserl 1971, p. 240.

certainty," realizing that, the *issue of certainty is crucial in the case of impossibility proofs*. Hilbert spelled out the rationale for this more fully:

> Occasionally it happens that we seek the solution under insufficient conditions or in an incorrect sense, and for this reason do not succeed. The problem then arises: to show the impossibility of the solution under the given hypotheses, or in the sense contemplated. Such proofs were effected by the ancients . . . when they showed that the ratio of the hypotenuse to the side of an isosceles right triangle is irrational. In later mathematics, the question as to the impossibility of certain solutions plays a preëminent part, and we perceive in this way that old and difficult problems, such as the proof of the axiom of parallels, the squaring of the circle, or the solution of equations of the fifth degree by radicals have finally found fully satisfactory and rigorous solutions, although in another sense than originally intended.[100]

Hilbert says that it is probably "this important fact along with other philosophical reasons" that encourages the belief of mathematicians that every "definite mathematical problem" is solvable, either by actual answer to the question or "by proof of the impossibility of its solution and therewith the necessary failure of all attempts." He also reminds us of the importance of such proofs in other sciences, e.g. the proof of the impossibility of a perpetual motion machine which led to the discovery of the law of the conservation of energy. This raises the question of whether it is "inherent in the nature of the mind, that all questions which it asks must be answerable?"[101] Hilbert probably has in mind here Kant's argument that every question raised by reason in transcendental philosophy must be answered by reason.

Hilbert proposed the solvability of mathematical problems in 1900, amid an outbreak of logical and set-theoretic paradoxes which complicated the issue of when they are "well posed." To motivate his second problem on the consistency of real arithmetic he refers to "the system of *all* cardinal numbers or of *all* Cantor's alephs, for which, as may be shown, a system of axioms, consistent in my sense, cannot be set up."[102] To motivate his doubt about Hilbert's thesis Richard argued for the paradoxical claim that Cantor's theorem implied the existence of real numbers that cannot be defined in a finite number of words. The contradiction, resulting from the application of the diagonal method used by Cantor to prove his theorem to the set of laws for finitely defining a real number, eventually obliges Hilbert to ask:

> What is a genuine law? e.g. for a sequence of numerals in the binary expansion of a real number? When is a mathematical problem correctly and clearly posed, so that we must demand the possibility of a clear answer? Why is it, for example, a clear question whether . . . $2^{\sqrt{2}}$ is rational or irrational, unlike the question of whether there is a totality of binary expansions expressed in a finite number of words?[103]

Hilbert found a more mathematical paradox but admitted that its use of Cantor's diagonalization called into question "one of the most beautiful proofs in set theory."[104] Still, Hilbert believed that the emergence of paradoxes in the youth of

[100] Ewald 1996, p. 1101–1102.

[101] loc. cit., p. 1102.

[102] loc. cit., p. 1105.

[103] Quoted in Hallet 1995, p. 161, from Hilbert's 1905 lectures.

[104] Peckhaus and Kahle 2002.

a science was a normal condition of its healthy development, and the first of his problems to be solved was a question about a pair of mathematical objects from which Kant had inferred a philosophical paradox.

For Kant a paradox is "an assertion that contradicts generally accepted opinion," which may be more fruitful than such opinion, for "paradox arouses the mind to attention and investigation, which often leads to discoveries."[105] Kant knows no paradoxes in mathematics itself but rather ones in which he sees mathematical facts in conflict with common opinion, e.g. that space and time are qualities of things in themselves. He says that if those who cannot free themselves from this conception reflect on his "paradox" on incongruent counterparts, they will see that "the demotion of space and time to mere forms of sensory intuition may indeed have foundation."[106] For we can intuit via their "outer relations to space an inner difference" between them, namely, that "one cannot be put in the place of the other," that "no understanding can specify as inner."[107] Thus does Kant support his paradoxical view of space. Hilbert shared his view of the fruitfulness of paradoxes, and in fact the heart of his third problem is whether incongruent symmetrical polyhedral of the same volume are equidecomposable. Dehn showed that they are not and hence that infinitesimal processes are necessary to calculate the volume of polyhedra. While Hilbert may not have thought of Kant's paradox when he formulated this problem, Dehn attributed his influence to "the characteristic Hilbertian spirit... combining logical power with intense vitality, disdaining convention and tradition, shaping that which is essential into antitheses with almost Kantian pleasure, taking advantage of the fullest freedom of mathematical thought."[108] We have already seen how Kant's ideality of space provided grist for phenomenological analyses, but his case for the ideality of time engenders his more challenging "paradox of inner sense" which will later engage Husserl's attention. But at this time Husserl is more concerned with what he regards as the "paradox" of the fruitfulness of imaginary concepts in mathematics, hitherto justified only by a principle of permanence.

Husserl's view of the state of logic and mathematics in 1900 is one of "towering thought piles and thought combinations intertwined in a thousand ways... moved about with the most sovereign freedom and spawned in ever increasing intricacy by our researches."[109] Here only art and *method*,

> Overcome the defects of our mental constitution, and permit an indirect achievement by way of symbolic processes from which the intuitive element, as well as all true understanding and inner evidence are absent, but which are rendered secure because a *general* proof of the efficiency of the method has been once and for all guaranteed.[110]

---

[105] Kant 2006, p. 17–18.

[106] Kant 2002, p. 81.

[107] loc. cit., p. 81.

[108] Reid 1970, p. 63.

[109] Husserl 1970a, b, p. 201.

[110] loc. cit., p. 201.

We shall come back to the question of just who might have supplied such a "general proof." As "towering thought piles" go, none stood taller that the object of Hilbert's first problem, which asks for a proof that the cardinality of the continuum is just the first uncountable cardinal in Cantor's "tower of powers." Another more unruly tower, not unlike nor entirely unrelated to Husserl's paradox, was the subject of another Hilbert problem; it was described with dismay by Pierce in 1900:

> Consider mathematics, as a field where, if anywhere, it might be supposed that machinery would be of little avail. In the early eighteenth century the greatest geometers in Europe were still 'stumping' one another with problems, and the discovery of a theorem might raise a man to greatness... Subsequently, it must be *method*, no longer a mere *theorem*, to impress the world so powerfully. Nowadays methods of the greatest power and profundity are turned out at such an astonishing rate that nobody but professional mathematicians ever hears of them... Herman Schubert's Calculus of Geometry, which enables us, for example by a brief computation, to determine that the number of cubic curves each of which shall touch any given twelve spheroids, is just 5 billion 819,539 million, 783,680 hardly makes a ripple in the ocean of modern mathematics.[111]

Schubert's "enumerative calculus" solved hitherto unsolved problems of this type in algebraic geometry, using his "principle of special position, or conservation of number" which however lacked any secure foundation. Thus Hilbert made it the subject of his 15th problem: "To establish rigorously and with an exact determination the basis of the so-called principle of special position, or conservation of number, by means of the enumerative calculus developed by him."[112]

Husserl was preparing his Double Lecture in which he argued that his own paradox was solved by the principle of permanence, provided that a formal system extended conservatively by imaginaries is itself definite or complete, something he purported to prove for all arithmetics. It may have been these proofs Husserl intended when he says that the "method of symbolic processes" is rendered secure by a "general proof" of its efficiency. In any case, by 1906 he can envisage a "supramathematics" of all possible such theory forms, providing a mathematical theory to fit "any pre-established purely logical or extralogical domain that in general admits of a mathematics, but is to be obtained by simple subsumption under the corresponding theory forms."[113] This ambitious vision of a pre-established harmony between mathematical thought and reality intersects with Hilbert's vision of such harmony at his sixth problem, calling for an axiomitization of physics, for which, "as he has in geometry, the mathematician will not merely have to take account of those theories coming near to reality, but also of all logically possible theories. He must be always alert to obtain a complete survey of all conclusions

---

[111] Peirce 1966, p. 172. Peirce praised Schubert's calculus as "the most extensive application of Boolean algebra which has ever been made."

[112] Gray 2000, p. 268. By the 1990s Schubert's calculus was finally provided with a rigorous foundation, through the combined efforts of some of the best mathematicians of the twentieth century.

[113] Husserl 2008, 90.

derivable from the system of axioms assumed."[114] The realization of Husserl's vision will require him to find a phenomenological home for Kant's transcendental idealism.

2.4 Idealism, Clarification, Intuitionism

In his winter lectures of 1910/1911 on phenomenology Husserl develops the theme that by "being in the *a priori* attitude, we grasp ideals, essences," and thereby avoid the temptation of mathematicians to interpret their pure universal truths in empirical terms. He explains that "the essence-attitude, ultimately that of *intuitive* ideation, brings a new *existence-free* sphere of givenness; and in a certain sense may indeed be characterized as the philosophical attitude."[115] Husserl contrasts "nature as fact," studied by empirical sciences, with "nature as Idea," the subject of pure natural science yielding "the sciences of the ideas which are constitutive of the idea of nature: geometry, pure theory of time, pure theory of motion and possible deformations of what is of the nature of things as such – this latter would correspond to Kant's idea of pure natural science."[116] But in addition to duration and shape, things also have *real* properties and *changes* standing in *causal* connections that are governed by *a priori* laws, which brings Husserl to "the Kantian 'pure science of nature' which, as is well known, is distinguished by him from geometry, pure chronometry, and the previously mentioned disciplines."[117] But the promise of this science, says Husserl, has not been realized, largely because philosophers have been unable to sustain the essence attitude against empiricist prejudices. Husserl soon makes a remarkable effort to sustain this attitude in his *Ideas*.

He begins by explaining that he uses the foreign word 'eiδos' in order to "keep the supremely important Kantian concept of the idea purely separate from the general concept of the (formal and material) essence."[118] Modern philosophy has been "secretly longing" for phenomenological clarification of the sense of 'being,' says Husserl, and

> The first to detect it correctly is Kant whose greatest Intuitions only become fully intelligible to us if we have elaborated in a fully conscious, clear way what is peculiar to the phenomenological domain. It then becomes evident that Kant's mental gaze rested on this field, although he was not yet able to appropriate it and to recognize it was a field of labor of a rigorous eidetic science of its own. Thus, for example, the transcendental deduction of the first edition of the *Critique of Pure Reason* is actually already moving on phenomenological ground. However, Kant misinterpreted it as psychological and, thus, gives it up in turn himself.[119]

---

[114] Gray 2000, p. 258.

[115] Husserl 2006, p. 18.

[116] loc. cit., p. 18.

[117] loc. cit., p. 19.

[118] Husserl 2014, p. 6.

[119] loc. cit., §62.

The main premise of Kant's transcendental deduction is that the unity of our experience can be ascribed to a single identical subject by means of its "transcendental unity of apperception" only if the elements of experience given in intuition are synthetically combined to present us with objects that are cognized by the categories. Everything centers on this identical transcendental subject or ego. And so it does for Husserl, who must ask whether "the phenomenological ego has become a transcendental nothing by means of the phenomenological reduction."[120] Husserl answers that there is a *flow* that survives the reduction as a "transcendental residuum" in which we find no pure ego as an experience among the other experiences, "but yet seems to be there." He says:

> The Pure ego lives itself out in a particular sense in every current cogito, yet all background experiences belong to it as well and it belongs to them. All of them, by virtue of belonging to the *one* stream of experience that is mine, *must* be capable of being converted into current cogitations or of being included in them immanently. In Kantian language: 'The "I think" must be able to accompany all my representations.'[121]

As for Kant's transcendental unity of apperception which anchors the unity of experience, Husserl insists that in the continuous progression of our apprehensions, we apprehend the *stream of our experience as a unity*, not as a single experience, "but in the manner of an *idea in the Kantian sense*. The stream of experience is nothing arbitrarily posited and maintained but instead an absolutely indubitable given – in a correspondingly wide sense of 'givenness.' This indubitableness, although it is grounded on intuition, has a source completely different from the one that obtains for the being of experiences, i.e. those that come to be purely given in immanent perception."[122] In fact, Husserl officially adopts the principle of "a things' givenness as an idea in the Kantian sense."[123] There are transcendental objects that cannot be given to any "isolated consciousness" with complete intuitiveness: "But the perfect givenness is nonetheless prefigured as an "idea" (in the Kantian sense) – as a system absolutely determined in terms of its type of essence, a system of endless processes of appearing ..." These processes proceed infinitely, but the apparent absurdity of their being given is avoided, since "The idea of an essentially motivated infinity is not itself an infinity; the insight that this infinity cannot in principle be given does not exclude, but instead far more demands the patently discernable givenness of the *idea* of this infinity."[124]

---

[120] loc. cit., §57.

[121] loc. cit., §57. Kant argued that we can find no substance behind this 'I think:' "For we cannot judge even from our own consciousness whether as soul we are persisting or not, because we ascribe to our identical Self only that of which we are conscious; and so we must necessarily judge that we are the very same in the whole of the time of which we are conscious. But from the standpoint of someone else we cannot declare this to be valid because, since in the soul we encounter no persisting appearance other than the representation "I," which accompanies and connects all of them, we can never make out whether this I (a mere thought) does not flow as well as all the other thoughts that are linked to one another through it." (A364)

[122] loc. cit., §83.

[123] loc. cit., §143.

[124] loc. cit., §143.

We have seen that Husserl shares Kant's optimism concerning the solvability of mathematical problems, and he certainly shares Kant's doctrine of the decisive importance of intuition for mathematical knowledge, having extended Kant's notion of pure intuition to categorial intuition. But we have also seen that Kant's case for solvability obligated him to accept as intuitive complicated impossibility proofs by contradiction that bore no resemblance to his paradigm cases of appeals to intuition. Let us now see how the same problem arises for Husserl's *Intuitionism* when he re-examines his old solution to the problem of meaningless expressions in Book III.

In the *Investigations* Husserl had encountered a "most unwelcome ambiguity" among absurd expressions: there he argued that if we judge the concept of 'round square' absurd because no object falls under it, we would have to judge as senseless not only concepts such as this whose absurdity is *immediate*, but also ones such as 'regular decahedron' whose absurdity was mediated by "lengthy indirect proofs" of their impossibility. So he distinguished truly meaningless words like 'abracadabra' from "the *a priori impossibility of a fulfilling sense*. An expression has meaning in this sense if a possible fulfillment, i.e. the possibility of a unified intuitive illustration, corresponds to its intentions."[125] And it is meaningless if "we apprehend the real impossibility of meaning fulfillment through an experience of the incompatibility of the partial meanings in the intended unity of fulfillment."[126] Husserl had taken the point of Marty's question of how we could understand how to present a proof of the impossibility of a regular decahedron if this concept is meaningless. Such questions are raised in legion by the Kant-Hilbert rationale for the solvability of all mathematical problems, as exemplified by the discovery of the impossibility of squaring the circle.

Husserl explains that in geometry we *posit* spatial formations as existent in *eidetic truth*, not in the world of experience, which posit is valid if such a formation is possible. This possibility guarantees the direct eidetic Intuition of the shape, bringing it to givenness:

> For every particular shape, insofar as it is not, like the fundamental shapes straight line, angle, and the like, taken from direct intuition, but rather is produced only by the connection of already formed and valid concepts of shape, the "proof of existence" is carried out on the ground of axioms, i.e. it is mediately shown that there truly is in space according to its essence a geometric essence corresponding to this shape-*concept* (a freely formed logical signification).[127]

Without such a proof of existence nothing can be said about regular bodies of ten surfaces. For those we need "negative existential propositions," which Husserl says, "have the function of separating out the invalid concepts, the expressions corresponding to no essence."[128] The proof of this for round squares would be almost immediate, but for decahedrons quite lengthy. That there is no essence of

[125] Husserl 1970a, b, p. 294.

[126] loc. cit., p. 294.

[127] Husserl 1980, p. 70.

[128] loc. cit., p. 71.

round square reminds us not to confuse noema with essence. Husserl also reminds us that since our intuitions of physical objects can be one sided and inadequate, "one must never overlook the fact that eidetic intuition in no way implies adequate intuition."[129] This brings us to the difference between phenomenology and ontology.

Husserl insists that despite the phenomenologically discernable eidetic relationships between intuition of noema and intuition of ontic essences, any intuition of the onta involves "an essential alteration of the judgmental position if we judge ontologically about the essences and eidetic particulars as such instead of judging phenomenologically about the noema."[130] In the phenomenological constitution of a thing we don't judge things as such, but rather reach a stage in which,

> The preconditions are fulfilled for being able to grasp the essence "physical thing" harmoniously in the greatest conceivable perfection and to reject dissonant accounts of essence and thereby for being able to obtain the most perfect evidentness of the axioms according to their pure and solely valid sense and to reject false axiomatic accounts as countersensical.

Now we have left phenomenology for ontology, but when we constitute, say a house, Husserl can explain how we know what 'house' means even though we don't know its essence: "Only in the clarifying intuition can it become apparent to us, through a purely expressive fitting of the word-signification, the logical one, to the essence given in the intuition, whether the expression with its sense actually fits onto that which is unclearly meant."[131] But what about all those countersensical judgments made throughout history about squaring the circle, a case where the corresponding essence cannot exist?

Husserl grants of course that in this case there is no harmonious essence, but says that we do "see the dissonant unity of the intended essences in the countersensical judgment, the incompatibility; in this discerned incompatibility we grasp instead of the essence pure and simple (the harmonious one) the non-essence, so to speak."[132] That we must recognize as ontological the insight that clarifies the meaning of this problem, which by then had come to require the algebraic construction of π, emerges, says Husserl from the fact that

> The clarifying of a concept with geometric sense, such as the squaring of the circle and, what is of equal value here, the demonstration that the thought-concept lying in these words could serve as descriptive for what is meant in an obscure act, still does not require that we carry out the geometric proof of impossibility of such a conceptual object: of that we would perhaps not even be capable.[133]

Husserl grants that the clarity of the problem hardly parallels the insight into its negative solution, *for which he is far from claiming an evident intuitive insight at*

[129]loc. cit., p. 74.

[130]loc. cit., p. 74.

[131]loc. cit., p. 74.

[132]loc. cit., p. 75.

[133]loc. cit., p. 75.

all. In the *Investigations* he has already described π as "unpresentable."[134] Still, perhaps viewing such problems as the exception which proves the rule of intuition, he proceeds to formulate his own homespun intuitionism, which answers "the 'dogmatic sciences' need for clarification."[135]

Husserl grants that we give birth to sciences while being unable to ground them rationally or have any insight into their empirical success, but is confident that:

> A remarkable teleology prevails in the development of human culture in general and thus also in scientific culture, in that valuable results can arise without insight or thought through a mixture of insight and instinct in the exercise of psychic powers. That the results are valuable is proved naturally by reasons subsequently produced that establish the claims to validity within the compass of perfect insight.[136]

But the sciences are still far removed from having all their claims so validated. As they develop their methodologies, the more their work "shifts into the symbolic sphere" where "concepts originally oriented towards Intuition are used merely symbolically, as counters with which one operates, without recourse to intuition in the style of the essentially symbolic method that has been developed."[137] What was at least *relatively* a matter of insight at a lower level of development is now symbolized on a higher level and "robbed of evidentness," turning the sciences into "factories" whose workers turn out valuable products of which they have no "inner understanding." Husserl's vision here of the plight of knowledge in this "century of technology," as he calls it, strikingly recalls Peirce's vision of modern mathematics in 1900. Without explaining exactly why the symbolic method must vanquish insight, Husserl argues that the treasures of knowledge buried in the

[134]Husserl 1970a, b, p. 653. Here π is classified with "mere concepts" for which "we have nominal expression inspired by signifigant intentions in which the objects of our reference are 'thought' more or less indefinitely, and particularly in the indefinite form of *an* A as the mere bearer of definitely named attributes. To mere thinking 'presentation' is opposed: plainly this means the intuition which gives fulfilment, and adequate fulfilment, to the mere meaning-intention." The unpresentability of π, referred to indefinitely as 'a transcendental number,' means that there is no intuition that adequately fulfills its concept; so one should not expect that the proof of the impossibility of constructing it algebraically would be intuitive. In fact, Lindeman's proof of this made crucial use of Euler's equation $e^{i\pi} = -1$.

[135]Husserl's intuitionism, unlike Brouwer's, is not strongly "revisionist," as van Atten 2002 explains. We saw Brouwer give what he called "an appropriate reworking of Lambert's negative irrationality proof of π" into a positive one. Kant's embrace of Lambert's proof as evidence of the solvability of mathematical problems appeared to contravene his own intuitionism, but he insisted that apagogic proofs by contradiction "carry more intuitiveness" than direct ones. Nor did Husserl evince any inclination to revise such a proof on principle. Van Atten also sees in Husserl's proposed "method of clarification" evidence of his "weak revisionism," but his efforts to clarify countersensical judgments about 'squaring the circle' seem too obscure to count as any kind of revision or foundation. A precise determination of his weak revisionism is important in view of van Atten's elegant argument that such revisionism would entail that he *should* have been a strong revisionist. All of which leads inescapably to Husserl's *pluralism* in philosophy of mathematics, as argued forcibly by Hartimo 2012.

[136]loc. cit., p. 81.

[137]loc. cit., p. 82.

symbolic trappings of the sciences must be exposed in intuition: "For knowledge is insight, is truth drawn from Intuition and thereby completely understood."[138] And so, Husserl says:

> *Intuitionism*, therefore, reacted with complete justification against the one-sided surrender of mankind to the expansion of the sciences as techniques of thought. What matters is to put an end to the plight, grown intolerable, of reason, which amidst all the riches of its theoretical possessions sees its proper aim, world understanding, insight into the truth, recede ever farther in the distance.

One might think that Husserl had followed Weyl in joining Brouwer's revolution, but Husserl demands intuition in *all* the sciences, not just mathematics, and in fact he immediately warns that: "Of course, Intuitionism must not degenerate into mysticism instead of approaching the sober tasks that are forthcoming from the situation described."[139] The phenomenological program of restoring intuition to science is combined with Husserl's optimistic vision of "how from the primal ground of Intuitive givenness the goal of an all around and complete knowledge that solves all reasonably posed problems can be attained."[140] This goes well beyond the thesis of Kant and Hilbert that all such problems of mathematics are solvable. In explaining what he calls "phenomenological intuitionism," Husserl warns that "the much misused word "intuition" makes it possible unfortunately to misplace phenomenology under the title of "intuitionistic philosophy" and thereby bring it under the heading of all sorts of mythical idolizations and unscientific raptures."[141] Nor does it provide for supernatural illumination of mysterious "intellectual intuitions." The "method of intuition" in phenomenology "has only a simple and sensible meaning; it says that I only pass judgment reliably, where I can show what I believe . . . and ultimate showing is seeing or something exactly analogous to seeing; and even that must still be shown in a pure visual display, and all forms of this seeing must be studied."[142] A more perfect intuitionism, says Husserl, is of itself a more perfect rationalism.

The pursuit of the ideal rational science leads to its exact mathematization, says Husserl, and thus to a modification of Kant's well known criterion that there is only so much "authentic science" in an empirical discipline as there is mathematics in it. Thereby such a rational science is an Idea in the Kantian sense. However, when we examine the most recent mathematical sciences, we see that they approach this ideal only in the direction of "deductive theorizing," says Husserl, while still wanting their critical reforms:

> The ardent pains taken to provide here authentic foundations, leads to known attempts at revolutionary reform, which are supposed to have profound changes in the content of science itself as a consequence. We think here of the conflict over the reshaping of

---

[138] loc. cit., p. 82.

[139] loc. cit., p. 83.

[140] loc. cit., p. 83.

[141] Husserl 2002, p. 288.

[142] loc. cit., p. 288. We wonder how analogous Husserl regards categorial intuition to seeing.

> geometrical and basic physical principles and Einstein's theory of relativity, and also of the strife over the reconstruction of the basic concepts of pure analysis (set, number, continuum) and of the revolutionary theories of Bouwer and Weyl. This state of affairs is the most paradoxical insofar as, for all that, it is still a question of axiomatics, which pretends to be consulting apodictic evidence. In fact, who could distance himself from, e.g. the evidence of geometrical axioms? And yet subsequent scientific reflective critique requires new foundations.[143]

Clearly then, Husserl's phenomenological intuitionism is his own agenda, however much of it might intersect with that of Weyl and Brouwer. Prima facie, the obvious such point would seem to be Kant, whose apparent view of the temporal basis of arithmetic was hailed by Brouwer. What about Husserl?

Husserl has more than once criticized Kant for allegedly trying to "bring arithmetic into an entirely inadmissible relationship to time."[144] While we may agree with Husserl, as I do, that such a relationship could give us no insight into arithmetic, he was wrong to suppose that Kant really tried to establish one, though understandably so. Some passages, including two cited by Husserl, do suggest that Kant was thinking along these lines, but the charge does not survive a wider look around.[145]

Even Husserl's own phenomenological intuitionism seems to have no special place for time in arithmetic when he declares that:

> It is now clear that the same holds for the research field of arithmetic as holds for nature, the universe of spaciotemporal facticities. Before I make a theoretical assertion about the domain of number, before I can believe anything concerning the lawfulness holding in it, it must itself be given to me. No general arithmetical proposition can be proved, that does not presuppose a cognitive consciousness at the deepest level, in which the number domain itself comes to originary givenness. To recite the numbers or their definitions is not to have given the number series in an originary way. If I call them infinite, so must the infinite be given to me.[146]

Let us examine Kant's view on arithmetic as well as his appeal to the ideality of space to demystify what he considers Plato's belief in pre-established harmony.

Kant mentions arithmetic only once in the *Critique*, to fix the terminology of "theorems" for "propositions of arithmetic and geometry."[147] But no theorem of arithmetic is mentioned, only "numerical formulas," by way of explaining why there are no axioms for "magnitude in general." Axioms must be synthetic, but such a plausible candidate as the immediately certain general proposition "equals added to

---

[143] loc. cit., p. 297–298.

[144] Husserl 2008, p. 109.

[145] Husserl cites the notorious claim in the Schematism chapter that while "space is the pure image of all magnitudes of outer sense," says Kant, "the pure *schema of magnitude* ... as a concept of the understanding, is *number*, which is a representation that summarizes the successive addition of one (homogeneous) unit to another. Thus number is nothing other than the unity of the synthesis of the manifold of an intuition in general, because I generate time itself in the apprehension of the intuition." (A142/B182) We shall try to put this curious claim in context later.

[146] Husserl 2002, p. 475.

[147] A736/B764. Kant would prefer not to call them "dogmata."

equals are equal" is according to Kant analytic: "The self-evident propositions of numerical relations, on the contrary, are, to be sure, synthetic, but not general, like those of geometry, and for that reason also cannot be called axioms, but could rather be named numerical formulas. That $7 + 5 = 12$ is not an analytic proposition."[148] For addition formulas Kant gives a categorial analysis of *Rechenkunst* he lectured on in Wolff, who based it on Euclid's definition of number as "a plurality of units." Kant appeals to his three categories of quantity; I have a number when I synthesize a plurality of units into a totality: "the concept of a *number* (which belongs to the category of totality)" illustrates his combination principle that the third of each triplet of categories arises from the first two by "a special act of understanding."[149] We see that Kant's conception of arithmetic develops from within the Wolffian framework of logic and mathematics, something Husserl points to by way of explaining the flaws in Kant's account. But I am not sure he appreciated the value of this framework.

In his *Mathematisches Lexicon* Wolff distinguished three levels of arithmetic: arithmetica practica (Rechenkunst), arithmetica theoretica (Zahlwissenschaft), and arithmetica infinitorum (Rechnung des Unendliche). In the *Critique* only arithmetica practica is at issue: as he told Schultz, $3 + 4 = 7$ is "an immediately certain practical judgment."[150] Arithmetica theoretica deals with "properties of numbers" and theorems about them as found in Euclid: *Zahlwissenschaft*, as he put it to Schultz, "is a pure intellectual synthesis that we represent to ourselves in thought."[151] For this Lambert formulated six axioms and four postulates in the *Architectonic* known to Kant.[152] His second postulate, that "any number can be taken as a larger unit", says Lambert, is "the basis of our characteristic number system, since in it one sees the numbers 10, 100, 1000, etc. as new units and interprets their value by their positions or order of precedence."[153] Kant bases this postulate on the "*pure synthesis*... which rests on a ground of synthetic unity *a priori*; thus our counting (as is especially noticeable in the cases of larger numbers) is *a synthesis in accordance with concepts*, since it takes place in accordance with a common ground of unity (e.g. the decad)."[154] Kant ignores arithmetical theorems until he later confronts Pythagorean mysticism.

Meanwhile he is interested in *intuitive* representations of the properties of numbers such as the "figured numbers" used by the Pythagoreans to prove the

---

[148] A164/B205 This controversial claim is convincingly defended by Anderson 2004, who shows independently of Kant's theory of intuition, that the non-conceptual content of $7 + 5 = 12$, and addition formulas generally, cannot be fitted into a hierarchy of concept containment for numbers comprising the criterion of analyticity drawn from traditional logic.

[149] B111.

[150] *Correspondence*, p. 284.

[151] loc. cit., p. 285. This was essentially his view of arithmetic already in the *Dissertation*.

[152] Indeed, it was Kant himself who managed to have the *Architectonic* published.

[153] Lambert 1771, §78.

[154] A78/B104.

theorems such as $1+3+5+\ \dots\ +(2n-1)=n^2$ for square numbers and $1 + 2 + 3 + \cdots + n = \frac{1}{2}n\,(n-1)$ for triangular numbers. These figured numbers and many more were explained by Wolff in his *Lexicon*, but the shortest and most telling entry was that on the prime numbers: "those which can be divided by no other number than 1, such as, 3, 5, 7, 29."[155] They are called "simple or linear" numbers, a reminder of the Pythagorean embarrassment at being unable to devise any figured representation of them. In Kant's terms, there appears to be no *schema* for the primes. Kant knew from Wolff that Leibniz believed that binary arithmetic can be used to investigate "hidden properties of numbers," above all the primes: as he explained to Locke:

> In the science of numbers themselves, great difficulties arise because so many things have to be taken into account: what we are looking for are short formulae, but we do not always know in a given case whether what we are seeking is there in nature to be found. For instance, what is simpler in appearance than the notion of a *prime number*? That is, a whole number divisible only by itself and unity. And yet we are still hunting for an easy positive criterion by which they can be identified with certainty, without having to try out all the prime divisors less than the square root of the prime in question.[156]

With Kant's grounding, quoted above of Lambert's second postulate in *a priori synthesis* in mind, Martin wrote that:

> In the 1771 *Architectonik* Lambert goes into questions of constructing numbers in considerable detail, also referring to the special significance of the dyadic system for Leibniz. Lambert had already indicated that there are properties of numbers independent of the number system, for example, the property of a number's being prime. Although Kant in dealing with numbers uses very careful and comprehensive formulations of the unity of apprehension in a concept, here we can really see his effort to include other concepts, for example, prime number, in the class of such arithmetical concepts as that of the base of the decimal system.[157]

Martin ably counters several of Husserl's early criticisms of Kant's theories of synthesis, but in his own studies of passive and active synthesis Husserl came to appreciate:

> Kant's brilliant insights that are expressed in his profound but obscure doctrine of the synthesis of productive imagination, above all in his transcendental deduction from the first edition of the Critique of Pure Reason. When Kant in his great work speaks of an analytic synthesis, he means cognition deployed there in explicit forms of concepts and judgments, and this points back for him, to a productive synthesis.[158]

Husserl is especially favorable to Kant's more psychologistic "threefold synthesis" of the A deduction. A striking example of the "intellectual synthesis" introduced

---

[155] Wolff 1965, p. 962. The usefulness of figured numbers for proving theorems is stressed in Lambert 1771.

[156] Leibniz 1981, p. 487–488. For the role of binary arithmetic in Leibniz' search for the secret law of primes, see Mahnke 1912, especially p. 44–47.

[157] Martin 1985, p. 91.

[158] Husserl 2001c, p. 410.

by Kant in the second edition, which he says *Zahlwissenschaft* is comprised of, is the investigation, alluded to by Martin, of Lambert in which he pushes forward the search for the law of primes into arithmetica infinitorum.

In the *Architectonic* Lambert formulates axioms and postulates to analyze positional number systems as well as properties of *numbers themselves*: "for considered in themselves the whole numbers have an absolutely necessary order which is entirely independent of characteristic numbering systems... In this sense the prime numbers are distinguished from those that are multiples of another number. Prime numbers are divisible by no other number than one. But one has not been able to find by any means the order in which they succeed one another, or determine by a simple criterion how each can be recognized for itself, or if a number is given, to find out if it has divisors or not."[159] In fact, says Lambert, "Euclid has proved of the prime numbers the theorem that, as many of them as we can think, there are still more. However, concerning their *order* one has not been able to find anything, since to all appearances, no local order at all occurs among them."[160] In a search for such an order Lambert formulated a new infinite series:

$$\text{(L)}\quad \frac{x}{1-x}+\frac{x^2}{1-x^2}+\frac{x^3}{1-x^3}+\frac{x^4}{1-x^4}+\frac{x^5}{1-x^5}+\ldots$$

which when resolved by "real division" yields the remarkable infinite polynomial:

$$\begin{aligned}\text{(S)}\; & x+2x^2+2x^3+3x^4+2x^5+4x^6+2x^7+4x^8+3x^9+4x^{10}+2x^{11}\\ & +6x^{12}+2x^{13}+4x^{14}+4x^{15}+5x^{16}+2x^{17}+6x^{18}+2x^{19}+6x^{20}+4x^{21}\\ & +4x^{22}+2x^{23}+8x^{24}+\cdots\end{aligned}$$

in which the coefficients of each term give the number of factors of its exponent. The prime numbers are thus the exponents of those terms whose coefficients are 2.[161] But Lambert denies that "these coefficients have any local order at all, making it difficult to deduce how the coefficients of a term can be determined without having its predecessor and successor."[162] This may be Lambert's admission that

[159]Lambert 1771, §875. The fundamental theorem of arithmetic enshrines the primes as its basic building blocks without providing any criterion for their recognition.

[160]loc. cit., §875. The order of prime numbers, said Euler, "is a mystery that the human mind has not been able to penetrate... which is all the more surprising since arithmetic provides us with sure rules for continuing their progression, without however, allowing us to perceive the least mark of an order." (*Opera Arithmetica*, additamenta, p. 639)

[161](L) is a special case of "Lambert's series," around which a fair literature has grown. G.H. Hardy and J.E. Littlewood: 1921, showed that a theorem about general Lambert series actually implies the prime number theorem. For their role in the mystic number theoretic research of Ramanujan, see R. P. Argawal: 1993.

[162]In denying "local order" among the primes, Lambert assimilates their apparent lawlessness to that of the non-repeating decimal expansions of irrational numbers, which cannot be calculated "der Stelle nach:" to calculate any one digit one must first calculate all its predecessors. He

(S) itself cannot provide the long sought law of primes. But his construction of (S) from (L) is a good example of what Kant called a "pure intellectual synthesis" in arithmetic. Each term of (L) becomes on division an infinite sequence itself, and Lambert synthesizes (S) from the resulting infinite sequence of infinite sequences by collecting the finite number of times that any nth power of x can occur in them.[163] Remarkably, if n is a prime number, x^n occurs only twice. For Lambert this synthesis is a prime specimen of symbolic cognition.[164]

The law of distribution of prime numbers is for Lambert one of the two fundamental problems of arithmetic, for which he institutes an *empirical* search by constructing factor tables for primes which show, however, only that, although infinite by Euclid's theorem, they seem to occur less frequently the further one extends their table. The other problem was to find faster methods of factoring composite numbers into primes. Lambert wrote to Kant that "now and then there are lovers of mathematics who are fond of calculation," telling him how any he knows in Konigsberg would be remunerated for working on his tables: "Perhaps the factor tables for numbers could rise to 1,000,000 and may yet make up two octavo volumes."[165] This request from "the leading genius in Germany" may have impressed on Kant the importance of even arithmetica practica for the future progress of mathematics. We shall come back to Gauss' discovery, using tables such as Lambert's.

While Kant's guarantee of the applicability of geometry to all objects of sensory experience stems from his exposition of space as the form of outer intuition, his claim that arithmetic applies only to objects of such experience derives no such explanation from his exposition of time as the form of inner sense. In fact, Kant freed

conjectured *all their digits will occur with equal frequency in the long run* so that each occurs with equal probability, their series comprising "ein absolutes Nicht Vorauswissen."(§311) Though determined by a law, we cannot *literally pre*dict them, but only wait for the law to generate them. If one *arbitrarily* draws a number from ten numbers one at a time, we cannot predict what number will be drawn on the hundredth trial: we could no more successfully *bet* on such draw, says Lambert, than we could on the digits of π. (§323) He claims that there are "uncountably more" such series than those with local order. (§318)

[163] Thus $\frac{x}{1-x} = x+x^2+x^3+\cdots+x^n$, $\frac{x^2}{1-x^2} = x^2+x^4+x^6+\cdots+x^{2n}$, $\frac{x^3}{1-x^3} = x^3+x^6+x^9+\ldots x^{3n}$, etc. and in general $\frac{x_n}{1-x^n} = x^n+x^{2n}+x^{3n}\ldots$ Number theorists I have consulted were amazed that Lambert ever thought of (L) to begin with, in connection with, prime numbers.

[164] As it would have been as well for Husserl, who observed that the division indicated in the very first term of Lambert's (L) "is not in a real sense feasible" (Husserl 1983, p. 238) any more than the calculation of the root of an irrational number. Indeed, he points out that this division along with other examples of infinite series easily leads to false results. (p. 239) This is a good example of what Husserl had in mind when he warned that it will take a "deeper analysis" to determine whether the inverses of his arithmetical operations are defined for given numbers or involve "an impossibility *a priori*" or that the computation problems they pose are "free of contradiction." (*PoA*, p. 293–294)

[165] AA10, p. 111. Kant was likely informed of the importance of Lambert's project by Schultz, who had stressed the need for tables of primes to simplify the theory of logarithms. See Schultz 1803, p. 34–38. Gauss will make use of Lambert's tables and those of others in his research on the frequency of primes that resulted in his conjecture of the prime number theorem.

arithmetic from dependence on time when he explained, in response to Kastner's critique of the infinite extendibility of the straight line in the theory of parallels, that:

> When the geometer says that a line, however far one has produced it, can always be extended further, that does not mean the same thing as when it is said in arithmetic of number that one can always increase it without end by the addition of others units or numbers (for these additional numbers and the magnitudes they express are *in themselves* possible, without having to be admitted with the previous ones only as parts of one magnitude); rather to say that a straight line can be extended into the infinite means no more than: the space in which I describe the line is greater than every space which I may describe in it, and so the geometer bases his task, of extending a space . . . into the infinite, on the original idea of a unique, infinite, subjectively given space.[166]

Whereas the geometer, says Kant, requires the philosopher's assurances of the existence of this infinite space in which he can extend his lines into its infinity, the arithmetician requires no such assurance of a subjectively given infinite time that would allow her to add units or increase numbers ad infinitum. Kant's "infinitude of possible or actual addition" comprises a manifold of intuition arising from the arbitrary spontaneity of the understanding which, he says, "generates time itself in the apprehension of the intuition." But her freedom to frame arbitrary definitions of properties of numbers is not guided by any intuition of 'natural kinds' the geometer has of his figures.

Kant's reflections on *Zahlwissenschaft* circle this issue: he writes that "All negations are limitations, for they can only be thought by real opposition, just as every line, figure, and solid is only thought as a delimitation of space. But why not numbers as well? Because their reality is not given, but are seen as arbitrarily thought up."[167] In fact, he says, "no other purely arbitrary concepts of pure reason can arise in us than those by means of repetition, hence of number and quantity."[168] But now, does not the primal arithmetic act of *counting* rest on *consciousness* as Kant invoked it in the threefold synthesis, even if not on "cognitive consciousness at the deepest level" as Husserl required for the proof of arithmetical theorems in his manifesto for phenomenological intuitionism?

For Kant the concept of number arises by a categorial synthesis as "nothing other than plurality considered as unity." For its *deduction* Kant subjects all cognition to time, the form of inner sense, where "the mere play of our representations" appear.[169] To represent a number "I must first grasp one of these manifold representations after another in my thoughts" and recall "the successively represented units."[170] This requires "the consciousness that that which we think is the very same as what we thought before . . . If in counting I forget that the units that now hover

---

[166] AA20, p. 420.

[167] AA17, p. 485.

[168] loc. cit., p. 371.

[169] A101.

[170] A102.

before my senses were successively added to each other by me, I could not cognize the generation of the multitude . . . and consequently I could not cognize the number, for this concept consists solely in the consciousness of this unity of synthesis."[171] But later Kant insists that "this cannot happen except by placing them side by side one another in space: for they must be conceived as given *simultaneously*, i.e. as taken together in one representation, otherwise, this multitude does not constitute a magnitude (number)."[172] Kant argues against Cartesian idealism that, although "the soul is the object of inner sense," this sense is *too amorphous to allow the determination of my existence in time*, for "inner sense is often dull, and its horizon is shrouded in fog, and it does not give much help."[173] How could anyone with Kant's dim view of inner sense envision its 'form' as the basis of arithmetic?

Kant wrote in the *Prolegomena* that "Geometry bases itself on the pure intuition of space. Even arithmetic forms its concepts of numbers through successive addition of units in time, but above all, pure mechanics can form its concepts of motion only by means of the representation of time."[174] It is mechanics that *require* the representation of time, not arithmetic where concepts of individual numbers such as 7 and 5 are formed by successive addition "in time." *Zahlwissenschaft* is *not* said to require pure intuition of time. In fact, to clinch his proof that time *is* a pure intuition, Kant had to fall back on its *spatial representation by a straight line.* As he assured Schultz, "time, as you correctly notice, has no influence on the properties of numbers."[175] Indeed, what influence could time possibly have on the properties of being odd, even, prime, or perfect? This brings us to Kant's case against what he sees as the mysticism of Plato and Pythagoras, who are initially inspired by the role of a perfect number in musical harmony.

Plato, writes Kant, "was led by such an original construction of things, in the discovery of which we can dispense with all experience, and by the mental capacity for drawing the harmony of things out of the supersensible principle (to which pertain the properties of numbers, with which the mind plays in music) to the enthusiasm that elevated him beyond the concepts of experience to the ideas which seemed to him explicable only by means of an intellectual communion with the origin of things."[176] He would derive that which Anaxagoras inferred from experience:

> from the pure intuition internal to the human mind. For in the necessity of that which is purposive and so constituted as if it were intentionally arranged for our use, but which nevertheless seems to pertain originally to the essence of things, without any regard to our use, lies the ground for the great admiration of nature, not outside of us so much as in our own reason; in which case it is surely excusable that through misunderstanding this admiration gradually rose to enthusiasm.[177]

---

[171] A103.

[172] Kant 2005, p. 359–360. This claim is essential for Kant's refutation of Idealism.

[173] Kant 1992c, p. 66.

[174] Kant 2002, p. 79.

[175] Kant 1999, p. 284.

[176] Kant 2000, p. 236.

[177] loc. cit., p. 237.

The unity of the rules for circles arousing his admiration, says Kant, derives not from the *concept* of circle but its exhibition in intuition, but "it thereby comes to seem as if this unity empirically possesses an external ground, distinct from our power of representation, for its rules," as if nature were in harmony with the purposes of the understanding. But,

> this very harmony, since it is, in spite of all this purposiveness, cognized not empirically, but *a priori*, should bring it home to us that space, by the determination of which... the object is alone possible, is not a property of the object outside me, but merely a kind of representation in me, and thus I *introduce* the *purposiveness* in the figure that I draw *in accord with the concept*.[178]

Having dissolved the mystery of the apparent purposiveness of Plato's geometrical forms in the ideality of space, Kant can make no such attempt to account for the purposiveness of Pythagoras's arithmetic. For this he is obliged to mount a critique of *the temptation to philosophize about mathematics over meaningless questions*.

In tracing all *a priori* knowledge to the intellectual sphere, both believed "they had stumbled on a *mystery*, where no mystery exists: not because reason can answer all the questions submitted to it, but its oracle falls silent, once the question has been elevated so high that it now no longer has any meaning."[179] Solvability excludes mysteries in mathematics by demanding purely mathematical solutions: "Anyone who wishes to solve a mathematical problem philosophically, thereby contradicts himself; for example, why is it that the *ratio* of the three sides of a right-angled triangle can only be that of the numbers 3, 4, and 5?"[180] Anyone "who *philosophizes* upon a mathematical problem believes that here he has stumbled on a mystery," says Kant, "which he can neither make intelligible to himself, nor communicate to others." But his example of such a question wrongly assumes the numbers 3, 4, 5 comprise the *only* Pythagorean triple, as Reimarus pointed out publically, much to Kant's embarrassment. But his extensive notes for answering Reimarus contain valuable points he rarely mentions.

Kant explains that to give "examples of the fanaticism that may be induced by attempts to philosophize about mathematical objects," I attributed to the Pythagorean number mystic the question why 3, 4, 5, was the only such triple. Reimarus proves there are infinitely many such triples, giving the formulas $a^2 + b^2$, $a^2 - b^2$, $2ab$ for them and proving (R): If $a > b$, then $(a^2 + b^2)^2 = (a^2 - b^2)^2 + (2ab)^2$. So it *appears* they are "embroiled in a truly mathematical dispute," but Kant explains that he only *meant* to say (K) that 3, 4, 5 were the only three *consecutive* numbers comprising such a triple.[181] In his notes Kant wrote the equation '$x^2 + (x+1)^2 = (x+2)^2$'

[178] loc. cit., p. 237. Later Kant grants that "Plato at least proceeds consistently. Before him there undoubtedly hovered, albeit obscurely, the question that has only lately achieved clear expression: How are synthetic propositions possible *a priori*?" (Kant 2002, p. 433). Had he realized that space and time were *sensuous* intuitions whose objects are only appearances whose *forms* are "determined *a priori* in mathematics," he would not have sought them in the divine understanding.

[179] loc. cit., p. 434.

[180] loc. cit., p. 434.

[181] loc. cit., p. 449.

calling its unique solution "a purely arithmetical problem."[182] Kant says both (R) and (K) are strictly provable, but what did he believe either proof owed to *intuition*? He writes that, "In pure mathematics there are really no controversies that can arise from misunderstanding, since there the object is given *a priori* in intuition and any ambiguity in the meaning of an expression can only last for an instant".[183] For Kant the object of (K) is a 3, 4, 5 right triangle, but can the intuition *of* such a triangle establish *that* $x = 3$ is the *only* solution of his "purely arithmetical" equation?[184] We can take away two things from Kant's discussion: that he knew there are infinitely many solutions for $x^2 + y^2 = z^2$ and is convinced that there are no real misunderstandings or conceptual disagreements among qualified mathematicians.

Transcendental idealism entails for Kant the solvability of all meaningful problems of mathematics as well as a guarantee of their applicability to sensory experience without invoking any mysterious pre-established harmony between the understanding and nature. There are no "strange and aesthetically remarkable properties of numbers," nor any "mystery" about them worthy of "surprise," if constructed *in intuition*. Euler regarded the distribution of primes as "a mystery that the human mind has not been able to penetrate." Reflection on the polynomial (S) that Lambert derived from his infinite series could have persuaded Kant that it was a meaningful problem to seek some formula for them. But Lambert's phenomenological observation, that "to all appearances there is no local order" among the primes, was a vague claim devoid of intuitive representation that qualified by Kant's lights as philosophizing about mathematical objects. The empirical study of their *average* distribution in the tables, however, invited Gauss' surprising conjecture that, despite their apparent disorder, the ratio of the number $\pi(n)$ of primes less than n to n is asymptotically equal to $\frac{1}{\log n}$. This was Gauss' *opinion* in Kant's sense, "taking something to be true, with the consciousness that it is subjectively *as well as* objectively insufficient."[185] But are opinions allowed in mathematics? What is their epistemic status? It is on this score that Husserl's idealism surpasses and is superior to Kant's, specifically in his analysis of the belief in solvability.[186]

---

[182] AA23, p. 200, p. 201.

[183] loc. cit., p. 201.

[184] In November 1796, Gustav Stark sent Kant a proof of (K) by a "geometrical construction" showing its truth from his figure (AA13, p. 121–124). The possibility that this could have figured in Kant's reply to Reimarus is explored in an intriguing paper: Marcucci 2001.

[185] A822/B850. We now have proof of Gauss' conjecture which he modeled on the tossing of a coin, where the probability that it would land prime on the Nth toss was $1/\log(N)$. If the primes were truly random, the error for the prime number coin should be of the order of $\sqrt{N}$. The Riemann hypothesis would imply just this error and hence explain why the primes *look so random*. So Lambert's phenomenological claim, that there is no apparent local order among the primes, touches on a problem at the forefront of mathematical research: the remaining unsolved Hilbert problem. See du Sautoy 2003, p. 166–167.

[186] In Husserl 1969 §79, which I have analyzed elsewhere in this volume.

Kant says we must never have an opinion without "*knowing* something by means of which the in itself mere problematic judgment acquires a connection with the truth which, although it is not complete, is nevertheless more than arbitrary intention".[187] The connection Gauss knew between his conjecture and the truths in his tables surely qualified as holding an opinion, but Kant is adamant that

> In judging from pure reason, to have an opinion is not allowed at all. For since it will not be supported on grounds of experience, but everything that is necessary should be cognized *a priori,* the principle of connection requires universality and necessity, thus complete certainty, otherwise no guidance to the truth is forthcoming at all. Hence it is absurd to have an opinion in pure mathematics: one must know, or else refrain from all judgment.[188]

Kant lectured that "in mathematics there is neither opinion nor believing, but only knowing – or nothing – and in merely speculative metaphysics it is the same."[189] But the great mathematician Leibniz believed that π was irrational: "Can't one also accept propositions *a priori* on belief? Yes, mathematics is also of this kind. One believes mathematicians because it is not possible that they can err, since they would hit upon false consequences at once."[190] Kant knew the equation $x^2 + y^2 = z^2$ had infinitely many solutions: had he known that the mathematician Fermat claimed to have a proof that it had no solutions at all for exponents greater than 2, he would believe this on the ground that Fermat would have hit upon a counter example to his theorem at once. Wiles finally found a proof, but it is so complicated that most philosophers will follow Kant in trusting the mathematicians' immunity from error, even if not inclined to invoke intuition to explain this.

That Kant could have brushed so close to such a difficult problem underscores what Riemann, Gauss, and Hilbert have all emphasized about number theory: that it is full of problems that seem *prima facie* very simple but which can be extraordinarily difficult to solve. Someone as mathematically trained as Husserl was more conscious of this than Kant, which should be factored into our assessment of the bearing of Gödel's incompleteness theorem on his "supramathematics" of complete formal systems, including those for number theory. I have elsewhere suggested that in formulating completeness proofs for what he called "arithmetics," Husserl was thinking of Kronecker's "general arithmetic" for algebraic fields only *including* the natural numbers, rather than Peano's arithmetic for *just* their recursive structure alone. Since Husserl was not nearly as explicit as we are now about the symbolism and axioms of systems, it is hard to be sure about this.[191] But consider first-order Peano arithmetic PA with just its axioms for the successor function, the four axioms for the recursive definitions of addition and multiplication, and the induction schema. What could have led Husserl to believe that Fermat's theorem

[187] loc. cit., A822/B850.

[188] A823/B851.

[189] Kant 1992c, p. 468.

[190] loc. cit., p. 469.

[191] I have been pressed on this point by Mirja Hartimo, whose grip on the Double Lecture is second to none.

could even be *expressed* in PA, let alone decided? The heart of Gödel's surprising discovery was that exponentiation can be defined in PA, for which Gödel had to laboriously construct pairing functions and his β-function and prove their properties by painstaking applications of the Chinese remainder theorem. In 1901 PA would have seemed to Husserl to be a rather impoverished system unable to even express outstanding problems such as Fermat's theorem or *most of this own totality of arithmetic operations*. On the other hand, had Husserl seen Gödel's paper in 1931 or after, it seems plausible to me that his first reaction would be: *Technisierung*.

Returning to Kant, we saw that his view of the mathematics of his time allowed him to believe that any disagreements or misunderstanding among mathematicians, due to ambiguities of meaning, could only "last for an instant," which seemed to be plausibly explained by his doctrine that mathematical objects are given in intuition. There were no paradoxes or contradictions in his mathematics but only antinomies that he argued arose only when its concepts of infinity and continuity were applied beyond the bounds of experience. With no disagreements of principle between mathematicians and no contradictions or paradoxes in mathematics, there was no urgent problem of explaining such intersubjective harmony for Kant. It was to be expected on his doctrine of intuition.

Husserl criticized Kant's idealism for leaving open, "at least as a limiting concept, the possibility of a world of things in themselves" which, he added later, "would signify for the Ego a realm in itself belonging to him mythically."[192] This falls short of phenomenology which discloses that what is constituted "in the meditating ego – as a world, a whole universe of being, we naturally could not avoid being mindful of "others" and their constitutings."[193] But Husserl's problem with intersubjectivity was made yet more difficult than Kant's by the emergence of sharp principled disagreements among the leading mathematicians of his time, as well as contradictions and paradoxes in logic and mathematics itself. "Consequently we have – and precisely at the high level attained by modern positive sciences – problems of fundamentals, paradoxes, and unintelligibilities."[194] This situation on the ground is compounded by widespread disagreement over how to solve these paradoxes or clarify the unintelligibilities, leaving Husserl to dream of "a community of Egos existing with each other and for each other – *ultimately a community of monads*, which moreover (in its communalized intentionality) constitutes the *one identical world*. *In this world* all Egos again present themselves but *in an Objectivating apperception* with the sense "*men*" or psychophysical men as worldly Objects."[195]

A decade earlier, before the outbreaks of the foundational crisis in mathematics and the unintelligibilities in quantum physics, Husserl had managed to combine his

---

[192]Husserl 1960, p. 86.

[193]loc. cit., p. 87.

[194]loc. cit., p. 153.

[195]loc. cit., p. 107.

vision of a supramathematics of all possible complete theory forms with the Kantian Idea of a rational ego, to claim that:

> The entire ideal sphere of rational objects and the sphere of objective ones knowable by every possible rational ego coincide with one another. It turns out that this sphere is a definite manifold, an infinite constructible system.[196]

This remarkable vision dates from a time when Husserl's hopes were rising for a phenomenological grounding of modern science, spearheaded by two of his students. Weyl had used infinitesimal geometry to extend Einstein's general theory of relativity to include electromagnetic phenomena, for which Becker pursued a phenomenological basis in his *Habilitation* on geometry. In 1922 Husserl wrote to Weyl that:

> It is nothing less than a synthesis between Einstein's and your discoveries with my phenomenological investigations on nature. It aims by deep and original means to prove that Einstein's theory, but only when completed and recast through your research into infinitesimal geometry, represents those forms of the 'structural lawfulness' of nature, which must be presumed necessary on the deepest *transcendental–constitutive* grounds... What will Einstein say when it is proved that nature seeks a relativistic structure on the *a priori* grounds of phenomenology and not on positivistic principles and that only in this way is a fully comprehensible and exact science possible.[197]

Given his belief that Becker had given such grounds for Einstein's theory, it seems that he saw nothing in 1922 of the crisis in physical science he will deplore a decade later. Nor in mathematical science, for he further informed Weyl that Becker was also able to:

> Portray a constitutive theory of the continuum... In this way he looks for a demonstration that only Brouwer's and Weyl's theories are coherent with the definite and fundamental demands of a program of research constituted in its sources phenomenologically.[198]

But by 1926 Weyl has rejected Becker's mantic version of phenomenology and decided that Brouwer's intuitionistic mathematics could not support modern physics, especially quantum mechanics in which Hilbert's non-constructive spectral theory of operators in infinite dimensional spaces had found what he later called a miraculous application. Becker then wrote to Mahnke that "we are facing here a certain crisis of phenomenological method itself. If Weyl is right that Brouwer's intuitionism cannot support theoretical physics, then phenomenology in Husserl's 'classical' sense must be seen as incapable of securing the modern form of knowledge of nature and making it understandable."[199] What did Husserl himself make of this?

---

196 Husserl 1974b, p. 388. This text is dated 1922/1923.

197 Tonietti 1988, p. 348. As for Einstein, he said Weyl's theory was beautiful but that it would not account for the stability and distinctness of the spectral lines. He argued that Weyl's theory made the radiation frequencies of two hydrogen atoms dependent on their history, thereby excluding stable spectra.

198 loc. cit., p. 369–370.

199 Mancosu and Ryckman 2002, p. 147.

In the *Crisis* Husserl considers "the possibility of a hidden truth in Kant's transcendental philosophy," represented as a hidden dimension that cannot be seen by those bound to the "life of the plane." He asks why positive sciences bound to the plane could have appeared so successful for so long. But above all, he wants to know, "Why did the later attempts at an 'intuitionistic' deepening, which in fact touch upon the higher dimension, and all attempts to clarify the situation in this way not lead to unanimously accepted, truly compelling scientific results?"[200] Here I believe Husserl means any kind of intuitionism, be it Brouwer's or his own very different version, whose least common denominator stems from Kant's doctrine that *a priori* mathematical knowledge derives from the construction of its concepts in intuition. But Brouwer rejects both Platonism and the solvability of all mathematical problems while Husserl apparently embraces both, while differing sharply with Kant on Platonism, who we have seen regards philosophizing about mathematical objects as fanatical and mystical. Which brings us to Husserl's final criticism of Kant.

Kant was never able to sufficiently free himself, "by means of a radical liberation from all scientific and prescientific traditions," says Husserl, to complete his Copernican revolution, and as a result "He does not penetrate to the absolute subjectivity which constitutes everything that is, in its meaning and validity, nor to the method of attaining it in its apodicticity, of interrogating it and of explicating it apodictically."[201] In short, Kant did not achieve "the genuine transcendental reduction," though Husserl does believe that a Kantian "way" to such a reduction is the most promising[202]. Another way of putting this critique of Kant might be this: Kant's Copernican proposal that we *a priori legislate* laws to nature was a step in the right direction, but he stopped short of the insight that we fully *constitute* nature in our absolute subjectivity that we reach only via the most radical reduction. But Husserl has often and eloquently expressed the prodigious difficulties that would confront such a reduction. So the question is: *can* anyone achieve it, in any way? In his classic authorized examination of this question, Eugen Fink wrote in 1933 that:

> Every discussion of the phenomenological reduction, no matter how incomplete, finds itself faced with the unavoidable difficulty of being compelled to speak about it as if it were an activity of knowledge which it is always possible to perform, and which from the very start lies within the horizon of our human possibilities. In truth, however, it does not at all present a possibility for our human existence. The unfamiliarity of the reduction is therefore not only an unfamiliarity with it as a fact but is also an unfamiliarity with its possibility.[203]

Of Fink's essay Husserl wrote that "I have carefully read through this essay at the request of the able editors of the *Kant-Studien*, and I am happy to be able

---

[200]CoESP, p. 120.

[201]loc. cit., p. 199. Earlier, Husserl had implied that Kant had not achieved the radical distinction between the actual world and "transcendental subjectivity, which, as constituting within itself the being-sense of the world, precedes the being of the world and accordingly bears within itself the world's reality, as an idea constituted actually and potentially within this same transcendental subjectivity." (Husserl 1969, p. 268)

[202]On this see Kern 1977

[203]Fink 1970, p. 104.

to state that it contains no sentence which I could not completely accept as my own or openly acknowledge as my own conviction."[204] Was his final judgment on Kant then that, although he was the first to detect the secret longing of modern philosophy for a phenomenological clarification of being, he never realized that the phenomenological reduction required for such clarification was humanly impossible? The meaning of such a judgment, however, would be unintelligible without reference to its intimate connection with Husserl's quest for the pure transcendental ego that Kant was the first to invoke.

Husserl had explained that "the world – whose conceivable non-being does not extinguish my pure being but rather presupposes it – is termed *transcendent*, whereas my pure being or my pure ego is termed *transcendental*. Through the phenomenological epoché the natural human ego, specifically my own, is reduced to the transcendental ego. This is the meaning of the phenomenological reduction."[205] His critique of Kant is essentially that he had not found his way from his empirical ego to the pure transcendental ego. Husserl's approval of Fink's efforts to sort the proliferation of egos spawned by a genuine reduction was an implicit admission that he himself had not yet limned a suitable transcendental ego in 1933, i.e. one that accommodated transcendental intersubjectivity.

In 1936 Husserl proposes that "As *primal* ego, I constitute my horizon of transcendental others as cosubjects within the transcendental intersubjectivity which constitutes the world."[206] Now all the others "are included for me, within my epoché, in the world-phenomenon which, in my epoché, is exclusively mine. The epoché creates a unique sort of philosophical solitude which is the fundamental methodical requirement for a truly radical philosophy."[207] The performance of an "absolutely universal epoché" delivers an absolute subjectivity to which,

> It is clear that there is no conceivable meaningful problem in previous philosophy, and no conceivable problem of being at all, that could not be arrived at by transcendental phenomenology at some point along the way. This includes problems that phenomenology itself poses, at a higher level of reflection, to the phenomenologist: that is, not only problems corresponding to the language, truth, science, and reason . . . which are constituted within the natural world, but the problems of phenomenological language, truth, and reason."[208]

This goes well beyond Kant's claim of the solvability of all problems of transcendental philosophy posed by reason, including those of mathematics. Kant admitted there are problems about nature raised by physics and chemistry that are unsolvable, e.g. how one piece of matter can attract another instantaneously at any distance. But he never managed a sufficiently "radical liberation" from scientific traditions, said Husserl, to complete his revolution.

---

[204] loc. cit., p. 74.

[205] Husserl 1967, p. 10.

[206] CoESP, p. 184. See Lohmar 2012 for an illuminating and sympathetic account of Husserl's efforts to conceive a primal ego that satisfies the lofty goals of his phenomenology.

[207] loc. cit., p. 184.

[208] loc. cit., p. 188.

Husserl believed that on reducing to the transcendental subjectivity of the pure ego,

> One becomes aware of standing within a sphere of self-evidence of such a nature that any attempt to inquire behind it would be absurd . . . All natural evidences, those of all objective sciences (not excluding those of formal logic and mathematics) belong to the realm of what is "obvious," what in truth has a background of incomprehensibility. Every kind of evidence is the title of a problem with the sole exception of phenomenological self-evidence, after it has been reflectively clarified itself and shown to be ultimate self-evidence.[209]

Once we are clear that transcendental phenomenology is not naïve Cartesianism, whose cogito provides premises from which the rest of knowledge is deduced, we can attain,

> The insight that no objective science, no matter how exact, ever explains or even can explain anything in a serious sense. To deduce is not to explain. To predict, or to recognize the objective forms of the composition of physical or chemical bodies and to predict accordingly – all this explains nothing but is in need of explanation. The only true way to explain is to make transcendentally understandable.[210]

Did Husserl really believe, e.g. that Bohr's deduction of the Balmer series for the hydrogen atom from his quantization premise, or Einstein's deduction of the correct advance of Mercury's perihelion from his field equations explained nothing in a serious sense? He says that his words do "not detract in the least from the greatness . . . of creative geniuses" or their deductive accomplishments. Did he believe that a genuine phenomenological reduction would lead to an absolute subjectivity for which their incomprehensible premises would be transcendentally understandable? Kant already acknowledged problems raised by physics and chemistry that were inscrutable in their current state, namely those of action at a distance and cohesion of matter. His final response to these challenges was to extend his transcendental deduction of the categories, supplemented by his principle of reflective judgement and *Selbstsetzungslehre*, to a transcendental deduction of an aether making action at a distance and cohesion comprehensible.[211] Though unsuccessful because of its dependence on Newtonian physics, Kant's attempt was every bit as heroic as Husserl's untiring pursuit of a phenomenological reduction to a transcendental ego, for which all questions about nature would be transparent.

Husserl also believes that when descriptive psychology is developed purely, "the radical and perfect reduction leads to the *absolutely single ego* of the pure psychologist" who emerges from his epoché of his human existence in this world as "the pure subject of his intentionality, which through the radical reduction is universal and pure, with all its intentional implications."[212] When "opened up," this

---

[209]loc. cit., p. 189.

[210]loc. cit., p. 189.

[211]This deduction is made in Kant 1993. An excellent account of Kant's final quest for a transcendental basis for a complete natural science can be found in Friedman 1992.

[212]loc. cit., p. 256. Kant's analysis of consciousness did not fully unravel its intentional implications, says Husserl, "although in his profound doctrine of synthesis he already discovered,

"apodictic ego" implies other egos.[213] Thus, says Husserl, "through the reduction [others] are transformed from human beings existing for me into alter egos existing for me, having the ontic meaning of *implicata* of my original intentional life."[214] If now, "in egological self-reflection" I withhold all my empathies, "a sort of epoché within the epoché," I attain "the essential structure of an original life" which, on according validity to my empathy, becomes the structure of "any alter ego I can conceive of."[215] Husserl concedes that in pursuing the phenomenological reduction the psychologist discovers that "no one actually arrives in [ordinary] self-knowledge at his true and actual self, the being which is his own as ego-subject and ... that pure psychology is nothing other than the infinitely toilsome way of genuine and pure self-knowledge."[216]

Husserl concludes that the inability of positive psychology to explain anything by itself reveals "the profound difference between mathematics, between every *a priori* science of the world, and phenomenology as *a priori* psychology, that is, as the theory of the essence of transcendental subjectivity."[217] The hypothesis that has dominated natural science for centuries is that,

> Nature is in itself what it is, and is in itself mathematical, no matter how much we know or do not of mathematics; everything is deduced in advance as pure mathematics and as nature itself.[218]

This absurd Laplacean theme is said to have led even Leibniz astray, but not Kant. Indeed, Husserl is clear that Kant gave us the "first system of a scientific transcendental philosophy. From the beginning certain individual problems guided him as a genuine scientist, those which had grown out of his philosophical consideration of mathematics and mathematical natural science and out of the critical awareness of the inadequacy of contemporary ontology."[219] In particular, says Husserl, Kant was

> the first who, in gigantic sketches, embarked on an attempt, which must be made again and again, until there is full success, of making nature, first of all the nature of intuition and that of mathematical science, theoretically understandable, as a formation constructing itself in the internality of transcendental subjectivity.[220]

---

basically, the peculiarity of intentional contexts and already practiced, in his own words, genuine intentional analyses." (Husserl 1974a, b, p. 15)

213 Husserl bases his claim on the extraordinary assumption that "all souls make up a single unity of intentionality with the reciprocal implication of the life-fluxes of the individual subjects, a unity that can be unfolded systematically through phenomenology; what is a mutual externality from the point of view of naïve positivity or objectivity is, when seen from the inside, an intentional mutual internality." (p. 257)

214 loc. cit., p. 258.

215 loc. cit., p. 259.

216 loc. cit., p. 261.

217 loc. cit., p. 264.

218 loc. cit., p. 264–265.

219 Husserl 1974a, b, p. 50.

220 loc. cit., p. 52. Husserl was presumably unaware of Kant's last such gigantic sketch in his *Opus Postumum*.

Husserl acknowledged more than once that his own quest for a reduction to transcendental subjectivity is as "infinitely toilsome" as that of the psychologist for pure self-knowledge. What remains unclear is whether he believed it to be humanly possible.

References

E. Adickes, *Kant als Naturforscher* (de Gruyter, Berlin, 1924)

R. Anderson, It adds up after all: Kant's philosophy of arithmetic in light of the traditional logic. Philos. Phenomenol. Res. **69**, 501–540 (2004)

R. Argawal, Lambert Series and Ramanujan, Indian Acad. Sci. (Math. Sci.) **103**, 269–293 (1993)

F.L. Bauer, Lamberts Kettenbruch. Informatik-Spektrum. **28**, 303–309 (2005)

L. Boi, Questions regarding Husserlian geometry and phenomenology. A study of the concept of manifold and spatial perception. Husserl Stud. **20**, 207–267 (2004)

M. Bullynck, Decimal Periods and their tables: A German research topic (1765–1801). Hist. Math. **36**, 137–160 (2009)

H. Busemann, *The Geometry of Geodesics* (Academic, New York, 1955)

J. da Silva, Husserl on geometry and spatial representation, in *The Road Not Taken*, ed. by C. Hill and J. da Silva, London 2013, 31–62

M. du Sautoy, *The Music of Primes. Searching to Solve the Greatest Mysteries of Mathematics* (Harper Perennial, New York, 2003)

W. Ewald, *From Kant to Hilbert: A Source Book in the Foundations of Mathematics* (Oxford University Press, Oxford, 1996)

L. Falkenstein, Kant, Mendelssohn, Lambert, and the subjectivity of time. J. Hist. Philos. **29**, 227–251 (1991)

E. Fink, The phenomenological philosophy of Edmund Husserl and contemporary criticism, in *The Phenomenology of Husserl,* ed. and translated by R. O. Elveton (Quadrangle books, Chicago, 1970)

M. Friedman, *Kant and the Exact Sciences* (Harvard University Press, Cambridge, 1992)

J. Gray, *The Hilbert Challenge* (Oxford University Press, Oxford, 2000)

M. Hallet, Hilbert and logic, in *Quebec Studies in the Philosophy of Science* (1995)

G. Hardy, L. Littlewood, On a Tauberian theorem for Lambert series, and some fundamental theories in analytic number theory. Proc. London Math. Soc. **19**, 21–29 (1921)

M. Hartimo, Husserl's pluralistic phenomenology of mathematics. Philos. Math. **20**, 86–110 (2012)

D. Hilbert, Mathematical problems (1900) in Gray 2000, pp. 240–282

D. Hilbert, *The Foundations of Geometry* (Open Court, LaSalle, 1971)

D. Hilbert, in *Lectures on the Foundations of Geometry*, ed. by M. Hallet, U. Majer (Springer, Berlin, 2004)

E. Husserl, *The Paris Lectures,* translated by P. Koestenbaum (Martinus Nijhoff, The Hague, 1967)

E. Husserl, *Formal and Transcendental Logic,* translated by D. Cairns (Martinus Nijhoff, The Hague, 1969)

E. Husserl, *Logical Investigations,* translated by J. Findlay, Humanities (1970a)

E. Husserl, [CoESP], *The Crisis of European Sciences and Transcendental Phenomenology*, translated by D. Carr (Northwestern University Press, Evanston, 1970b)

E. Husserl, *Cartesian Meditations*, translated by D. Cairns (Martinus Nijhoff, The Hague, 1960)

E. Husserl, Kant and the idea of transcendental philosophy. Southwest. J. Philos.**9**, 9–56 (1974a), translated from Husserl 1956 by T. Klein and W. Pohl

E. Husserl, in *Formale und transzendentale Logik. Versuch einer Kritik der logischen Vernunft. Mit ergänzenden Texten*, ed. by P. Jansson. Husserliana 17 (Nijhoff, The Hague, 1974b)

E. Husserl, *Phenomenology and the Foundations of the Sciences. Third Book of Ideas*, translated by T. Klein and W. Pohl (Martinus Nijhoff, The Hague, 1980)
E. Husserl, *Studien zur Arithmetik und Geometrie, Texte aus dem Nachlass 1886–1901*, Husserliana, XXI (1983)
E. Husserl, *On the Phenomenology of the Consciousness of Internal Time*, translated by J. Brough, (Kluwer, 1991)
E. Husserl, *Early Writings in the Philosophy of Logic and Mathematics*, translated by D. Willard (Kluwer, Dordrecht 1994)
E. Husserl, *Thing and Space*, translated by R. Rojcewicz (Kluwer, Dordrecht, 1997)
E. Husserl, in *Logik. Vorlesung 1896*, ed. by E. Schuhmann, *Husserliana*, Materialienbände 1 (Kluwer, Dordrecht, 2001a)
E. Husserl, in *Logik. Vorlesung 1902/03*, ed. by E. Schuhmann, *Husserliana*, Materialienbände 2 (Kluwer, Dordrecht, 2001b)
E. Husserl, *Analyses Concerning Passive and Active Synthesis: Lectures on Transcendental Logic*, translated by A. Steinbock. *Collected Works*, vol. 9. (Kluwer, Dordrecht, 2001c)
E. Husserl, *Einleitung in die Philosophie, Vorlesungen 1922/1923*, Husserliana XXXV (2002)
E. Husserl, [PoA], *Philosophy of Arithmetic. Psychological and Logical Investigations with Supplementary Texts from 1887–1901*, translated by D. Willard (Kluwer, Dordrecht, 2003)
E. Husserl, *The Basic Problems of Phenomenology.* From the Lectures, Winter Semester, 1910–1911, translated by I. Farin and J. Hart, (Springer, 2006)
E. Husserl, *Introduction to Logic and Theory of Knowledge.* Lectures 1906/07, translated by C. O. Hill (Springer, Dordrecht, 2008)
E. Husserl, *Ideas for a Pure Phenomenology and Phenomenological Philosophy*, translated by D. Dahlstrom/Hackett, Indianapolis/Cambridge, 2014
I. Kant, *Universal Natural History and Theory of the Heavens*, translated with an introduction and notes by S. L. Jaki (Scottish Academic Press, Edinburgh, 1981)
I. Kant, *Theoretical Philosophy 1755–1770*, translated and ed. by D. Walford and R. Meerbote (Cambridge University Press, Cambridge 1992b)
I. Kant, *Lectures on Logic*, translated and ed. by J. Young (Cambridge University Press, Cambridge 1992c)
I. Kant, *Opus Postumum*, translated by E. Förster and M. Rosen (Cambridge University Press, 1993)
I. Kant, *Critique of Pure Reason*, translated and ed. by P. Guyer and A. Wood (Cambridge University Press, Cambridge, 1997)
I. Kant, *Correspondence,* translated by A. Zweig (Cambridge University Press, Cambridge, 1999)
I. Kant, *Critique of Aesthetic Judgment,* translated by P. Guyer and E. Matthews (Cambridge University Press, Cambridge, 2000)
I. Kant, in *Theoretical Philosophy after 1781,* ed. by H. Allison, P. Health, translated by G. Hatfield, M. Friedman, H. Allison, and P. Health (Cambridge University Press, Cambridge, 2002)
I. Kant, *Notes and Fragments*, translated by P. Guyer, C. Bowman, F. Rauscher (Cambridge University Press, Cambridge, 2005)
I. Kant, *Anthropology from a Pragmatic Point of View*, translated and ed. by R. Louden with an introduction by M. Kuehn (Cambridge University Press, Cambridge, 2006)
I. Kant, [AA], *Kant's Gesammelte Schriften*, ed. by the Königlich Preussischen Akademie der *Wissenschaften*, subsequently the Deutsche Akademie der Wissenschaften, 29 volumes
I. Kern, *Husserl und Kant. Eine Untersuchung über Husserls verhältnis zu Kant und zum Neukantianismus* (Martinus Nijoff, Den Haag, 1964)
I. Kern, The three ways to the transcendental phenomenological reduction in the philosophy of Edmund Husserl, in *Husserl. Expositions and Appraisals,* ed. by F. Elliston, P. McCormick (University of Notre Dame Press, Notre Dame, 1977)
J. Lambert, *Neues Organon oder Gedanken über die Erforschung und Bezeichnung des Wahren und des Unterscheidung vom Irrthum und Schein* (Leipzig, 1764)
J. Lambert, Memoire sur quelque propiétés remarquables des quantités transcendentes circulaires et logarithmeques, Akademie de Berlin. **17**, 265–322 (1768)

J. Lambert, Solution général et absolut du problem de Trois Corps moyen des suites infinites. Memoires de l'Academie. **23**, 353–364 (1769)

J. Lambert, Vorläufige Kenntnisse fur die, so die Quadratur und Rectification des Circuls suchen, Beiträge zum Gebrauche der Mathematik und der Anwendung. Bd. **2**, 140–169 (1770)

J. Lambert, *Anlage zur Architectonic,oder Theorie des Einfachen und des Ersten in der philosophischen und mathematischen Erkenntniss,* Riga (1771)

J. Lambert, Theorie der Parallelinien, in *Die Theorie der Parallellinien von Euklid bis auf Gauss,* ed. by P. Stäckel, F. Engel (Tuebner, Leipzig, 1895)

G. Leibniz, in *New Essays on Human Understanding*, translated by P. Remnant and J. Bennett (Cambridge University Press, Cambridge, 1981)

D. Lohmar, Husserl's concept of categorial intuition, in *One Hundred Years of Phenomenology*, ed. by D. Zahavi, F. Stjernfelt (2002), pp. 125–145

D. Lohmar, Intuition in mathematics: On the function of eidetic variation in mathematical proofs, in *Phenomenology and Mathematics,* ed. by Mirja Hartimo (Springer, 2010)

D. Lohmar, Ego and Arch-Ego in Husserlian phenomenology, in *Life, Subjectivity, and Art: Essays in Honor of Rudolph Bernet* (Springer, 2012), pp. 277–302

E. Mach, in *Space and Geometry in the Light of Physiological, Psychological, and Physical Inquiry,* translated by T. McCormak (Open Court, LaSalle, 1960)

D. Mahnke, Leibniz auf der Suche nach einer allgemeinen Primzahlgleichung, *Biblioteca Mathematica*, 13 (1912/1913), 3 Folge, pp. 29–61

P. Mancosu, *Philosophy of Mathematics and Mathematical Practice in the Seventeenth Century* (Oxford University Press, Oxford, 1996)

P. Mancosu, *From Brouwer to Hilbert* (Oxford University Press, Oxford, 1998)

P. Mancosu, T. Ryckman, Mathematics and phenomenology: The correspondence between O. Becker and H. Weyl. Philos. Math. **10**, 130–202 (2002)

S. Marcucci, Kant, Gustav von Stark, und die Pythagorischen Ternen, in *Vernunftkritik und Aufklarung*, 2001, ed. by M. Oberhausen, pp. 251–263

G. Martin, *Arithmetic and Combinatorics, Kant and his Contemporaries*, translated by J. Wubnig, (Southern Illinois University Press, Carbondale, 1985)

E. Orth, Can 'Phenomenology' in Kant and Lambert be connected with Husserlian Phenomenology, *Kant and Phenomenology*, ed. by T. Seebohm, J. Kockelmans (Center for Advanced Research in Phenomenology and University Press of America, Washington, DC 1984)

C. Parsons, Mathematical Intuition. Proc. Aristotelean Soc. **80**, 145–169 (1980)

V. Peckhaus, R. Kahle, Hilbert's paradox. Hist. Math. **29**, 157–175 (2002)

C. Peirce, in *Collected Papers of Charies Sanders Peirce,* vol. 7, ed. by A. W. Burks (The Belknap Press of Harvard University, Cambridge, MA, 1966)

C. Reid, *Hilbert* (Springer, New York/Heidelberg/Berlin, 1970)

J. Schultz, *Sehr leichte und kurze Entwicklung einiger wichtigsten mathematischen Theorien* (F. Nicolovius, Königsberg, 1803)

F. Schur, Johann Heinrich Lambert als Geometer. Jahresberichte der deutsche Mathematiker Vereinigung. **14**, 186–198 (1905)

J. Sommer, Review of Hilbert's *Festschrift*. Bull. Am. Math. Soc. **6**, 289–295 (1900)

T. Tonietti, Four letters of E. Husserl to H. Weyl and their context, in *Exact Sciences and Their Philosophical Foundations*, ed. by W. Deppert (1988)

M. Van Atten, Why Husserl should have been a strong revisionist in mathematics. Husserl Stud. **18**, 1–18 (2002)

J. Vuillemin, Sur la demonstration de l'irrationalité de π chez Leibniz, Lambert et Kant, Revue philosophique.**43**, 417–431 (1961)

C. Wolff, in *Mathematisches Lexicon*, ed. by J. E. Hofmann (Georg Olms, Hildesheim, 1965)

W. Zage, The geometry of binocular visual space. Math. Mag. **53**, 157–175 (1980)

Chapter 3
Husserl and Bolzano

Ettore Casari

Abstract The paper examines the all too often neglected role of the Czech philosopher and mathematician Bernard Bolzano for Husserl's work, from ca. 1893–1894 onwards. Husserl himself finds it important to stress in an appendix to chapter 10 of the *Prolegomena to Pure Logic* that his investigations are not "in any sense mere commentaries upon, or critically improved expositions of, Bolzano's thought patterns", but that they "have been crucially stimulated by Bolzano ...". The paper examines early Bolzano's ideas on the ground-consequence relation, Bolzano's logical universe as presented in his masterpiece, the monumental *Wissenschaftslehre*, the role of Hermann Lotze in making Husserl receptive for Bolzano and, finally, a lecture course on logic held by Husserl at the University of Halle in 1896, working out just what Husserl is taking, and not taking, from Bolzano.

Keywords Husserl • Bolzano • Lotze • Ground-consequence • Lectology

3.1 The First Contacts

Husserl himself says us that his first contacts with Bolzano concerned his mathematical work and, to this regard, he mentions three occasions[1]: in the time he was a student of Weierstrass in Berlin, an article of the Austrian mathematician Otto Stolz (1842–1905) on the *Significance of B. Bolzano in the History of Infinitesimal Calculus*[2]; in the time he was in Vienna, a course of Brentano[3] in *Winter Semester* 1884/1885 in which the *Paradoxes of the Infinite*[4] were analyzed in depth; and, later on, in the time he was in Halle, through his familiarity with

[1]Husserl *LU Umarbeitung*, 297.

[2]Stolz 1881.

[3]Husserl *Erinnerungen an Franz Brentano*, in Kraus 1919, 307–8.

[4]Bolzano 1851.

E. Casari (✉)
Scuola Normale Superiore, Pisa, Italy
e-mail: ettore.casari@sns.it

S. Centrone (ed.), *Essays on Husserl's Logic and Philosophy of Mathematics*, Synthese Library 384, DOI 10.1007/978-94-024-1132-4_3

Cantor. It is worth to be remarked, however, that the course of Brentano had also discussed in depth "the differences of the 'intuitive and non intuitive', 'clear and unclear', 'distinct and indistinct', 'proper and improper', 'concrete and abstract' presentations," i.e., as can be easily seen, almost exactly, the issues discussed in the §§ 276–81 of Bolzano's *Wissenschaftslehre.*[5] His first contacts, however, with Bolzano's philosophical views, were not particularly convincing for him. Although he had "the chance" to discover a copy of the *Wissenschaftslehre* "in a secondhand book store"[6] and had made "a point of looking through" that long-forgotten work "and of making use of it from time to time with the help of its copious index," he, however, "misinterpreted as metaphysical abstruseness his original thoughts about ideas, propositions and truths 'in themselves.'"[7]

3.2 Bolzano's Starting Point

But what had Bolzano said? A basic starting point of the logical reflection of this thinker was the idea, already here and there emerging in the Leibniz-Wolffian tradition in the discussion on the 'Principle of Sufficient Reason', that among the truths there is a *connection*, according to which some truths are *ground* (*Grund*) of other and these in turn *consequence* (*Folge*) of the former, and that such a connection is *objective*, i.e., subsisting independently of every cognitive activity of the subject. Bolzano's idea was that science has the aim to identify and expose this connection. The exposition of the grounds of a truth constitutes its *proof* intended as its *foundation* (*Begründung*), in opposition to a proof of it as *certification* (*Gewissmachung*), i.e., as an argumentation aiming to convince of its truth. He found support to this distinction in the history of mathematics[8]:

> I believe that *Euclid* and his predecessors were in agreement with me and they did not regard the mere *increase in certainty* as any part of the purpose of their method. This can be seen clearly enough from the trouble which these men took to provide many a proposition (which in itself had complete certainty) with a proper [*eigener*] *proof*, although it did not thereby become any more certain. Whoever became more *certain* after reading *Elements* Bk. I Prop. 5 that in an isosceles triangle the angles at the base are equal?

This idea, already orienting his first essay *Considerations on Some Objects of Elementary Geometry*,[9] led him to a new conception of logic which received a first, still unripe, but already very innovatory exposition in his *Contributions to a Better-Grounded Presentation of Mathematics I*[10] and later on an almost definitive

---

[5]Bolzano *WL*.

[6]Schumann 1977, 463.

[7]Loc. cit.

[8]Bolzano 1810, II, § 2.

[9]Bolzano 1804.

[10]Bolzano 1810.

assessment in his monumental *Theory of Science* in which, the recognition of the theoretical primacy of foundation over certification was however accompanied by a more realistic view of its use in scientific expositions.

It was just the attempt to account for that distinction between the subjective and the objective levels of knowledge which led Bolzano to accept and articulate the idea that the reference of the subject to the object was mediated by a realm of entities without existence or efficient reality that, recalling the Stoic *lekta*, will be called here 'lectological.' Of the two main ways through which that reference may take place, however, psychic activity on the one hand, and linguistic activity, on the other hand, Bolzano favored the first and traced back to it the problems of the second one, i.e., he considered those intermediate entities firstly as possible *contents* (*Inhalt*) or *matters* (*Stoff*) of psychic phenomena and only subordinately, on the basis of a complex theory of signs, as *senses* (*Sinne*) or *meanings* (*in the strict sense*) (*Bedeutungen*) of linguistic phenomena.

3.3 Bolzano's Logical Universe

The view of the logical universe emerging from the *Theory of Science* may be summarized as follows. It is made up of *objects* (*Gegenstände*) among which two basic relations may occur. The first one is *being a proper part* (*Theil*)*:* an object may be part of another object. The second one is *having* (*haben*) or, conversely, *belonging* (*zukommen*)*:* an object may *have* another object which then, in turn, *belongs* (*kommt zu*) to it. An object, which belongs to an object is a *quality* (*Beschaffenheit*) (of it). There are objects that are not qualities: the *pure* objects. Moreover, there is a basic partition of the objects according to which some *exist* or *have reality* (*Dasein* or *Wirklichkeit haben*) whereas the others simply *are there* (*es gibt*). Besides God, the only existing objects are those which possess space-temporal determinations.

3.4 Lectological Objects: Ideas

Among the *pure non existing objects* there are the *lectological ones.* They are partitioned into *(objective) ideas* [*(objective) Vorstellungen*], or *ideas in themselves* [*Vorstellungen an sich*], and *propositions in themselves* [*Sätze an sich*]—here simply called 'ideas' and 'propositions.' From the point of view of their structure, ideas are subdivided into *simple* and *composed* [*zusammengesetzt*]; in a composed idea it must be distinguished between its *content* [*Inhalt*], i.e., the collection of simple ideas that enter in its composition, and its *type of composition* [*Zusammensetzungsart*]. The basic relation between ideas and objects in general, is *referring to* [*sich beziehen auf*] or, conversely, *underlying* [*unterstehen*]. From the point of view of their possible references Bolzano distinguishes, on the one side, between *object ideas* and *quality ideas*, and, on the other side, between *objectual* [*gegenständlich*] ideas, that refer to at least one object and *objectless* [*gegenstandlos*] ideas, referring to none. An objectual idea has an *extension* [*Umfang*] consisting of the set of the objects to which it refers and "this [set], as any [set], has a certain *magnitude*, which is called the *width* [*Weite*] of the extension." A nearer consideration of the width leads to classify ideas into *finite* and *infinite*, and the finite into *singular* and *plural* and plural into *at most, exactly, at least n-element* ideas. An object idea that is simultaneously simple and singular is an *intuition* [*Anschauung*] and an object idea in the content of which no intuitions occur is a *pure concept* [*reiner Begriff*]. The notion of intuition is the tool by means of which Bolzano deals with the issue of the relationship *knowledge—experience.* As far as the type of composition is concerned, Bolzano considered mainly three operations by means of which composed ideas are produced. If A is an object idea and b a quality idea, he considered the basic composition *'an A which has* b', which refers to all objects underlying A and having a quality underlying b. As he accepts that there is an idea *'Something* [*Etwas*]' referring to all objects, he has as a special case *'Something which has* b', which obviously refers to all objects having a quality underlying b. So, for instance, 'a men' is interpreted as 'something having humanity.' Moreover, he also considered the operation which combines the references of two quality ideas *a* and b, that he wrote '$a+$b' and called 'sum of *a* and b', as well as the negation of a quality idea *a* that he often wrote 'n*a*.' Departing from the tradition, Bolzano emphasized that there is in general no correspondence among the parts of an object and the parts of an idea of it. Disregarding its systematic nature, Bolzano's theory of relations between two ideas is apparently not very different from the traditional one. Indeed, he distinguishes firstly the relations of *compatibility* [*Verträglichkeit*] and *incompatibility* [*Unverträglichkeit*]; the first obtains when there is at least one object to which both ideas refer; the second in opposite case. He then considers the special cases of both. For compatibility: *inclusion* [*Umfassung*] and its still more special cases: *equivalence* [*Gleichgültigkeit*] and its opposite *subordination* [*Unterordnung*]. For incompatibility: *contradictoriness* [*Widerspruch*], when the two ideas are *exhaustive*, and its opposite *contrariety* [*Widerstreit*]. Not very new appears also, at least in principle, its extension to situations in which more than two ideas are involved. The real novelty is that those relations are first introduced only for *objectual* ideas and only later extended to the objectless ones on the basis of his famous theory of variation [*Veränderlichkeit*].

3.5 Lectological Objects: Propositions and Truths

As to the second component of the lectological realm—the propositions—Bolzano was convinced that all propositions or at least all true propositions have the atomic form 'A *has* b', where A is an object idea and b a quality idea, and gave for them the following fundamental truth definition:

A *has b* is true iff A *is objectual and for each object underlying* A *there is a quality underlying b that belongs to* it.[11] It is worth to be mentioned that Bolzano introduced the notion of 'proposition' by what he used to call an *'agreement* [*Verständigung*]', i.e., a procedure apt to transmit the intended meaning of a term when it refers to a notion that does not admit *definition.* In fact §19 in which they are introduced bears the title: *What the Author Means by a Proposition in Itself.*[12]

> I wish to show as clearly as possible what I mean by a *proposition in itself* [*Satz an sich*]. In order to accomplish this, I want to define first what I mean by a *spoken* [*ausgesprochen*] proposition or a *proposition which is expressed in words* [*durch Worte ausgedrückt*]. With this name I wish to designate any ... *speech act* [*Rede*], if through it anything is asserted or expressed; that is to say, whenever it is one of the two, either true or false in the usual sense of these words.... Given that it is understood what I mean by a spoken proposition, I should like to note that there are also propositions which are not presented in words but which somebody merely thinks, and these I call *mental propositions* [*gedachte Sätze*].... A *proposition in itself* ... is that very entity which one thinks of as being a proposition when one asks whether or not somebody has articulated it, or whether or not somebody has thought it.... In other words, by proposition in itself I mean any assertion that something is or is not the case, regardless whether or not somebody has put it into words, and regardless even whether or not it has been thought.

As alluded to above, Bolzano was convinced that at least all true propositions have the atomic form 'A *has* b.' In *Bolzano's Theory of Science and Science of Religion in a Critical Overview*, a sort of self-review of his two main works, Bolzano summarizes as follows this conviction[13]:

> Turning now to the *second chapter "On the propositions in themselves"* and entering into its *first section* on the "*general qualities of the propositions*," we find one of the least complete parts of B.'s book. In §127, Bolzano goes so far to posite the opinion that all propositions underlie one and the same form and precisely the following 'A *has (the quality)* b.' ... B. doesn't trust himself to regard this opinion as definitely decided; because he doesn't know how to prove it other than by an *induction*, which is by its own nature incomplete, so he tries to show, partly here and partly in the following, that whenever he comes across a seemingly deviant form, this expression could also be traced back to the form 'A *has* b', or better that it should be traced back if what one is here properly saying must be really understandable.

To rightly understand Bolzano's conviction, that may, at first glance, appear to be nothing more than a naive assumption able to completely trivialize the system, let us first of all remark that nobody before Bolzano has ever had such a great awareness of the variety of the ways in which a proposition may result by composition of other propositions. Indeed, we may say that it was his opinion that any relation among any number of propositions univocally determines one of those ways. His conjecture was that *despite* the fact that the *verbal expression* of a composition of this kind is in general certainly not of the form 'A *has* b', it is nonetheless thinkable that its *sense* is always (or at least equivalent to) a *proposition* (in itself) of that form. In this formulation, his conviction appears to be more properly a 'thesis', an

[11] Bolzano 1837, §131, II, 26–7; Bolzano 1969ff. IIA 12/2, 61 and 12/2, 105.

[12] Bolzano 1837, §19.

[13] Bolzano 1841, 30–31.

assertion which cannot be proven, but only made plausible and hence, first of all, empirically verified. The conceptual kernel of the Bolzanian strategy to this aim may be summarized as *the sense of a composed verbal proposition is a 'semantical' proposition*, i.e., a proposition which attributes 'semantic qualities' to lectological things. Bolzano's conviction, indeed, was that not only both ideas and propositions are *objects* and that there are *ideas*, called *'symbolic* [*symbolisch*]*'*, referring to them, but also that both *objectuality of an idea* and *truth of a proposition* are qualities and that there are ideas referring to them so that, in particular, there are propositions (in themselves) enunciating that a certain idea has (or not) objectuality or that a certain proposition has (or not) truth. So, to make some simple examples, the *negation* of a proposition p is simply the proposition enunciating that p has not truth; the *conjunction* of the propositions p and q is the proposition enunciating that the objects falling under the idea that refers to p and q have truth; the *disjunction* of p and q is the proposition enunciating that the idea referring to someone among p and q having truth has objectuality.

3.6 Variation

As alluded to above, the possibly most renowned Bolzano's innovation is his method of *variation.* In its most simple case and in modern terms it consists in the consideration, for a given idea a or proposition p and some other idea b (possibly) occurring in them, of the totality of ideas, respectively of propositions, obtained from a or p by replacing in them the idea b by some other idea. The natural generalizations of this method consider the result of the simultaneous replacement of more ideas and this not only in a single idea or proposition but in a plurality of them. In its applications some constraints are imposed: for instance, that the replacement do not destroy the objectuality or even the truths of the ideas or propositions to which it is applied.

On this basis he, first of all, extends to objectless ideas the theory of the relations among ideas (compatibility, inclusion, etc.) previously introduced only for objectual ones: a relation holds among objectless ideas iff it holds for all replacements that render them objectual. His most important applications, however, concern propositions. Given a objectual proposition p and an idea a occurring in it, he considers the totality of objectual propositions obtained from p by variations of a and the totality of them which are true. If these totalities coincide he says that p is *analytically true* (with respect to a) whereas if they are disjoint he says that *p* is *analytically false* (with respect to a). In all other case he tries to reduce the problem in such a way that the ratio between the two totalities (which are in general infinite) may be seen as a fraction: the *degree of validity* of p (with respect to a). By far more important, however, are the applications of this method to the relations among propositions and, in particular, to the relation we would call *'(logical) consequence*' but which he called *'(logical) derivability* [*Ableitbarkeit*].*'* He states: the propositions $q_i, \ldots, q_m$ are derivable from the propositions $p_i, \ldots, p_m$ with respect to the ideas $a_i, \ldots, a_m$ if and only if: (1) the $p_i, \ldots, p_m$ are compatible with respect to the ideas $a_i, \ldots, a_m$ (i.e., there is at least a replacement of the $a_i, \ldots, a_m$ that makes simultaneously true all the $p_i, \ldots, p_m$); (2) any replacement of the $a_i, \ldots,$a that makes simultaneously true all $p_i, \ldots, p_m$ makes also simultaneously true all $q_i, \ldots, q_m$.

To the systematic treatment of the *inferences* [*Schlüsse*], i.e., of those propositions "which enunciate a relation of derivability," with respect to certain ideas, between certain propositions called *premises* [*Vordersätze*] and certain propositions called *conclusions* [*Schlusssätze*], Bolzano devotes the §§223–268 of the *Theory of Science*, for a total of 178 pages. His treatment does not develop axiomatically, but according to the explicative-foundational model that had already been examined and theorized in the *Contributions: "from simple to complex; from general to particular."* So, he begins with the examination of the conclusions that may be inferred from a premise of the simplest and most general form 'A *has* b', and continues with the conclusions from a premise with a composite subject, then from the negation of such propositions and so on.

3.7 Consecutivity

In §§162 and 168 Bolzano introduces and in §§ 198–222 he tries to outline the formal properties of the relation of *consecutivity* [*Abfolge*] that, as mentioned at the beginning, was the real staring point of his logical reflection. Aware of the complexity of this issue his treatment is very cautious and he preliminary remarks "My investigations are, in the best case, only fragments and suggestions which will reach their aim if they will induce somebody else to further reflections on these matters."

A main problem to be faced in this context is obviously the exact relationship between the relation of derivability and the relation of consecutivity because it is immediate to see that often this relation does not occur where the first one occurs. For example let us consider the three truths:

(A) When it is warmer, then the thermometer reading is higher;
(B) In summer, it is usually warmer than in winter;
(C) In summer, the thermometer reading is usually higher than in winter.

Clearly, from (A) and (B) we can derive (C) but also from (A) and (C) we can derive (B); but whereas we can say that (A) and (B) are the ground of (C) nobody would say that (A) and (C) are the ground of (B). Bolzano concludes the discussion of this issue with the words[14]:

> This should be, moreover, the most suitable place to confess to the reader that sometimes I am assailed by the doubt that the notion of consecutivity, which I claimed above to be simple, is possibly, on the contrary, composed and after all nothing else than the notion of an ordering of the truths by means of which from the smallest number of simple premises one can derive the greatest possible number of other truths as mere conclusions.

---

[14]Bolzano 1837, §221, II, 388.

It is worth mentioning that Bolzano uses the relation of consecutivity to explain the *causality* relation. For him, indeed, that *x* causes y means simply that the proposition asserting the existence of b is a ground of the proposition asserting the existence of y.

3.8 Theory of Knowledge

Among the existing qualities occurring in the universe there are the *psychic phenomena*, that are qualities of those pure existing objects that are the *thinking beings* [*denkenden Wesen*], and among such phenomena, besides *sensations* [*Empfindungen*], *wishes, volitions*, etc., there are those we may call '*cognitive*', that are characterized by having to some lectological object the relation Bolzano called *'having as matter* or *content* [*als Stoff oder Inhalt haben*].' A psychic phenomenon the matter of which is an idea is a *(subjective) representation* [*subjective Vorstellung*] whereas one having as material a proposition is a *judgment* [*Urtheil*], representations and judgments are considered equal if their materials are identical; moreover they inherit their 'logical properties' from their material. So, for example, a representation is *objectless* or *represents at least* n *objects* if its material is objectless or refers to at least n objects and a judgment is *true* (or a *knowledge* [*Erkenntnis*]), respectively *false* (or an *error* [*Irrthum*]) if such is the proposition constituting its matter.

Besides such properties, however, representations and judgments may have properties deriving from their *cognitive* and *physical* nature. So, for instance, a representation may be *clear* or *obscure, distinct* or *confused* and, on the other side, by its physicality, it may have different degrees of *effectiveness* [*Wirksamkeit*] due to its *intensity* or *power* or *liveliness* [*Stärke* oder *Lebhaftigkeit*]; an half- or a three-meter representation of a snake have a remarkably different power. *Mutatis mutandis* such properties go over also to judgments where, however, another property emerges: *trust* or *approval* [*Beifall, assensio*]. Even equal judgments having the same intensity may deeply differ as to the *degree of trust* [*Grad der Zuversicht*] with which they are formulated, the *degree* with which they *are kept as true* [*Grad des Fürwahrhaltens*]. It is worth to be mentioned that Bolzano in § 320, bearing the title *How Different Degrees of Trust May Be Presented trough Calculus*, outlines a possible mathematical development of this strictly subjectivist conception of probability stating, however, in advance that[15]

> [a]lthough the knowing how to present with great exactness, through calculus, the degree of trust with which we judge may rarely be of a particular utility, nevertheless the problem of the way this degree should be calculated may not be regarded as a useless thinking exercise; and only with this intention, I will pay here to it some attention.

[15]Bolzano 1837, §320, III, 283.

To the treatment of the psychic cognitive phenomena is devoted—under the title *Theory of Knowledge* [*Erkenntnislehre*]—the whole third *Book* of the *Theory of Science.*

Among the issues which find extensive treatment in this Book, besides those alluded to above, we simply mention the formation of judgments, the possible origins of errors, the difference between knowing and believing.

3.9 Semiotics

In the same Book Bolzano also develops (§ 285) an important theory of signs. The basis of his construction his what he calls the law of awakening [*Erweckung*], renewal [*Erneuerung*] connection [*Verknüpfung*] or association of our ideas[16]:

> If a pair of ideas has ever been produced simultaneously in our soul, and afterwards an idea is produced in us which is equal to one of them (has the same matter), then, if not always, at least quite often, the other will also be renewed, i.e., an idea will be produced which has the same matter as the other.

It is just this peculiarity of our mind "that ideas that were once connected with each other also mutually stimulate each other" that leads us to use "certain ideas which can be easily stimulated [. . .] as means to produce others (either in ourselves or in other people) that are more difficult to generate, but are associated with the former".[17] According to Bolzano the communication process through signs runs as follows. When a thinking being A wants to transmit to a thinking being B a certain *idea* a, he *produces* a certain object (for instance a voice) i; B perceives it – i.e., the object i *stimulates* [*anregen*] in B a *representation* r which *represents* i (i.e., a *representation* r, which has as *matter* the *idea* that *refers* to i) –; in B the representation r is *associated* to a *representation* s the *matter* of which is the idea a. In this manner A has transmitted to B, through the *sign* [*Zeichen*] i, the idea a that is *denoted* [*bezeichnet*] by that sign and constitutes the *meaning* [*Bedeutung*] of that sign, its *sense* [*Sinn*] or its *intention* [*Verstand*]. A *sign* [*Zeichen*] is thus, by definition, "an object ... through the representation of which we want to know renewed in a thinking being a representation associated to it," whereas the *meaning in the strict sense* (or the *sense*, or the *intention*) of the sign is the matter of the representation produced by association (in the interpreting being) of the representation of the sign. If the idea meant in a strict sense from the sign is objectual, then the object or the objects referred to by it constitute the *meaning in a broad sense* of the sign.

---

[16]Bolzano 1837, §284, III, 58–59.

[17]Bolzano 1837, §285, III, 67.

The meaning in a strict sense of a sign is thus always an idea. This idea may of course refer to a vast range of things; in particular, it may refer to another sign (i.e., it may be the *idea of a sign*). This second sign has, as such, a meaning in the strict sense (and possibly also a meaning in a broad sense). In these cases, we say that the first sign means *mediately* in a strict sense (in a broad sense) the meaning in a strict sense (in a broad sense) of the second sign. This is, in general the case of the written signs. A *vox scripta* means *immediately* in *the strict sense* the idea of the 'corresponding' *vox prolata* and the *vox prolata* to which that idea refers means *immediately*, in *the strict sense*, a certain idea; this idea constitutes the *mediate* meaning, in *the strict sense*, of the first sign and its referents (if occurring) the mediate meaning *in broad sense* of the first sign. A difference between Bolzano's and Frege's theory is that for Bolzano the sense of a linguistic proposition is the idea of a proposition (in itself), not the proposition (in itself) itself. Surprisingly, however, Bolzano himself remarks: "However, even if things were different here, the question should be of no importance."

3.10 Lotze's Ideas

As mentioned above, this 'realm of abstruseness' was made accessible to Husserl by his study of Lotze. In the first book of his *Logic*,[18] Lotze had attempted to describe the way thought converses an impression into an idea by distinguishing two main steps. In the first one the impression is transformed into an object of thought having a proper individuality and this happens through two operations. The first operation described "as the beginning of an objectification of the subjective" consists in giving a name to the impression. As soon as we have given a name,

> we have separated something before unseparated, our sensitive act from the sensible matter to which it refers. This matter we now present to ourselves ... as a something which has its being and its meaning in itself, and which continues to be what it is and to mean what it means whether we are conscious of it or not.

The second operation, called *'position* [*Setzung*]*'*, completes the first by giving singular identity to the impression, "for no matter can have a name made for it unless it has been thought of as identical with itself, as different from others, and as comparable with other." In the second step the abstraction of the universal concept is accomplished. Here again Lotze distinguished two operations. The starting point for the recognition of the first operation is the remark that "in a number of different impressions there is something common which can be thought apart from their differences." And this common element is grasped by intuition:

> Such a common element is usually considered by logic only in the form of a universal concept, and in this shape it is a product of more or less numerous acts of thought. It is

[18] Lotze 1874.

> therefore important to point out that this *first Universal* [*erstes Allgemeine*] which we find here involved in the comparison of simple ideas, is of an essentially different kind; that it is the expression of an *inward experience* [*innere Erfahrung*] which thought has merely to recognize.

Just the fact that it not thought, but inward experience which grasps the common element, is for Lotze the proof that "this first universal, therefore, is no product of thought, but something which thought finds already in existence." The passage from the first to the *second Universal* [*zweites Allgemeine*], the final act of the *formation of concept* [*Bildung des Begriffes*], is accomplished by various logical means, in particular, by "comparison of different instances and observation of the same instance under different circumstances."

> [T]he "'first Universals' ... come to us without logical effort as simple facts of observation in our mental life; and just for this reason they can be applied in building up this second Universal, which we do produce by logical effort.

Lotze made also a radical distinction among four independent kinds of meanings of the term *Reality* [*Wirklichkeit*] "one or other of which it must necessarily assume, and of which no one is reducible to or contained in the other": *Being* [*Sein*], which may or not belong to a *thing* [*Ding*]; *Occurring* [*Geschehen*], which may or not belong to an *event* [*Ereignis*]; *Obtaining* [*Bestehen*] which may or not belong to a *relation* [*Verhältnis*]; *Validity* [*Geltung*] which may or not belong to a proposition and he emphasized that no one of this four meanings can be explained.

> As little as we can say how it happens that anything *is* or *occurs*, so little can we explain how it comes about that a truth has Validity; the latter conception has to be regarded as much as the former as ultimate and underivable, a conception of which everyone may know what he means by it, but which cannot be constructed out of any constituent elements which do not already contain it.

With respect to ideas, he distinguished between the (subjective) *representations* [*Vorstellungen*] which may *occur* in us and their (objective) *content* which may possess *validity:*

> Now Ideas, in so far as they are present in our minds, possess of an activity of presentation they are never a Being at rest but a continual Becoming; their content on the other hand, so far as we regard it in abstraction from the mental activity which we direct to it, can no longer be said to occur, though neither again does it exist as things exist, we can only say that it possesses Validity.

On this basis Lotze had given, in the third book of his *Logic*, a reinterpretation of Plato's theory of ideas. Referring to his conceptions sketched above, he recalled that

> perception shows us the things of sense undergoing changes in their qualities. But while black becomes white and sweet sour, it is not blackness itself which passes into whiteness, nor does sweetness become sourness; what happens is that these several qualities, each remaining eternally identical with itself, succeed each other in the thing, and the conceptions through which we think the things have themselves no part in the mutability which we attribute on account of their changes to the things of which the qualities are the predicates.

Indeed his conviction was that

> a knowledge may be possible the truth of which is wholly independent of the question of Skepticism as to its agreement with a world of things outside it. If the current of the outer world had brought before us only once in a transient appearance the perception of two colours or two sounds, our thought would immediately separate them from the moment of time at which they appeared, and fix them and their affinities and their contrasts as an abiding object of inner contemplation, no matter whether they were ever presented to us again in actual experience or not.

According to him, the great significance of Plato, results from his

> endeavour to bind together the predicates which are found in the things of the external world in continual change, into a determinate and articulated whole, and how he saw in this world of Ideas the true beginnings of certain knowledge; for the eternal relations which subsist between different Ideas, and through which some are capable of association with each other and others exclude each other, form at all events the limits within which what is to be possible in experience falls; the further question what is real in it, and how things manage to have Ideas for their predicates, appeared to Plato not to be the primary question, and was for the time reserved.

3.11 Husserl's Logic Lectures 1986

Of primary relevance for the comprehension and valuation of Bolzano's impact on Husserl's thought is the text of the *Logic Lectures* that Husserl held at Halle in Summer Semester 1896 and for the edition of which in 2001 we are indebted to Elisabeth Schuhmann. As she right remarks in her preface, this work "is strongly dependent on Bolzano's *Theory of Science* both in the discussions of the introductory lectures and the subdivision and conceptualizations of the principal part."

The introductory lectures, indeed, that furnish amongst other things a refusal of psychologism that will be resumed in the *Prolegomena*, correspond to the 16 initial sections of the *Theory of Science* in which Bolzano had introduced and justified his *prima facie* rather bewildering characterizations of a *science*—"an aggregate of truths of a certain kind ... important enough to be set forth in ... a treatise"—and of the *theory of science* or *logic*—"the science which instructs us in the representation of sciences in adequate treatises"—and had discussed the questions: [§11] whether logic is an art or a science; [§12] whether logic is purely formal science; [§13] whether logic is an independent science. It is worth to be remarked that Husserl tried at first to dignify his position by adjoining:

> Hence it is clear the notion of a systematical collective order and connection of truths does not exhaust the notion of a science. Sciences are systematical connections of truths serving a certain aim, and precisely the most possible enrichment and extension of knowledge within the field they have carved out for investigation,

but that later he himself crossed this remark.

Also the introductory lectures end with the discussion of the four questions: (1) whether logic is an independent or a dependent science; (2) whether logic is an art or a science; (3) whether logic has to do only with the forms of knowledge or it must take into account also its materia; (4) whether logic has the character of a demonstrative or of an empirical science. These questions are answered basing essentially on the Bolzanian distinction between the objective and the subjective dimensions of logic—"The source of most errors in logic has been the lack of distinction between thought truths and truths in themselves, and between thought propositions and thought concepts on one hand, and propositions and concepts in themselves on the other"—and on his basic distinction between logic in the strict sense—also called 'theory of inferences [*Schlusslehre*]'—and logic in the broad sense—also called 'theory of methods [*Methodenlehre*].'

Also the systematic development of logic in the following three parts—*Concepts in the broadest sense and their objects*; *Propositions*; *Inferences*—often closely parallels the Bolzanian exposition.

This does not mean, however, that the *Lectures*, may be simply regarded as a rewriting of Bolzano's work; there are indeed some non secondary differences between the twos. Some of them concern more directly the logical content. Let us mention two particularly relevant differences of this kind. The first one is that now in the sphere of objects enters also a notion that, after his introduction in the modern philosophical analysis by Lotze and Stumpf, was familiar, even if with rather different interpretations, within the Brentano school: the *state of affairs* [*Sachverhalt*]. Indeed, whereas Bolzano had distinguished ideas from propositions attributing to the first ones the (possible) referring to objects and to the second ones the being true or false, now in parallel to the triad <(*subjective*) *representation—idea—object* > the triad <(*subjective*) *proposition—proposition—state of affairs* > is introduced.[19] The second difference of this kind is that whereas Bolzano insisted on the fact, that, in spite of current verbal formulations, at least all atomic propositions have the form 'a *has* b', Husserl devoted a pair of sections to discuss, besides those, also the propositions of the form 'a *is* b' and introduces the distinction between *quality* and *predicate:* "In our sense, qualities are objects just as well as abstracts or individuals. They have however the peculiarity that, unlike those, they are, so to say, merely objectifications of predicates."[20]

This last difference, however, depends on one of a more general type and is a significant indication of the further development of Husserl's thought. Bolzano, indeed, was certainly very careful to point out the ambiguities of the linguistic formulations, but its attention was not directed to the meanings as such, but separately to the lectological and ontological spheres. Husserl, on the contrary, is more interested in the meanings as such. This is possibly due not only to Husserl's

---

[19]Apart from terminology, however, looking at Bolzano 1837, §225, II, 399, where Bolzano discusses the notion of '*circumstance* [*Umstand*]', this novelty appears possibly not so remarkable.

[20]Husserl 1896, 64.

experience within the Brentanian school but also to his Lotzean studies; Lotze, indeed, has sometimes been considered as the first 'philosopher of language.' In the successive re-elaboration of this ideas Husserl will come to the identification of the entities of the objective world with the 'meanings of "ideal" unities', a perspective very far from the Bolzanian.

3.12 Husserl on His Debt to Bolzano

Besides some rand remark to be found here and there, we have two main passages in which Husserl give us his acknowledgment of his debt to and at the same time his differentiation from Bolzano's thought. The first one is in an Appendix to the Tenth Chapter of the *Prolegomena*, entitled *Hinweise auf F. A. Lange and B. Bolzano*; the second one is to be found in the draft of a *Preface*, prepared in 1913, to the second edition of the *Logical Investigation* that after some complicated vicissitudes was published in 1939 by Eugen Fink in the Leuven's *Tijdschrift voor Philosophie.*

In the first passage, Husserl, referring to Lange's aspirations to a purely formal logic, remarks that his "Idea had already been realized to a relatively high degree" in the *Theory of Science* "a work which, in its treatment of the logical 'theory of elements', far surpasses everything that world-literature has to offer in the way of a systematic sketch of logic." In Husserl's opinion, indeed, Bolzano, although without any explicit demarcation of the pure logic in his proper sense, had "de *facto* provided one in the first two volumes of his work, in his discussions of what underlies a *Theory of Science* in the sense of his conception" and he adjoins a judgment that only after several decades has begun to acquire some popularity: "he did so with such purity and scientific strictness, and with such a rich store of original, scientifically confirmed and ever fruitful thoughts, that we must count him as one of the greatest logicians of all time."

In Husserl's eyes Bolzano's "wonderful work" is characterized by the "mathematical straightforwardness and plainness" as well as the "mathematical clearness and strictness" of its thought-patterns and he seizes the opportunity to take a position on the flourishing beginnings of mathematical logic:

> Logic as a science must, however, be built upon Bolzano's work, and must learn from him its need for mathematical acuteness in distinctions, for mathematical exactness in theories. It will then reach a new standpoint for judging the mathematicizing theories of logic, which mathematicians, quite unperturbed by philosophic scorn, are so successfully constructing.

The main defects Husserl thinks to find in Bolzano's work are on the epistemological plain; moreover his acting on logic as usually a mathematician does "without having to bother himself about questions of underlying principles," has the consequence that this investigations lacks to give "genuine philosophical intelligibility to logical thought-achievements, and so provide a philosophical estimate of logic as a discipline." Worried however that this praise of Bolzano's work could overshadow the originality of his own work he closes the *Appendix* with the claim:

> Our comparison of these present *Logical Investigations* with Bolzano's work is meant to make clear, not that our *Investigations* are in any sense mere commentaries upon, or critically improved expositions of Bolzano's thought-patterns, but that they have been crucially stimulated by Bolzano (as also by Lotze).

In the second passage Husserl renews his appreciation and the acknowledgment of his debt to Bolzano but this time he distances himself much more explicitly both on the logical and the epistemological level.

It is worth to be mentioned that in a note of his draft he remarks: "The practice (which now seems to be so popular) of projecting my own views back upon Bolzano (obviously without taking the trouble of a real study of the *Wissenschaftslehre*) produces a picture of Bolzano that is totally distorted historically."[21]

Having mentioned again that it suddenly occurred to him that

> the first two volumes of Bolzano's *Theory of Science* ... were to be looked upon as a first attempt at a unified presentation of the area of pure ideal doctrines—in other words, that here a complete plan of a 'pure' logic was already available,

he emphasizes that this insight offered him "an immense benefit." With its help he could step by step verify his 'Platonic' interpretation and became conscious of the necessity to completely revise his lectures on logic. But just doing this he became aware of "the incompleteness of Bolzano's plan: It lacked the idea of a purely formal mathematics or, correspondingly, a 'universal theory of manifolds'"; he concedes, however, that this idea "was at that time by no means in any way familiar to mathematicians as it is today." Always on the logical level he objected that in Bolzano's work

> there was not even a beginning of a discussion on the relation between formal ontological and formal semantical considerations, which in its own right is connected with the lack of any clarification of the concepts 'proposition in itself' and 'idea in itself' There was missing both the contrast between proposition (as logical judgment) and state of affairs as well as all of other fundamental distinctions which pertain thereto).

Whereas the distancing from Bolzano on the epistemological level was in the first occasion limited to the kind remark mentioned above, it is now much more radical. With reference to a passage of Bolzano[22] in which he had attributed the certainty of logic, arithmetic, geometry, and pure physics to their having

> the advantage, that the most important of their theories can be themselves very easily and in various ways verified through experience.... We are so certain of the correctness of the rules 'Barbara', 'Celarent', etc. only because thousands of tests in our conclusions which we carried out according to them certify them. As long as we have not convinced ourselves of the correctness of a proposition either by experiment or by repeated testing of its method of derivation, we do not yet, if we are at all smart, grant it any unconditional trust in spite of all that critical philosophy might recite to us about the infallibility of the pure intuitions upon which our judgment here is supposed to be based,

[21] Husserl 1913, 37.

[22] Bolzano 1837, III, 244f.

he concludes that in spite of the fact that "no one has taken notice of the fact that Bolzano's epistemology rests upon the foundation of an extreme empiricism" in reality "there is here no difference between Bolzano and his contemporary, John Stuart Mill."

References

C. Beyer, *Von Bolzano zu Husserl. Eine Untersuchung über den Ursprung der phänomenologischen Bedeutungslehre* (Kluwer, Dordrecht, 1996)

B. Bolzano, [BGA] *Gesamtausgabe*, ed. by E. Winter, J. Berg, F. Kambartel, J. Louzil, E. Morscher, B. van Rootselaar (Stuttgart-Bad Cannstatt, Frommann, 1969ff)

B. Bolzano, [1804] *Betrachtungen über einige Gegenstände der Elementargeometrie* (Barth, Prague). Reprint: L. Novy (ed.), *Acta historiae rerum naturalium necnon technicarum* 12, Prague 1981. Engl. transl. in: S. Russ (ed.), *The mathematical works of Bernard Bolzano* (Oxford University Press, Oxford, 2004), pp. 25–81

B. Bolzano, [1810] *Beyträge zu einer begründeteren Darstellung der Mathematik. Erste Lieferung* (Caspar Widtmann, Prague). Reprint: L. Novy (ed.), *Acta historiae rerum naturalium necnon technicarum* 12, Prague 1981. Engl trans. in: S. Russ (ed.), *The mathematical works of Bernard Bolzano* (Oxford University Press, Oxford, 2004) pp. 82–137

B. Bolzano, [*WL*] *Dr. Bolzanos Wissenschaftslehre. Versuch einer ausführlichen und größtentheils neuen Darstellung der Logik mir steter Rücksicht auf deren bisherige Bearbeiter* (Her. von mehren seiner Freunde, Sulzbach, 1837). 2nd ed. by A. Hofler, W. Schultz (Meiner, Leipzig, 1928). Engl. trans. by R. George and P. Rusnock, *Theory of Science, Vols. 1–4* (Oxford University Press, Oxford, 2014)

B. Bolzano, [1841] *Bolzano's Wissenschaftslehre und Religionswissenschaft in einer beurtheilenden Uebersicht. Eine Schrift für Alle, die dessen wichtigsten Ansichten kennen zu lernen wünschen* (Seidel, Sulzbach)

B. Bolzano, [1851] *Paradoxien des Unendlichen*, her. aus dem schriftlichen Nachlasse des Verfassers von Fr. Prihonsky, Reclam, Leipzig. Reprint: A. Hoefler, with Notes by H. Hahn, Reclam, Leipzig, 1921. Reprint with Notes by B. van Rootselaar (Meiner, Hamburg, 1975)

P. Bucci, *Husserl e Bolzano. Alle origini della fenomenologia* (Edizioni Unicopli, Milano, 2000)

E. Casari, Sull'origine dell'«oggettivo» in Bolzano, in *Logica e teologia*, ed. by A. Fabris, G. Fioravanti, E. Moriconi. (ETS, Pisa, 1997), pp. 93–115

E. Casari, *Bolzano's Logical System* (Oxford University Press, Oxford, 2016)

S. Centrone, *Begründungen bei Bolzano und beim frühen Husserl*, in S. Centrone, *Studien zu Bolzano*, (Academia, Sankt Augustin, 2015), pp. 117–143

E. Husserl, *Logische Untersuchungen*, Bd I, [*PR*] *Prolegomena zur reinen Logik*, Max Niemeyer, Halle, 1900 (^{2}1913); Bd II, [*LU*] *Untersuchungen zur Phänomenologie und Theorie der Erkenntnis*. Max Niemeyer, Halle a.d.S., 1901 (^{2}1913–1921, ^{7}1993). English translation: [*PRe*/*LI*] *Logical Investigations*, ed. by J.N. Findlay (Routledge, New York, 1970)

E. Husserl, [*LU Umarbeitung*] *Logische Untersuchungen. Ergänzungsband. Erster Teil. Entwürfe zur Umarbeitung der VI. Untersuchung und zur Vorrede für die Neuauflage der Logischen Untersuchungen* (Sommer 1913), ed. by U. Melle. HGW XX/1, 2002. English translation: *Introduction to the Logical Investigations. A Draft of a Preface to the Logical Investigations* (1913), ed. by E. Fink (Tijdschrift voor Philosophie, 1939). English translation with Introductions by P.J. Bossert and C.H. Peters, (M. Nijhoff, The Hague, 1975)

E. Husserl, *Aufsätze und Rezensionen (1890–1910)*, ed. by B. Rang. HGW XXII (1979)

E. Husserl, *Aufsätze und Vorträge 1911–21*, ed. by Th. Nenon, H.R. Sepp. HGW XXV (1986)

E. Husserl, [LV'96] *Logik: Vorlesung 1896*, ed. by E. Schuhmann. *Husserliana* Materialienbände I (Kluwer, Dordrecht, 2001)

E. Husserl, [HGW] *Husserliana, Gesammelte Werke,* I–XXVI: Martinus Nijhoff, Den Haag, 1950–, XXVII–XXXVII: Kluwer, Dordrecht, 1989–; XXXVIII– Springer, New York, 2005
O. Kraus, *Franz Brentano. Zur Kenntnis seines Lebens und seiner Lehre* (Beck, München 1919) (includes: E. Husserl, *Erinnerungen an Franz Brentano*)
W. Künne, Analyticity and logical truth. From Bolzano to Quine, in Textor (2006), pp. 184–249
S. Lapointe, Bolzano e Husserl. Discipline filosofiche. **XXI**(2), 228–244 (2011)
R.H. Lotze, *Logic, in three Books: of Thought, of Investigation, and of Knowledge* (1874), ed. by B. Bosanquet (Clarendon Press, Oxford, 1884, 2nd edn., 1887)
K. Schuhmann, *Husserl-Chronik, Denk- und Lebensweg Edmund Husserls,* (M. Nijhoff, Den Haag, 1977)
J. Sebestik, Husserl Reader of Bolzano, in *Husserl's Logical Investigations Reconsidered*, ed. by D. Fisette (Kluwer, Dordrecht, 2003), pp. 59–81
P. Simons, Austrian philosophers on truth, in M. Textor (2006), pp. 159–83
O. Stolz, B. Bolzanos Bedeutung in der Geschichte der Infinitesimalrechnung. Mathematische Annalen. **18**, 255–279 (1881)
M. Textor, *The Austrian Contribution to Analytic Philosophy* (Routledge, London, 2006)

Chapter 4
Husserl and Grassmann

Jan von Plato

Abstract Husserl's *Philosophie der Arithmetik* came out in 1891. In this short essay, its place at the crossroads of two traditions in the philosophy and foundations of arithmetic is described; what preceded it and could thus have influenced Husserl. A brief look at what came later is also taken; where the choices at the crossroads led to.

Keywords Husserl • Grassmann • Number • Arithmetic • Recursive Definition

4.1 Husserl's Philosophy of Arithmetic at the Crossroads

Husserl's *Philosophie der Arithmetik* came out in 1891. In this short essay I shall describe its place at the crossroads of two traditions in the philosophy and foundations of arithmetic; what preceded it and could thus have influenced Husserl. I shall also have a brief look at what came later; where the choices at the crossroads led to.

The main question in the philosophy of arithmetic has been the *ontological* one: What are the numbers? In the nineteenth century, attempts at defining the notion of a number led to what was called *Anzahllehre*, a term apparently coined by Paul Bernays that stands for something like "the doctrine of numerical quantity." The best-known names today that belong to this tradition are Frege and Russell. In the late 1920s, *Anzahllehre* was still considered a chapter of pure logic as shown by its presentation in the first book devoted to pure modern logic, i.e., first-order and higher-order logic, namely the *Grundzüge der theoretischen Logik* published under the names of Hilbert and Ackermann in 1928, but practically all of it written by Bernays on the basis of his elaboration of Hilbert's lectures 10 years earlier.

A second main trend in the philosophy of arithmetic is one in which numbers are undefined basic concepts and the aim is to define instead the arithmetic operations so that their properties could be proved by induction. German writers tend to locate the beginning of this tradition in Dedekind's *Was sind und was sollen die Zahlen?*

J. von Plato (✉)
Department of Philosophy, University of Helsinki, Helsinki, Finland
e-mail: jan.vonplato@helsinki.fi

S. Centrone (ed.), *Essays on Husserl's Logic and Philosophy of Mathematics*,
Synthese Library 384, DOI 10.1007/978-94-024-1132-4_4

(What are numbers and what are they for?, 1888). Others consider Skolem the initiator of this *recursive arithmetic*, in the *Begrüngung der elementaren Arithmetik durch die rekurrierende Denkweise ohne Anwendung scheinbarer Veränderliche mit unendlichem Ausdehnungsbereich* (Foundation of elementary arithmetic through the recurrent mode of thought without use of bound variables with an infinite domain of extension, 1923). In a recent study of mine, *In search of the roots of formal computation*, I have been able to show that this tradition traces back to Kant and especially his disciple, a certain Johann Schultz.

It became clear by the 1930s, chiefly through the efforts of Bernays as exposed in the *Grundlagen der Mathematik* of 1934, that steps of formal computation are also steps of formal deduction as defined by recursion equations and other principles of arithmetic. It is a development that began with attempts at giving a foundation to the laws of commutativity and associativity of sum: Followers of Kant's doctrine of the *synthetic a priori* in arithmetic, the mentioned Johann Schultz as foremost, missed by a hair's breadth the proper recursive definition of addition that appeared instead first in a book of Hermann Grassmann: his 1861 *Lehrbuch der Arithmetik für höhere Lehranstalten* contains the first explicit *recursive definitions* of the basic arithmetic operations, ones that go hand in hand with *inductive proofs* of properties of the recursively defined operations.

Schultz in his two-volume *Prüfung der Kantischen Critik der reinen Vernunft* (Examination of the Kantian critique of pure reason, 1789 and 1792) asks (vol. 1, p. 220):

> How do I know, then, that this willful procedure [of changing the order of a sum] that lies in no way in the concept of addition itself, provokes no difference in the sum $7 + 5$?

For a proof, Schultz gives first the axioms of commutativity and associativity of addition, then states that they are "indispensable for arithmetic" (p. 219). In his book *Anfangsgründe der reinen Mathesis* (1790), Schultz found it necessary to pose the commutativity and associativity of sum as axioms, as can be seen from the comparison:

$$\textit{Schultz}\ 1790: 7 + 5 \overset{\text{def}}{=} 7 + (4 + 1) \overset{\text{comm}}{=} 7 + (1 + 4) \overset{\text{assoc}}{=} (7 + 1) + 4$$

$$\text{Grassmann}\ 1861: 7 + 5 \overset{\text{def}}{=} 7 + (4 + 1) \overset{\text{def}}{=} (7 + 4) + 1$$

Schultz thus missed the inductive proofs of commutativity and associativity, and could not do better than claim that any attempted proof of the commutativity of addition would be circular. He gives instead an inductive proof of the commutativity of product in which the right recursion equations appear as "corollaries," a reverse of the conceptual order recursive definition – inductive proof. This order was found by Grassmann and a line can be followed from it to Hankel, Schröder, von Helmholtz, Dedekind, Peano, and Skolem, the last mentioned marking the birth of what is now called *primitive recursive arithmetic*, a formal system for finitary number theory characterized by the restriction of the principle of proof by induction to quantifier-free formulas.

Grassmann's approach to the foundations of arithmetic is found explained in detail in the first of Hermann Hankel's two-volume treatise *Vorlesungen über die complexen Zahlen und ihre Functionen* of 1867. Grassmann's recursion equation for sum is $A + (B + 1) = (A + B) + 1$ and Hankel now states that "this equation determines every sum," (p. 37):

> In this way one finds through a recurrent procedure, one that goes on purely mechanically without any intuition, unequivocally every sum of two numbers.

This prophetical proclamation about the nature of computation is followed by a mechanical computation of the Kantian formula $7 + 5 = 12$ through the writing of 17 equations that begins with $7 + 5 = 7 + (4 + 1) = (7 + 4) + 1$ and ends with $7 + 5 = 11 + 1 = 12$.

Peano writes in his 1889 *Arithmetices Principia, Nova Methodo Exposita* that he has followed in logic amongst others Boole, and for proofs in arithmetic the book by Grassmann (1861) [in arithmeticae demonstrationibus usum sum libro: H. Grassmann], with the pattern recursive definitions – inductive proofs.

One who added to the knowledge of Grassmann's achievement was Hermann von Helmholtz, through his widely read essay *Zählen und Messen, erkenntnistheoretisch betrachtet* (Counting and measuring, epistemologically considered) of 1887. Grassmann's approach is explained right from the beginning: von Helmholtz writes that one has until now posed axioms in arithmetic that include the associativity and commutativity of sum, whereas *Grassmann's axiom*, as he calls the recursion equation for sum, leads to inductive proofs of the mentioned axioms, by which "the right foundation has been gained for the doctrine of sum of pure numbers." What this foundation is, and how the basic laws of sum and product follow from Grassmann's axiom is presented with admirable clarity in von Helmholtz, mostly in reference to Schröder's *Lehrbuch der Arithmetik und Algebra* of 1873. The axiom itself is described as an equation that captures the process of counting from a number a on, with transparent inductive proofs of commutativity and associativity.

Husserl discusses Helmholtz' "nominalistic attempt" in detail in his book (appendix to part I, pp. 190–196): after giving Grassmann's axiom he admits that "the basic formulas of the computational algorithm of the positive integers can be shown, i.e., those formulas from which all its rules can be derived purely deductively," but then adds: "What shall we do with this empty game of signs, one asks in astonishment" (p. 192).

The essential tension between the two competing schools in the philosophy of arithmetic, and in the *Philosophie der Arithmetik* as well, is a very precise one: Should one try to define the numbers, to find a proper ontology for them, or should one leave the notion implicit and concentrate instead on the properties of numbers and how they are proved?

4.2 The "Anzahllehre" in Frege

The *Begriffsschrift* of 1879 resulted as a side-effect from Frege's attempt at a logical definition of an ordered sequence. He begins there with a definition of an expression that contains a property $F(b)$ and a two-place relation $f(b, a)$, the latter described as a "procedure" that gives an element a from a given b. Thus, it can be, as the lower case letter suggests, a proposition defined by a function, as in $f(b) = a$. Frege's notion has the intuitive reading that "property F is inherited in the f-sequence" (p. 58). The notion Frege defines is, in a modern linear writing, an abbreviation for the following:

$$\forall x(F(x) \supset \forall y(f(x,y) \supset F(y)))$$

Various consequences of this formula are given, all of them hard to read, but Frege gives verbal formulations, such as (p. 64):

> *If x has property F that is inherited in the f-sequence, if y follows x in the f-sequence, then y has property F.*

An added footnote gives that "the Bernoullian induction rests on this." The last result of the *Begriffsschrift* is (pp. 86–87):

> *If the procedure f is unique, and if m and y follow x in the f-sequence, then y belongs to the f-sequence that begins with m or appears in the f-sequence before m.*

It is seen that Frege tries to express in logical terms properties of an order relation such as the linear ordering of y and m in the last theorem.

In the *Grundlagen der Arithmetik* 5 years later, in 1884, it is seen what the point of all of the above is, for there Frege defines the concept of *number series* (§79, p. 89). The book, though, is written in such a verbose and polemical way that it is difficult to wade one's way through it. For example, §52 tries to derive foundational results from "German usage," quite contrary to what Frege says elsewhere about the harmful effect of ordinary language. In §70, he writes that "definitions must prove themselves through their fruitfulness." Two, three pages later we come to a series of involved definitions in which the basic concepts are "a is subsumed under concept F" and "a stands in relation φ to b." The idea is that there can be a definite number of objects subsumed by a predicate, its numerical quantity that Frege calls *Anzahl;* something like "numerosity" would be a one-word analogy to the German. There can be further a one-to-one relation between two such predicates F and G, which condition defines *equinumerosity*. In general, a numerical quantity is the scope of the concept "to be equinumerous with F" (p. 85). Note that no definition of numbers in any usual sense has been given so far. No formulas appear anywhere, but the number 0 is now defined through the predicate "to be unequal to oneself" (p. 88, and let's write $a \neq a$ for that).

Next comes a tricky passage in which Frege tries to define the relation of immediate succession between numbers (pp. 89–90).

> There is a concept F and an object x under its scope such that the numerical quantity that belongs to the concept F is n and that the numerical quantity that belongs to the concept "under the scope of F but not equal to x" is m.

Frege adds at once that the definition means the same as "n follows immediately m in the natural number sequence," a plain error because Frege's definition does not give the passage from m to its successor n, but the other way around: *given n*, we get its *predecessor m* by excluding one of the objects from the scope of F by the "not equal to x" clause. To wit, Frege defines the numerical quantity 1 as that which "belongs to the concept *equal to* 0." Let's write this $x = 0$, and there is just one object that falls under the predicate, namely 0. The predecessor of 1 is obtained by adding to clause $x = 0$ the negation $x \neq 0$ and indeed the number of such x is 0. §79 gives various paraphrases and results about 0 and 1, among them (p. 91):

> 6. Each numerical quantity except for 0 follows immediately a numerical quantity in the natural number sequence.

Despite the wording, the result is about the predecessor, not the successor of a number: It says that *for every x except* 0, *there is a y such that x follows immediately y*, and there is no way to reverse this order of dependence in which x comes first, then y.

No way is given for writing the numbers 2, 3, and so on, though it is clear how these should go: The "twoity" of a predicate F means that there exist x and y that belong to F, they are distinct, and any z that belongs to F is equal to x or y, etc. Still, whenever in a work on the foundations of arithmetic the expression "etc" or equivalent appears, we should be on guard: The writing of a few initial cases amounts to a recursive definition.

4.3 Are Numerical Formulas Provable?

Frege's discussion of the foundations of arithmetic in the *Grundlagen* begins properly with the question (p. 5): Are numerical formulas provable? After a presentation of Kant's view in which they are unprovable he illustrates the opposite view through Leibniz' proof of $2+2 = 4$: With the definitions $2 \equiv 1+1, 3 \equiv 2+1$, and $4 \equiv 3 + 1$, we have $2 + 2 = 2 + 1 + 1 = 3 + 1 = 4$. Frege points out a gap in the proof that remains "hidden by the leaving out of parentheses" (p. 7). The Erdmann edition of 1840 of Leibniz' *Nouveaux essais* that Frege used has in fact parentheses of sorts, namely horizontally drawn curly brackets as detailed out in my (2016). These are not properly reproduced in Frege, even if he puts Leibniz' proof in quotation marks. Frege writes that the proof should be:

$$2 + 2 = 2 + (1 + 1), 2 + (1 + 1) = (2 + 1) + 1 = 3 + 1 = 4$$

Thus, Frege contends that the associative law of sum needs to be added, i.e., that an axiom is indispensable. Let us assume this to have been done with the proof. Then (pp. 7–8):

> If one assumes this law, one sees easily that each formula of one-plus-one can be thus proved. Each number can be then defined from the preceding one... The infinite set of numbers is reduced back through such definitions to one and the increase by one and each of the infinitely many number formulas can be proved from a few general propositions.

Pursuing his discussion, Frege mentions Grassmann's aim to arrive at the law $a + (b + 1) = (a + b) + 1$ "through a definition" (p. 8), but states at once that this definition is circular, a definition of sum through sum. Even putting that objection aside, writes Frege, it could still be said that $a + b$ is "an empty sign" if there is no element in the sequence of natural numbers such that associativity holds. Frege contends that Grassmann merely assumes such not to be the case, by which "the rigour is only apparent."

4.4 Grassmann's Formalization of Computation

In his *Ausdehnungslehre* of 1844, Grassmann wrote (p. xix):

> Proof in the formal sciences does not go into another sphere, beyond thinking itself, but resides purely in the combination of the different acts of thought. Therefore the formal sciences must not begin with axioms unlike the real ones; their foundation is made up of definitions. If axioms have been introduced into the formal sciences, such as arithmetic, this is to be seen as a misuse that can be explained only through the corresponding treatment in geometry.

Fifteen years after the above passage was published, Grassmann put his *credo* about avoiding axioms in arithmetic to full effect through the application of the "recurrent procedure" of the science of combinatorics to the most elementary parts of arithmetic, namely the basic arithmetic operations. His 1861 *Lehrbuch* contains the first explicit *recursive definitions* of arithmetic operations, ones that go hand in hand with *inductive proofs* of properties of the recursively defined operations. The base case of induction is 0, or, in Grassmann's times, usually 1, and the step case is that of n to $n + 1$, the *successor* of n [Nachfolger]. Grassmann's definition of sum is for an arbitrary "basic sequence" with a unit e and a succession of units as in $e, e + e, e + e + e, \dots$, with the addition of a unit always intended at right, in the sense of $(e + e) + e$. He actually constructs a doubly infinite series of integers with a positive and negative unit, but the following is for brevity the part on natural numbers that begin with the positive unit denoted e. Here is Grassmann's explanation of the positive integers (p. 2):

> **7.** *Explanation*. Let a sequence of quantities be built out of a quantity e [Grösse] through the following procedure: One sets e as *one* member of the sequence, $e + e$ (to be read e plus e) as the successive [nächstfolgende] member of the sequence, and one continues in this way, by deriving from the member that is each time last the successive one through the joining of $+e$.
>
> ⋮
>
> When one assumes each member of this sequence to be different from all the other members of the sequence, one calls this sequence the *basic sequence* [Grundreihe], e the *positive* unit.

It is remarkable that Grassmann does not use the word "addition" in this characterization of the number sequence. The addition of the unit e is explained as follows (p. 3):

8–9. *Explanation.* If a is any member of the basic sequence, one means by $a + e \ldots$ the member of the sequence that follows next to $a \ldots$ that is, if b is the member of the sequence next to a, we have

(8) $b = a + e$

One calls this operation [Verknüpfung] the addition of a unit.

Addition of a unit is a clearly defined separate operation explained as the taking of a successor in the basic series. Grassmann now sets out to prove that "the members of the sequence that follow e are sums of positive units." The proof goes as follows:

Proof The members of the basic series that follow e have (by **7**) resulted from e by a progressive addition of positive units, are therefore sums of positive units.

There is a clear awareness in Grassmann that the sequence of natural numbers is generated from the unit through repeated application of the successor operation. The only fault is that there is no separate notation for the successor; it is denoted by $+e$. Grassmann's definition of addition should be read with the overloading of the symbol + and the convention about dropping parentheses from $(a + b) + e$ in mind (p. 4):

15. *Explanation.* If a and b are arbitrary members of the basic sequence, one understands with the sum $a + b$ that member of the basic sequence for which the formula

$$a + (b + e) = a + b + e$$

holds.

In Grassmann's terms, this equation gives a procedure for reducing arbitrary sums into members of the "basic sequence" that by the above proof are "sums of positive units."

The recursive definition of sum is put into use in Grassmann's "inductory" (inductorisch) proofs of the basic properties of addition, such as associativity and commutativity. Anyone who tries to prove the commutativity of addition, $a + b = b + a$, will notice that the proof requires as a preliminary a proof of associativity, $(a + b) + c = a + (b + c)$. This must have happened to Grassmann who offers the following proof (p. 8):

22. $a + (b + c) = a + b + c.$

"Instead of adding a sum one can add the summands step by step," or "instead of adding two quantities step by step, one can add their sums."

Proof (inductorily in relation to c) Assume formula 22 to hold for whatever value of c, then we have

$$\begin{aligned} a + [b + (c + e)] &= a + [b + c + e] \text{ (by15).} \\ &= a + (b + c) + e \text{ (by15).} \\ &= a + b + c + e \text{ (by assumption).} \\ &= a + b + (c + e) \text{ (by15).} \end{aligned}$$

> Therefore, if formula 22 holds for whatever value c, it holds even for the one following next, thereby for all values that follow.

The use of parentheses could have made the steps a bit clearer. Next, as result 23, Grassmann proves the commutativity of addition.

The treatment of addition is followed by a similar recursive definition of multiplication, with accompanying inductive proofs of its properties, then division and prime factorization and so on, i.e., a development of elementary arithmetic through rigorous inductive proofs.

4.5 Frege's Blindness

Reading Frege's *Grundgesetze der Arithmetik* of 1893, one cannot but think: how is it possible not to see, some thirty years after Grassmann and in full knowledge of his work and the further explanations of Hankel, Schröder, and von Helmholtz, that the natural numbers have a beginning, 0 or 1, and a step ahead, the successor, and that arithmetic operations are defined recursively and their properties proved inductively? One who stated the matter very clearly was Schröder (1873, pp. 64–65):

> The concept of addition has been determined for two numbers the second of which is equal to 1, and then one can make clear this concept for two arbitrary numbers, by setting by way of definition for arbitrary values of a and b the equation:
>
> (6) $$a + (b + 1) = (a + b) + 1.$$
>
> ⋮
>
> This equation contains in fact a *recurrent* definition of the sum of any two numbers. It does not, though, state at once what such a sum means; it traces instead back the understanding of a sum of a and $b + 1$ to that of a and b. For when $a + b$ is a known number of our series of numbers, we have already seen that even $(a + b) + 1$ must be one; then the left side of equation (6) is really explained by its right side.

By rejecting the recursive definition of sum and other arithmetic operations, Frege had condemned his attempt at a foundation for arithmetic to fail, on page 8 of the first volume of his 500-odd page *Grundgesetze* in two volumes. There is no answer to such questions as the basic laws for arithmetic operations.

In the *Begriffsschrift*, not a single arithmetic result is proved formally, and the principle of proof by induction especially is left untouched. The same is true of the *Grundgesetze*: it does not bring proofs in formal arithmetic an inch further even if that was the very purpose of his "concept writing." As impeccable as was Frege's insight in getting the principles of quantificational logic exactly right, the more disastrous was his failure with arithmetic. If he had just followed his advice to Schröder, namely to put the formalism into actual use instead of trying to decide philosophically in advance what a definition is, how numerical quantity is defined, and so on, he would have been steered a course past the rocks that sunk his attempt. Now, instead, he put decisive weight to objections such as that Schröder's approach does not define what numbers are but just represents them (the German word is *abbilden*), as in the *Grundlagen*, §43: "It follows that he explains just the *number sign*, not number... what is expressed by the sign, what I had called the number, he assumes to be known."

The main reason for Frege's blindness about the possibility of recursive definitions and a formal development of arithmetic on their basis within his quantificational logic seems to be a ridiculous little detail, namely that in the nineteenth century, natural numbers began with 1 instead of 0: With the latter, there are two recursion equations for sum. The first one, $a+0=a$, shows that to add 0 reduces to doing nothing, with the symbol + for sum eliminated. Secondly, to add 1 reduces to the taking of a successor, with the notation $a+1=a'$ as in Schröder, and, finally, to add n reduces to the taking of n successors – nothing circular about that and the symbol + completely eliminated after so many recursions that give as a result an expression of the form $0'^{\cdots\prime}$. But Frege was *folgerichtig*, upright and consequential, and would not proceed just formally and later fix the philosophy behind.

In the approach through recursive definitions, one can actually develop elementary arithmetic to the full: all the arithmetic operations and their properties, relations of equality and order, the notion of divisibility, and the fundamental theorem of arithmetic by which every composite natural number has a unique factorization into a product of prime numbers. A comparison to what the *Anzahllehre* has accomplished in the way of actually developing arithmetic is crushing:

4.6 The Last Word on the *Anzahllehre*

The last word on the *Anzahllehre, the logicist definition of number* is found in Hilbert-Ackermann of 1928. There, in Chapter 4, Section 4.2 is dedicated to the definition. Here is how it goes:

Numbers are defined as properties of predicates: 0 is the property of predicates F given by the formula $\overline{(Ex)}F(x)$, "there is no x that has F" as Hilbert-Ackermann explain, and handily abbreviated as $0(F)$. 1 is defined as a property of those F for which we have, with $\equiv(x,y)$ the two-place equality relation (p. 86):

$$(Ex)[F(x)\ \&\ (y)(F(y)\ \rightarrow\ \equiv(x,y))]$$

This formula is abbreviated as $1(F)$. There follows another definition, that of $2(F)$, with a negation on top of the first $\equiv$:

$$(Ex)(Ey)\{\overline{\equiv}(x,y)\ \&\ F(x)\ \&\ F(y)\ \&\ (z)[F(z)\ \rightarrow\ \equiv(x,z)\vee\equiv(y,z)]\}$$

Next the *equinumerosity* (Gleichzahligkeit) of two predicates F and G, notation $\mathrm{Glz}(F,G)$, is defined as the existence of a one-to-one relation R between those objects for which F resp. G hold, a rather involved formula in second-order logic (p. 87):

$$(ER)\{(x)[F(x) \rightarrow (Ey)(R(x,y) \,\&\, G(y))] \,\&$$
$$(y)[G(y) \rightarrow (Ex)(R(x,y) \,\&\, F(x))] \,\&$$
$$(x)(y)(z)[(R(x,y) \,\&\, R(x,z) \rightarrow \equiv (y,z)) \,\&$$
$$(R(x,z) \,\&\, R(y,z) \rightarrow \equiv (x,y))]\}$$

Two predicates F and G are defined to be *incompatible* (unverträglich), a second-order relation found already in §1:

$$\mathrm{Unv}(F,G) \text{ is to mean the same as } (x)(\overline{F}(x) \ v\ \overline{G}(x)), \text{ i.e.,} (x)(\overline{F(x) \,\&\, G(x)}).$$

The equation $1 + 1 = 2$ amounts now to the following provable second-order formula (p. 87):

$$(F)(G)([\mathrm{Unv}(F,G) \,\&\, 1(F) \,\&\, 1(G)] \rightarrow 2(F \vee G))$$

The next level of abstraction introduces the general notion of a number, the *third-level* predicate Φ for which the following is posed (p. 87):

$$(F)(G)\{(\Phi(F) \,\&\, \Phi(G) \rightarrow \mathrm{Glz}(F,G)) \,\&\, [\Phi(F) \,\&\, \mathrm{Glz}(F,G)) \rightarrow \Phi(G)]\}$$

With the abbreviation $\mathfrak{Z}(\Phi)$ (a fraktur letter Z for *Zahl*) for this formula, we have (p. 88):

A number is a predicate of predicates Φ *that has the property* $\mathfrak{Z}(\Phi)$.

Written out without abbreviations, this definition would be close to half a page in length; One cannot but bring into mind what Hilbert had stated at the beginning of his Paris problem list of 1900, namely that "what is clear and easily comprehensible attracts us, what is wicked frightens us away" (das Klare und leicht Fassliche zieht uns an, das Verwickelte schreckt uns ab). Moreover, there is no reward for the labour in the form of easy proofs, on the contrary.

The attempted definition of the concept of number is a failure: When applied to a finite domain of n members, all numbers beyond n will be equated, a fact described as an "Übelstand" (a bad state of things). The only remedy suggested is an axiom of infinity, by which the original aim of a purely logical foundation of arithmetic is defeated.

There is a blind spot in this whole affair of the logicist attempt at defining numbers and the rest of the foundations of arithmetic in purely logical terms, one that can be exposed as follows:

Say, we want to prove $n + m = m + n$ and to this effect, let n and m be any given numbers. How do their definitions look like? There is no definition of $n(F)$ and $m(F)$ that would let us start to work with the expression $n+m$. We saw above how to define

$0(F)$, $1(F)$, and $2(F)$, but Hilbert-Ackermann failed to add the required "etc" after these three first cases. Some little pondering will lead to the insight that the ability to write down a definition of $n(F)$ for any given n is based on an implicitly understood *recursive procedure*, the missing "etc" in Hilbert-Ackermann. Given $n(F)$, $n'(F)$ is formed by taking a fresh variable $x_{n'}$, then adding to the formula for $n(F)$ the parts $(Ex_{n'})$ and $F(x_{n'})$ and the negations $\overline{\equiv}(x_i, x_{n'})$ for $1 \leq i \leq n$ and so on in the appropriate places. When the inductive definition is supplied, the whole enterprise of a purely logical definition of the natural numbers collapses, for the properties of numbers are proved by an induction that corresponds to their build-up. Therefore, in comparison to the approach through recursive definition, an unnecessary logical detour has been taken.

Today, the *Anzahllehre* has been justly forgotten, and theories of arithmetic and computation are based exclusively on recursion and induction, but with Husserl, we must not judge his accomplishment just against a wisdom that arrived much later.

4.7 Reading *Philosophie der Arithmetik* Against the Two Traditions

Husserl's references in the *Philosophie der Arithmetik* contain of the above-mentioned, as they appear in the original edition, Frege's 1884 *Grundlagen* (p. 12), Hankel's 1867 *Zahlensysteme* (p. 31), Grassmann's 1861 *Lehrbuch* (p. 104), von Helmholtz' 1887 *Zählen und Messen* (p. 105), Schröder's 1873 *Lehrbuch* (p. 106), and Dedekind's 1888 *Was sind und was sollen die Zahlen?* (p. 138). He discusses some of these in detail and we can safely assume that he knew the *Anzahllehre* and the recursive way well enough.

I will not say much about Husserl's book; the groundwork has been done by Centrone (2010), and it will not be too difficult to find out how the two traditions here presented appear in Husserl. Neither is fully endorsed, but it is clear that the *Anzahllehre* dominates Husserl's thinking. Frege's definition of *Anzahl* is given, as in (p. 133): "The *Anzahl* that belongs to a concept F is the scope of the concept 'equinumerous with the concept F'." Husserl makes his own adjustment to this definition, and then discusses the "nominalistic attempts" as he describes the more formal approach to arithmetic. He definitely does not present in any way the scheme recursive definition – inductive proof in the building up of elementary arithmetic, even though he clearly must have recognized the pattern in von Helmholtz.

In reading through Husserl's book in the light of the two traditions, the lack of any formal development is striking. Husserl tries to decide on the basis of general criteria what is reasonable and defensible, what not. Chapter VI is dedicated to the definition of equinumerosity. Grassmann's Leibnizian position is given as one extreme: The criterion is substitutability *salva veritate*, a condition that already in Leibniz' times had been improved in arithmetic into substitutability *salva quantitate*. Husserl, though, does not see this as a criterion that is posed on equality, but as an attempted definition that in his view turns out circular. He is unable to see that arithmetic is given a formal representation in Grassmann, one that works to perfection.

In chapter VII, a notion of *equivalence of sets* is described, similar to one in Otto Stolz' *Vorlesungen über allgemeine Arithmetik* of 1885 (p. 121, cf. also p. 105 for Stolz): "One takes the sets to be compared, connects them element by element and checks if any elements were left over or not." Sets can be brought together on this basis into classes; their common property is the numerical quantity or *Anzahl*. Next, this quantity is possessed by any one of the sets, and Husserl puts up a *representation* through a stroke, the symbol 1, and its repetition as in 11, 111, 1111, ... conceived as concrete sets of as many strokes (p. 124). Now Husserl observes that the mixing with the decimal notation can be avoided by writing instead $1+1, 1+1+1, 1+1+1+1, \ldots$. This discourse is taken over directly from Schröder and is a sign of his careful reading of the Schröderian *Vorlesungen* of 1873. Again, Husserl gives no weight to what Schröder is able to accomplish in arithmetic, but discards the approach: The equivalence of sets is just a criterion, not a "nominal definition" (p. 126). Further, "We must take it as totally amiss if these sets of strokes are taken as 'natural numbers' and the names two, three, etc. as naming *them*" (p. 128). The approach "attempts to construct the elementary arithmetic concepts from their ultimate definitional characteristics" but results in the end "in an alien concept formation that is useless for practice and science" (p. 129).

What is wrong here? Husserl's judgement of the uselessness of the formal development of arithmetic would apply rather to the sterile *Anzahllehre* as above described. It must be that the whole idea of a formal arithmetic was alien to him, whereas we today would think that that is the very point: computation has to proceed according to fixed rules that can be applied without thinking so to say, for otherwise there would have to be a second level of rules for how the computational rules are applied; no computer would tick even once if there were such a regress of rules. Similarly with the proofs of the properties of elementary arithmetic: those rules have to be applicable without a second level of rules, a point emphasized in Bernays (1927). To check the correctness of the application of a rule is not to apply another rule.

Husserl does discuss formal arithmetic, in an appendix to part I of his book with the title: *The nominalistic attempts of von Helmholtz and Kronecker.* Of von Helmholtz' approach, as in the 1887 essay, he writes (p. 192):

> The signs of the fixed sequence that begins with one serve to signify sequences of arbitrary objects, among these also whatever sections of the "number sequence" itself. In this way, Helmholtz gains the possibility to define addition as a purely signifying operation, by which follow theorems, or more precisely said, equivalences in signs such as
>
> $$(a + b) + 1 = a + (b + 1)$$
>
> In general, one succeeds in proving all the basic formulas of the computational algorithm of positive entire numbers, i.e., those formulas from which all of its rules can be deductively inferred. One has just to observe that the whole of it appears only as mere equivalences of certain complexes of symbols (that hold in the sense of the signifying initial definitions).

Here we have Grassmann's recursive definition of addition, written as in von Helmholtz. Hankel, Schröder, and Frege gave long discussions about this definition, and von Helmholtz writes that Grassmann's axiom leads to inductive proofs of the axioms that include the associativity and commutativity of sum. "Thereby, as we hope to show in what follows, the right foundation has been gained for the doctrine of addition of pure numbers."

The passage in Husserl after Grassmann's recursion equation would be magnificent were it not for the belittlement that follows: Formal arithmetic is described as an "empty game with signs" that however has, and here Husserl quotes von Helmholtz, "extraordinarily useful applications." One can "give by the system of signs for numbers descriptions of the conditions of objects in reality, whenever they apply, to any required degree of accuracy." Then, by the known laws of nature, "one can compute in a great number of cases in advance the numerical values that measure the course of events," obviously a reference to the classical mechanics of moving bodies, say the astonishingly precise computations of planetary positions ahead in time (p. 192).

Chapter XIII is dedicated to "the art of computing and arithmetic." There are some suggestive passages, as one about computation (p. 293):

> It can be taken as *any regulated art of derivation of signs out of signs within whatever algorithmic system of signs, by the "laws" — or better: conventions — of combination, separation, and transposition.*

It becomes even better further below (pp. 293–294):

> If we free the number signs from their conceptual correlates and build up, disregarding any conceptual application, the technical methods that the system allows, we have delimited the pure mechanics of computation [die reine Rechenmechanik] that lies at the basis of arithmetic and that forms the technical side of its methods. This art of computation, now, is clearly not identical anymore to the art of *arithmetic* knowledge.

Husserl has taken in the point of Grassmann, Hankel, and Schröder about mechanical computation. That of Peano about formal proof was still to come; Peano's treatise that made even steps of arithmetical proof formal came out in 1889 but was unknown to Husserl. With Peano, steps of computation became just a special kind of steps of deduction. The full understanding of the connection came as said with Bernays who in the *Grundlagen* of 1934 wrote (p. 290, p. 292):

> If c is a numeral, the computation of the value of $f(a, \ldots, t, x)$ is already completely formalized through the derivation of the equation
>
> $$f(a, \ldots, t, x) = c$$
>
> We can reconstruct completely the recursive procedure of computation of finitary number theory in our formalism, through the deductive application of the recursion equations.

4.8 Husserl's *Philosophie der Arithmetik*, Part II

Husserl summarizes his *Philosophie der Arithmetik*, part I in a closing paragraph, "our result expressed in a brief form" (pp. 322–323):

> The fact that we are limited in the vast majority of cases into a *symbolic construction of numbers* forces us into a regulated construction of the domain of numbers in the form of a *system of numbers* (be it the natural number sequence or a system in a narrower sense of the word), one that grabs after a fixed principle out of the totality of symbolic constructions one such as belongs to each real number-concept and is equivalent to it, and gives it at the same a systematic place. There arises then for all other thinkable number-forms the problem of determination of value, i.e., its classificatory reduction to a number of the system equivalent to it. An overview of the thinkable forms of number formation taught, though, that the discovery of apposite methods of value determination depends on the construction of a *general arithmetic* in the sense of a general theory of operations.

This suggestive closing of the first volume of the *Philosophie der Arithmetik* finds its continuation in Husserl's sketches for the second volume that never saw the light of the day in a finished and printed form. In fact, he promises in the preface of the first part, concerning the second part, to "deliver it to print within a year." Sketches for the second part have been published in the collected works of Husserl, in volume XII to which I refer. From the point of view taken here, not much needs to be added:

A short text on "arithmetic as an aprioristic science" written around 1890–1891 begins with a discussion of the contents of a "scientifically conceived textbook on arithmetic," in German precisely the title of Grassmann's and Schröder's *Lehrbuch der Arithmetik* (p. 380). The preface of the former is written in such terms of "scientific conception." There will be, writes Husserl, *basic operations*, the standard ones of sum, subtraction, product, and division (ibid.):

> One defines these operations and puts up their laws, in which an organization is followed that lies in the nature of things. With each operation, one considers combinations with two members to start with, then with several members, which can be seen as compositions or nestings of combinations with two members.

The repetition of operations gives rise to "an unlimited multitude of determinations of numbers."

Husserl writes in the quotation that one *defines* the basic operations, certainly the crucial starting point in any development of arithmetic, but says not a word more about how it is done. Grassmann had given the recursive definitions of these operations, and Schröder had discussed their nature in detail, as we saw above. Husserl's section on arithmetic as an aprioristic science is preceded by a similar short text of 1890 on "general arithmetic," an expression found in the title of Stolz' book of 1885. Stolz mentions just in one place the "recurrent procedure" for the determination of the value of a sum and gives there a detailed example of computation by the recursion equation for sum (p. 9). For the rest, he mostly lists elementary properties as postulates without discussing their status either as theorems or axioms. Husserl, in comparison, does not state a single thing on how the properties of the basic operations are grounded.

4.9 An Echo of Husserl: Weyl's Predicative Arithmetic of 1918

Hermann Weyl was the most prominent follower of Husserls' philosophy in mathematics and exact science. The introduction of his widely read book on relativity theory, the *Raum-Zeit-Materie* of 1918, echoes Brentano's and Husserl's ideas in relation to the mathematical physics to follow, in particular, the idea of a symbolic construction of the world. Weyl's treatment of arithmetic in his well-known booklet *Das Kontinuum*, also published in 1918, has a similar tone. In reference to Husserl, Weyl writes in the preface:

> Our treatment of the continuum problem offers a contribution to the epistemological question of the relations between what is given immediately (in intuition) and the formal concepts (of the sphere of mathematics) through which we in geometry and physics attempt to construct that given.

The notation in *Das Kontinuum* is as follows: There are free-variable atomic formulas, such as $E(x)$ and $U(xy)$ for a property and a two-place relation, respectively. There are next the operations of negation, conjunction, disjunction, substitution in argument places, and existential quantification. The last one is written as in $E(*)$ and read as *there is an object x such that* $E(x)$. Universality is defined as the inexistence of a negating counterexample. Weyl's notation for the propositional part is exemplified by his writing of the laws of excluded middle, contradiction, and double negation (p. 10): $U + \overline{U}$, $\overline{U \cdot \overline{U}}$, $\overline{\overline{U}} = U$. His example of a quantified judgment reveals the awkwardness of his variable-free existential quantifier notation: The task is to write a typical mathematical statement, *there is for every x a y such that* $U(xy)$. He posits (p. 7):

$$U(x\,*) = A(x)$$

$$\overline{A}(x) = B(x)$$

$$B(*)$$

$$\overline{B(*)}$$

The last formula, an abbreviation for something like $\overline{\overline{U}(*\,*)}$, can be read with some good will as $\neg\exists x \neg\exists y U(xy)$.

Arithmetic is based on a single relation written as $\mathcal{F}(n, n')$, a relation that holds precisely when n' is the *successor* of n. With the natural numbers beginning with 1, the number 1 is characterized by the property that it is not the successor of any number (p. 9):

$$\overline{\mathcal{F}(*, x)} = I(x)$$

The uniqueness of 1 is just given "as a fact." Next 2 is characterized by:

$$I(y) \cdot \mathcal{F}(y, x) = \mathcal{F}_2(y, x) : \mathcal{F}_2(*, x) = II(x)$$

"Analogously 3, 4, etc" writes Weyl, not seeing that the "etc" hides the inductive step in his definition of numbers.

The logical definitions of numbers in the style of the *Anzahllehre* are not put to any real use in Weyl: later in the book, arithmetic operations are defined recursively, with a reference to Dedekind's *Was sind und was sollen die Zahlen?* (1888), and the basic laws proved inductively (pp. 39–44). Weyl writes that "the elementary truths about numbers" can be proved logically through complete induction from the two axioms: "there is to every number a unique successor, and every number except 1 has a unique immediate predecessor" (p. 44).

Dedekind's book appeared too late for Husserl to make real a use of it in the *Philosophie der Arithmetik*, a book based on his *Habilitationsschrift* of 1887, titled *Über den Begriff der Zahl. Psychologische Analysen*. With some bending of mind, we can take his follower Weyl's eclectic combination of both approaches after the model of Dedekind, the *Anzahllehre* and that of recursive definitions, as belonging to Husserl, too.

After this paper was finished I came to know a recent study by Hartimo and Okada (2016) that discusses in detail Husserl's remarks on Hankel (1867).

References

P. Bernays, Probleme der theoretischen Logik (1927). As reprinted in Bernays (1976), pp. 1–16

P. Bernays, *Abhandlungen zur Philosophie der Mathematik*. Wissenschaftliche Buchgesellschaft (1976)

S. Centrone, *Logic and Philosophy of Mathematics in the Early Husserl* (Springer, Dordrecht, 2010)

R. Dedekind, *Was sind und was sollen die Zahlen?* (1888)

G. Frege, *Begriffsschrift, eine nach der arithmetischen nachgebildete Formelsprache des reinen Denkens* (Nebert, Halle, 1879). Reprinted in *Begriffsschrift und andere Aufsätze*. Ed. I.Angelelli, Olms 1964. English translation in Van Heijenoort

G. Frege, *Grundlagen der Arithmetik* (F. Meiner, Hamburg, 1884)

G. Frege, *Grundgesetze der Arithmetik. Begriffsschriftlich abgeleitet*, vol. 1 (Pohle, Jena, 1893)

H. Grassmann, *Die Wissenschaft der extensiven Grösse oder die Ausdehnungslehre* (Wigand, Leipzig, 1844)

H. Grassmann, *Lehrbuch der Arithmetik für höhere Lehranstalten* (Enslin, Berlin, 1861)

H. Hankel, *Vorlesungen über die complexen Zahlen und ihre Functionen I* (Leopold Voss, Leipzig, 1867)

M. Hartimo, M. Okada, Syntactic reduction in Husserl's early phenomenology of arithmetic. Synthese **193**, 937–969 (2016)

D. Hilbert, W. Ackermann, *Grundzüge der theoretischen Logik* (Springer, 1928)

D. Hilbert, P. Bernays, *Grundlagen der Mathematik I* (Springer, 1934)

G. Leibniz, *Nouveaux essais sur l'entendement humain* (1704). Edition by J. Erdmann (1840)

G. Peano, *Arithmetices Principia, Nova Methodo Exposita* (1889). (Partial English translation in Van Heijenoort)

J. van Heijenoort (ed.), *From Frege to Gödel, A Source Book in Mathematical Logic, 1879–1931* (Harvard University Press, Cambridge/London, 1967)

H. von Helmholtz, Über Zählen und Messen, erkenntnistheoretisch betrachtet, in *Philosophische Aufsätze, Eduard Zeller zu seinem fünfzigjährigen Doctorjubiläum gewidmet* (Fues' Verlag, Leipzig, 1887), pp. 17–52

J. von Plato, In search of the roots of formal computation, in *Proceedings of the 2015 HaPoC Conference*, ed. by F. Gadducci, M. Tavosanis (Springer, 2016), pp. 300–320

E. Schröder, *Lehrbuch der Arithmetik und Algebra für Lehrer und Studirende. Erster band. Die Sieben algebraischen Operationen* (Teubner, Leipzig, 1873)

J. Schultz, *Prüfung der Kantischen Critik der reinen Vernunft.* Part I (Hartung, Königsberg, 1789)

J. Schultz, *Anfangsgründe der reinen Mathesis* (Hartung, Königsberg, 1790)

J. Schultz, *Prüfung der Kantischen Critik der reinen Vernunft.* Part II (Hartung, Königsberg, 1792)

T. Skolem, Begrüngung der elementaren Arithmetik durch die rekurrierende Denkweise ohne Anwendung scheinbarer Veränderliche mit unendlichem Ausdehnungsbereich (1923). As reprinted in Skolem 1970, pp. 153–188 (English translation in Van Heijenoort)

T. Skolem, *Selected Works in Logic*, ed. by J.E. Fenstad (Universitetsforlaget, Oslo, 1970)

O. Stolz, *Vorlesungen über allgemeine Arithmetik* (Teubner, Leipzig, 1885)

H. Weyl, *Das Kontinuum. Kritische Untersuchungen über die Grundlagen der Analysis* (Veit, Leipzig, 1918a)

H. Weyl, *Raum, Zeit, Materie. Vorlesungen über allgemeine Relativitätstheorie* (Springer, 1918b)

Chapter 5
Husserl and Boole

Stefania Centrone and Pierluigi Minari

Abstract We aim at clarifying to what extent the work of the English mathematician George Boole (1815–1864) on the algebra of logic is taken into consideration and discussed in the work of early Husserl, focusing in particular on Husserl's lecture "*Über die neueren Forschungen zur deduktiven Logik*" of 1895, in which an entire section is devoted to Boole. We confront Husserl's representation of the problem-solving processes with the analysis of "symbolic reasoning" proposed by George Boole in the *Laws of Thought* (1854) and try to show how and why Husserl, while praising Boole's calculus, strongly criticizes his attempt at a philosophical clarification and justification of it.

Keywords Boole • Algebra of Logic • Formal-algorithmic processes • Symbolic reasoning

5.1 Introduction

The present chapter aims at clarifying to what extent the work of the English mathematician George Boole (1815–1864) on the algebra of logic is taken into consideration in the work of early Husserl, focusing in particular on Husserl's 1895 lecture "*Über die neueren Forschungen zur deduktiven Logik*", in which an entire section is devoted to Boole.[1] We proceed from the assumption that the distinctive trait of early Husserl's work can be identified in the simultaneous presence in his logical and mathematical reflections of two different directions of research, (1) the project of a substantial mathematization of logic and (2) a conception of logic as

[1]In Husserl (2001), Appendix, 305–328 (henceforth *LV 96,* App.).

S. Centrone (✉)
Institut für Philosophie, Carl von Ossietzky Universität Oldenburg, Oldenburg, Germany
e-mail: stefania.centrone@uni-oldenburg.de

P. Minari
Department of Letters and Philosophy, Università di Firenze, Florence, Italy
e-mail: pierluigi.minari@unifi.it

S. Centrone (ed.), *Essays on Husserl's Logic and Philosophy of Mathematics*,
Synthese Library 384, DOI 10.1007/978-94-024-1132-4_5

the study of objective relations occurring among certain abstract logical entities. As regards (1), Husserl succeeds in grasping with great clarity and insight the implications of the *formal-abstract* trend in mathematics and, in particular, of its tendency toward *algebrization*, which he is able to transfer to and elaborate at the logico-theoretical level. As regards (2), Husserl develops a philosophy of logic and mathematics focused on the systematic investigation of the properties and relations that occur among certain abstract semantical entities: a source of inspiration for this project is the theory of *ideas* (*Vorstellungen an sich*) and *propositions* (*Sätze an sich*) in Bolzano's *Wissenschaftslehre,* and one of its more remote ancestors is the Stoic doctrine of *Sayables* (*lektá*). A detailed analysis of claim (2) is to be found in Chap. 3 of this volume. In this chapter we address claim (1).

Let us preliminary observe that in the work of the English algebraists in Cambridge (notably C. Babbage, G. Peacock, J. W. Herschel) in the period 1830–1840, a tendency can be seen to emerge consisting in considering the abstract properties of arithmetical operations while leaving aside their numerical substrate. Operations began to be studied as 'autonomous'. The so-called symbolic algebra became an algebraic theory of magnitudes in general: on the one hand, it assumed as principles (and hence as rules of calculus) the laws that apply to the usual arithmetical operations, while, on the other hand, it eliminated the restriction concerning their exclusive applicability to natural numbers. A distinction began to be made between (i) the symbolic aspect of algebra, that is a system of formal laws of combinations ($+, \times, \ldots$) and of abstract algorithms of computation, by means of which conclusions are drawn in a deductive-algorithmic way,[2] and (ii) the possible systems of entities that can satisfy such formal conditions. With the same symbolic system a unitary treatment could be provided for systems of heterogeneous entities manifesting a similar structural behavior.

With the contributions of scholars such as W. R. Hamilton, H. Grassmann, and A. Cayley there was a progressive distancing from the idea of algebra as "symbolic algebra of magnitudes," which will culminate in the explicit disengagement of algebraic research from the quantitative dimension with George Boole and his creation of an algebra of logic. Algebra no longer treats numbers or magnitudes only, but also propositions, concepts and, in general, *qualitative data*. The laws under which they fall are independent from any specific interpretation of the symbolism, and the structural properties of the operations that are reflected in such laws are unleashed from numerical elements and assume the character of abstract algorithmic procedures for "calculations" performed with symbolic expressions.[3] The explicit separation between laws of calculus – purely *formal* laws – and their interpretations, is more or less the distinctive trait of modern abstract algebra and mathematics. In Boole's words[4]:

---

[2]Casari 1973, 8–9.

[3]See Cantini 1979, 41 ff.

[4]Boole 1847, Introduction. Cp. also Webb 1980, 79; Hartimo 2007, 285 ff.

> [T]he validity of the process of analysis does not depend upon the interpretation of the symbols which are employed, but solely upon the laws of their combination. Every system of interpretation which does not affect the truth of the relations supposed, is equally admissible, and it is thus that the same process may, under one scheme of interpretation, represent the solution of a question on the properties of numbers, under another, that of a geometrical problem, and under a third, that of a problem of dynamics or optics.

Synopsis In Sect. 5.2 we present Husserl's idea of a 'system of numeration in a given base' (as we would say nowadays), to be found in chapter XII of the *Philosophy of Arithmetic,*[5] entitled "*The Symbolic Representations of Numbers*", and in particular in paragraph 4, "*The system of Numbers*".[6] All number systems, so Husserl, present a very peculiar feature: they perfectly function even when unleashed from their conceptual interpretation. In Sect. 5.3 we consider Husserl's analysis of problem-solving processes in chapter XIII of the *Philosophy of Arithmetic* ("*The Logical Sources of Arithmetic*"),[7] in particular in paragraph 1, "*Calculation, Calculational Technique and Arithmetic*".[8] In Sect. 5.4 we present the analysis of "symbolic reasoning" proposed by George Boole in chapter V of his masterpiece, *The Laws of Thought* (1854).[9] In Sect. 5.4 we compare Husserl's and Boole's view of problem-solving algorithms as well as Husserl's comprehensive judgment both on Boole's calculus and on Boole's interpretation of the *calculus*, focusing, in particular, on the mentioned section of Husserl's lecture "*Über die neueren Forschungen zur deduktiven Logik*" (1895) that is entirely dedicated to Boole.

5.2 Number Systems in Husserl's *Philosophy of Arithmetic*

In the *Philosophy of Arithmetic* Husserl proceeds from the observation that arithmetic does not operate with proper numerical concepts. As a (at that time) logical psychologist he takes proper numerical concepts to be the product of abstraction, the latter viewed as the process of freeing an object of its peculiar features. The lowest numerals, say 2, 3, 4, are taken to designate immediately recognizable differences between sets of 2, 3, 4 elements, respectively, where a sort of cantorian abstraction from the nature and order of the elements is presupposed. Since Husserl believes that "in the authentic sense one can barely count beyond three"[10] he introduces symbolic number concepts that stand for proper ones. However, to conceive the

---

[5]Husserl 1891, henceforth *PdA* (Husserliana edition) and *PoA* (English translation). Here *PdA* 222–255; *PoA* 335–269.

[6]*PdA* 228–233; *PoA* 241–246.

[7]*PdA* 257–283; *PoA* 270–299.

[8]*PdA* 257–259; *PoA* 270–274.

[9]Boole 1854.

[10]*PdA* 339; *PoA* 357.

system of proper number concepts to be replaced by a system of symbolic number concepts that is completely isomorphic to it, so Husserl, still presupposes a strong idealization of our representational capacity. What is needed is an efficient method to construct the totality of positive integers in a manner that it is adequate for the purposes of the arithmetic. And, what is more, the number system should be able to solve sufficiently complex calculatory problems. The ultimate organization of the numerical field must be such that (i) we can effectively calculate with all conceivable numbers; (ii) these are univocally classified on the base of the order relation $\leq$; and (iii) the following conditions are met: speed in calculations, ease of distinguishing numerical constructs and univocity in the principle of construction (the same number must not be present more than once, i.e. all must have different names in the construction). So Husserl turns to what in current terminology is called 'a system of numeration in a given base'. The essence of such a system consists in the fact that it constructs all numerical concepts using a few elementary concepts and rules for operations.

The base ($1, \ldots, X$) is constituted by the numbers that are properly given to us. We then construe new number concepts by means of the operation of "successor". We obtain[11]:

$X + 1, X + 2, \ldots, X + X,$
$X + X + 1, X + X + 2, \ldots, X + X + X,$
$\ldots \; \ldots.$

Yet,[12]

> [T]his mode of designation also does not suffice. The further we go the more tedious becomes the designation by the accumulating (*sich anhäufende*) sums of X's. A new means of abbreviation presents itself at this point: the simple enumeration (*Abzählung*) of the X's leads to the multiplicative symbolization in thought and sign; that is, to: $2X, 3X, 4X \ldots$

Using the same kind of argument, Husserl justifies the introduction of the operation of exponentiation: the number constructs formed by the iterative multiplication of the X (XX, XXX, XXXX) proliferate to such an extent that new abbreviations become necessary; counting the factors leads to the operation of exponentiation (X^2, X^3, X^4, ...). We see that the series continues and that the iteration of the last introduced operation always leads to a new operation; however, *for practical reasons and ends* – Husserl maintains – it is *sufficient to stop at the operation of exponentiation.* Thus mathematically, each number is a "whole, whole-number function (*ganze, ganzzahlige Funktion*)" of a determinate fundamental number (*Grundzahl*) X established conventionally, i.e. it is symbolically presentable as a *finite* set with the form:

$\{a_0, a_1X^1, a_2X^2, \ldots, a_nX^n\}$

with a_i between 0 and X.

[11] *PoA* 244; *PdA* 230–231.

[12] *PoA* 244; *PdA* 231.

All numerical constructs outside the number system (e.g. those, in decimal notation, of the form 10+5, 3·7, etc.) form a "problem" that "awaits" a solution, i.e. they must be *reduced* to the corresponding number in the system. Husserl regards systematic numbers as *normal forms* to which all the others must be reduced through the operations.[13] Calculating means *reducing a numerical construct to its normal form*, to the number of the system that corresponds to it. For example, in '49+17 = 66', '49+17' is one of the possible symbolic forms that serve to represent that given number, and '66', the result, is the normal form, the number of the system to which all other representations of that number are to be reduced.

Thus, according to Husserl, arithmetical operations are procedures to reduce complex numerical expressions to the corresponding number in normal form, and, *vice versa*, to construct complex expressions starting from numbers given in normal form. One hardly needs to emphasize the depth and "modernity" of this view.

Moreover, the number system presents a peculiar feature that deserves particular attention. On the one hand, the system produces all symbolic number concepts, using only the base (1, 2, ..., X) and the arithmetical operations of addition, multiplication and exponentiation. On the other hand, it produces for all number–constructs the corresponding notation using the signs '1', '2', ... 'X' and the signs for the operations of addition, multiplication, and exponentiation.[14] Husserl observes that if the signs are separated from their conceptual correlates, the symbolic aspect of the system keeps working *autonomously*. The system consists of two correlative structures, a conceptual and a signitive one. The *conceptual* structure is a way to generate new concepts by combining elementary concepts in accordance with certain laws. The *signitive* structure is a way of producing signs from signs according to pre-established rules, without any need to refer to their conceptual content.[15]

> Let us abstract from the signification of the numerals "1," "2," ..., "X," as well as from the significations of the notations for the operations of addition, multiplication, and exponentiation, and take them as totally arbitrary symbols without signification (as, for example, the counters in a game). Let us replace number definitions and operation rules which are the regular medium of systematic procedure, with corresponding, conventionally fixed formulas expressing the equivalences of sign combinations. One will then recognize that, in this way, there actually originates an independent system of symbols which permits the derivation of sign after sign in a uniform pattern without there ever turning up – nor could there ever, as such, turn up – other sign formations that appear in other circumstances, accompanying a conceptual process, as designations of the concepts here formed.

The point is, so Husserl, that calculating is *not* an *activity with concepts, but with signs*. Husserl's account of the signs-system as an autonomously functioning

---

[13]Webb 1980 observes that "Husserl's theory of calculation has some of the flavor of Church's calculi of λ-conversion: 'systematic numbers' (e.g. arabic numerals) result from a series of rule governed 'reductions' of 'unsystematic numbers' (terms compounded out of numerals with function symbols), also called 'symbolische Bildungen'" (25).

[14]Cp. Hartimo 2007, 288.

[15]*PoA* 251–252; *PdA* 237–238.

mechanism indicates that he understood very well the results of the process of transformation of algebra that, during the nineteenth century, led to the birth of abstract algebra. However, at a variance with the algebraists of logic, in particular with Boole, Husserl takes the symbols to always describe a well-determined reality, in other words, the objective referent is not eliminated. Husserl never forgets the conceptual basis of the signs-system and maintains that there must be a conceptual content that guarantees the meaningfulness of mechanical operations with signs, even if *we can manipulate the signs* without attending to the concepts that are their meanings.[16] What happens on the level of signs is legitimated by what happens on the level of proper concepts, and for Husserl this depends on the fact that the signs are construed as expressing concepts and that the system of concepts, as well as the operations on them, are consistent (*konsequent*). It is the exactness of the conceptual operations that determines the exactness of the system of signs, even if the latter can then be considered completely independently from its substratum.

5.3 Husserl's Concept of Computation in the *Philosophy of Arithmetic*

In the last chapter of the *Philosophy of Arithmetic* ("*The Logical Sources of Arithmetic*"),[17] Husserl distinguishes *arithmetic* from *the art of computation* (*Rechenkunst*, computistic). The latter is conceived as the technical side of the methods of the former, on the basis of the distinction between concepts and signs and, moreover, on the basis of the peculiar property of the autonomous functioning of the system of signs.

Arithmetic is characterized as the science of *numbers*, or, more precisely, as the science of *numerical relations*, as it does not have as object the specific properties of single numbers, but numbers in so far as they are identified through certain relations or complexes of relations with other numbers. The task of suitably characterizing the art of computation is more delicate and complex. It is necessary, first, to consider the various meanings of the concept of *computation* (*rechnen*).

Husserl distinguishes between three meanings of 'calculation', one broader and two narrower ones. In the *first*, broader sense, calculating is *any kind of derivation of numbers from numbers*. Within this characterization, the *method of derivation* is not qualified, since it can be *conceptual-abstract* as well as *signitive-concrete*. If it is conceptual, new number concepts are derived from (proper or symbolic) number concepts on the basis of certain operations *that are conceptual as well*. The signs, in this sense of calculation, have only a subsidiary role.

In a *second*, more restricted sense of calculation, the method of derivation is required to be sensuous, i.e. it operates on signs according to pre-established rules.

[16] See e.g. Tieszen 1996, 312–313 & 2005, 9.

[17] *PoA* 271–299; *PdA* 256–286.

Under this reading, "calculating" means "deriving numbers from numbers through operations on sensuous signs". In comparison with this sense of calculation, the *former*, that is, the method of conceptual derivation, is *less general*, and, moreover, makes calculation a long and complex operation. By contrast, the method of sensuous signs, besides being more practical and more functional, is also more general and *all-inclusive*, i.e. *universally applicable* and potentially able to solve any conceivable problem. The method of sensuous signs seems therefore to be "*the* logical method for arithmetic." Ultimately, it makes the method of conceptual derivation superfluous. However, so Husserl, "[m]ost researchers – guided by the general prejudice that every scientific methodology operates with the respective intended concepts – have also held the arithmetical operations to be abstract-conceptual, in spite of all clear indications."[18] He maintains that for a general acknowledgment of the view that calculating is a signitive activity we still lack a well–founded epistemological account of symbolic methods of knowledge.

Calculating can also be interpreted in a *third* sense, different from these other two, i.e. as deriving signs from signs following formal rules. In Husserl's words[19]:

> One can ... conceive of calculation as *any rule-governed mode of derivation of signs from signs within any algorithmic sign-system according to the "laws" – or better: the conventions – for combination, separation, and transformation peculiar to that system.*

Here, at a variance with the previous sense of calculating, not only the algorithmic computation is abstract-symbolic but also the manipulated entities are nothing but signs.

With this third sense of 'calculation' we have obtained a true and proper characterization of the *formal-algorithmic* method. Hereby the notion of algorithm is bound up with that of a 'mechanical process.[20] An algorithm is in fact a mechanical procedure that operates on configurations of (sensuous) signs according to certain formal rules. Calculating, now, means deriving signs *from signs* according to pre-established formal rules. Husserl attributes great importance to this third concept of 'calculation', since it makes possible an exact separation of the various *"logical" moments* that are involved in every derivation of numerals from numerals. "Each solution obviously decomposes into one calculational part and two conceptual parts: *conversion of the initial thoughts into signs – calculation – and conversion of the resulting signs back into thoughts*."[21]

---

[18] *PoA* 272; *PdA* 257.

[19] *PoA* 273; *PdA* 258 (italics in the original). Cp. Hartimo 2007, 289 f.

[20] We borrow here terminology from Webb 1980. As Webb 1980 puts it: "Husserl ... attempted a complete development of the algorithmic conception of arithmetic, which required "*die logische Untersuchung des arithmetischen Algorithmus*". The notion of algorithm, Husserl felt, had to be bound up with that of a 'mechanical process'" (24–25).

[21] *PoA* 273; *PdA* 258 (italics in the original). Husserl does not fail to stress how important a good *choice of the system of signs* is, in terms of efficiency, for all three of these phases (encoding – calculation – decoding) of the solution of a problem.

The importance of this concept of computation is determined also by the fact that it is sensitive to the developments in logical-abstract algebra. The system of signs is indeed *polyvalent*: it allows a uniform treatment of "similar" situations or, equivalently, a single computational system permits the mastery of more than one conceptual system[22]:

> It is a fact highly significant for the deeper understanding of mathematics that one and the same system of symbols can serve in *several* conceptual systems which, different as to their content, exhibit analogies solely in their structural form. They are then, as we say, governed by the same calculational system.

It is important to stress again that, according to Husserl, and at a variance with Boole, the abstract algorithm of computation always has a conceptual ground. It is the correct functioning of the underlying number system that guarantees the correct functioning of the algorithm.

5.4 Husserl and Boole

As we have seen in the above Section Husserl arrives in chapter XII of the *Philosophy of Arithmetic* at a characterization of the notion of *computation* (*Rechnen*) or *calculation* as "any rule-governed mode of derivation of signs from signs within any algorithmic sign-system according to the 'laws' – or better: the conventions – for combination, separation and transformation peculiar to that system."[23]

What is really interesting in this characterization lies mainly in the explicit acknowledgement of the *generality of the concept* under scrutiny, more precisely, in the fact that the notion of computation is completely uncoupled from that of *number*, and more generally from that of *quantity*, in which the whole issue had originated. In Husserl's words, "there are higher logical interests than those of *arithmetica numerosa* (with which we currently have to do) which require this delimitation of the concept."[24]

Thanks to the generality of this characterization it is possible to analyze and represent the structure of *any* 'problem-solving' process – and not just of those of numerical kind, but independently from the specific nature of the domain in which we are operating – in three distinct and sequential moments, the first and third of which are *conceptual*, while the second is *purely formal-algorithmic*:

[22] *PoA* 273; *PdA* 258 (italics in the original).

[23] *PoA* 273; *PdA* 258.

[24] loc. cit.

formal encoding of the problem
("conversion of the initial thoughts into signs")
↓
calculation
↓
solution as decoding of the result of the calculation
("conversion of the resulting signs back into thoughts")

A careful reader cannot fail to notice the similarity (with just one important difference, to which we will return below) of this abstract representation of the problem-solving processes with the analysis of "symbolic reasoning" proposed by George Boole in the *Laws of Thought* (1854). Diagrammatically (as can be seen from chapter V, esp. §4), according to Boole a (correct) *symbolic reasoning* can be represented by the composition of three moments or steps A, B, C:

- The first one (A) consists in the encoding of the data in symbols: these must have a well-defined *interpretation*, and their laws of combination must be determined by that interpretation.
- The second one (B) consists in the application of *symbolic-formal processes* conforming to the laws of combination, and it is completely independent from the requirements of interpretability: Boole explicitly allows the *possibility of non-interpretability of the intermediate steps in calculation.*
- The third one (C) finally consists in the *interpretation* (decoding) of the result of the symbolic process on the basis of the 'coding system' chosen for the symbolization of the data.

Admittedly, in chapter XIII of the *Philosophy of Arithmetic* Boole is not mentioned (indeed, he is not mentioned at all in the entire work). However, in Husserl's lecture "*Über die neueren Forschungen zur deduktiven Logik*" of 1895, we find an entire section (circa 20 pages) devoted to Boole. The main part of this section (which is the one we want to consider here) is focused on the notion of *calculus*.

Husserl's considerations begin with the observation that the real strength of Boole's work does not reside in his proposed logical analysis of language (on the contrary: "he has been less concerned about a possibly complete analysis of the forms of judgment"[25]), but rather in the development of a *logical calculus, i.e. the development of logic as a computational discipline*:

> He took ... and accepted from his predecessors, from Hamilton and in part also from De Morgan, what he could use. Use for what? The answer is: for the development of calculus (*für die kalkulatorische Entwicklung*). For this is the goal that he had set for himself from the beginning and that he pursued with powerful genius and achieved with full certainty:

[25] *LV 96*, App. 305.

> establishing formal logic as a mathematical science. As arithmetic is the computational discipline about numbers, so formal logic is to be developed as the computational discipline about concepts and states of affairs (*kalkulatorische Disziplin von den Begriffen und Sachverhalten*).[26]

At this point, Husserl puts Boole aside (returning to him only in the final pages, where he summarily describes Boole's class calculus) and develops an articulated and very interesting discussion of the concept of *calculus*, using arithmetic as a paradigm.

Of the four elementary arithmetical operations on natural numbers (greater than zero), addition and multiplication are total, i.e. defined for every natural number, while subtraction and division are partial ("subtraction: a−b. Defined, if $a > b$; and then also univocal"[27]). Consider the following nine fundamental laws (*Fundamentale Gesetze* or *Grundgesetze*):

(1) $a + b = b + a$
(2) $(a + b) + c = a + (b + c)$
(3) $(a - b) + b = a$
(4) *if* $a + c = b + c$ *then* $a = b$
(5) $ab = ba$
(6) $a(bc) = (ab)c$
(7) $a(b + c) = ab + ac$
(8) $(a/b)b = a$
(9) *if* $cb = db$ *then* $c = d$.

The laws of addition, subtraction, multiplication, division, *are grounded* in the *concept* of number; and the nine laws listed above *derive a priori from these concepts*. A law is grounded in a concept if and only if it is *analytically included in that concept*. All remaining true propositions of arithmetic that exclusively concern these four operations of the calculus are *purely formal* deductions from the principles (1)–(9). More exactly, they can *be proved without any appeal to the concepts* (of number, of the four operations), by manipulating the "sensuous expressions on paper" corresponding to them in accordance with the *rules of symbolic transformation* corresponding to the principles (1)–(9). For example, principle (1) corresponds to the rule of transformation that allows the equivalent substitution of every sign construction of the form 'a + b' with the construction 'b + a'.

> To prove any further proposition I do not need at all to fall back on the concept of number or on the concepts of addition, multiplication, etc. The proof rather consists entirely of steps in which we do not need to do anything but apply one of the nine propositions, i.e. we have to do nothing but subsume. The nine basic propositions (*Grundsätze*), however, are formally independent from each other... In order to grasp (*einsehen*) them we must appeal again to

[26] *LV 96*, App. 306.
[27] *LV 96*, App. 307.

> the concepts of number or addition etc. We cannot represent any of these propositions as a special case of the others or derive it from them through transformations.[28]

Calculating is a *non-conceptual* process that operates on complexes of signs and proceeds formally according to pre-established rules of transformation. And it is clear, Husserl claims, that there is no *a priori* reason to limit calculating to the numerical domain: in general it is possible to apply the calculus, an algorithmic system, to any domain that allows an 'algebraic' structuring[29]:

> It is immediately clear that there is no a priori reason why calculating should be restricted to the arithmetical domain. Wherever we find a domain of concepts in which obtain analogous relations as in that of arithmetic, i.e. wherever we can find uniform ways of constructing new concepts from given ones, in such a way that the results of the constructions can always serve as elements for new constructions, and where there is a limited number of laws for these kinds of construction, an infinite manifold of pure theorems (*Folgesätze*) is deducible from the axioms (*Grundsätze*), and this in the way of a purely formal deduction. And the computational way of proceeding will also be possible that makes falling back on the concepts superfluous and relies only on the external forms of the process.

Here we have in a nutshell the essence of calculation[30]:

> What is characteristic of calculation? Nothing but the fact that it is a procedure for deducing equivalent propositions from certain given propositions in a research domain, without falling back on the specific concepts and relations. How is this possible? What are we dealing with, if not concepts and relations? The answer is: the concepts are thought by means of certain terms, the relations by means of corresponding connecting signs. When we calculate in arithmetic, we only care about the signs and the rules of their connection.

This characterization is perfectly in line with what emerged at the end of the discussion on computation in chapter XIII of the *Philosophy of Arithmetic*. But in Husserl's reflections on Boole's approach we also find further interesting remarks, two of which are worth considering in the present context.

Husserl underscores repeatedly, and with abundant examples, the importance of generality and flexibility of systems of calculation for their application: an algorithmic system obtained – as the one considered at the beginning – by *abstraction* from a specific conceptual domain will also be interpretable in *different* though structurally similar conceptual domains. This results into obvious "savings in deductive labour": a formal proposition, once deduced by calculation, translates into a true conceptual proposition in every domain that falls under the algorithm. But the most significant aspect here is that Husserl explicitly takes into account also the process opposite to one exemplified above (schematically: determinate conceptual domain → abstraction → algorithmic system), i.e. the process (which is the essence of the "axiomatic revolution" of the late nineteenth century) that consists in *starting* from the constitution of an algorithmic system, and *then to* look for possible *interpretations* (possible *models*).[31]

---

[28] *LV 96*, App. 308.

[29] *LV 96*, App. 312.

[30] *LV 96*, App. 309.

[31] *LV 96*, App. 314.

> But also another way is open. One develops the algorithm for itself and says: every conceptual domain which is such that we can designate its basic concepts (*Grundbegriffe*) by the basic signs, its concepts of combination (*Verbindung*) by the signs for combinations of the algorithm – obviously to the basic rules (*Grundregeln*) in the domain there correspond basic laws (*Grundgesetze*) in the algorithm – falls under the algorithm with respect to all of its deductions.

While in the case of algorithmic systems obtained by abstraction we have to do with *synthetic a priori* laws, founded *in given concepts*, in this second case the laws are *purely analytic*, and the concept is obtained through reflection on the form of the law.

Let us proceed to the second point that we deem relevant here. Suppose we set up an algorithm (i.e. a system of signs plus rules of transformation): do the signs still have a meaning? And if so, what is it? Husserl's answer to this question is affirmative: the signs do (still) have a meaning in the algorithm, namely an *operational* meaning, a "*Spielbedeutung*", which is determined by the complex of formal rules that govern its manipulation.[32]

> Suppose a kind of given signs is set and memorized and so is a certain number of rules that like the rules of a game determine how we are allowed to operate with the signs, in such a way that every other way of proceeding is considered unacceptable. Then an arbitrary connection of signs can, on the basis of the rules, be replaced by various equivalent connections of signs. And a derivation is correct if all of its steps are in accordance with the rules, i.e. no step is taken which does not have a justification by a simple subsumption under one of the rules. Hence if I consider the signs in this way by themselves, they are not merely doodles on the paper, they clearly have a certain meaning. What, then, is their meaning? It is no longer the corresponding arithmetical meaning, because I have wholly abstracted from it. Clearly the meaning now lies in the rules of the game. It is exactly like as in the game of chess: the bishop, castle etc. Now I maintain: all calculating consists in the fact that the original concepts, the concepts of number and the concepts of relation and connection belonging to them, are replaced by their mere symbols and these are now considered only as such purely conventional game-concepts. The game-meaning (*Spielbedeutung*) of these symbols then lies in certain rules which are nothing but the exact counterparts of the fundamental laws to which all arithmetical deduction can be reduced by mere subsumption.

Incidentally, there is a stark contrast here with Frege's attitude in his paper *Über formale Theorien der Arithmetik*[33](1885). In the course of his criticism of Heine's and Thomae's theory of irrational numbers in the second volume of his later *Grundgesetze*[34] Frege concedes that the *Begriffschrift*, too, can be conceived as a *Spiel* (§ 90) but he doubts that a theory of *Rechenspiele* is possible (§ 93).

Husserl appends a brief description of the Boolean calculus to his general reflections on the concept of calculus. This calculus is presented as a calculus for the domain of classes in general: "Boole originally constructs his calculus as a calculus of classes. As arithmetic is the calculus for the field of numbers (*Anzahlgebiet*),

---

[32] *LV 96*, App. 310.

[33] Frege 1985.

[34] Frege 1903.

so the Boolean calculus is the calculus for the field of classes in general (*im Allgemeinen*)." While this expository part is not really worth discussing a remark should be made on Husserl's (and not only on Husserl's) fundamental critique of Boole, as regards the *correctness* of the Boolean method. One can summarize the entire discussion as follows. In order to avoid losing the generality of the calculus, Boole explicitly admits (and freely avails himself of) the possibility that not all the steps in the symbolic-formal process that leads from the encoded data to the to-be-decoded result are *interpretable*; in other words, he does not require that there be a parallelism between the conceptual and the symbolic-algorithmic level *along the entire symbolic process*: the important factor is that this parallelism exists at the beginning (*input*) and at end (*output*). Husserl's position appears to be the opposite, i.e. that the parallelism between the two levels should be *constant*: "Finally, to every principle corresponds a certain rule for operating with the signs, and every derived proposition is obtained by mere stepwise subsumption under these rules. Thus there is a one-to-one corresponding parallelism between the game-system and its rules and the number-system and its laws. Hence there is no mechanically-symbolically derivable proposition that does not have its counterpart in the domain of arithmetic."[35] Similar considerations are to be found in "*On the concept of the operation*", in the context of Husserl's discussion of algorithmic systems[36]:

> Whoever has ... attained clarity about the algorithmic methods that run precisely parallel to the ... objective [*sachlichen*] methods (operating on the concepts themselves ...) will see that all immediate equivalences that mediate (*vermitteln*) in the objective methods must have their counterpart in formulae that mediate in the algorithmic methods.

Husserl's conclusion is that in Boole the strength of the idea of extending the notion of calculation beyond the sphere of quantity is coupled with an intrinsic weakness at the level of the foundations of the computational process.[37]

> Hence Boole's method must have appeared like shadow boxing (*Spiegelfechterei*, i.e. a sham), where, however, one had to register the unexplainable miracle that Boole's calculations always led to correct results: somebody who trusts the Boolean method and solves a logical task by calculatorial means, would in fact find a true solution, while one would have expected that a meaningless method would also deliver meaningless or at least false results. However, Boole himself did not have an entirely clear conception of the reasons for the validity of his method. In his case we are dealing with a brilliant intuition rather than with a conceptual insight. The logical principles[38] of the calculatorial method remained completely precluded to him as well as to later researchers.

---

[35] *LV 96*, App. 311.

[36] *PdA* App. 414; *PoA* 391.

[37] *LV 96*, App. 322–323.

[38] Husserl means here: "the epistemological ground of the calculatorial method".

References

G. Boole, *The Mathematical Analysis of Logic* (Macmillan & Co., Cambridge, 1847)

G. Boole, *An Investigation of the Laws of Thought* (Macmillan & Co., Cambridge, 1854)

A. Cantini, *I fondamenti della matematica da Dedekind a Tarski* (Loescher, Torino, 1979)

E. Casari, *La filosofia della matematica del '900* (Sansoni, Firenze, 1973)

G. Frege, Über formale Theorien der Arithmetik (1885), in G. Frege, *Kleine Schriften*, ed. by I. Angelelli (Olms, Hildesheim, 1990), pp. 103–111

G. Frege, *Grundgesetze der Arithmetik, begriffschriftlich abgeleitet,* Band 1, Pohle, Jena, 1893; Band 2, Pohle, Jena, 1903

M.H. Hartimo, Towards completeness: Husserl on theories of manifolds 1890 – 1901. Synthese. **156**, 281–310 (2007)

E. Husserl, [*LV 96*] *Logik. Vorlesung 1896*, ed. by E. Schuhmann, *Husserliana, Materialienbände* vol. I (Kluwer, Dordrecht 2001)

E. Husserl, *Philosophie der Arithmetik. Logische und psychologische Untersuchungen*, Bd. I, Pfeffer, Halle a.d.S. 1891. Critical Edition: [*PdA*] *Philosophie der Arithmetik. Mit ergänzenden Texten (1890–1901)*, ed. by L. Eley. HGW XII, 1970. English translation: [*PoA*] *Philosophy of Arithmetic, Psychological and Logical Investigations with Supplementary texts from 1887 – 1901*, ed. by D. Willard (Kluwer, Dordrecht, 2003)

E. Husserl, [HGW] *Husserliana, Gesammelte Werke,* I–XXVI: Martinus Nijhoff, Den Haag, 1950–; XXVII–XXXVII: Kluwer, Dordrecht, 1989–; XXXVIII–: Springer, New York, 2005–

R. Tieszen, Logicism, Impredicativity, Formalism. Some remarks on Poincaré and Husserl (1996), in Tieszen (2005), pp. 294–313

R. Tieszen, *Phenomenology, Logic and the Philosophy of Mathematics* (Cambridge University Press, Cambridge, 2005)

J. Webb, *Mechanism, Mentalism and Mathematics* (Reidel, Dordrecht, 1980)

Chapter 6
Husserl and Schröder

Stefania Centrone and Pierluigi Minari

Abstract We aim at clarifying to what extent the work of the German mathematician Ernst Schröder (1841–1902) on the algebra of logic is taken into consideration and rehashed in the work of the early Husserl, focusing on Husserl's 1891 *Review* of the first volume of Schröder's monumental *Vorlesungen über die Algebra der Logik* (*Exakte Logik*) and on Husserl's text *Der Folgerungskalkül und die Inhaltslogik* written in the same year. We will try to show how and why Husserl, while praising Schröder's calculus, strongly criticizes Schröder's attempt at a philosophical clarification and justification of it.

Keywords Algebra of logic • Calculating vs Inferring • Extensional vs Intensional logic

6.1 Introduction

Before turning to philosophy under the influence of Franz Brentano, the father of phenomenology Edmund Husserl had studied mathematics in Berlin with Kronecker, Kummer and Weierstrass, and had been Weierstrass' assistant, working with him until about 1883. At least from the publication of Husserl's *Philosophy of* Arithmetic along with some unpublished early texts from the *Nachlaß* in Volume XII of the *Husserliana* edition in 1975 and from Dallas Willard's translation of the same[1] Husserl's early, pre-phenomenological works have increasingly attracted the

[1]Husserl 1891b. Henceforth cited as *PdA* from vol. XII of the *Husserliana* Edition (Husserl 1950+; abbreviation: *Hua*). English translation cited as *PoA*. Responsibility for translations from German is ours, even when we refer to, benefit from, or simply echo published translations.

S. Centrone (✉)
Institut für Philosophie, Carl von Ossietzky Universität Oldenburg, Oldenburg, Germany
e-mail: stefania.centrone@uni-oldenburg.de

P. Minari
Department of Letters and Philosophy, Università di Firenze, Florence, Italy
e-mail: pierluigi.minari@unifi.it

S. Centrone (ed.), *Essays on Husserl's Logic and Philosophy of Mathematics*, Synthese Library 384, DOI 10.1007/978-94-024-1132-4_6

attention of many scholars.[2] It has been stressed, i.a., how Husserl's early writings are characterized by the project of a substantial mathematization of logic and how Husserl enters in a very lively and stimulating debate with important interlocutors, some of whom made definitive contributions to the development of formal logic as an autonomous discipline in the last two centuries, in particular with the algebraists of logic as well as with Bernard Bolzano, Gottlob Frege and David Hilbert.

The present paper pursues the goal of clarifying to what extent the work of the German mathematician Ernst Schröder (1841–1902) on the algebra of logic is taken into consideration and rehashed in the work of the early Husserl, focusing on Husserl's 1891 *Review*[3] of the first volume of Schröder's monumental *Vorlesungen über die Algebra der Logik* (*Exakte Logik*)[4] and on Husserl's text *Der Folgerungskalkül und die Inhaltslogik*[5] written in the same year of *RevS* and first published in the *Vierteljahrschrift für Wissenschaftliche Philosophie*. We will try to show how and why Husserl, while praising Schröder's calculus, strongly criticizes Schröder's attempt at a philosophical clarification and justification of it.

Schröder's *Vorlesungen* consist of three bulky volumes, preceded by a *Foreword* and a long *Introduction*. The ensuing 14 *Lectures* of the first volume, the one Husserl is reviewing in *RevS*, deal, roughly speaking, with the calculus of classes; more precisely, they are devoted to the elaboration and the formal investigation of an axiomatic calculus (characterizing boolean algebras, as we might now say) having the former calculus as the privileged interpretation. The second volume (in two parts, the second of which was published posthumously by Karl Müller, using material from the *Nachlass*) is concerned, by and large, with the syllogistics and the propositional logic, treated from the algebraic point of view set up in the first volume. The third volume (whose planned second part never appeared), the one containing Schröder's major contributions, is devoted to the algebra of relations.

Husserl's criticism in *RevS* is mostly directed to Schröder's philosophical justifications of the calculus, in particular his conception of the "essence and value" of the calculus and his views on the relationship between old and modern logic (the latter being intended by both authors as the logic after Boole, cp. *RevS* 4), and focuses therefore especially on the *Foreword*,[6] the *Introduction*[7] and the *1st Lecture*[8] of Schröder's masterwork. This is the part of *RevS* we will concentrate here. Notice however that Husserl's *Review* provides also an accurate exposition of the content of

---

[2]Cp. at least Eley's Introduction to *Hua XII* (1970); Rang's Introduction to *Hua XXII* (1979); Miller 1982; Willard 1984; Rosado-Haddock 1997; Majer 1997; Da Silva 2000; Ortiz Hill & Rosado-Haddock 2000; Tieszen 2005, 314–336; Hartimo 2007; Ierna 2005, 2006 & 2012; Centrone 2010a.

[3]Husserl 1891a (henceforth cited as *RevS*). English translation in Husserl 1994, 52–91.

[4]Volume I: Schröder 1890 (henceforth cited as *EL I*); Volume II part 1: Schröder 1891 (henceforth cited as *EL II*); Volume II part 2 (*posth.*): Schröder 1905; Volume III part 1: Schröder 1895.

[5]Husserl 1891c. Henceforth cited as *FK*.

[6]*EL I*, iii-iv; *RevS* 3–4.

[7]*EL I*, 1–125; *RevS* 4–23.

[8]*EL I*, 126–167; *RevS* 23–29.

the remaining lectures,[9] and that some criticism is also directed to certain specific, more technical points (more on this in Sect. 6.6 below).

The title *Vorlesungen über die Algebra der Logik*, so Schröder in the *Foreword*, indicates that the object of the investigation is *deductive logic*. More precisely, Schröder calls "exact logic" the computational treatment of deductive logic[10] and claims that such an exact logic reflects the laws of correct thinking (*folgerichtiges Denken*).[11] He sets himself as a goal to present a completely reformed deductive logic in form of an *algebra of logic*.[12] At the background is the work of George Boole and Charles S. Peirce—to this tradition Husserl refers in *RevS* as "extensional logic" (*Umfangslogik*) or "logic of classes."[13]

The *Introduction* is divided into three parts, which bear the following titles:

(A) "Preliminary considerations on the nature and the delimitation of the problem to be solved, with reference to induction, deduction, contradiction and sound reasoning. The rational subject, his representations and the things. (*Vorbetrachtungen über Charakter und Begrenzung der zu lösenden Aufgabe mit Bemerkung über Induktion, Deduktion, Widerspruch und folgerichtiges Denken. Denkendes Subjeckt, seine Vorstellungen und die Dinge.*)"
(B) "Preliminary considerations on signs and names. (*Vorbetrachtungen über Zeichen und Namen.*)"
(C) "On concepts, classification, definition and categories, pasigraphy, logic of content or logic of extension? On judgements, inferences and their correctness. Why an algebra of logic. (*Über Begriffe. Einteilung, Definition und Kategorien, Pasigraphie. Logik des Inhaltes oder des Umfanges? Über Urteile, Schlüsse und deren Folgerichtigkeit. Warum Algebra der Logik.*)"

Husserl's discussion and criticism relative to each of the above items (A) - (C) are presented and evaluated in Sects. 6.3, 6.4 and 6.5 of our paper, in the order, preceded by a cursory exposition of Schröder's calculus in Sect. 6.2. Finally, Sect. 6.6 is about some more technical points raised by Husserl, and by Frege as well in his *Kritische Beleuchtung einiger Punkte in E. Schröders Vorlesungen über die Algebra der Logik,*[14] concerning Schröder's calculus.

---

[9] *RevS* 30–31 on Schröder's *2st Lecture* (*EL I*, 168,184,188); *RevS* 31–34 on Schröder's *3rd Lecture* (*EL I*, 196,199,202,204,207,211); *RevS* 34–37 on Schröder's *4th Lecture* (*EL I*, 241–242,245–248); *RevS* 37 on Schröder's *5th Lecture* (*EL I*, 254–281); *RevS* 37–38 on Schröder's *6th Lecture* (*EL I*, 282–298); *RevS* 38–39 on Schröder's *7th Lecture* (*EL I*, 299–341); *RevS* 39–40 on Schröder's *Lectures 8–11* (*EL I*, 342–477); *RevS* 40–42 on Schröder's *12th Lecture* (*EL I*, 378–520); *RevS* 42–43 on Schröder's *13th Lecture* (*EL I*, 521–558); *RevS* 43 on Schröder's *14th Lecture* (*EL I*, 558–592).

[10] *EL I*, iii; *RevS* 3.

[11] *EL I*, 4; *RevS* 4.

[12] *EL I*, 118ff.; *RevS* 3.

[13] See, for instance, *RevS* 4, 17, 19.

[14] Frege 1895.

6.2 Schröder's "Identischer Kalkul"

The calculus presented in the first volume of the *Vorlesungen* (which Schröder calls "*identischer Kalkul*," or also "*Gebietekalkul*") is set up gradually, along four stages. The starting point is the 'base' calculus exposed in the *2nd Lecture*, to which further postulates are added in the *3rd*, the *6th* and the *7th Lecture*. It eventually amounts to a fully axiomatic characterization—indeed the first (non equational) complete definition—of the notion of *Boolean Algebra*, as we would now say. Here is a very quick and rough overview of the system.[15]

The most primitive notion of the calculus is the binary relation of *subsumption* (*Subsumtion*), denoted by the symbol "€." Intuitively, it corresponds to the copula "is" of the categorical statements. Schröder talks of "judgements" and writes "*Subjekt* € *Prädikat*."[16] This form covers both the relation of *subordination* (*Unterordnung*) and the relation of *extensional equality* between classes. The latter is exemplified by statements like e.g. "*Kochsalz* = *Chlornatrium*." The former is intended in a wide sense, as usual in pre-fregean logic, thus not only as the relation of inclusion between concept-extensions, as exemplified by "*Gold* ⊂ *Metall*", but also as the relation of membership between an individual and a concept under which this individual falls, as exemplified e.g. by "*Erde* ⊂ *Planet*."[17]

The basic properties of the relation € are singled out in the *2nd Lecture*, in the abstract setting of a "calculus with letters" (*Buchstabenrechnung*[18]). The letters "*a*", "*b*", "*c*", ... are generically referred to as staying for "domains" (*Gebiete*) of a "multiplicity of elements" (*Mannigfaltigkeit von Elementen*). The formulas (*Formeln*) are of the form "*a* € *b*" and admit diverse "applications" (*Anwendungen*) according to a selected *interpretation*—e.g. in terms of classes (*Klassen*), or in terms of concepts (*Begriffe*), of statements (*Urteile*), of inferences (*Schlüsse*) among others, "kurzum, bei geeigneter Auslegung der Zeichen so ziemlich alles Denkmögliche," as Schröder expresses himself.[19]

To this aim two axioms are first of all stated:

[*Prinzip I*]: € is reflexive, in symbols: *a* € *a*;
[*Prinzip II*]: € is transitive, that is: if *a* € *b* and *b* € *c* then also *a* € *c*.

The relation of identity (*Identität*), denoted by "=", is not taken as primitive, being instead explicitly *defined* as mutual subsumption:

---

[15] See also (among others) Chapter 7 of Brady 2000, and Peckhaus 1994.

[16] *EL I*, 132.

[17] *EL I*, 131. Husserl will adopt Schröder's symbol "€" in his own lecture course on logic held at the University of Halle in Summer 1896 (known as *Logikvorlesung* 1896, in: Husserl 2001; henceforth cited as *LV'96*), while presenting a "Theory of conceptual and propositional inferences" (*LV'96* 254–264). Cp. Centrone 2010a, 128–145.

[18] *EL I*, 157.

[19] *EL I*, 160.

$a = b$ if and only if $a \in b$ and $b \in a$.

Next, on the basis of this definition it is proved, using the two Principles I and II, that identity is an equivalence relation and in fact a congruence with respect to $\in$. Finally, two constants *0* (*der identischen Null*) and *1* (*der identischen Eins*) are introduced (oddly enough, through a Definition!), satisfying for an arbitrary element a the conditions

$0 \in a$, resp. $a \in 1$,

and their uniqueness is proved.

All in all, this first stage of the elaboration of the calculus provides a neat axiomatization (in the signature $\in$, *0*, *1*, as we would say) of those structures now currently called *bounded partial orderings*.[20]

In the *3rd Lecture* the binary operations of multiplication (*identische Multiplikation*) and addition (*identische Addition*) are introduced (again, officially through a Definition!) by postulating the existence, for arbitrary elements a and b, of a *greatest lower bound* and of a *least upper bound* of a and b with respect to $\in$, that is the existence of two elements—denoted respectively by ab and $a + b$—which satisfy the conditions:

$c \in ab$ if and only if $c \in a$ and $c \in b$,
$a + b \in c$ if and only if $a \in c$ and $b \in c$,

for any element c. The uniqueness of such elements is then proved.[21]

All in all, this second stage of the elaboration of the calculus provides an axiomatization of the notion of a *bounded lattice*, as we are used to say after Birkhoff.[22] The basic properties of such kind of structures, as well as the privileged model, the algebra of classes w.r. to inclusion, intersection and union, are thoroughly investigated by Schröder in the *4th* and the *5th Lecture*.

The *6th Lecture* is devoted to Peirce's problem[23] concerning whether the *distributive* law

$a(b + c) = ab + ac$, or equivalently[24] $(a + b)(a + c) = a + (bc)$

is a consequence or not of what has been postulated up to this stage, that is of the (bounded) lattice axioms. More precisely, the question is whether

$a(b + c) \in ab + ac$, or dually $(a + b)(a + c) \in a + (bc)$

[20] A similar axiomatic characterization, as acknowledged by Schröder, is found in Peirce 1880, 24.

[21] *EL I*, 210.

[22] Cp. Birkhoff 1940. In fact the characterization of *infs* and *sups* via inclusion goes back again to Peirce 1880, 32–33.

[23] Raised in Peirce 1880.

[24] In a lattice, each of the two identities entails the other one. Indeed, Schröder is fully aware of the duality between multiplication and addition, and makes extensive and systematic use of this fact throughout his treatise.

is derivable (the converse inclusions are so).[25] Schröder is actually able to solve the question *negatively* by skilfully providing a suitable counter-example.[26]

Yet the distributive laws are valid in the algebra of classes, the privileged model of the calculus. This prompts a third extension of the calculus through the postulation of distributivity. Or, more precisely, of a *restricted* form of distributivity

[*Prinzip* $III_\times$]: if $bc = 0$ then $a(b + c) \in ab + ac$,

amounting to the distribution of multiplication over *boolean* addition,[27] which is sufficient to get the full distributivity law in presence of complementation (see below).

The final stage of the construction of the calculus is found in the *7th Lecture*, and consist in the addition of negation (*Negation/Verneinung*), that is *boolean complementation*. First of all Schröder introduces, by definition, the notion of negation: given a domain a, a negation of a is any domain a' (this is his own notation) satisfying both the conditions

$a\,a' = 0$ and $a + a' = 1$.

Next, he proves *Hülfstheorem* 29,[28] indeed by making essential use of the *restricted* distributivity law (Principle $III_\times$ above), that any domain a has *at most one* negation. And finally, this time by an explicit axiom, he postulates

[*Postulate (3)*] For every domain a there is at least one negation a' of a (and so exactly one).

The final step is the verification that the full distributivity law(s) follow from their restricted form together with Postulate (3).[29]

Summing up, Schröder's full *identischer Kalkül* amounts to an elegant, *non equational* axiomatization of boolean algebras, conceived as bounded distributive lattices with complementation. The remaining *Lectures* (*8–14*) of the first volume are devoted to the derivation of further consequences of the addition of negation, and then to the study of the boolean polynomials and the method of expansion, to the question concerning the inversion of the operations of addition and multiplication, and to the comparison with the work of Peirce and other scholars as well.

---

[25] *EL I*, 280.

[26] *EL I*, 617 ff.

[27] According to Boole, the operation + of addition is defined *only* when the addends are disjoint.

[28] *EL I*, 300.

[29] *EL I*, 310.

6.3 "Is Calculating the Same as Inferring?"

Right at the beginning of Section (A) of the *Introduction*, Schröder claims that "*logic*, in the broader sense of the word, deals with all the rules, the following of which promotes the knowledge of the truth"; more exactly, logic tries to answer the question "*how* do we get knowledge, how do we get truth?"[30] It is natural to ask whether Schröder is claiming here that logic is a normative science that prescribes how we *ought* to think if we want to get knowledge, or whether he is regarding logic as a branch of epistemology, namely as the complex of procedures through which we get knowledge from knowledge. Since he lacks the distinction between mental acts-or-states and their contents,[31] he characterizes the object of logic as "reasoning in so far as it has knowledge as ... its ultimate purpose."[32] About 10 years later, in the first volume of his *Logical Investigations*, the *Prolegomena to Pure Logic*,[33] Husserl will characterize positions like that of Schröder as (variants of) logical psychologism: according to the more mature Husserl logical laws are to be sharply distinguished from the mental acts-or-states in which they are grasped. The former have all the features which are generally attributed to abstract objects: they have no spatial and temporal determinations, they cannot stand in causal relations, they are mind-independent, they are objects that *are there* before we become aware of them and that would be there even if they weren't experienced by any subject. Mental acts-or-states, on the contrary, are spatial and temporally located and are individually and generically bearer-dependent, i.e., they are dependent on their particular bearer as well as on the very fact that there are bearers.

At the time in which *RevS* appears—1891 is the year in which Husserl's youthful work, *The Philosophy of Arithmetic*, is published—Husserl attitude towards logic is still a psychologistic one. He proceeds from the assumption that psychology is the science from which logic has to hope for foundation and philosophical clarification.[34] Later on, in the *Prolegomena* he will tell us that, while writing the *Philosophy of Arithmetic*, he had become "more and more disquieted by doubts of principle, as to how reconcile ... the subjectivity of knowing and the objectivity of the content known."[35] He will write:

> Where one was concerned with questions as to the origin of mathematical concepts, or with the elaboration of those practical methods, which are indeed psychologically determined,

[30]"Die *Logik*, im weiteren Sinne des Wortes, beschäftigt sich mit all' den Regeln, durch deren Befolgung die Erkenntnis der Wahrheit gefördert wird.... Sie sucht die Frage zu beantworten: *wie* gewinnen wir Erkenntnisse, auf welchem Wege gelangen wir zur Wahrheit?" (*EL I*, 1).

[31]Cp. e.g. *EL I*, 126.

[32]The full text reads as follows: "Mithin, da Erfassen der Wahrheit ein Akt des Denkens ist, dürfen wir als Gegenstand der Logik überhaupt bezeichnen: das *Denken*, sofern es das *Erkennen* zum Endzweck hat" (*EL I*, 1).

[33]Husserl 1900, henceforth cited as *PR*. English translation cited as *PRe*.

[34]Cp. *PR* vi; *PRe* 42.

[35]*PR* vi; *PRe* 42.

> psychological analysis seemed to me to promote clearness and instruction. But once one had passed from the psychological connections of thinking, to the logical unity of the thought-content ..., no true continuity and unity could be established.[36]

To understand Husserl's criticism of Schröder in *RevS* it is important to stress that he conceives problems like those mentioned in the quoted passage, i.e. problems such as "the origin of mathematical concepts" or "the elaboration of practical methods," as "problems of *general logic* and epistemology."[37] He thinks (and he never changed his view about this) that logic in its broader sense has an epistemic function.[38] The point is that such a general logic with epistemic tasks *is not* and *cannot* be identified with *deductive logic* in Schröder's sense. For, Schröder takes his own calculus to be a theory of formal inferences. Also in this assumption he is, according to Husserl, wrong, but even if he were right, "deductive sciences not only infer (*schließen*): they also operate, they construct and they calculate."[39]

Schröder takes his own calculus to be "exact" or "deductive logic" presented in a purely formal manner. "One can hardly be more mistaken about his own aims"—so Husserl—"than [Schröder] is here."[40] Moreover, this misinterpretation is characteristic of the whole extensional logic.[41]

Husserl observes that Schröder's calculus (see Sect. 6.2 above) can be viewed, *prima facie*, as a "formal theory of inferences (*formale Schlußlehre*)."[42] Such view turns out to be plausible when one considers that (i) all propositions occurring in the calculus are analytical, and that (ii) the calculus permits to completely enumerate all propositions that are purely derivable for any given terms $S, P, \ldots Z$.[43] (i) and (ii) imply namely that only formal inferences, like "if all A are B and S is an A, then S is a B" or "if something is a S, then it is not a non-S" or either "if something is S and P, then it is either S or P" are at issue, but not also material inferences like "if a > b and b > c, then a > c", for inferences of the latter kind make no sense if one does not already know that the letters a, b, c stand for magnitudes. Under this reading the algebra of classes treats those inferences that are valid independently of the nature of the domain[44]:

---

[36] *PR* vii; *PRe* 42.

[37] *PR* vi; *PRe* 42.

[38] See for instance *RevS* 6: "Deductive Logic has as its object ... the logical activities employed in all deductive sciences. The latter make up the essence of its methods. (Die deduktive Logik geht ... auf die logischen Betätigungen, die in allen deduktiven Wissenschaften zur Geltung kommen und das Wesen ihrer Methoden ausmachen.)" Cp. also *PR*. 12–26, *PRe* 60–70.

[39] "Die deduktiven Wissenschaften schließen nicht allein: Sie operieren auch, sie konstruieren und rechnen" (*RevS* 6).

[40] "[M]ann kann sich kaum mehr über die eigenen Ziele täuschen, das der Verf. Es hier tut" (*RevS* 5).

[41] *RevS* 5.

[42] *RevS* 5.

[43] *RevS* 5.

[44] "Daher die Möglichkeit im konkreten Falle, die Termini durch allgemeine Zeichen zu ersetzen, nach dem korrespondierenden allgemeinen Schema zu schließen und erst dann die Zeichen durch

> Therefore it is possible to replace the terms by ... signs in a concrete instance, to infer after the corresponding general scheme and only then to substitute the signs by their particular meaning. Whatever is derivable from a system of premises on grounds of the mere form of the premises falls in the field—but also no more than this.

However, if one conceives of a deductive science as a complex of methods to obtain new truths proceeding from given truths, or from truths that have been already established—as Husserl does—it is clear that inferring (*Schließen*) is only *one* of such methods. Algorithms, for instance, and computational procedures in general are another kind of methods that are typically used in deductive sciences to extend knowledge.

Furthermore, Schröder's calculus not only is no deductive logic, but also no formal theory of inference in the proper sense. Husserl asks, and this is a fundamental question for him: "Is calculating the same as inferring (*Ist aber Rechnen ein Schließen*)?" His own answer is "No." Full stop. We have quoted above a passage from the *Prolegomena* in which Husserl, when talking of his own investigations in *The Philosophy of Arithmetic*, claims that psychological analysis turns out to be clear and instructive when it is a matter of explaining "the origin of mathematical concepts" and "the practical elaboration of computing methods"—see the first quotation in this Section. Psychological analysis teaches us that numerical algorithms have a natural origination in the process of counting.[45] Since we are able to directly intuit only the smallest numerals, say, 2, 3, and 4, because of the contingent restrictions of human nature, we substitute proper number-concepts with *symbolic* ones. The concept of something is *proper* when it is given to us, "as it were, in persona;"[46] it is *symbolic* when it presents its object through signs that univocally identify it.

Before going on we recall that Husserl, in his early writings, uses the words "concept" and "presentation" in quite the same way.[47] Indeed, he tacitly works with Leibniz's distinction between "cognitio intuitiva" and "cognitio caeca vel symbolica."[48] By the way, these ideas by Leibniz are well known to Schröder, who often refers to Leibniz in his *Introduction*,[49] if only through Trendelenburg's *Über Leibnizens Entwurf einer allgemeinen Charakteristik*.[50] Now, Husserl's rendering of Leibniz's thoughts about symbolic thinking can be presented as follows: We tend to reproduce simple concepts/presentations "as they were, in persona" and complex

---

ihre besondere Bedeutung zu ersetzen. Was immer aus irgendeinem System gegebener Prämissen auf Grund ihrer bloßen 'Form' gefolgert werden kann, fällt in diesen Bereich—aber auch nicht mehr" (*RevS* 6).

[45]Cp. *PoA* 258–267; *PdA* 244–252.

[46]Husserl 1901, henceforth cited as *LU* followed by *I*, *II*, ... or *VI*; English translation cited as *LI*. Here *LU VI*, §45, 144; *LI VI*, 785–786.

[47]Cp. Willard 1984, 26.

[48]See Leibniz 1684, 586.

[49]*EL I*, 38, 40. Cp. *RevS* 9.

[50]Cp. Trendelenburg 1856, 45.

concepts by means of signs. Compelled by the finitude of human nature we erect algorithms in a scientifically systematic way. Algorithms present a characteristic feature: they perfectly function also when unleashed from the conceptual correlates of the signs employed.

What is then the essence of algorithms? They are blind symbolic procedures set up with the purpose of making effectively performed inferences superfluous. They proceed "after mechanically-reproduced rules of conversion and transformation of signs (*nach mechanisch-repoduzierten Regeln der Umwandlung und Umsetzung von Zeichen*)."[51] Moreover, neither the encoding of the data in symbols, "*das in Gleichung setzen*," as Husserl puts it, nor the final interpretation of the result of the symbolic process are a kind of inferring[52]:

> The whole procedure spares and replaces a number of pure deductions, without being itself a deduction. The theory of all these intellectual operations that, notwithstanding the fact that they are not themselves of inferential kind, serve the deduction, belongs to the field of deductive logic, but not to that of pure inferences. The belief of achieving the aims of logic, by restricting itself to this narrow domain [of pure inferences], was a fundamental mistake of the old formal logic. And, algorithmic logic, its direct descendant, has inherited this mistake....

To summarize, the first objection Husserl addresses to Schröder in *RevS* is that the calculus *is not*, as the author of the *Vorlesungen* believes, a deductive logic or *the* deductive logic as such. It is therefore not what it seems to be *prima facie*, namely a theory of purely formal inferences; it is only a well-working method to spare deductions.[53]

6.4 Semiotic Issues

Section (B) of Schröder's *Introduction*, "On Signs and Names," sets out to tackle semiotic questions. Aimed at is the explanation of the function of signs for cognition. Schröder tries to present a plausible account of the semiotic relations and at the same time to clarify the "principles of denotation" ("*Prinzipien der Bezeichnung*"[54]) that should preside over the set up of the calculus. However, he is not very clear about the difference that subsists between giving an account of

---

[51]*RevS* 7. Cp. *PoA* 273; *PdA* 258.

[52]"Das ganze Verfahren erspart und ersetzt mannigfaltige reine Deduktionen, ist aber selbst keine solche. Die Theorie aller dieser Geistesbetätigungen, welche, obschon selbst keine folgernden, der Deduktion wissenschaftlicher Wahrheiten dienen, gehört offenbar in die deduktive Logik, aber sie gehört nicht in das Gebiet der reinen Folgerungen. Es war ein Hauptirrtum der alten formalen Logik, daß sie, auf dieses enge Gebiet sich einschränkend, dennoch glaubte, die Ziele der Logik zu erreichen. Die algorithmische Logik, in der Tat ihr direkter Abkömmling, hat diesen Irrtum übernommen ..." (*RevS* 7).

[53]Cp. *RevS* 7–9. Cp. *FK* 45, fn.

[54]*EL I*, 42.

the semiotic relations, i.e. of the relations between the natural language and the world, on the one side, and explaining the specific linguistic conditions that are to be observed when erecting an algorithm, on the other side. As Husserl rightly stresses, semiotic relations "are in nowise yet the principles that govern those peculiar notations that enter into algorithms."[55]

Schröder treats altogether under the heading "On Signs and Names" (i) the question of the "surrogative function"[56] of signs in rational thinking, and particularly in algorithms,[57] (ii) the question of a good choice of the signs when erecting algorithms,[58] and (iii) the issue of the semiotic relations.

As to (i), both Husserl and Schröder work with Leibniz' idea[59] that a relation subsists between a sign and the concept expressed by that sign,[60] say, for instance, between the numeral "1" and the concept *1*, in virtue of which the sign can stand for the concept under all circumstances, and both employ this idea to explain the essence of algorithms. Husserl calls this relation "equivalence" or "logical equivalence."[61] Schröder calls it sometimes "univocal correspondence."[62] It is precisely this relation that makes the surrogative fuction of signs achieved in all algorithms possible.

As to (ii), both Husserl and Schröder gather from the observation of functioning algorithms the conviction that the signs employed should be "suitably short," should avoid "circuitousness" and superfluous "complication,"[63] which is, at root, Leibniz' requirement that the *lingua characteristica* should be "easy to learn, to utter and to write."[64]

As to (iii), the semiotic relations are the target of Husserl's critique. What Schröder would like to say is that the semiotic relations are as follows. A sign *has* a meaning, *denotes* (in certain cases) an object and has the function to *arouse* in the hearer the representation of the object it is about. However, as we saw above, he misses a clear distinction between mental acts-or-states and their contents and confuses the presentation aroused by the sign in the mind with the meaning of the sign as well as this meaning itself with the denoted object. So, for instance, in the attempt to explain the relation between "sign and denoted object (*Zeichen und*

55 "Aber Prinzipien der Benennung—und darum allein dreht sich alles Weitere—sind noch keineswegs die Prinzipien derjenigen eigentümlichen Bezeichnungen, die sich zu *Algorithmen* zusammenschließen." (*RevS* 10).

56 Husserl's terminology in *LU I*, §20, 68; *LI*, 304.

57 *EL I*, 40–41.

58 *EL I*, 44–45. Cp. *PoA* 249–250; *PdA* 235–237.

59 Cp. e.g. Leibniz 1666, 5.

60 Leibniz would not speak *tout court* of "expressed concept". See, for instance, Leibniz 1677b.

61 *PoA* 205–206; 247–240; *PdA* 193–194; 237–239. Hereto also cp. Husserl, *Zum Begriff der Operation* (*On the Concept of Operation*), in: *PoA* 385–408; *PdA* 408–429. Here: *PoA* 395; *PdA* 418.

62 *EL I*, 48; 49.

63 *EL I*, 44–45.

64 Leibniz 1678a and 1678b. Cp. Trendelenburg 1856, 43.

Bezeichneten)"[65] he writes that the relation between the name and "the referred-to object (*was der Name benennen soll*)" is the same as the relation between that name and "the representational content (*Inhalt der Vorstellung*)" conveyed by the utterance of that name.[66]

Moreover, in the attempt to distinguish univocal from equivocal names, he confuses what Frege would call "sense" of a name with the ambiguity (*Mehrdeutigkeit*) of a name.[67] One may refer to Socrates as to the founder of the definition theory or as to Plato's teacher, but this is quite different from the lexical ambiguity of a word having more than one meaning, as it is, for instance, the case for the word "bank" that may refer to a financial institution as well as to the edge of a river. Husserl omits to dwell upon the latter point, but observes that a consequence of Schröder's mistaken identification of the meaning of a sign with the representational content conveyed by that sign is that all general names (*Gemeinnamen*) become equivocal.[68] Namely, the representational content associated with a general name, like, say, "horse", may be sometimes this, sometimes that particular object belonging to its extension, i.e., sometimes this, sometimes that particular horse, but such instances are anything but the meaning of the general name "horse."[69]

Another consequence of Schröder's misinterpretation of the meaning of a name is that he misses to get the difference between *meaningless* and *objectless* names.[70] "*Asa-nisi-masa*" in the children's mouth in Fellini's *Otto e mezzo* is a meaningless expression. "Round square", at a variance with it, is objectless, but not meaningless.

In his *A System of Logic* (1864) John Stuart Mill had given a picture of semiotic relations bound to last. Names *denote* objects and *not* our ideas of them. *General—singular* and *abstract—concrete* are pairs of conceptual determinations that are opposite. They can be combined, so that we obtain four kinds of names. Consider the two examples (i) "Socrates is courageous" and (ii) "courage is a virtue." In (i) "Socrates" is a concrete singular term: it purports to denote only one object and this object is a concrete one (it is spatio-temporally located). "Courageous", on the other side, is a concrete general term, it refers to many objects, namely all those that share courage, and these objects are concrete ones. In (ii) "courage" is an abstract singular term: it purports to denote only one object and this object is an abstract one (it has no spatial and temporal determinations, it does not stand in causal relations). "Virtue" is an abstract general term, it refers to many objects, namely all those with respect to which the predication "(. . .) is a virtue" expresses a truth, and these objects are abstract ones.[71]

---

[65] *EL I*, 46.

[66] *RevS* 11.

[67] *EL I*, 49.

[68] *RevS* 11.

[69] About this issue, cp. the later Husserl in *LU I*, §12, 47; *LI*, 288.

[70] *RevS* 12.

[71] Mill 1884, 27–30. Cp. Quine 1960, §§20, 25, 49. Also cp. Künne 2007, 23ff.

This leads to the consideration of another important division of names, that between *connotative* and *non-connotative* names. "A non-connotative term is one which signifies a subject only, or an attribute only. A connotative term is one which denotes a subject, and implies an attribute".[72] Concrete singular terms, like "Socrates", "George", "Germany", ... *denote* only *one object*. Abstract singular terms, like "whiteness", "length", "courage", ... *denote* only *one property*. They are *non-connotative*. Concrete general terms, like "white", "long", virtuous", "courageous" ... are, instead, *connotative*. "Courageous", for instance, denotes all courageous things, and "implies, in the language of the schoolmen, *connotes*"[73] the property of being courageous. "Courageous" is not predicated of the property, but of the subjects (Socrates, ...), "but when we predicate it of them, we convey the meaning that the attribute [courageous] belong to them."[74]

There is, according to Mill, "another kind of names, which, although they are individual names, that is predicable only of one object, are really connotative."[75] After Russell, we call names of this kind definite descriptions. "The present prime minister of England", "the first emperor of Rome", "the father of Socrates" are connotative. They *denote* one individual *as* bearer of one or more properties. It is important to stress that, for Mill, the meaning of a name is not the denoted object, but what is conveyed by the name. "The meaning [of a name] resides not in what [it] *denote*[s], but in what [it] *connote*[s]. The only names of objects which connote nothing are *proper* names; and these have, strictly speaking, no signification."[76]

In his *Introduction* Schröder dashes Mill's distinction between *concrete* and *abstract* names with the one between *dependent* and *independent* parts of a whole. The latter is a mereological distinction. Consider the following example: The head of a horse is an *independent* part of the horse, for it can be thought of *separated/abstracted* from its whole. Colour and extension, on the contrary, are *dependent* parts, for they cannot be thought separated from each other. Schröder writes[77]:

> We may abstract certain elements ... from the representation of a concrete thing.... Such representations of the parts ... we call "*marks*" of the thing (nota, mark ...). When this separation succeeds completely, the mark is called a part (pars, part) of the thing

[72]Mill 1846, 31. Note that Mill writes here "signifies", but in the light of what he is going to say next he had better written "it purports to denote."

[73]*Loc. cit.*

[74]*Loc. cit.*

[75]Mill 1846, 33.

[76]Mill 1846, 34.

[77]"Aus der Vorstellung eines konkreten Dinges vermögen wir ... gewisse Elemente abzusondern.... Solche Teilvorstellungen ... nennen wir "Merkmale" desselben (nota, mark ...). Gelingt solche Isolierung vollkommen, so heißt das Merkmal ein Teil (pars, part) des Dinges und wird sich auch seinerseits wieder als ein konkreter Gegenstand in's Auge fassen lassen.... Gelingt jene Isolierung (Absonderung, ...) nicht vollkommen, so nennen wir das vorgestellte Ding etwas Abstraktes, seinen (Eigen-)Namen ein nomen abstractum" (*EL I*, 57–58). Cp. *RevS* 12.

> and is considered itself as a concrete object. . . . When this separation does not succeed completely, we call the represented thing something abstract and his (proper) Name a nomen *abstractum*.

Moreover, Schröder doesn't exactly catch Mill's distinction between connotative and non-connotative names and in fact considers it as superfluous. What Mill says is that proper names "denote the individuals who are called by them; but they do not indicate or imply any attributes as belonging to those individuals."[78] They are, as we saw, non-connotative, they do not denote an individual *as* bearer of a property, but "are simply marks used to be made those individuals subjects of discourse."[79] Schröder, however, translates "non-connotative" into "nicht bezeichnend" (!), instead of "nicht *mit*-bezeichnend", wherefrom follows that, according to Mill, proper names are *non-denoting* names (!). Husserl comments thereupon[80]:

> *Mill's* classification of names into connotative and not connotative is treated very badly in *Schröder*. Insofar as it does not coincide with the separation between common names and proper names, it is "irrelevant" and "dissolves into thin air" (62).—What has occasioned the author, of whom we may indeed presuppose that he has read *Mill's* Logic, to such strong assertions, is inexplicable to me. For *Mill* (Book I, Ch. 2, § 5) expressly gives examples of singular names (in *Schröder's* manner of speaking: proper names) that are connotative. Moreover, *Schröder* ascribes to Mill's "*not-connotative*" the sense of "not-denotative", as if the not-connotative names where senseless and as if *Mill* had never written his correct and precise explanations of this!

6.5 Inhaltslogik or Umfangslogik?

Husserl's comments on Section C of Schröder's *Introduction* contains the sticking point of his critique, which is of an epistemological nature and is directed against Schröder's choice of taking the extensional interpretation as the privileged interpretation of the calculus, as well as against the resulting interpretation of traditional logic.

Husserl interprets Schröder so, as if he were saying that deductive logic is, at root, extensional logic (*Logik der Umfangs*), i.e. a calculus of classes, even if it admits different interpretations. Husserl on the contrary believes that deductive logic is, in

[78] Mill 1846, 33.

[79] Mill 1846, 33. Cp. *RevS* 12.

[80] "[D]ie Millsche Klassifikation der Namen in mitbezeichnenden und nicht mitbezeichnende [wird] bei Schröder gar über mitgespielt. Soweit sie nicht übereinkomme mit derjenigen in Gemeinnamen und Eigennamen, sei sie "belanglos" und "lose sich in eitel Dunst auf" (EL I, 62).—Was den Verf., von dem wir doch voraussetzen dürfen, daß er Mills Logik gelesen hat, zu so starken Behauptungen veranlaßt hat, ist mir unerfindlich. Gibt doch Mill ausdrücklich (Buch I., Kap. 2, §5) Beispiele von Einzelnamen . . . , die nicht mitbezeichnend sind. Schröder unterlegt überdies dem Millschen 'not-connotative' den Sinn von 'nicht-bezeichnend' als ob die nicht-mitbezeichnenden Namen sinnlose wären, und als ob Mill seine trefflichen und präzisen Erklärungen hierzu nie geschrieben hätte!" (*RevS* 12).

its essence, intensional logic (*Logik des Begriffsinhalt*), i.e. a calculus of concepts-contents,[81] even if it admits different interpretations. In the background, Husserl's critical attitude towards *the use of defining a concept by defining its extension* is tacitly at work.

Schröder distinguishes between *actual* and *ideal* content of a concept. Under "actual content" he understands "those marks common to the things referred-to by the common name ... those on which it has been reflected in building the concept,"[82] under "ideal content of a concept" he understands those marks that are common to all things falling under that concept, in his own words: "all common marks in general that can be recognized as such, although perhaps it is not always possible to conceive them completely."[83] Under "extension (*Umfang*)" of a concept Schröder understands, more or less, what is also meant today, namely "the class of individuals (distributively) comprehended under the common name."[84] (Oddily enough, Husserl does not remark that Schröder's definition of the *actual* content of a concept is not a good one, for if the content of a concept is not, as in the tradition, a conjunction of conceptual marks, but the conjunction of those marks on which it has been reflected in the construction of the concept, then the concept expressed by the word, say, "man", would have different contents depending on the different individuals who contruct it.)

A concept, so Schröder, can be identified either by determining its content or by determining its extension. He understands all definitions corresponding to those Wittgenstein would have called "family resemblances (*Familienähnlichkeiten*)",[85] together with all definitions corresponding to those Wittegenstein would have called "ostensive Definitions"[86] as concepts-determinations by extension.[87] We have no proper grasp of the extension of a concept (*EL I*, 85; *RevS* 16), thus "the specification of the extension appears ... to be the most imperfect of the two ways we have at our disposal to determine a concept."[88] And Husserl remarks: "'[A]ppears', he says, for [Schröder] takes the imperfections of the content-specification to be still bigger and more insurmountable and draws the conclusion that the specification of

[81] *RevS* 17.

[82] "die gemeinsamen Merkmale der mit dem Gemeinnamen bezeichneten Dinge ..., auf welche bei seiner Bildung reflektiert wurde" (*EL I*, 83). Cp. *RevS* 17.

[83] "die sämtlichen gemeinsamen Merkmale überhaupt, welche als solche erkannt werden könnten, die es aber vielleicht niemals vollständig auszudenken möglich" (*EL I*, 83 ff.). Cp. *RevS* 17; *FK*, 171; 172. Note the difference with Husserl's understanding of "ideal content of a concept" in his *LV'96*, 70–71.

[84] "Klasse der unter dem Gemeinnamen (distributive) zusammengefassten Individuen." (*EL I*, 83).

[85] Wittegenstein 1953, §§ 65–67. Cp. Mill 1843, Book 1, Ch. 8.

[86] In Wittgenstein's words, "hinweisende Erklärung." See Wittgenstein 2001, §§2, 8.

[87] Cp. *EL I*, 84.

[88] "[d]ie Umfangsangabe erscheint ... als das unvollkommenere der beiden Mittel, einen Begriff zu bestimmen" (loc. cit.).

the extension alone shall be taken as basis for Logic."[89] It is worth noticing that Schröder gives as reason for his privileging definitions by extension the fact that it is not possible to have a *proper* grasp of all conceptual marks that make up the ideal content of a concept (!),[90] and not the difficulties that are connected with the intensional treatment of negation. Indeed, it appears, *prima facie*, plausible to give both an extensional and an intensional interpretation of the calculus on the basis of the *Canon of Reciprocity*, i.e. the traditional principle that intension and extension of a concept stand in an inverse relation. However, there are well known difficulties with negation, that is with the intensional interpretation of the extensional operation of complementation: while the extension of the concept *non-man* is simply the class of all things that are not in the extension of the concept *man*, it is difficult to say what is the intension of the concept *non-man*, given the intension of the concept *man*. Husserl too seems not to be aware of the latter point, for he replies that definitions have, traditionally, never been conceived as the enumeration of the conceptual marks that make up the *ideal* content of a concept, but just as lists of those marks that make up the *actual* content of a concept, and that Schröder is wrong in taking traditional logic as a science that has as its own theme the relations that occur between *ideal* contents of concepts.[91] It deals instead, according to Husserl, with relations between *actual* concept-contents. In *FK* he adds that: (i) such a calculus of *actual* concept-contents is, at root, useless, for it does permit to treat formally analytic judgments only, and not also synthetic ones, and (ii) that it is very well possible to set up a calculus of *ideal* concepts-contents. For this, one does not need to have a proper grasp of ideal concepts-contents more than one needs having a proper grasp of extensions by setting up a calculus of concepts-extensions (*FK*, 171–172). He writes[92]:

> The best confutation of Schröder's argument is provided by the following fact: Exactly in the same sense in which so-called extensional logic is a valid theory, also a "logic of ideal contents" can be erected, by means of a technique, which is identical with that of the former. Indeed there's more to come. Also the logical tasks that they are able to solve are identically the same. I show this at another place.

The "other place" is *FK*. Against Schröder's claim that a calculus of classes can be established setting aside the concept-contents, Husserl objects in *RevS* that precisely

---

[89]"'Erscheint', heißt es; denn die Unvollkommenheiten der Inhaltsangabe sollen noch größere und unüberwindliche sein, daher die Umfangsangabe allein der Logik zugrunde gelegt werden müsse" (*RevS* 16).

[90]"Es lassen die Merkmale, welche den unter den Begriff fallenden Individuen gemeinsam sind, und welche in ihrer Verbindung den idealen Inhalt ausmachen, sich überhaupt nie vollständig aufzählen" (*RevS* 16; cp. *EL I*, 86).

[91]*RevS* 16–17; 18.

[92]"Die beste Widerlegung der Schröderschen Argumentation bildet die folgende Tatsache: Genau in demselben Sinne, als die sog. Umfangslogik eine gültige Theorie ist, kann auch eine "Logik idealer Inhalte" aufgebaut werden, mit einer Technik, die sogar identisch ist mit derjenigen der ersteren. Ja noch mehr, auch die von beiden beherrschten logischen Aufgaben sind identisch dieselben. Ich beweise dies an anderem Orte" (*RevS* 18).

the opposite is true. An *Umfangslogik* cannot be erected without considering the corresponding relations between concept-contents, while the opposite is well possible. How this is possible, Husserl tries to substantiate in *FK*. Indeed, this text sets out to show how a calculus of logic can be set up on the basis of an intensional interpretation of the employed symbols, and precisely "by following exactly the same formal pathways (*Formelbahnen*) as any given calculus of classes (be it that of Boole or of Schröder or of Peirce and so on)."[93] If we conceive contents-concepts as sets of conceptual marks, Husserl maintains, it is natural to assume that they can stand in the same relations ($=$, $\cap$, $\cup$, $\subset$, ...) extensions can stand. Thus, we can apply the calculus to content-concepts simply by assuming that the employed letters are placeholders for *ideal* concepts-contents and not for extensions. What we obtain by changing the interpretation is, according to Husserl, "a truthfully logical calculus. It encompasses the whole field of pure consequences."[94] Indeed, each categorical proposition can be equivalently read both as denoting a relation between concepts-extensions and a relation between ideal concepts-contents. We may interpret, "All men are mortal" both as saying that the class of men is included in the class of mortals and as saying that "the ideal content [of the concept] *of man* includes the ideal content [of the concept] of mortality" (loc. cit.). Furthermore, we can interpret it as saying that all objects of the concept *man* are objects of the concept *mortal*. We can also set aside the interpretation in terms of extensions and concepts, respectively, and take the letters A, B, ... to be placeholders for propositions—again, the calculus is a system of signs that is polyvalent, it admits several interpretations.

6.6 Husserl (and Frege) on "Definitions" and Subsumption

We know from Frege's letter to Husserl dated 24.5.1891[95] that Husserl had sent his *RevS* (together with his *Philosophie der Arithmetik*) to Frege. "I have read your notice of Schröder's work with great interest, and it has prompted me to write down my own thoughts now, rather than wait for the second volume to appear ...",[96] Frege says in the letter, going then on by indicating two main points raised in Husserl's *Review* he agrees with. As we observed in the Introduction, Husserl's criticism in *RevS* is mainly directed to Schröder's philosophical justifications of the calculus, but it also tackles other more specific issues, like the two mentioned by Frege, which we are now going to consider briefly.

---

[93] *FK* 169.

[94] *Loc. cit.*

[95] Frege 1976, 92–94 (English edition: 61–64).

[96] Frege 1976, 61 (henceforth from the Engl. edition). Frege's *Kritische Beleuchtung einiger Punkte in E. Schröders Vorlesungen über die Algebra der Logik* (Frege 1895) will actually appear 4 years later.

The first one concerns "the manner in which the *crucial* signs '0' and '1' are introduced into the calculus"—and "similar objections concern the crucial definitions of *addition* and *multiplication*."[97] Indeed, as we remarked in Sect. 6.2, Schröder officially introduces "0", "1", and the operations of addition and multiplication through corresponding "Definitions." This sounds of course odd for our formal standards concerning the presentation of an axiomatic theory,[98] but also Husserl cannot agree to such a use of the word "Definition", although for slightly different reasons[99]:

> Expressly rejecting any question about existence, we are from here on 'to count the *symbols* '0' and '1' ... as among the domains of the manifold.' ... The author here proves to be a member of that school of formalist arithmeticians who suppose that the oft-mentioned freedom of definition gives them the right 'creatively to introduce' an arbitrary number x of concepts into a deductive discipline, provided only that one takes care that the definitions are free from contradiction. But besides contradiction there is yet another thing which is to be logically avoided: opposition to subject matter.
>
> With what justification may the *0* nonetheless be added, since in the system of the calculus as presented above [i.e. the system given by the first two Principles], the 'creative' definition of *0* yields no legitimation of existence.... For this addition I find not even the shadow of a proof. The *0* of the identity calculus presents the same problem as the $\sqrt{-1}$ of the arithmetical calculus.

Analogous objections apply to the definitions of addition and multiplication. Husserl's point here—which interestingly anticipates his later, more elaborate reflections and theorizations on the *Imaginary in Mathematics*[100]—is, to put it shortly, that the extension of a theory/manifold T (here, the theory given by Schröder's first two *Principles*) with a new operation (possibly a nullary one) satisfying a certain condition φ is legitimate only if we are able to provide a proof

---

[97] *RevS* 31.

[98] What Schröder is actually doing is, we would now say, to add existential postulates of the form $\forall x \forall y \exists z\ \varphi(x,y,z)$—in the case of addition and multiplication, and analogously for "0", "1." Then it is shown that the resulting system proves $\forall x \forall y \exists ! z\ \varphi(x,y,z)$, and so one can expand the language with a binary function letter f and add $\varphi(x,y,f(x,y))$ to the axioms. The resulting system is a conservative extension of the original one.

[99] "Mit ausdrücklicher Ablehnung der Existenzfrage sollen von da ab 'die Symbole 0 und 1 ... mit zu den Gebieten der Mannigfaltigkeit gezählt werden'.... Der Verf. dokumentiert sich hier als Glied jener Schule formalistischer Arithmetiker, die da meinen, die vielberedete Definitionsfreiheit gäbe das Recht, inmitten einer deduktiven Disziplin x-beliebige Begriffe 'schöpferisch einführen' zu können, sofern man nur Sorge trage, daß die Definitionen widerspruchslose seien. Aber neben den Widerspruch ist logisch noch ein anderes zu vermeiden: der Widerstreit" (*RevS* 31). "Da jene 'schöpferische' Definition der 0 ihr im System des Kalküls, wie oben dargelegt worden, noch kein Existenzrecht gibt; ...—mit welchem Rechte darf sie dennoch adjungiert werden? Ich finde auch nicht den Schatten eines Beweises hierfür. Die 0 des identischen Kalküls bietet dasselbe Problem wie die $\sqrt{-1}$ des arithmetischen" (*RevS* 33).

[100] This was the subject of the lecture (known as *Doppelvortrag*) presented by Husserl to the *Mathematische Gesellschaft* in Göttingen in the winter semester of 1901/02. For an analysis of this text (and related ones) see Ch. 3 of Centrone 2010a.

not only of the fact that the theory T' = T + φ is consistent, but also of the fact that it is *conservative* over T.

On this point, Frege's comment in the letter is much shorter (and perhaps more sharp):

> I agree with you when you find that Schröder's definitions of 0, 1, a+b, a.b are faulty. Strictly speaking, instead of defining a+b, Schröder defines the sign € a second time together with the sign +. One and the same definition must not be used to define two different things, and least of all, something that had already been defined previously. The sign should have been introduced at once for all possible cases.[101]

The second issue concerns Schröder's rejection of the Boolean conception of 1 as the class of all possible objects of thought, more precisely Schröder's alleged proof that the "universal interpretation" of 1 leads to a contradiction.[102] His argument runs as follows. Let *a* be the class of all classes contained in 1, which are equal to 1. If 1 contains all possible objects of thought, then 1 contains *a*. On the other side, 0 is contained in each class contained in 1. Thus 0 is contained in *a*, hence 0 = 1, which is absurd (from 0 = 1, Schröder observes, one can derive any equation). Schröder's way out, delineated in *EL I*, 246 ff. has been interpreted (by Church and Quine, among others) as a kind of anticipation of the simple theory of types. In any case, it is fairly obvious that the argument above is fallacious. As Husserl observes,[103]

> This argument, which at first glance is quite impressive, is sophistical. Let us form the class K, whose *elements* are themselves classes, and indeed are those which are = 1. Then K obviously would contain the class 1 as an element, since 1 = 1. But does this K also then contain the class 0 as an *element*? Absolutely not! Certainly it contains 0, as does every other class. But it contains it as a subordinate class, not as an element.

In other words, Husserl appears here to be aware that the relation of "standing under" (*stehen unter*) that holds between an object and a concept, that is, extensionally, the relation of membership of an element in a class, has to be distinguished from the relation of subordination between concepts or, extensionally, of class inclusion.[104] The mistake in Schröder's argument lies exactly, according to Husserl, in the confusion between these two different relations and so, ultimately, in Schröder's very notion of subsumption, symbolized by "€", that dangerously mixes those two relation into a single one (see Sect. 6.2 above).

---

[101] Frege 1976, 62.

[102] *EL I*, 245–246.

[103] "Diese im ersten Moment verblüffende Argumentation ist eine sophistische. Bilden wir die Klasse K, deren *Elemente* selbst Klassen sind, und zwar diejenigen, die = 1 sind; dann enthält K selbstredend die Klasse 1 als Element, da 1 = 1. Enthält dieses K aber auch die Klasse 0 als *Element*? Ganz und gar nicht. Allerdings enthält es, wie jede andere Klasse, die 0, aber als subordinierte Klasse und nicht als Element" (*RevS* 36).

[104] In his *Review* of Husserl's *Philosophy of Aritmetic* (Frege 1894), Frege criticizes Husserl i.a. for not having distinguished between subsumption and subordination. Husserl seems to benefit from this criticism in his *LV'96*, 163 ff. (Hereto cp. Centrone 2010b, 332–334). The fact that the distinction is, in some way, already present in *RevS* suggests however that Husserl had come to it independently from Frege's remarks.

It goes without saying that Frege agrees with Husserl's criticism! Indeed, in Frege 1895 he will once again reaffirm one of the cornerstones of his position, namely "the relation of an object (an individual) to the extension of a concept when it falls under the concept" must be kept separate from "the relation between the extension of one concept and that of another when the first concept is subordinate to the second", deriving as a consequence that "the domain-calculus, in which the fundamental relation is that of part to whole, must be wholly separated from logic."[105]

References

G. Birkhoff, *Lattice Theory*, 3rd edn. (AMS Colloquium Publications, Providence, 1940)

B. Bolzano, *Beyträge zu einer begründeteren Darstellung der Mathematik. Erste Lieferung*, Prag, 1810. Reprint: Wissenschaftliche Buchgesellschaft, Darmstadt, 1974

G. Brady, *From Peirce to Schroeder. A Neglected Chapter in the History of Logic* (North Holland, Amsterdam, 2000)

E. Casari, Anche la Filosofia dell'Aritmetica compie cent'anni, in *Atti del Convegno: Peano e i fondamenti della matematica*, (Mucchi, Modena, 1991), pp. 35–49

S. Centrone, *Logic and Philosophy of Mathematics in the Early Husserl* (Springer, Berlin, 2010a)

S. Centrone, Functions in Frege, Bolzano and Husserl. Hist. Philos. Log. **31**, 315–336 (2010b)

J. da Silva, Husserl's two notions of completeness: Husserl and Hilbert on completeness and imaginary elements in mathematics. Synthese. **125**, 417–438 (2000)

G. Frege, Rezension von Dr. E. G. Husserl: Philosophie der Arithmetik. Zeitschrift für Philosophie und Philosophische Kritik. **103**, 313–332 (1894)

G. Frege, *Wissenschaftlicher Briefwechsel*, ed. by G. Gabriel et al. (Meiner, Hamburg 1976). Abridged English edition: *Philosophical and Mathematical Correspondence* (Basil Blackwell, Oxford, 1980)

G. Frege, Kritische Beleuchtung einiger Punkte in E. Schröders Vorlesungen über die Algebra der Logik, Archiv für systematische Philosophie **1**, 433–456 (1895). English translation in *G. Frege: Collected Papers on Mathematics, Logic and Philosophy* ed. by B. McGuinness (Basil Blackwell, Oxford, 1984), pp. 210–228

M. Hartimo, Towards completeness: Husserl on theories of manifolds 1890–1901. Synthese. **156**, 281–310 (2007)

C.O. Hill, G.E. Rosado Haddock, *Husserl or Frege? Meaning, Objectivity, and Mathematics* (Open Court, La Salle, 2000)

E. Husserl, Besprechung von: E. Schröder, *Vorlesungen über die Algebra der Logik*, vol. I, Leipzig 1890, Göttingische gelehrte Anzeigen (1891a), pp. 243–270. English translation in Husserl (1994), pp. 52–91

E. Husserl, *Philosophie der Arithmetik. Logische und psychologische Untersuchungen*, vol. I, Pfeffer, Halle a.d.S. 1891b. In Husserl 1950+, vol. XII. English translation: D. Willard (ed.), *Philosophy of arithmetic, psychological and logical investigations with supplementary texts from 1887–1901* (Kluwer, Dordrecht, 2003)

E. Husserl, Das Folgerungskalkül und die Inhaltslogik, Vierteljahrsschrift für wissenschaftliche Philosophie. **15**, 168–189 (1891c). English translation in Husserl (1994), pp. 92–114

E. Husserl, *Logische Untersuchungen*, vol. I: *Prolegomena zur reinen Logik*, Max Niemeyer, Halle a.d.S. 1900 (^{2}1913; ^{7}1993 Max Niemeyer, Tübingen). English translation in J. N. Findlay (ed.), *Logical Investigations*, 2 vols. (Routledge and Kegan Paul, London, 1970)

[105] Frege 1895, 456–457.

E. Husserl, *Logische Untersuchungen,* vol. II/1–2: *Untersuchungen zur Phänomenologie und Theorie der Erkenntnis,* Max Niemeyer, Halle a.d.S. 1901 (^{2}1913; ^{6}1993 Max Niemeyer, Tübingen). English translation in J.N. Findlay (ed.), *Logical Investigations*, 2 vols. (Routledge and Kegan Paul, London, 1970)

E. Husserl, *Husserliana, Gesammelte Werke* [HuA], I–XXVI: Martinus Nijhoff, Den Haag, 1950+, XXVII–XXXVII: Kluwer, Dordrecht, 1989–; XXXVIII–: Springer, New York, 2005–

E. Husserl, *Aufsätze und Rezensionen. 1890–1910,* ed. by B. Rang (Martinus Nijhoff, The Hague, London, 1979)

E. Husserl, *Early Writings in the Philosophy of Logic and Mathematics*, translated by D. Willard (Kluwer, Dordrecht, 1994)

E. Husserl, *Logik: Vorlesung 1896*, ed. by E. Schuhmann, *Husserliana,* Materialienbände vol. I. (Kluwer, Dordrecht, 2001)

C. Ierna, The Beginnings of Husserl's Philosophy (Part 1: From *Über den Begriff der Zahl to Philosophie der Arithmetik)*, New Yearb. Phenomenol. Phenomenol. Philos. **5**, 1–56 (2005)

C. Ierna, The Beginnings of Husserl's Philosophy (Part 2: Philosophical and Mathematical Background), New Yearb. Phenomenol. Phenomenol. Philos. **6**, 23–71 (2006)

C. Ierna, Husserl's psychology of arithmetic. Bulletin d'Analyse Phénoménologique. **8**, 97–120 (2012)

W. Künne, *Abstrakte Gegenstande. Semantik und Ontologie* (Klostermann, Frankfurt, 2007)

G. Leibniz, *Dissertatio de Arte combinatoria* (1666), in Leibniz 1936+, vol. VI, 1, pp. 163–230

G. Leibniz, *La vraie méthode* (1677a), in Leibniz 1936+, vol. VI, 4/A, pp. 3–7

G. Leibniz, *Quid sit Idea* (1677b), in Leibniz 1936+, vol. VI, 4/B, pp. 1369–1371

G. Leibniz, *Lingua generalis* (1678a), in Leibniz 1936+, vol. VI, 4/A, pp. 65–67

G. Leibniz, *Lingua universalis* (1678b), in Leibniz 1936+, VI,4/A, pp. 67–68

G. Leibniz, *Meditationes de cognitione, veritate et ideis* (1684), in Leibniz 1936+, vol. VI,4/A, pp. 585–592

G. Leibniz, *Philosophische Schriften*, in G. Leibniz, *Sämtliche Schriften und Briefe*, Series VI, vols. VI.1, VI.2, VI.3, VI.4, VI.6 (Akademie Verlag, Darmstadt/Berlin 1936+)

U. Majer, Husserl and Hilbert on completeness: A neglected chapter in early twentieth century foundations of mathematics. Synthese **110**, 37–56 (1997)

J.S. Mill, *A System of Logic. Ratiocinative and Inductive*, 2nd edn. (Harper & Brothers, New York, 1846)

J.P. Miller, *Numbers in presence and absence. A study of Husserl's philosophy of Mathematics* (Martinus Nijhoff, The Hague, 1982)

V. Peckhaus, Wozu Algebra der Logik? Ernst Schröders Suche nach einer universalen Theorie der Verknüpfungen. Mod. Log. **4**, 357–381 (1994)

C.S. Peirce, On the algebra of logic. Am. J. Math. **3**, 15–57 (1880)

W.V.O. Quine, *Word and Object* (MIT Press, Cambridge, MA, 1960)

G. Rosado Haddock, Husserl's relevance for the philosophy and foundations of mathematics. Axiomathes. **8**, 125–142 (1997)

E. Schröder, *Vorlesungen über die Algebra der Logik (Exakte Logik)*, vol I (Teubner, Leipzig, 1890)

E. Schröder, *Vorlesungen über die Algebra der Logik (Exakte Logik)*, vol. II, part 1 (Teubner, Leipzig, 1891)

E. Schröder, *Vorlesungen über die Algebra der Logik (Exakte Logik)*, vol. III, part 1: *Algebra und Logik der Relative* (Teubner, Leipzig, 1895)

E. Schröder, *Vorlesungen über die Algebra der Logik (Exakte Logik)*, vol. II, part 2 (Teubner, Leipzig, 1905)

R. Tieszen, *Phenomenology, logic and the philosophy of mathematics* (Cambridge University Press, Cambridge, 2005)

F. A. Trendelenburg, Über Leibnizens Entwurf einer allgemeinen Charakteristik, in *Philosophische Abhandlungen der Königl Akadademie der Wissenschaften zu Berlin*, 1856, pp. 37–69

D. Willard, *Logic and the objectivity of knowledge: A study in Husserl's philosophy* (Ohio University Press, Athens, 1984)

L. Wittgenstein, *Philosophical Investigations* (Blackwell, Oxford, 2001)

Chapter 7
The Brentanist Philosophy of Mathematics in Edmund Husserl's Early Works

Carlo Ierna

Abstract A common analysis of Edmund Husserl's early works on the philosophy of logic and mathematics presents these writings as the result of a combination of two distinct strands of influence: on the one hand a mathematical influence due to his teachers is Berlin, such as Karl Weierstrass, and on the other hand a philosophical influence due to his later studies in Vienna with Franz Brentano. However, the formative influences on Husserl's early philosophy cannot be so cleanly separated into a philosophical and a mathematical pathway. Growing evidence indicates that a Brentanist philosophy of mathematics was already in place before Husserl. Rather than an original combination at the confluence of two different streams, his early writings represent an elaboration of topics and problems that were already being discussed in the School of Brentano within a pre-existing framework. The traditional account understandably neglects Brentano's own work on the philosophy of mathematics and logic, which can be found mostly in his unpublished manuscripts and lectures, and various works by Brentano's students on the philosophy of mathematics which have only recently emerged from obscurity. Husserl's early works must be correctly placed in this preceding context in order to be fully understood and correctly assessed.

Keywords Franz Brentano • Edmund Husserl • School of Brentano • Early phenomenology • Philosophy of mathematics

7.1 Introduction

A common analysis of Edmund Husserl's early works on the philosophy of logic and mathematics presents these writings as the result of a combination of two distinct strands of influence: on the one hand a mathematical influence of his teachers is Berlin, such as Karl Weierstrass and Leopold Kronecker, and not to

C. Ierna (✉)
Faculty of Philosophy, University of Groningen, Groningen, The Netherlands

Institute of Philosophy, Czech Academy of the Sciences, Prague, Czech Republic
e-mail: carlo.ierna@gmail.com

S. Centrone (ed.), *Essays on Husserl's Logic and Philosophy of Mathematics*, Synthese Library 384, DOI 10.1007/978-94-024-1132-4_7

forget his dissertation advisor in Vienna, Weierstrass' student Leo Königsberger, and on the other hand a philosophical influence from his later studies in Vienna with Franz Brentano and then in Halle an der Saale with Brentano's student Carl Stumpf, who supervised his habilitation.[1] In this picture it is Husserl's eclectic combination of these independent sources of inspiration that resulted in his original approach to mathematics and logic in the 1880s and 1890s, laying the foundations for his later development of phenomenology.[2] However, this picture is somewhat simplistic, since it does not sufficiently take into account how much the nineteenth century philosophers and mathematicians that inspired Husserl had already inspired one another. The formative influences on Husserl's early philosophy cannot be so cleanly separated into a philosophical and a mathematical pathway. Growing evidence indicates that a Brentanist philosophy of mathematics was already in place when Husserl started working on his habilitation and his first book.[3] Rather than an original combination at the confluence of two different streams, his early writings represent an elaboration of topics and problems that were already being discussed in the School of Brentano within a pre-existing framework. Besides the mathematical literature on the foundations of mathematics, particularly on the concept of number, that originated in the 1880s and 1890s,[4] there is also a surprising number of works from the School of Brentano on the philosophy of mathematics in the same timeframe, as we will see later on. This reduces, or at least contextualizes, Husserl's originality considerably. The standard account about the School of Brentano tends to underscore how each of the students elaborated a specific topic according to Brentano's philosophical and psychological method: Stumpf a Brentanist philosophy of sound and music, Marty of language, etc. and Husserl would then have been chosen to supply a Brentanist philosophy of mathematics. Brentano himself would merely have provided the method and metaphysics of these inquiries, and hence in the historiography of the School of Brentano it is often observed that Brentano remains somewhat "invisible"[5] and was overshadowed by his students (and the schools and movements they contributed to found and influence). Based on such reasonings, it is sometimes called into question whether we can even speak of a

[1]See i.a. Miller 1982, 4; Haddock 1997, 127: "*Über den Begriff der Zahl* and later *Philosophie der Arithmetik* are in some sense the result of such a marriage between the mathematician formed under the guidance of Weierstrass and the philosopher formed in the school of the philosopher-psychologist Brentano," Centrone 2010, 5: "From Weierstrass, one could say, Husserl inherits the project of founding analysis on a restricted number of simple and primitive concepts, and from Brentano he inherits the method for identifying these primitive concepts, namely by describing the psychological laws that regulate their formation."

[2]Hartimo 2006, 319: "The paper examines the roots of Husserlian phenomenology in Weierstrass's approach to analysis.... Husserl's novelty is to use Brentanian descriptive analysis to clarify the fundamental concepts of arithmetic in the first part."

[3]For a first outline of the idea of a Brentanist philosophy of mathematics, see Ierna 2011a.

[4]E.g. Frege 1884; Kronecker 1887; Dedekind 1888.

[5]Albertazzi et al. 1996, 3 ff.; Schuhmann 2004, 278.

"school" of Brentano at all[6]: after all heterodoxy was rampant, each of the students went his own way, and came to disparate and incompatible results.[7] Yet, this account understandably neglects the extent and detail of Brentano's own work on the philosophy of mathematics and logic, which can be found mostly in his unpublished manuscripts and lectures. Moreover, only recently has the content of various works by Brentano's students on the philosophy of mathematics emerged from obscurity through, among other things, the critical edition of Stumpf's habilitation work on the foundations of mathematics,[8] a reconstruction of Husserl's original habilitation essay,[9] in part based on the content of his previously unknown lectures on the concept of number,[10] and Ehrenfels' dissertation on *Relations of Magnitude and Numbers*.[11] These constitute a growing body of evidence regarding the existence and content of a Brentanist philosophy of mathematics. Husserl's early philosophy of mathematics must be correctly placed in this preceding context in order to be fully understood and correctly assessed. At the same time, the existence of such a shared framework in one specific field of inquiry, also suggests a stronger unity in the school of Brentano and a more prominent role of Brentano himself in the formation of the basic ideas that shaped the positions of his students.

7.2 Husserl's Mathematical Background

Husserl began his academic career as a mathematician, having written a technical-mathematical dissertation on the calculus of variations.[12] He certainly did receive an important formation and inspiration in this regard from his mathematical teachers, among whom Husserl indicates Weierstrass as his most important mentor.[13] Specifically regarding Weierstrass, we know from Husserl's annotations of his lectures "*Introduction in the Theory of Analytical Functions*" that Weierstrass began his lecture with a philosophical introduction into the concept of number and how all

[6] Moran 2000, 24 "one can speak loosely of a 'Brentano school'."

[7] Albertazzi et al. 1996, 6.

[8] Stumpf 2008.

[9] Ierna 2005.

[10] Husserl 2005.

[11] Ehrenfels' Dissertation and an introductory article are forthcoming in Meinong Studies 8 (Ehrenfels 2017, Ierna 2017).

[12] Beiträge zur Variationsrechnung, Husserl 1883.

[13] Schuhmann 1977, 7. It is often repeated in the secondary literature that Husserl worked for Weierstrass as an "assistant." The source for this nugget of information is Malvine Husserl's brief biographical sketch of Edmund Husserl, written in 1940. Karl Schuhmann, the editor of this text, however, indicates that this expression could not have meant any kind of "official or semi-official function" and that Malvine Husserl probably anachronistically projected the more recent institution of assistantships back on Husserl's Berlin period (Schuhmann 1988, 121, n. 33).

higher mathematics was founded on natural numbers and the basic operations.[14] In essence, this represents the core of the project of arithmetization of analysis: an ongoing project of the Berlin School at the time when Husserl studied there.[15] Yet "Weierstrass' (mathematical and philosophical) thought moves exactly in the same conceptual space described extensively by Bolzano."[16] Hence, it is not just the case that Husserl was first "made aware" of Bolzano when attending Weierstrass lectures, but he was implicitly taught a philosophy of mathematics based on Bolzanian elements. It is no wonder, then, that Husserl a few years later was so highly receptive to Brentano's extensive discussion of Bolzano in his lectures on logic. Indeed, also other elements of Husserl's thought that we tend to associate with Brentano, probably had antecedents in Husserl's mathematical formation, such as the distinction (though not necessarily the technical terminology) of proper ("*Eigentliche*") and improper or symbolic ("*Uneigentliche*" or "*Symbolische*") presentations, despite Husserl's explicit acknowledgement of an inspiration by Brentano.[17] In order to single out the original influences that are specifically due to Husserl's Brentanist background, we first need to identify which fundamental concepts were instead due to his mathematical background, beyond and besides his technical-mathematical knowledge.

The idea that "all philosophy of mathematics must start with the analysis of the concept of number,"[18] that the concept of number is based on the operations of collecting and counting, and the "usual definition" that "number is a multiplicity of unities"[19] can all be found in the introductory parts of Weierstrass lectures. Based on the published version of Königsberger's lectures,[20] we also know that he shared these views and likely conveyed them in the lectures that Husserl attended while working on his dissertation. Additionally, Weierstrass had already pointed out that there is a difference between equality (as relation between two different objects) and identity (in Leibniz' sense), a distinction which was then exploited by Husserl in his polemic with Frege in the *Philosophy of Arithmetic*, but also occurs independently elsewhere in the School of Brentano (e.g. Ehrenfels' dissertation).

Regarding Bolzano's influence on Husserl's early works in the philosophy of mathematics and logic, preceding the 1900 *Prolegomena* and 1901 *Logical Investigations*, it is mainly limited to the *Paradoxien des Unendlichen*. Husserl derives a fair few elementary definitions and examples quite directly from the opening paragraphs of this booklet. Exactly like Bolzano, Husserl defines collections through

[14]Husserl 1891, 5 n.; 1970, 12 n., see manuscript Q 3, 1, quoted in Ierna 2006, 36 f.

[15]Cf. Miller 1982, 1–4.

[16]Spalt 1991, 351 f., also see 355; compare Kline 1972, 953; for a more extensive account, see Ierna 2006, specifically 40 f.

[17]See Ierna 2006, 38 ff. and 46 ff.

[18]Husserl 1887, 8; 1970, 295.

[19]Husserl 1887, 12; 1970, 297.

[20]Königsberger 1874, 1.

the conjunction "and,"[21] used between completely arbitrary objects,[22] which again leads to the conception of quantities as multiplicities of units.[23] Moreover, Husserl takes over the term "*Inbegriff*" from Bolzano's *Paradoxien*. In the 1891 *Philosophie der Arithmetik* we also see echoes of Bolzano's main topic, infinity, in the paragraph Husserl dedicates to this subject.[24]

Given the shared background of Georg Cantor and Husserl in Weierstrass and, at least implicitly, Bolzano, and the fact that Brentano extensively discussed (and criticized) not only Bolzano's *Paradoxien*, but also Cantor's *Mannigfaltigkeitslehre* in his 1884/85 lectures on elementary logic, it is both difficult and pointless to try to extrapolate an autonomous influence of Cantor on Husserl's early works. While they were bound by a personal friendship and were colleagues in Halle, it is doubtful whether Cantor had much influence on Husserl's theories at that point. Indeed, precisely one pivotal concept of Husserl's *Prolegomena*,[25] likely was not inspired by Cantor at all: the concept of manifold (*Mannigfaltigkeit*). While Cantor obviously wrote about "manifolds" in one sense,[26] the way Husserl employs the term is decidedly closer to the usage of Bernard Riemann.[27] Cantor generally used manifold simply as synonym for *Menge* ("quantity"), or *Inbegriff* ("collection"), and in the mid 1890s actually started calling his work *Mengenlehre* ("Theory of Quantities" or set-theory), instead of theory of manifolds. Husserl explicitly distinguishes Cantor's conception of manifold from that of Gauss and Riemann:

> By manifold, Cantor means a simple collection of elements that are in some way united. . . . However, this conception does not coincide with that of Riemann and as used elsewhere in the theory of geometry, according to which a manifold is a collection not of merely united, but also ordered elements, and on the other hand not merely united, but continuously connected elements.[28]

In the 1880s and 1890s Husserl extensively lectured on Riemann's theories of space, possibly due to Brentano's emphasis on the problems regarding the continuum in his logic lectures. In the *Prolegomena* Husserl will then link his conception of the theory of manifolds to Riemann's, bringing it closer to topology than to (Cantorian) set-theory.[29]

---

[21] Husserl 1887, 58 f.; 1970, 334; Bolzano 1889, 2 f.

[22] Husserl 1887, 13; 1970, 289; Bolzano 1889, 3.

[23] Bolzano 1889, 4 f.

[24] Husserl 1891, 246–250; 1970, 218–221; also see Ierna 2003.

[25] Husserl 1900, § 70.

[26] See Cantor 1883.

[27] Also see Haddock 2006 and Ierna 2012.

[28] Husserl 1983, 95 f.

[29] Husserl 1900, 156 n. 2.

7.3 The Brentanist Philosophy of Mathematics

Already a few times I had occasion to mention that some of the mathematical influences on Husserl were also mediated through Brentano's and later Stumpf's lectures. Beyond and besides the presence of theories developed by other mathematicians in Brentano's lectures, there certainly also is a philosophy of mathematics originating within the School of Brentano itself. Based on Brentano's teachings, nearly all Brentanists produced works on the philosophy of mathematics, often specifically on the concept of number as foundational element. In itself this is not a feature specific of the Brentanist approach, since in the same period there were multiple works by mathematicians on this topic: famously Frege's *Grundlagen* were subtitled "*Eine logisch mathematische Untersuchung über den Begriff der Zahl*,"[30] as well as the equally well-known works by Kronecker[31] and Dedekind,[32] which all took the concept of number as central to a discussion on the foundations of mathematics. Additionally, this also shows that such a tendency was not exclusive to the Berlin School and their project of artihmetization of analysis either. There are, however, some typical features of the Brentanist approach that need to be taken into account, if we want to correctly determine Husserl's position in the context of the School of Brentano with respect to the philosophy of mathematics.[33]

In his early works, Husserl identifies the concept of number as the fundamental issue in the philosophy of mathematics. Whether this is due to his exposure to Weierstrass' project of the arithmetization of analysis in Berlin, to the context of the School of Brentano, or simply to the generally accepted views among mathematicians at the time, is not as relevant as the precise terminology he employs in the formulation of the problem and the methods he uses to tackle it. These are all decidedly Brentanist: Husserl's goal is to secure a foundation for mathematics through a psychological analysis of the origin and content of the concept of number.[34]

The *Philosophie der Arithmetik* is divided in two parts, one dealing with "The Proper Concepts of Multiplicity, Unity, and Cardinal Number," the other with "The Symbolic Number-Concepts and the Logical Sources of Cardinal-Arithmetic."[35] In other words, the first part contains the "psychological analyses" of the subtitle to

[30]Frege 1884.

[31]Kronecker 1887.

[32]Dedekind 1888.

[33]While we can (and should) distinguish Husserl's three early works on the philosophy of mathematics, i.e. his (unpublished, lost) habilitation work, the partial print thereof by the same title Über den Begriff der Zahl, and the Philosophie der Arithmetik, here we will discuss his early position as a comprehensive whole. Various historical sub-periodizations can be found in Miller (1982), Willard (1984), and Ierna (2005).

[34]Husserl 1887, 7; 1970, 294.

[35]Husserl uses Anzahl ("amount") interchangeably with *Zahl* ("number"), so the straightforward translation of *Anzahl* with "cardinal number" is not entirely uncontroversial or unproblematic.

his habilitation essay, while the second contains logical investigations. In the first, Husserl approaches the concept of number as multiplicity of units and provides an extensive account of the "origin and content" of the proper presentations of numbers. The basic elements of this approach involve the idea of "something in general" (*Etwas überhaupt*) and the conjunction "and," by which a collective connection or *Inbegriff* is established. The operation of collecting is essentially based on those of analyzing and distinguishing,[36] which Husserl derives from Stumpf: "Noticing a plurality we want to call analysis."[37] We then consider every object as "something" by completely abstracting from its concrete contents and taking it merely as the foundation of a relation. Numbers are then generated by counting,[38] much like Weierstrass already argued. One and zero would therefore not be numbers properly speaking, in the sense of being multiplicities of units,[39] as they do not really answer the question "how many?" ("none" and "not many" wouldn't be acceptable answers).[40] Beyond this (contingent, psychological) limit, we can only operate with the symbolic concepts of number, to which the second part of the *Philosophie der Arithmetik* is dedicated.

In part two, Husserl does not address merely the symbolic conception of numbers, but also that of quantities (*Mengen*). For this latter, he famously introduced the idea of "quasi qualities" that would immediately provide an indication of the quantity-character of a whole.[41] While he termed these "higher order qualities" as "figural moments" in the *Philosophie der Arithmetik*, these actually are none other than *Gestalt* qualities. Husserl himself had already used precisely this term more than a year earlier in his lecture "On the Concept of Number" of January 1890.[42] The symbolic concept of number, then, would be due to the parallel generation of names and concepts from a numeral system. Arithmetic, rather than the "science of numbers" would be the "science of relations among numbers"[43] since its man task would be to find all possible symbolic number constructions and the methods of reducing them to their normal forms, provided by the numeral system.[44]

Husserl's eclectic combination of the examples of collectives from Brentano's lectures, the notion of fusion (*Verschmelzung*) from Stumpf's *Tonpsychologie* and

[36] Husserl 1891, 49, 59, 66 n., 241.

[37] Stumpf 1883, 96.

[38] Husserl 1887, 24; 1970, 307.

[39] Husserl 1891, Ch. 8.

[40] If we combine this with his 1887 habilitation thesis that "In the proper sense we cannot count beyond three," then the only numbers, properly speaking, would be two and three.

[41] Husserl 1891, 225, 1970, 201.

[42] Husserl 2005, also see Husserl 2002, 295; Ierna 2005, § 3.2.

[43] Husserl 1891, 290.

[44] Husserl 1891, 297.

lectures,[45] and the notion of *Gestalt* gleaned from his early reading of Mach's 1886 *Beiträge zur Analyse der Empfindungen*, is among his most original innovation in his first book. However, obviously, none of the constituents are. If we comb through Husserl's sources and consider parallel developments in other authors, there is little originality left in Husserl's early philosophy of mathematics.

Among Husserl's sources and inspirations in the School of Brentano, we will consider mainly three here. First and foremost, Brentano himself, especially with respect to the lectures on elementary logic of the Winter Semester 1884/85 that Husserl attended and that "gave the breakthrough" in his choice to pursue philosophy rather than mathematics. These logic lectures also contained an extensive engagement with the theories of Bolzano and Cantor, specifically regarding the continuum. As further evidence for the influence these lectures had on Husserl at the time, we might adduce the fact that his very first research manuscript dates from shortly after this period and addresses precisely the issues surrounding the continuum.[46] Indeed, in his letter to Stumpf containing his recommendation of Husserl, Brentano explicitly mentions these writings and points out that Husserl originally wanted to use them as basis for his habilitation.[47] Husserl then transferred to Halle and came under the tutelage of Brentano's first student, Carl Stumpf, who was also the first to produce a work in the tradition of the Brentanist philosophy of mathematics, i.e. his 1870 habilitation *Über die Grundsätze der Mathematik* (*On the Foundations of Mathematics*).[48]

Regarding the mediation of Brentano's influence on Husserl with respect to the philosophy of mathematics, Stumpf plays a very interesting role. First of all, exactly like Husserl, he wrote his habilitation work on the philosophy of mathematics, after having been influenced by Brentano's lectures. Secondly, through his lectures Stumpf taught Husserl the basic method of Brentano's descriptive psychology, while often using the very same examples Brentano used, particularly when speaking of mathematics. Hence, Stumpf's works and lectures represent an important source to clarify what the content of the Brentanist philosophy of mathematics was and how it was conveyed to Husserl.[49]

---

[45] We find a first definition in Stumpf 1883, 101, which was underlined by Husserl in his personal copy, conserved at the Husserl Archives Leuven with signature BQ 472, and then a more extensive discussion in Stumpf 1890, 64 ff.

[46] See Schuhmann 1977, 16. The manuscript has the signature K I 50/47. A picture of the first page of the manuscript (with title and date) can be found in Sepp 1988, 157. The surrounding pages in the manuscript contain excerpts and critical discussions of Riemann and Helmholtz, which also were the topic of Husserl's early lectures at Halle.

[47] Also see Ierna 2005, 7.

[48] Stumpf's habilitation work has only recently been published (Stumpf 2008) and I discuss its main points more extensively elsewhere (Ierna 2011a and esp. Ierna 2015). Since it cannot yet be conclusively established that Husserl had access to this material, the notions shared between Stumpf's and Husserl's habilitation works at least suggest a shared background in Brentano.

[49] I will not go into Stumpf's discussions of space and geometry or the theory of probability here, which would go well beyond the scope of the present contribution.

Despite the general agreement in the views of Brentano, Stumpf, and Husserl that I will substantiate below, this would still merely point to a continuity from teacher to pupil: an original, significant influence with respect to the philosophy of mathematics, proceeding from Brentano himself, first on Stumpf and then, jointly with him, on Husserl. However, I actually wish to make an even stronger claim: that there is a Brentanist philosophy of mathematics in the School of Brentano. For this we need more that the mere vertical lineage of Brentano, Stumpf and Husserl, namely horizontal connections and parallel among independent pupils and members of the School. These are not at all hard to find. Indeed, there is one particular author from the School of Brentano that presents several amazing parallelisms to Husserl's development. Exactly like Husserl, he was a student of Brentano, influenced by Meinong's works on relations, developed a technical notion of *Gestalt* inspired by Mach around 1890, wrote an academic work containing psychological analyses of the concept of number in the mid-1880s, and then published another one on the philosophy of mathematics in 1891. I speak, of course, of Christian von Ehrenfels.

Ehrenfels represents another important piece of the puzzle of Brentanist philosophy of mathematics. Based on the theories developed by Ehrenfels we can assess more readily which notions at the time were actually "in the air" and common knowledge in the wider School of Brentano, beyond the master-pupil lineage of Brentano-Stumpf-Husserl. This provides a fuller context and background for Husserl's early works, since Ehrenfels and Husserl developed many central notions in parallel. By "in parallel" I do not mean in any way collaboratively, but rather independently, though based on the same background. Husserl's introduction of the term "*Gestalt*" and "*Gestaltmoment*" prior to Ehrenfels in a lecture of January 1890[50] is less relevant here, though it underscores how similar their common theoretical background must have been, to be receptive to the same kinds of ideas. Ehrenfels' anticipation in his 1885 dissertation *Über Größenrelationen und Zahlen. Eine psychologische Studie* of some of the doctrines set forth in Husserl's 1887 habilitation is rather more interesting here. However, a fuller and more detailed side by side comparison will have to wait for the edition of Ehrenfels' dissertation, which remained unpublished until now.[51]

So what did Brentano et al. have to say about mathematics? In his *Psychologie vom Empirischen Standpunkte* Brentano already identified mathematics as the first science to obtain a mature theoretical unity[52] and as the most general and hence most fundamental science[53]:

> Mathematics considers the most simple and independent phenomena.... Consequently, mathematics reveals in a clear and understandable way the fundamental nature of all true

---

[50] Husserl 2005; Ierna 2005, 2009.

[51] Ehrenfels' Dissertation and an introductory article are forthcoming in Meinong Studies 8 (Ehrenfels 2017, Ierna 2017).

[52] Brentano 1874, vi.

[53] The following discussion of Brentano's theories is partially based on Ierna 2006 and Ierna 2011a.

> scientific investigation. There is no better field of study for gaining one's first clear view of laws, deduction, hypothesis, and many other important logical concepts.[54]

Indeed, since mathematics does not only treat numbers as such, but magnitudes in general, for Brentano it is "necessary for the exact treatment of all sciences only because we now in fact find magnitudes in every scientific field."[55] Magnitudes imply measurement, which presupposes mathematics. It falls to mathematics to define what "magnitude" is and how it can be treated with the required degree of exactness. It is in this sense that "mathematics proves to be the indispensable adjunct of all sciences, at every level of exactitude and in all kinds of different circumstances."[56]

Likewise, in his habilitation work, Stumpf critiques Mill's empirical approach to number as well as Kant's conception of mathematics as synthetic a priori:

> The critique of Mill's theory has shown that the mathematical axioms (as well as its propositions) cannot be founded on induction, but that they must be necessary a priori. The critique of Kant's theory has shown that they cannot be synthetic a priori. The only remaining possibility is: that they must be analytic.[57]

For Stumpf, the "basic propositions" of mathematics are completely analytic and can be reduced to tautologies.[58] In fact, the axioms of arithmetic are so general that they are applicable "simply everywhere" and "presupposed by all sciences with the same right and the same need" (Stumpf 2008, 20–2). Mathematics itself does not presuppose any "truths, judgments, facts," but instead "presentations, objects, definitions" (Stumpf 2008, 35–4).

In this context, number is conceived as a "sum of units," counting is a matter of distinguishing different units, and then the concept of number is obtained by "grasping these acts [of differentiating] together in thought: one and one and etc."[59] For this we need the concept of a "thing in general" i.e. "something" (*Etwas*), so we can count "a thing and an other thing etc." Greater numbers can be conceived only on the base of a system of numerals, i.e. a rule-based system for the construction of signs. Indeed, Stumpf, very similarly to Husserl, argues that "by constructing a numeric system, we have constructed the higher numbers themselves."[60] Higher numbers are then conceived through relations to lower numbers: 100 as 10 times 10. However, in describing this process of construction, Stumpf does not use the technical terminology of "proper" and "improper" (or "symbolic") presentations or concepts, yet.

---

[54]Brentano 1874, 29, 34.

[55]Brentano 1874, 86.

[56]Brentano 1874, 93.

[57]Stumpf 2008, 18–2.

[58]Stumpf 2008, 19–1.

[59]Stumpf 2008, 19–3.

[60]Stumpf 2008, 24–2.

Nevertheless, mathematics turns out to be the prime example of a science based on symbolic presentations, i.e. presentations through signs. In the *Philosophie der Arithmetik*, Husserl extensively thanks Brentano for the "deeper understanding of the eminent significance of improper presenting" and the awareness of the importance of the distinction between proper and symbolic presenting in general.[61] This distinction, as we have seen, underlies the division of the *Philosophie der Arithmetik* in two parts. Brentano made this distinction in his logic lectures:

> We improperly present that of which we have no precisely corresponding presentation and often can have none. We name it, but we do not quite even understand the name when we name it.... It is, however, also similar when we name objects, the single features of which we could presumably grasp, but which are for us no longer presentable due to their complication. A million, a billion, we cannot properly present any longer and we name them without understanding the name precisely.[62]

These improper or symbolic presentations are presentations through relations.[63] Relations play a central role in Brentanist mathematics, not only as mediating sign and signified in symbolic presentations, but also on the elementary level of proper presentations of numbers. Indeed, as Husserl already argues in *Über den Begriff der Zahl*, in order to obtain a countable quantity, we have to collect the objects-to-be-counted together, establishing a "collective connection" between them.[64] This kind of relation is taken straight from Brentano's lectures,[65] who uses it precisely to define numbers as multiplicities of units. Nevertheless, the more important case is that of symbolically conceived numbers, since after all we can hardly count beyond three[66]:

> A *symbolic* or improper presentation is, as the name already says, a presentation through signs. When content is not given directly to us as what it is, but only indirectly *through signs that characterise it unambiguously*, we have, instead of a proper presentation, a symbolic presentation of it.[67]

While the requirement of being "unambiguous" has been indicated as an element that Husserl adds to Brentano's and Stumpf's theories, possibly based on his study of Schröder, it can nevertheless already be found in Brentano's logic lectures: "Mathematics is just as abstract as metaphysics. But it does not confuse us as easily, because the signs exclude equivocation."[68] Indeed, Brentano connects this explicitly to the Leibnizian project of a *characteristica universalis*, though built bottom up and not top down. For Brentano, the signs aid our cognitive capacities, allowing us to

61 Husserl 1891, 215; 1970, 193.

62 Brentano EL 80/13060, quoted from Rollinger 2009, 81 f.

63 Husserl 2002, 296.

64 Husserl 1887, 50 f.; 1970, 328.

65 see Brentano 1884/85, Y 2, 107 ff.

66 Stumpf 2008, 24–1. Echoed by Husserl: "In the proper sense one can hardly count beyond three" (Husserl 1887, Theses; 1970, 339)

67 Husserl 1891, 215; 1970, 193.

68 Brentano 1884/85, Y 2, 32.

think concepts that would otherwise be inaccessible to us, i.e. “Who could conceive of a million in any other way but as of a great number, as a 1 with six zeroes?”[69] However, this is not simply a matter of creative definition, but stepwise construction: higher numbers are built up out of lower numbers. Ultimately, for Brentano, we can reduce the inconceivable to the conceivable: “A multiplicity, a number is composed of smaller numbers, 12 from 6 + 6.” Indeed, Brentano continues: “But if you go further back, you will reach the unity. Is that still a number? It is not a number at all. Thus, ultimately, multiplicity is composed of what is not-multiple [*Vielheit aus Nichtvielheit*].”[70] Brentano quotes Gauss on this point: Unity is not a magnitude, because it is not further divisible (“*Die Einheit ist nach Gauß keine Größe, weil nicht teilbar*”). Several of Brentano's numerical examples point out that number in essence is a sum of units (1 + 1 + 1 etc.) to which we assign a new, unambiguous name, to sharply, unambiguously distinguish one sum from another.[71]

We find the same doctrines repeated in Stumpf's lectures on psychology from the Winter Semester 1886/87, which Husserl attended. When Stumpf introduces the idea of symbolic presentations “which occur only as signs for others,” he again immediately uses the familiar mathematical example as illustration:

> [I]n the case of larger numbers, the adequate presentation [*adäquate Vorstellung*] is altogether impossible for us and we think, instead of it, the indeterminate concept of a large number together with certain relations of the number we mean (we intend) [*der gemeinten (intendierten) Zahl*] to other numbers. E.g. $1000 = 10 \times 100$, $100 = 10 \times 10$.[72]

A few pages later, Stumpf also quite literally repeats Brentano's observation that “A million: easily one operates in thought with these contents that in themselves are enormous and cannot be fully thought out.” In his lectures on logic, Stumpf then further clarifies the use of signs in mathematics. Stumpf argues, as he had already done in his habilitation essay, that mathematics is a completely analytic science, contra Kant's view that it would be based on synthetic a priori judgements. We can always analyze, i.e. decompose, higher numbers into lower ones, which is actually illustrated by Kant's example $7 + 5 = 12$. We cannot fully and properly conceive 7 or 12, but can analyze these number concepts into simpler ones, e.g. $7 = 5 + 2$, again and again, until we reach the basic level of $3 = 1 + 2$, $2 = 1 + 1$. So actually 3 means $1 + 1 + 1$. Since these propositions are completely analytic, all the propositions for higher numbers based on these must also be analytic.[73] All higher numbers, beyond our cognitive capacity, in the end would reduce to sums of units, turning mathematics essentially into a kind of abbreviated counting.[74]

As we already saw in his habilitation work, Stumpf again points out that in order to count objects, we must first have distinguished them, which also means

[69] Brentano 1884/85, Y 2, 28 f.

[70] Brentano 1884/85, Y 2, 47.

[71] E.g. Brentano 1884/85, Y 2, 29.

[72] Stumpf 1886/87, Q 11/II, 504. Translation from Rollinger 1999, 301.

[73] Stumpf 1887, Q 14, 86 ff.

[74] Stumpf 1887, Q 14, 114a.

that relations play an essential role for the construction of numbers.[75] Indeed, also relations among relations, i.e. higher order relations, are required to distinguish the objects to be counted from each other: we can distinguish not only a from b and b from c, but also the relations of difference between those objects, through a higher order distinction. This idea, including its symbolism, is then literally taken over by Husserl in his habilitation work.[76]

After Brentano's own lectures, Stumpf's lectures and works certainly represent one of the main proximal sources for the Brentanist philosophy of mathematics. Husserl did not only find a workable philosophical and psychological toolkit in them, regarding the notions of intentionality, proper and symbolic presentations, consciousness-*nebenbei*, fusion, the limits of our cognitive capacities or "the narrow mind of man," etc.[77] Indeed, at all stages of his works, from his 1870 habilitation to the chapter on the concept of number in his *Erkenntnislehre*,[78] Stumpf endorses the central tenets of the Brentanist philosophy of mathematics: number is a multiplicity of units, given by the mental operation of counting and "*Zusammenfassung*" ("grasping together," i.e. collecting) operated on "somethings," in which case one and zero are not to be considered numbers properly speaking. Also higher numbers, beyond our presentational capacity are to be understood as presented improperly, i.e. symbolically, through signs provided in a systematic way by a (positional) numeral system.[79]

Let us now consider the broader context of the School of Brentano and the horizontal connections I mentioned earlier. Especially the works of Ehrenfels about the philosophy of mathematics reveal that these basic elements were current beyond the vertical lineage of Brentano-Stumpf-Husserl.

Much like Husserl, Ehrenfels begins his 1885 dissertation with the question "what is number?" and reviews the options: is it given in sensation, is it a relation, or is it something new? Just like Stumpf had done and Husserl would do soon thereafter, he opposes the view that number would be given through sense perception. For Ehrenfels, as for the other Brentanists, numbers are given by counting, which requires establishing relations of difference. When we abstract from all concrete contents of the foundations of a relation of difference and only consider them as foundations of such a relation, we obtain the concept of "twoity." We can obtain the concept of "unity" by abstracting away one foundation from the "twoity." This process of distinguishing becomes increasingly difficult for greater numbers, as Husserl will of course underscore in the second section of his *Über den Begriff der Zahl*, and hence we need to analyze greater numbers, of which a complete presentation of all their foundations and relations of difference is impossible, into

[75] Stumpf 1886/87, Q 11/II, 494.

[76] Husserl 1887, 41 f.; 1970, 321.

[77] See Ierna 2006 for a more detailed reconstruction of the provenance and role of these concepts in Husserl's early works.

[78] Stumpf 1939.

[79] From the conclusion to Ierna 2015.

their constituent partial numbers, i.e. $9 = 3 \times 3$. Ultimately, besides the fact that numbers can be considered as multiplicities of other numbers, they essentially are multiplicities of units. Given the complete abstraction from the particularities of the counted objects, Ehrenfels provides the same kinds of examples as Husserl and Bolzano for countable objects: the moon and a grain of sand, but also the moon and hope.

Most of these basic elements are taken over in his 1891 article "*Zur Philosophie der Mathematik*." Ehrenfels provides an extensive critique of Kant and J.S. Mill, in the familiar style of Brentano and Stumpf, and again describes the generation of numbers based on counting by successive addition of unities and the need for symbolic presentations for greater numbers.[80] Ehrenfels also defines the concept of "something" as "the abstract presentation, the most empty of content,"[81] which is quite similar to Husserl's account of "something" as "most general and emptiest of content of all concepts"[82] or "the presentation, empty of content, of the counted something."[83]

This brief and superficial sketch of Ehrenfels' earliest works in the philosophy of mathematics[84] is meant to further contextualize Husserl's early works in the School of Brentano. The mere existence of the (relatively) independent works of Ehrenfels, still of course based on Brentano's original influence and prominently also Meinong's, still do not yet fully reveal the horizontal connections in all their strength. What we would need is an acknowledgement of their existence within the School of Brentano, otherwise they would remain a mere historical curiosity. Such acknowledgement is found first of all in the abundance (for the time) of references made in Ehrenfels' 1891 article to other writings in the Brentanist philosophy of mathematics. Ehrenfels refers i.a. to Husserl's 1887 habilitation thesis, Benno Kerry's series of articles, which had appeared in the *Vierteljahrsschrift für wissenschaftliche Philosophie* between 1885 and 1891, and Konrad Zindler's 1889 prize winning essay "*Zur Theorie der mathematischen Erkenntniss*." That such works belonged together in a common tradition is moreover underscored by the survey of the Brentanist philosophy of mathematics provided in Höfler's 1894 comprehensive review of Kerry's articles, Husserl's *Philosophie der Arithmetik* and Ehrenfels' article. This certainly shows that there was a wider awareness of the cross-connections in the School of Brentano with respect to their common interests and theories with respect to the philosophy of mathematics. Hence, we can truly speak of a criss-crossing network of connections and therefore a Brentanist philosophy of mathematics.

[80]Ehrenfels 1891, 291 f.

[81]*Op. cit.*, 295.

[82]Husserl 1891, 91.

[83]*Op. cit.*, 186.

[84]His much later work on the law of primes based on the concept of Gestalt is less relevant in this context.

Husserl's early works are not only part of this network, but in large measure due to it. He is clearly and openly indebted for many elementary notions to his teachers (in particular Brentano and Stumpf) and was preceded by other students of Brentano in working out a philosophy of mathematics (specifically Stumpf and Ehrenfels). Neverheless, Husserl's early works represent the most extensive and systematic engagements with the philosophy of mathematics in the School of Brentano, both in sheer quantity as well as in detail. The originality of Husserl's early works lies mostly in the eclectic combination and systematization of the theories in the School of Brentano. Moreover, Husserl addresses potential objections and alternative positions (besides the favorite punching-bags Kant and Mill) that were not extensively discussed in the previous literature and he does go deeper into mathematical technicalities. However, as we have seen, even if he would not have had such an extensive mathematical background, there is little in his position that he could not have obtained in the School of Brentano alone.

However, the *Philosophie der Arithmetik* represents but the beginning of Husserl's philosophy, not its end. In what respects and for which reasons did Husserl then go beyond this initial framework of the Brentanist philosophy of mathematics?

7.4 Beyond the *Philosophie der Arithmetik*?

Husserl was very ambivalent about the *Philosophie der Arithmetik*. On the one hand, he kept positively quoting it in his later works, on the other, he privately remarked in 1906 that it was "unripe, naïve, and almost childish," that he was "already beyond it" when it was published,[85] and in 1890 before even publishing it, that it was mistaken in its most fundamental assumption:

> The opinion, by which I was still lead during the elaboration of the *Habilitationsschrift*, that the concept of amount [*Anzahl*] would constitute the foundation of general arithmetic, soon proved itself wrong.[86]

If we take this to mean the fundamental assumption of both the Brentanist philosophy of mathematics as well as the project of arithmetization of analysis, then Husserl's *Philosophie der Arithmetik* would indeed be a jarring patchwork of conflicting theories. However, though taking this into account, they still seems to fit coherently together. Where in the first part Husserl provides psychological analyses of the proper concept of number, i.e. $1 + 1 + 1$ etc., the second part complements this with logical analyses of the symbolic concept of number. These are not necessarily in opposition or mutually exclusive. The key to the correct interpretation that connects his "newer" insights to the original Brentanist framework can be found

[85] Husserl 1956, 294.

[86] Husserl 1994, 158.

in certain remarks spread over various writings. First of all, in his lecture on the concept of number from January 1890 he already pointed out that:

> The fundamental error we made was to overlook the fact that all presentations of number that we have at all beyond about 5, are not given to us as real ones, but only as symbolic ones. This fact, which determines the whole character, sense and purpose of arithmetic, has been generally over looked by logicians and mathematicians, and this is to a considerable extent why they could not arrive at a real understanding of this science. And the goal of our preceding deliberations was to vividly portray to them that an understanding of this science based on the proper concepts of number is impossible.[87]

Then soon after in his letter to Stumpf from February 1890 he claimed that "Already the analysis of the ordinal number led me to this,"[88] i.e. that the idea that amounts would provide the ultimate foundation for mathematics was wrong. We can find such an analysis of ordinals in his unpublished materials for the second volume of the *Philosophie der Arithmetik*, although this particular text, *Arithmetik der Reihen*,[89] was probably originally meant to be included in the first volume.[90] Here, Husserl comes to the insight that indeed one cannot "deduce" other kinds of numbers from amounts, i.e. cardinals, by any "tricks" of "improper presenting,"[91] but that instead:

> They [all number forms developed up to now] can be conceived independently from each other, and even, when they are not, each area requires a special arithmetic for itself. The fact that this arithmetic possesses the same formal rules everywhere, i.e., that the algorithm is identical, that is a fact in itself.[92]

You cannot start out with the proper concept of number, proper presentations of cardinals, and then by mentally squinting end up with ordinals, i, π, or $\sqrt{2}$: "The concept of amount does not admit of any expansion; what is expanded and admits of expansion, is only the *arithmetical technique*."[93] This is a story we know very well from his alter writings: how the difficulties of calculating with contradictory concepts (i.e. $\sqrt{-1}$) led him to the position developed in his famous Göttinger *Doppelvortrag*[94] on the "Passage through the Imaginary."[95] Yet, this very lecture was based "on older mathematical manuscripts," and is therefore still significantly connected to and dependent on his analyses from the last phases of the *Philosophie der Arithmetik*. What, then, is his ultimate position in this phase? As he expresses it in his letter to Stumpf, echoing his analysis of the ordinals:

---

[87] Husserl 2005, 297.

[88] Husserl 1994, 158.

[89] Husserl 1983, 154–214.

[90] See Ierna 2005, 47–48, n. 170.

[91] Husserl 1994, 158.

[92] Husserl 1983, 175.

[93] *Op. cit.*, 42 f.

[94] Schuhmann & Schuhmann 2001.

[95] Also see Ierna 2011b.

> Since I originally only saw signs in relation to signified concepts, signs such as e.g., $\sqrt{2}$, $\sqrt{-1}$, and similar ones, had to be representations of "impossible" concepts in the case of amounts. I tried thus at first to understand clearly, how thought operations with contradicting concepts could lead to correct sentences.... Finally, I noticed that through the calculation itself and through its rules, as defined for those fictional numbers, the impossible vanishes and a correct equation remains.... Thus the signs and their rules must do it.[96]

Since "All higher mathematical signs are indirect, are highly stacked signs of signs of signs,"[97] we can certainly simply operate with the signs according to "blind psychological rules."[98] However, Husserl then still would have the problem that there would be "signs that don't designate anything, signs without any interpretable meaning," which seems absurd. Husserl is still against the kinds of nominalism of Helmholtz and Kronecker, and indeed claims to have reached a position that "formally harmonizes" with that of Helmholtz, though being founded completely differently. This foundation implies the possibility of an ultimate reduction:

> Higher numbers appear at first as symbolic presentations which are composed from symbolic number formations, these again, which perhaps are likewise [composed of symbolic number constructions], but in such a way that ultimately we reach the small but for the conception of the concept of number indispensable field of real [i.e. properly conceived] numbers.[99]

For, e.g. imaginary numbers, this would imply that we treat them as temporary assumptions: "Every imaginary relation is, regarded for itself, a fictional one. However, in so far as it is reducible on a correct equation, it contains a piece of factual truth."[100] So if we follow his account in the letter to Stumpf, that mathematics is a system of concentric circles,[101] which anticipates the later theories in his *Doppelvortrag*, we see that this is not in essence so very different from the Brentanist philosophy of mathematics. Higher numbers are reducible to lower ones, until we reach a properly-presentable rock bottom. We do need a number system to generate the signs and concepts of these higher numbers systematically and unambiguously, which at the same time provides the means for a reduction, i.e. for solving equations. Husserl's key insight is that the really important element here are the rules of calculation, the algorithm, and not what you feed into it. If the rules are right, the result will be right ("*die Rechnung bleibt richtig, wenn sie regelrecht ist*"): proper or improper concepts of number or even impossible ones like $\sqrt{-1}$ don't matter.

Bringing together these remarks, spread over various writings, but forming a coherent whole, explains Husserl's position in the preface and in the conclusion

[96] Husserl 1994, 160.

[97] Husserl 1970, 344. This is from his treatise on Semiotik, which was meant as an appendix to the Philosophie der Arithmetik as a whole, also see 358, 368.

[98] Husserl 1970, 357.

[99] Husserl 2005, 301.

[100] *Op. cit.*, 307.

[101] Husserl 1994, 161.

of the *Philosophie der Arithmetik.* As he claims, his observations in the second volume, would have shown that it is not the amount concept (or any specific concept of number) that is fundamental, but the algorithm: the *arithmetica universalis*. In the last line of the book he then points forward again that general arithmetic would need to be developed as "a general theory of operations,"[102] i.e. actually as part of a more encompassing formal logic as an "art of signs."[103]

While the *Philosophie der Arithmetik* can be seen as a work in the Brentanist philosophy of mathematics, what will take Husserl beyond this framework is the recognition of the ideality of the objects and truths implied by the ideal validity of the sign system and its operations.[104] This, however, did not happen overnight. Nevermind his own retrospective rhetoric or facile historiography, Husserl's progress did not proceed by sudden breakthroughs that exchange one paradigm for the next, but by painstaking stepwise conquest. Still in his 1896 logic lectures, what makes the *arithmetica numerosa* as well as the *arithmetica universalis* work, is the parallelism between signs and concepts:

> To every correct derivation [*regelrechten Herleitung*] thus corresponds a result that, when interpreted conceptually, yields a correct proposition. This is due to the precise parallelism between mental operations and operations on signs.[105]

Still in 1896, like in the *Philosophie der Arithmetik*, we see Husserl claim that all defining comes to an end with the most primitive concepts[106] and that "one" technically is not a number, but "two" would be the first one.[107] Signs here still serve the role that Brentano identified for them: to avoid the ambiguities of natural language.[108] After introducing the *Inbegriff* again as collective unity of "something and something etc."[109] Husserl here explicitly connects the notion of *Inbegriff* to that of *Mannigfaltigkeit*,[110] establishing an important link between his earlier and later works. Indeed, considering his later recollections about the development of his *Logische Untersuchungen*, we can see a strong continuity with his work on the second volume of the *Philosophie der Arithmetik*, which was to deal i.a. with the arithmetical algorithm in other domains.[111]

Of course, his position becomes more Bolzanian and less Brentanian, and increasingly anti-psychologistic, but then again Husserl's *Philosophie der Arithmetik* wasn't really a psychologistic work either, and I doubt whether the Brentanist

---

[102] Husserl 1891, 323.
[103] Husserl 1994, 161.
[104] See Husserl 1900, 175.
[105] Husserl 2001, 247–248.
[106] *Op. cit.*, 52.
[107] *Op. cit.*, 102.
[108] *Op. cit.*, 252.
[109] *Op. cit.*, 86 f.
[110] *Op. cit.*, 100.
[111] Husserl 2002, 296 f.; compare Haddock 2006, 194.

philosophy of mathematics as such could be called psychologistic. After all, from the very first works in this tradition, the attempt was made to establish mathematics as a wholly presuppositionless, analytic, and a priori science. Husserl's innovation is not to have developed a formal mathematics or logic out of the Brentanist psychologistic framework, but rather to have adapted the Brentanist philosophy of mathematics to logic. Indeed, his main innovation is to have been able to separate the practical logic as "art of signs" from theoretical logic as pure and foundational discipline. This distinction was enabled by the already existing distinction between proper and symbolic mathematics and the recognition that the latter was epistemically independent of the former: the independence of the algorithm with respect to the concepts it is applied to.

From his 1890 lecture on the concept of number and the letter to Stumpf, Husserl's position does not appear to change radically. Compare his pronouncements on "the signs and their rules must do it," to his statement in his 1895 lectures on deductive logic: "when we calculate in arithmetic we only care about the signs and the rules of their connection [*Verknüpfung*]."[112] This would be a kind of nominalism or formalism if Husserl would then not add, as echo to the "indispensable field of properly conceived numbers," that "when I then go over from the symbols and their conventional meaning to the proper and original concepts, then I have a proposition about numbers and this proposition is then always correct."[113] What is new is his additional quest for the justification of the application of the signs and rules, not the separation between psychology and logic in the domain of formal sciences. Husserl does not wholly substitute Bolzano for Brentano, but simply recognize that descriptive psychology cannot give any foundational insight into such a justification. For that we will need a pure logic. In 1895, however, the justification lies still in the parallelism between concepts and signs and between actual and mechanical deduction.[114] As in his treatise on ordinals, he remarks that we can develop the algorithm starting from either cardinals, ordinals, etc. and then apply it to all others, but then adds much more clearly and explicitly that we could also develop the system straightaway *in abstracto*.[115] This additional possibility, of defining formal numbers directly, leads straight to his discussion of the theory of manifolds in the *Prolegomena*. When we start from a given concept of number, the rules we would end up with would actually be synthetic a priori. It can be purely analytic only if we construct the concept in reflection on the form of the rules.[116] Due to this switch in perspective, any reduction to a "proper" domain becomes arbitrary and the foundational irrelevance of any pre-given concept of

[112]Husserl 2001, 309. Husserl elsewhere (318) suggests that instead of Verknüpfung we could also say Operation.

[113]*Op. cit.*, 311.

[114]*Op. cit.*, 311 f.

[115]*Op. cit.*, 314 f.

[116]*Op. cit.*, 315.

number becomes obvious.[117] The solution to the problem of giving an analytic a priori foundation of mathematics in the Brentanist philosophy of mathematics can only be found by going beyond its framework to a larger more encompassing account, which subsumes it without invalidating it.

Acknowledgements This chapter is an outcome of the project "From Logical Objectivism to Reism: Bolzano and the School of Brentano" P401 15-18149S (Czech Science Foundation), realised at the Institute of Philosophy of the Czech Academy of Sciences.

References

L. Albertazzi, M. Libardi, R. Poli (eds.), *The School of Franz Brentano* (Kluwer, Dordrecht, 1996)
B. Bolzano, *Paradoxien des Unendlichen* (Mayer & Müller, Berlin, 1889)
F. Brentano, *Psychologie vom empirischen Standpunkte* (Duncker & Humblot, Leipzig, 1874)
F. Brentano, Die elementare Logik und die in ihr nötigen Reformen I. Manuscript Y 2, Vienna 1884/85
G. Cantor, *Grundlagen einer allgemeinen Mannichfaltigkeitslehre. Ein mathematisch-philosophischer Versuch in der Lehre des Unendlichen* (Teubner, Leipzig, 1883)
S. Centrone, *Logic and Philosophy of Mathematics in the Early Husserl* (Springer, Dordrecht, 2010)
R. Dedekind, *Was sind und was sollen die Zahlen?* (Vieweg & Sohn, Braunschweig, 1888)
G. Frege, *Grundlagen der Arithmetik. Eine logisch mathematische Untersuchung über den Begriff der Zahl* (Koebner, Breslau, 1884)
G.R. Haddock, Husserl's relevance for the philosophy and foundations of mathematics. Axiomathes **1–3**, 125–142 (1997)
G.R. Haddock, Husserl's philosophy of mathematics: Its origin and relevance. Husserl Stud. **22**, 193–222 (2006)
M. Hartimo, Mathematical roots of phenomenology: Husserl and the concept of number. Hist. Philos. Log. **27**, 319–337 (2006)
A. Höfler, Besprechung von Kerry, 'Über Anschauung und ihre psychische Verarbeitung', Husserl, Philosophie der Arithmetik, von Ehrenfels, 'Zur Philosophie der Mathematik', Zeitschrift für Psychologie und Physiologie der Sinnesorgane VI (1894)
E. Husserl, *Beiträge zur Variationsrechnung*, Vienna 1883, Unpublished dissertation
E. Husserl, *Über den Begriff der Zahl: Psychologische Analysen*, Heynemann'sche Buchdruckerei (F. Beyer), Halle (1887)
E. Husserl, *Philosophie der Arithmetik. Psychologische und Logische Untersuchungen* (C.E.M. Pfeffer (Robert Stricker), Halle, 1891)
E. Husserl, *Logische Untersuchungen. Erster Theil: Prolegomena zur Reinen Logik* (Niemeyer, Halle, 1900)
E. Husserl, Persönliche Aufzeichnungen. Philos. Phenomenol. Res. **XVI**, 293–302 (1956)
E. Husserl, in *Philosophie der Arithmetik. Mit ergänzenden Texten* (1890–1901), ed. by L. Eley Husserliana XII. (Nijhoff, Den Haag, 1970)
E. Husserl, *Studien zur Arithmetik und Geometrie*, ed. by I. Strohmeyer Husserliana XXI (Nijhoff, Den Haag, 1983)
E. Husserl, *Briefwechsel.* in *Die Brentanoschule*, vol. I, ed. by K. Schuhmann, E. Schuhmann (Kluwer, Dordrecht, 1994)

[117] Something Husserl apparently didn't fully appreciate yet in 1891, 292–294.

E. Husserl, in *Logik: Vorlesung* 1896 ed. by E. Schuhmann, *Husserliana*, Materialienbände I. (Kluwer, Dordrecht, 2001)
E. Husserl, in *Logische Untersuchungen* (Ergänzungsband: Erster Teil), ed. by U. Melle, *Husserliana* XX/1. (Kluwer, Dordrecht, 2002)
E. Husserl, *Lecture on the Concept of Number* (WS 1889/90). C. Ierna (ed. & tr.), The new yearbook for phenomenology and phenomenological philosophy V (2005)
C. Ierna, Husserl and the infinite. Studia phaenomenologica III No. 1–2 (2003), 179–194
C. Ierna, *The beginnings of Husserl's Philosophy (Part 1: From Über den Begriff der Zahl to Philosophie der Arithmetik).* The new yearbook for phenomenology and phenomenological philosophy V (2005), 1–56
C. Ierna, *The beginnings of Husserl's Philosophy (Part 2: Philosophical and mathematical background).* The new yearbook for phenomenology and phenomenological philosophy VI (2006), 23–71
C. Ierna, Husserl et Stumpf sur la Gestalt et la fusion. Philosophiques **36**(2), 489–510 (2009)
C. Ierna, Brentano and mathematics. Rev. Roum. Philos. **55**(1), 149–167 (2011a)
C. Ierna, Der Durchgang durch das Unmögliche. *An Unpublished Manuscript from the Husserl-Archives*. Husserl Stud. **27**(3), 217–226 (2011b)
C. Ierna, *La notion husserlienne de multiplicité : au-delà de Cantor et Riemann.* Methodos 12 (2012)
C. Ierna, Carl Stumpf's philosophy of mathematics, in Philosophy from an Empirical Standpoint: Essays on Carl Stumpf, ed. by D. Fisette, R. Martinelli (Rodopi, Amsterdam, 2015)
C. Ierna, On Christian von Ehrenfels' Dissertation, Meinong Studies 8 (2017)
M. Kline, *Mathematical Thought*, vol 1 (Oxford University Press, New York, 1972)
L. Königsberger, *Vorlesungen über die Theorie der elliptischen Functionen nebst einer Einleitung in die allgemeine Functionenlehre*, vol 1 (Teubner, Leipzig, 1874)
L. Kronecker, Über den Zahlbegriff. Crelle's J. Reine Angew. Math. **101**, 337–355 (1887)
E. Mach, *Beiträge zur Analyse der Empfindungen* (Gustav Fischer, Jena, 1886)
J. P. Miller, *Numbers in Presence and Absence*, Phaenomenologica 90 (Nijhoff, Den Haag, 1982)
D. Moran, *Introduction to Phenomenology* (Routledge, London, 2000)
R. Rollinger, *Husserl's Position in the School of Brentano*, Phaenomenologica 150 (Kluwer, Dordrecht, 1999)
R. Rollinger, Brentano's logic and Marty's early philosophy of language. Brentano Studien **XII**, 77–98 (2009)
K. Schuhmann, *Husserl-Chronik (Denk- und Lebensweg Edmund Husserls)* (Nijhoff, Den Haag, 1977)
K. Schuhmann, Malvine Husserls "Skizze eines Lebensbildes von E. Husserl". Husserl-Studies **5**, 105–125 (1988)
K. Schuhmann, Brentano's impact on twentieth-century philosophy, in The Cambridge Companion to Brentano, ed. by D. Jacquette (Cambridge University Press, Cambridge, 2004), pp. 277–297
E. Schuhmann, K. Schuhmann, Husserls Manuskripte zu seinem Göttinger Doppelvortrag von 1901. Husserl Stud. **17**(2), 87–123 (2001)
H. R. Sepp (ed.), *Edmund Husserl und die Phänomenologische Bewegung. Zeugnisse in Text und Bild* (Alber, Freiburg München, 1988)
D. Spalt, Die mathematischen und philosophischen Grundlagen des Weierstraßschen Zahlbegriffs zwischen Bolzano und Cantor. Arch. Hist. Exact Sci. **41**(4), 311–362 (1991)
C. Stumpf, *Tonpsychologie*, vol 1 (Hirzel, Leipzig, 1883)
C. Stumpf, *Tonpsychologie*, vol 2 (Hirzel, Leipzig, 1890)
C. Stumpf, *Vorlesungen über Psychologie*, manuscript Q 11/I, Halle WS 1886/87 (unpublished lecture notes)
C. Stumpf, *Logik und Enzyklopädie der Philosophie*, manuscript Q 14, Halle SS 1887 (unpublished lecture notes)
C. Stumpf, *Erkenntnislehre* (Barth, Leipzig, 1939)
C. Stumpf, in *Über die Grundsätze der Mathematik*, ed by W. Ewen (Königshausen & Neumann, Würzburg, 2008)

C. von Ehrenfels, Zur Philosophie der Mathematik. Vierteljahrsschrift für wissenschaftliche Philosophie **15**, 285–347 (1891)
C. von Ehrenfels', Größenrelationen und Zahlen: eine psychologische Studie, ed. by C. Ierna, Meinong Studies **8** (2017)
D. Willard, *Logic and the Objectivity of Knowledge* (Ohio University Press, Athens, 1984)
K. Zindler, Beiträge zur Theorie der mathematischen Erkenntnis. Sitzungsberichte der Wiener Akademie. *Philos-Histor. Abth.*, Bd. 118 (1889)

Chapter 8
Husserl and Cantor

Claire Ortiz Hill

Abstract Husserl and Cantor were colleagues and close friends during the last 14 years of the nineteenth century, when Cantor was at the height of his creative powers and Husserl in the throes of an intellectual struggle during which he drew apart from people and writings to whom he owed most of his intellectual training and drew closer to the ideas of thinkers whose writings he had not been able to evaluate properly and had consulted too little. I study ways in which Husserl and Cantor might be said to have been alike, while pointing to dissimilarities between them. In particular, I discuss how their ideas overlapped and crisscrossed with regard to mathematics and philosophy, Platonic idealism, abstraction, empiricism, psychologism, actual consciousness and pure logic, Frege's reviews of their works, metaphysics and mysticism, sets, arithmetization, strange and imaginary numbers and manifolds. I conclude that Cantor was among those of his mentors from whose ideas Husserl drew away and Lotze and Bolzano were among those to whose ideas he drew closer.

Keywords Abstraction • Idealism • Consciousness • Metaphysics • Sets

8.1 Introduction

In 1886, Edmund Husserl arrived at the University of Halle to prepare his *Habilitationsschrift*, which would be entitled *Über den Begriff der Zahl* (*On the Concept of*

C.O. Hill (✉)
Independent Scholar, Paris, France
e-mail: claire.hill@orange.fr

S. Centrone (ed.), *Essays on Husserl's Logic and Philosophy of Mathematics*, Synthese Library 384, DOI 10.1007/978-94-024-1132-4_8

Number).[1] He studied the calculus of probabilities with Georg Cantor,[2] the creator of set theory, who then served on his *Habilitation* committee and approved the mathematical portion of the work.[3]

Cantor took a liking to his younger colleague, and the two became close friends during the last 14 years of the nineteenth century, when Cantor was at the height of his creative powers and Husserl in the throes of an intellectual struggle during which his ideas changed considerably and definitively. During those years, he published the *Philosophy of Arithmetic*,[4] numerous articles and reviews[5] and the groundbreaking *Logical Investigations*,[6] where he began laying the foundations of his phenomenology that went on to shape the course of philosophy in Continental Europe.

While Husserl labored as a *Privatdozent*—meaning that he gave courses, but did not have a regular position—Cantor multiplied efforts to find him a professorship. Surviving books of Cantor's correspondence show him "most warmly" recommending Husserl in "good conscience" for a position of full professor. He praised Husserl's "exceptional academic qualifications beyond all question" and cited the great respect generated by his writings, the *Philosophy of Arithmetic*, in particular. Cantor described Husserl as "a person highly-valued and generally loved by us because of his peaceable and sterling character", as thoroughly honest and irenically disposed and assured that "his personality and character justify the belief that he would discharge his duties with desirable tact". He said that he knew for certain from courses that Husserl had given on the proofs for the existence of God and against Darwinism that he was a theist. In response to reservations about the fact that Husserl was a Jewish convert to Protestantism, Cantor appealed to the "holiest

[1] *Über den Begriff der Zahl* is published in Husserl, *Philosophie der Arithmetik. Mit ergänzenden Texten (1890–1901)*, *Husserliana* vol. XII, henceforth cited as *PdA*, 289–338, and in English as "*On the Concept of Number: Psychological Analyses*" in *Philosophy of Arithmetic, Psychological and Logical Investigations with Supplementary Texts from 1887–1901*, henceforth cited as *PoA*, 305–357.

[2] Hereto cp. Hill & da Silva 2013, 368.

[3] Husserl's time in Halle is studied in Gerlach and Sepp 1994.

[4] *PdA* 1–283; *PoA* 1–299.

[5] Many of the articles and reviews were published in *Aufsätze und Rezensionen (1890–1910)*, *Husserliana* vol. XXII. They are translated and published Husserl's *Early Writings in the Philosophy of Logic and Mathematics*, vol. V of Husserl's *Collected Works*, which also includes translations of short works published in *PdA* and *Einleitung in die Logik und Erkenntnistheorie, Vorlesungen 1906/07*, *Husserliana* vol. XXIV.

[6] Husserl, *Logische Untersuchungen*, Bd I: *Prolegomena zur reinen Logik*, Max Niemeyer, Halle, 1900 (^{2}1913), henceforth cited as *PRe*. Bd II: *Untersuchungen zur Phänomenologie und Theorie der Erkenntnis*, Max Niemeyer, Halle a.d.S., 1901 (^{2}1913–1921). English translation: *Logical Investigations* (J. N. Findlay, translator), Routledge and Kegan Paul, New York 1970, henceforth cited as *LI*.

traditions of the Church" and to the fact that "the most holy person of our Savior, his twelve apostles were all Jews."[7]

In 1901, when he was in his forties, Husserl was appointed associate professor at the University of Göttingen. Cantor died in 1918 after bouts of mental illness. Of the relationship between the two men, Husserl's wife recalled that "Cantor, the greatest mathematician since Gauss, the creator of set theory" loved her husband tenderly ("*liebte H. zärtlich*"). She added: "They were alike in many ways, but with otherwise great dissimilarity. The Cantors' house was like home ...".[8] In what follows, I discuss respects in which Husserl and Cantor might be said to have been alike while pointing to dissimilarities between them.

8.2 On the Borderline Between Mathematics and Philosophy

In the 1880s, Husserl and Cantor (along with Frege) figured among the small number of their contemporaries intent upon marrying mathematics and philosophy. A mathematician by training,[9] Husserl had just been transformed into a philosopher by Franz Brentano.[10] In *On the Concept of Number*, he wrote of definitively removing the real and imaginary difficulties on the borderline between mathematics and philosophy. He regretted that mathematicians had failed to examine the logic of the concepts and methods they were introducing and using and he saw the need for logical clarification, precise analyses.[11]

According to its 1882 foreword, Cantor's *Mannigfaltigkeitslehre* was a work, "written with two groups of readers in mind—philosophers who have followed the developments in mathematics up to the present time, and mathematicians who are familiar with the most important older and newer publications in philosophy."[12] Joseph Dauben has noted that in the *Mannigfaltigkeitslehre*, "Cantor made philosophy an equal and intentional partner to mathematics.... In the German version issued as a separate monograph by Teubner in 1883, a simple introduction was added in which he stressed that the mathematical and the philosophical sections were inextricably connected."[13]

During the late 1880s, the embattled creator of set theory was in fact hard at work trying to put the new numbers he was inventing on solid foundations and

[7]See Cantor letter's 240, 264, 291 as translated and published in Hill & da Silva 2013, specifically pages 367–369, 374–375, 377.

[8]See M. Husserl 1988, §E.

[9]Husserl's training in mathematics is discussed at length in my essay "On Husserl's Mathematical Apprenticeship and Philosophy of Mathematics" in Hill 2002a.

[10]See Hill 1998.

[11]See *PdA* 294; *PoA* 310.

[12]Quotes after Hallett 1984, 7.

[13]Dauben 1979, 120.

philosophically justifying the claims he was making about them.[14] Those attempts found expression in his correspondence and most particularly in the "Mitteilungen zur Lehre vom Transfiniten",[15] which he said he published in the *Zeitschrift für Philosophie und philosophische Kritik* because he had grown disgusted with mathematical journals.[16]

Over the years, Cantor's thoughts had been turning increasingly to philosophy and he was ready to abandon mathematics for philosophy. He had come to see himself as a sort of philosopher *manqué*. He tried to teach philosophy[17] and was seasoning his writings with philosophical reflections and references to Democritus, Plato, Aristotle, Augustine, Boethius, Aquinas, Descartes, Nicolas von Cusa, Spinoza, Leibniz, Kant, Comte, Francis Bacon, Locke and so on. In 1883, Gösta Mittag-Leffler saw fit to warn him that his work would be much better received in the mathematical world "without the philosophical and historical explanations"[18] and in 1885 that he risked shocking most mathematicians and damaging his reputation with his philosophical way of expressing himself.[19] Despite those warnings, Cantor continued to clothe his theories in philosophical garb.

8.3 Cantor's Platonic Idealism

In the *Mannigfaltigkeitslehre*, Cantor emphasized that the idealist foundations of his theories were essentially in agreement with the basic principles of Platonism according to which only conceptual knowledge in Plato's sense afforded true knowledge.[20] By "manifold" or "set", he said, he was defining something related to the Platonic *eidos* or *idea*.[21] He once wrote to Giuseppe Peano that he conceived of numbers as 'forms' or 'species', general concepts, of sets and that in essentials that was the conception of the ancient geometry of Plato, Aristotle, Euclid etc.[22] His talk of awakening and bringing to consciousness the knowledge, concepts and numbers

---

[14]See Dauben 1979, Chapter 6, 120–148, "Cantor's Philosophy of the Infinite" for an in depth discussion.

[15]Cantor 1887/8.

[16]See Dauben 1979, 139, 336 n. 29.

[17]See his letter of October 20 and 28, 1884 to Gösta Mittag-Leffler as published in Cantor 1991, 210 and 218 n. 3; Dauben 1979, 282, 337 n. 31.

[18]See page 118 of his letter of March 3, 1883 to Gösta Mittag-Leffler as published in Cantor 1991, 118.

[19]*Op. cit.*, 244.

[20]Cantor 1883, 181, 206 n. 6.

[21]*Op. cit.*, 204 n. 1.

[22]See Cantor's letter of September 21, 1895 to Giuseppe Peano as published in Cantor 1991, 365.

slumbering in us[23] was an unmistakable allusion to Plato's theory of recollection and Socratic theories of concept formation.

In the "Mitteilungen", Cantor described the realm of the transfinite as "a rich, ever growing field of ideal research" and his transfinite numbers as a special form of Plato's *arithmoi noetoi* or *eidetikoi*.[24] There, he was particularly intent upon proving that his theorems about transfinite numbers were firmly secured "through the logical power of proofs" which, proceeding from his definitions were "neither arbitrary nor artificial, but originate naturally out of abstraction, have, with the help of syllogisms, attained their goal."[25]

He considered that his technique for abstracting numbers from reality provided the only possible foundations for his Platonic conception of numbers.[26] In the "Mitteilungen", he repeatedly gave the same recipe for extracting cardinal numbers from reality through abstraction[27]: by abstracting from both the characteristics of the elements of the set and the order in which they are given, one obtains the cardinal numbers; abstracting only from the characteristics of the elements and leaving their order intact, one obtains the ideal numbers or *eidetikoi*.[28]

Cantor saw his theory of abstraction as the distinctive feature of his number theory and believed it was an entirely different method for providing the foundations of the finite numbers than was found in the theories of his contemporaries. He envisioned it as a technique for focusing on pure, abstract arithmetical properties and concepts that would divorce them from any sensory apprehension of the particular characteristics of the objects figuring in the sets and so free mathematics from psychologism, empiricism, Kantianism and insidious appeals to intuitions of space and time. He believed that with it he was laying bare the roots from which the organism of transfinite numbers develop with logical necessity. A good measure of the freedom he felt was his as a mathematician derived from his distinguishing between an empirical treatment of numbers and Plato's pure, ideal *arithmoi eidetikoi*, which by their very nature are detached from things perceptible by the senses.[29]

In his 1885 review of the *Foundations of Arithmetic*, Cantor praised Frege for requiring that all psychological factors and intuitions of space and time be banned from arithmetical concepts and principles because that was the only way their strict

[23]Cantor 1883, 207 n. 6, 7, 8; Cantor 1887/8, 418 n. 1.

[24]Cantor 1887/8, 406, 420.

[25]*Op. cit.*, 418.

[26]See Cantor's letters of September 14 and 21, 1895 to Giuseppe Peano as published in Cantor 1991, 363, 365 respectively; Cantor 1887/8, 380, 411.

[27]Cantor 1887/8, 379, 387, 411, 418 n. 1.

[28]Cantor 1887/8, 379–380; Cantor 1883, 180–181; also see Cantor's letters of November 13 and 17, 1890 to Giuseppe Veronese as published in Cantor 1991, 329, 330 respectively.

[29]Cantor 1883, 191–192; Cantor 1887/8, 380–381 n. 1, 411; see Cantor's letters of September 14 and 21, 1895 to Giuseppe Peano as published in Cantor 1991, 363, 365, respectively. Hereto cp. Hill 1999.

logical purity and validity might be secured.[30] In so doing, Cantor was in step with Karl Weierstrass' aims, which Husserl also shared. Section 2 of *On the Concept of Number* and Chapter 2 of the *Philosophy of Arithmetic* were devoted to discussions aimed at obtaining pure arithmetical concepts by detaching the concept of number from any spatio-temporal intuitions and so also from any taint of Kantianism in keeping with goals Weierstrass had set.[31] The early years of the twentieth century would find Husserl teaching that numbers could not concern what happens in or to the real temporal matters of fact that we call mental experiences of experiencing individuals. He stated unequivocally that Kant had brought pure arithmetic into an entirely inadmissible relationship to time.[32]

8.4 Husserl's Theory of Abstraction

Husserl owned offprints of the "Mitteilungen" and marked and underlined precisely (and almost exclusively) those passages in which Cantor explained the abstraction process. Calling Cantor a mathematician of genius, Husserl further commended him for having written with a great deal of precision in the "Mitteilungen" that for "the formation of the general concept 'five' one needs only a set (for example all the fingers of my right hand) which corresponds to this cardinal number; the act of abstraction with respect to both the properties and the order in which I encounter these wholly distinct things, produces or rather awakens the concept 'five' in my mind."[33]

In *On the Concept of Number*, Husserl himself characterized the distinctive abstractive process yielding the concept of number of a concrete set of objects as follows[34]:

> It is easy to characterize the abstraction which must be exercised upon a concretely given multiplicity in order to attain to the number concepts under which it falls. One considers each of the particular objects merely insofar as it is a "something" or a "one," simultaneously retaining the collective combination; and in this manner there is obtained the corresponding general form of multiplicity, one and one and ... and one, with which a number name is associated. In this process there is total abstraction *from* the specific characteristics of the particular objects. But this neither means nor implies that the concrete objects have to disappear from our consciousness. To "abstract" from something merely means to pay no special attention to it. Thus, also in our case at hand, no special interest is directed upon the peculiarities of content in the separate individuals....

[30] Cantor 1885, 440.

[31] "On the Concept of Number: Psychological Analyses" as published in *PdA* 303–304; *PoA* 320–321.

[32] Husserl 1906/07 §§11, 13b, 23.

[33] Cantor 1887/8, 418 n. 1; *PdA* 116 n.; *PoA* 121 n. 3.

[34] *PdA* 294, 337; *PoA* 310, 354–355.

In *Philosophy of Arithmetic*, he studied this same process in greater detail, especially underscoring the uniqueness of the abstraction process that yields the number concept. He made a point of disassociating his theory from the better known abstraction theories of Locke and Aristotle, something it is important to call attention to because philosophers have been all too wont to assimilate the process Husserl advocated to theories more familiar to them.[35]

Since Cantor was propounding this theory of abstraction at the very time Husserl was writing *On the Concept of Number* and the *Philosophy of Arithmetic* where a similar theory was advocated, it is worth noting here that in the latter work, Husserl approvingly noted that the definition of number in the "Mitteilungen" was profoundly different from that of the *Mannigfaltigkeitslehre*.[36] He specifically pointed to two passages of the "Mitteilungen." In the first, Cantor had written[37]:

> By the *power* or *cardinal number* of a set *M* (which is made up of distinct, conceptually separate elements *m*, *m*', . . . and is to this extent determined and limited), I understand the general concept or species concept (universal) obtained by abstracting from the properties of the elements of the set, as well as from all the relations which the elements may have, whether to themselves or to other things, but especially from the order reigning among the elements and only reflect upon what is common to all sets *equivalent* to *M*.

8.5 Psychologism

However, as similar as Husserl's and Cantor's theories of abstraction might seem to have been in the late 1880s, there was a major difference between them that reflected deep contradictions in both men's philosophies of arithmetic and soon brought Husserl to transform his entire approach to philosophy. For, however psychologistic Cantor's mysterious references to inner intuition[38] or to experiences helping produce concepts in his mind[39] may seem, he opposed empiricism, naturalism, sensualism, skepticism, psychologism and Kantianism which, he argued, mistakenly locate the sources of knowledge and certainty in the senses or in the "supposedly pure forms of intuition of the world of presentation." He maintained that certain knowledge could "only be obtained through concepts and ideas which, at most stimulated by external experience, are on the whole formed through inner induction and deduction as something which in a way already lay within us and was only awakened and brought to consciousness."[40]

[35] *PdA* 86; *PoA* 89–90. Hereto cp. Hill 1999.

[36] *PdA* 116 n. (see above); *PoA* 121 n. 3

[37] Cantor 1887/8, 387; also see Cantor's letter of February 15, 1884 to Kurd Lasswitz, as published in Cantor 1991, 178.

[38] Ex. Cantor 1883, 168, 170, 201.

[39] Cantor 1887/8, 418 n. 1.

[40] Cantor 1883, 207 n. 6.

Husserl, on the other hand, began writing *On the Concept of Number* as a committed empirical psychologist *à la* Brentano, whose philosophical ideal was most nearly realized in the exact natural sciences.[41] There and in the *Philosophy of Arithmetic*, he tried to anchor arithmetical concepts in direct experience by analyzing the actual psychological processes to which he believed the concept of number owed its genesis. He considered the analysis of elementary concepts to be one of the more essential tasks of psychology. "In truth", he declared in *On the Concept of Number*, "not only is psychology indispensable for the analysis of the concept of number, but rather this even *belongs within* psychology."[42] "In view of my entire training", he later recalled, "it was obvious to me when I started that what mattered most for a philosophy of mathematics was a radical analysis of the 'psychological origin' of the basic mathematical concepts."[43]

However, Husserl's confidence in empirical psychology eroded quickly and the enthusiastic espousal of it in *On the Concept of Number* does not appear in the *Philosophy of Arithmetic*, —a sign that change was on the horizon. In his 1913 preface to the *Logical Investigations*, he confessed to having been "unsettled – even tormented" by doubts about the psychological analysis of sets from the very beginning. He came to believe that the concept of number had to be something basically different from the concept of collecting, which was all that could result through reflection on acts. He said that his doubts extended to all categorial concepts and to concepts of objectivities of any sort whatsoever, ultimately to include modern analysis and the theory of manifolds, and simultaneously mathematical logic and the entire field of logic in general. He could find no way to reconcile the objectivity of mathematics with psychological foundations for logic.[44]

8.6 Husserl's Idealistic Turn

Franz Brentano taught his students to despise metaphysical idealism. So, Husserl came to Halle quite disinclined to traffic in the kind of idealism that pervades Cantor's writings,[45] and any such considerations are conspicuously absent from *On the Concept of Number* and *Philosophy of Arithmetic*. Notwithstanding, he left Halle persuaded that there was "an essential, quite unbridgeable difference between the sciences of the ideal and the sciences of the real", the correct assessment of which presupposed "the complete abandonment of the empiricistic theory of abstraction, whose present dominance renders all logical matters unintelligible."[46] He began

[41] Husserl 1919, 344–345. Hereto cp. Hill 1998.

[42] *PdA* 295, *PoA* 311.

[43] Husserl 1913, 33.

[44] Husserl 1913, 16–17; 34–35.

[45] Husserl 1919, 344–345.

[46] *PRe* 185.

to accord idealist systems the highest value and to see them as shedding light on totally new, radical dimensions of philosophical problems. The ultimate and highest goals of philosophy, he came to believe, are only opened up when the philosophical method that those particular systems call for is clarified and developed.[47] Of his *Logical Investigations*, he stated that every possible effort had been made "to dispose the reader to the recognition of this ideal sphere of being and knowledge ... to side with 'the ideal in this truly Platonistic sense', 'to declare oneself for idealism' with the author".[48] He had become a committed Platonic idealist persuaded that pure mathematics was a strictly self-contained system of doctrines to be cultivated by using methods essentially different from those of natural science.[49] He wrote to Brentano in 1905[50]:

> The empirical sciences—natural sciences—are sciences of 'matters of fact'.... Pure Mathematics, the whole sphere of the genuine Apriori in general, is free of all matter of fact suppositions.... We stand not within the realm of nature, but within that of Ideas, not within the realm of empirical ... generalities, but within that of the ideal, apodictic, general system of laws, not within the realm of causality, but within that of rationality.... Pure logical, mathematical laws are laws of essence....

Phenomenology was to be an "eidetic" discipline. The "whole approach whereby the overcoming of psychologism is phenomenologically accomplished", Husserl explained, "shows that what ... was given as analyses of immanent consciousness must be considered as a pure *a priori* analysis of essence."[51]

However, although Cantor's ideas probably helped pry Husserl away from psychologism and turn him in the direction of Platonic idealism, he always maintained that Hermann Lotze's work was responsible for his conscious, radical rejection of psychologism and the accompanying Platonism. Lotze's interpretation of Plato's doctrine of Ideas, Husserl said, lent him his first major insight, became a determining factor in all his further studies, and gave him the idea to transfer all of mathematics and a major part of traditional logical into the realm of the ideal. Lotze's work further helped Husserl discover a complete plan of a pure logic and an initial attempt to provide a unified presentation of the domain of pure ideal doctrines in Bolzano's work.[52]

---

[47] Husserl 1919, 345.

[48] Husserl 1913, 20.

[49] *Op. cit.*, 29.

[50] Husserl 1905, 37.

[51] Husserl 1913, 42.

[52] Husserl 1903b, 'Review of Melchior Palágyi's *Der Streit der Psychologisten und Formalisten in der Modernen Logik*', as translated in Husserl 1994, 201–302; Husserl 1913, 36–38, 46–49.

8.7 Actual Consciousness and Pure Logic

Husserl saw that empirical psychology could not solve the epistemological problems he was facing, but he never considered utterly forsaking the world of actual consciousness. He said that it was troubling puzzles about the interrelationship and intrinsic unity of what he called the "incredibly strange" worlds of the purely logical and of actual consciousness that he saw opening up all around him while he was trying to understand the logic of mathematical thought and calculation and achieve clarity regarding the true meaning of the concepts of set theory and the theory of cardinal numbers in the *Philosophy of Arithmetic*[53] that launched him on his phenomenological voyage of discovery. He strove until the end of his life to solve those puzzles. He considered that only those who, like himself, deeply distressed by the issues in the most intense way,

> compelled by the critical dissolution of the blinding prejudices of psychologism to recognize the purely logical ideal ... but ... at the same time ... compelled by the revealing emphasis upon the essential relationships between the ideal and the psychological ... not to abandon by any means the psychological entirely but rather to keep it within view as somehow belonging with the ideal ... can also have the insight that such psychological critiques are indispensable for forcing recognition of the ideal as something given prior to all theories ... can realize that the being-in-itself of the ideal sphere in its relation to consciousness brings with it a dimension of puzzles which remain untouched by all such argumentation against psychologism and hence must be solved through special investigations ... through phenomenological ones.[54]

Now, in Cantor's writings, the world of consciousness and the world of pure mathematics mingle together in confusing and frustrating ways that cry out for clarification,[55] so that the naive epistemological theorizing in which Cantor was so earnestly engaging while Husserl was grappling with analogous questions could easily have impressed upon his younger colleague an urgent need to develop the more sophisticated logical and epistemological tools needed to attain a deeper, clearer understanding of how the human mind interacted with the world of numbers.

For instance, Husserl might have fairly wondered in just what way the cardinal number belonging to a set is an abstract image in our intellect[56] or exactly how the act of abstraction awakened the number concepts in Cantor's mind.[57] In the same passage of the *Mannigfaltigkeitslehre* in which Cantor explicitly rejected the belief that "the source of knowledge and certainty is located in the senses or in the so-called form of pure intuition of the world of presentation", he wrote that "certain knowledge ... can only be obtained through concepts and ideas, which are at best only stimulated by outer experience, but which are principally formed

[53]Husserl 1906, 'Personal Notes', as translated in Husserl 1994, 491–492.

[54]Husserl 1913, 21–22.

[55]See Hallett 1984, 16–18, 34–35, 121, 128–133, 146–158.

[56]Cantor 1887/8, 416.

[57]*Op. cit.*, 418 n. 1.

through inner induction, like something which, so to speak, already lay within us and is only awakened and brought to consciousness."[58]

Frege was perfectly justified in qualifying Cantor's appeals to direct inner intuition[59] as "rather mysterious."[60]

So given the particular nature of Cantor's experiments, it is not surprising to find Husserl asking in those years how rational insight was possible in science,[61] how the mathematical in itself as given in the medium of the psychical could be valid, how logicians penetrated an objective realm entirely different from themselves, how objective, mathematical and logical relations constituted themselves in subjectivity, how symbolic thinking was possible,[62] how abandoning oneself completely to thought that is merely symbolic and removed from intuition could lead to empirically true results, or how mechanical operations with mere written characters could vastly expand our actual knowledge concerning number concepts.[63] These are all questions that Cantor's theories raise.

8.8 Husserl Teaches Metaphysics

Husserl communicated the new vision of metaphysics, epistemology and the natural sciences that he developed during the 1890s to a new generation of students in search of a scientific metaphysics that could stand up to the challenges of the natural sciences. He called for a science of metaphysics to study problems lying beyond empirical investigation, to engage in the exploration of what is *realiter* in the ultimate and absolute sense, and so provide ultimate and deepest knowledge of reality. He believed that such a science of metaphysics was possible, justifiable, and that human beings were ultimately capable of attaining knowledge of reality.[64]

He defined metaphysics as the science of absolute Being in contrast to the individual sciences, which he saw as merely sciences of Being in the relative, provisional sense sufficient for practical orientation in the phenomenal world and for the practical mastery of nature.[65] Metaphysics was to be understood in a broad sense as the radical science of Being in the absolute sense, instead of the science

---

[58] Cantor 1883, 207 n. 6; Hallett 1984, 15.

[59] Cantor 1883, 168, 170, 201.

[60] Frege 1884, §86.

[61] Husserl 1894, 'Psychological Studies in the Elements of Logic', as translated in Husserl 1994, 167.

[62] Husserl 1913, 35, 222.

[63] Husserl 1890b, 'On the Logic of Signs (Semiotic)' as translated in Husserl 1994, 50; Husserl 1894, 'Psychological Studies in the Elements of Logic' as translated in Husserl 1994, 167.

[64] Husserl 1898/99, 232, 233, 252; Husserl 1906/07, §§ 20, 21.

[65] Husserl 1902/03, 12–13.

of Being in the empirical sense, which we think we know so well, but which upon closer inspection at times turns out to be deceptive and an illusion.[66]

Husserl considered a science of metaphysics so necessary for science that even natural scientists could not do without it, for as soon as they began reflecting on the principles of their science, they fell into metaphysics, though they most certainly did not want to call it by that forbidden name. It is certain, he maintained, that knowledge of the world of the natural sciences is not definitive knowledge of reality. Through them, the practical mastery of nature, is attained, a far-reaching orienting of empirical reality, the possibility of formulating laws by which we exactly foresee and foretell the course of empirical processes, redirect the course of the processes. But, we are not then in possession of definitive, ultimate, conclusive knowledge of the essence of nature. One undoubtedly arrives at worthwhile results, but the lack of critical insight into the meaning of the fundamental concepts and principles makes it impossible to be clear about what is thereby ultimately achieved and consequently about the sense in which the results may be taken as expressions of ultimate Being.[67]

The empirical sciences, he taught, are subject to principles which govern thinking and research in the natural sciences and make them possible, and consequently which cannot in turn be searched for again by thinking and research in the natural sciences.[68] Wherever it is a question of reality, in life and in all empirical sciences, he explained, we apply certain concepts like thing, real property, real relation, state, process, coming into being and passing away, cause and effect, space and time, that seem to belong necessarily to the idea of a reality. Whether or not all these concepts are actually intrinsic to the idea of reality, there surely are such concepts, the basic categories, in which what is real as such is to be understood in terms of its essence. Thus, investigations must be possible that simply reflect everything without which reality in general cannot be conceived. There must be a science of real Being as such in the most universal universality, an a priori metaphysics.[69]

Husserl lamented the fact metaphysics had come to be seen as a relic of scientifically backward times on a par with alchemy and astrology. He considered the fight against it and most of the contemptuous remarks against it to be directed at a kind of a hobgoblin (*eine Art Popanz*) that people had concocted. He saw the metaphysical needs of his time going unmet and gave this as an explanation as to why spiritism and the occult were thriving and superstition of every kind was spreading.[70] Moreover, he wished to make clear that he himself was "far from any mystico-metaphysical exploitation of 'Ideas', ideal possibilities and such"[71] and

---

[66]Husserl 1906/07, §20.

[67]Husserl 1898/99, 233.

[68]Husserl 1906/07, §20; Husserl 1898/99, 233.

[69]Husserl 1906/07, §21.

[70]Husserl 1898/99, 225–255.

[71]Husserl 1905, 39.

that he "meant anything but a dialectical spinning of the concrete results of these sciences out of some abstract conceptual mysticism."[72]

8.9 Cantor, Metaphysics, Mysticism and the Occult

What remains of Cantor's correspondence reveals a great deal about his metaphysical views. For example, in 1894, he wrote to the mathematician Charles Hermite that "in the realm of the spirit" mathematics had no longer been "the essential love of his soul" for more than 20 years. Metaphysics and theology, he "openly confessed", had so taken possession of his soul as to leave him relatively little time for his "first flame."[73]

Given Husserl's views on metaphysics as expounded above, he could have conceivably agreed with the Cantor who wrote to Fr. Thomas Esser in 1896[74] that the

> grounding of the principles of mathematics and natural science is a matter for metaphysics. Metaphysics has therefore to look upon these two sciences not only as its servants and helpers but also as its children which it should not let out of its sight, but must watch over and control ...,

or with Cantor's conviction that the

> general *Mengenlehre* ... belongs thoroughly to metaphysics. You can easily convince yourself of this by examining the basic concepts of *Mengenlehre*, the categories of cardinal number and ordinal type, and noticing not only the degree of their generality, but also how thinking [*Denken*] with them is fully pure, so that there is not the slightest room for fantasy.[75]

And Dauben's interpretation of what Cantor meant by metaphysics fits in with Husserl's later phenomenology[76]:

> Whenever Cantor spoke of metaphysics he meant the philosophical study of the relations between the constructs of mind and the objects of the external world. Thus the study of the abstract theory of the transfinite numbers was the business of mathematics, but the study of the realization or embodiment of the transfinite numbers in terms of the objects of the phenomenological world was the concern of metaphysics. And so metaphysics assumed its place in Cantor's continuing program to establish the legitimacy of his new theory....

More mystical, however, was Cantor's conviction that "the whole numbers both separately and in their actual infinite totality exist in that highest kind of reality as

[72]Husserl 1896, 5; Husserl, 1902/03b, 12–13; Hill 2008, 2009 and 2012.

[73]See Cantor's letter to Charles Hermite of January 24, 1894 as published in Cantor 1991, 350.

[74]Cited in Hallett 1984, 10.

[75]*Loc. cit.*

[76]Dauben 1979, 125.

eternal ideas in the Divine Intellect",[77] or that contradictions in his own theories were but apparent and that one must distinguish between the numbers which we are able to grasp in our limited ways and "numbers as they are *in and for themselves, and in and for the Absolute intelligence*."[78] To Fr. Ignatius Jeiler, he wrote in 1888[79]:

> Each individual *finite* cardinal number is in God's intellect both a representative idea and a unified form for the knowledge of innumerably many things, that is, those which possess the cardinal number in question. All *finite* cardinal numbers are thus distinct and simultaneously present in God's mind. They form in their *totality a manifold, unified thing for itself* [*Ding für sich*] *delimited* from the remaining content of God's intellect, and this thing is itself again an object [*Gegenstand*] of God's knowledge.

And in another letter of the same year, Cantor wrote that he knew that his set theory stood "as firm as a rock", because he had studied it from all sides for many years, had examined all objections ever made against the infinite numbers and above had "followed its roots to the infallible cause of all created things."[80]

But Cantor went rather further than that. In Dauben's opinion,[81]

> [t]here can be no mistake about Cantor's identification of his mathematics with some greater absolute unity in God. This also paralleled his identification of transfinite set theory with divine inspiration.... Cantor ... told Mittag-Leffler that his transfinite numbers had been communicated to him from a more powerful energy.... Cantor believed that God endowed the transfinite numbers with a reality making them very special. Despite all the opposition and misgivings of mathematicians in Germany and elsewhere, he would never be persuaded that his results could be imperfect.

Such considerations may explain why Husserl wrote in the *Logical Investigations* of banishing "all metaphysical fog and all mysticism" from mathematical investigations into numbers and manifolds like those of Cantor.[82] Cantor's philosophy was mystical to say the least.[83]

It is important in this respect to realize that, while the end of the nineteenth century witnessed attempts to rehabilitate the respectability of metaphysical inquiry and place it centrally on the philosophical agenda along with rigorous, rational, scientific thinking, alongside this there was an occult revival, a rise in cults, spiritism, Satanism, the occult, magic, witchcraft, irrationalism and superstitious practices. In fact, the same letter books that show Cantor multiplying efforts to obtain Husserl a satisfactory job reveal that, alongside interests in traditional

---

[77]Cited in Hallett 1984, 149.

[78]See Cantor's letters of September 7, 1890 to Giuseppe Veronese, of November 3, 1886 to Axel Harnack, and of March 26, 1887 to Aloys Schmid as published in Cantor 1991, 326, 267, 282 respectively.

[79]Cited in Hallett 1984, 36.

[80]Cited in Dauben 1979, 298.

[81]Dauben 1979, 290–291.

[82]*PRe* 242.

[83]See Dauben 1979, ch. 6, 236–239; Hallett 1984, 9–11, 35–36.

theology and metaphysics, Cantor was rather heavily involved in occultism and occultist groups, Rosicrucianism in particular.[84]

8.10 Frege's Reviews of Husserl and Cantor

In 1894, Frege published a damaging, abusive review of Husserl's *Philosophy of Arithmetic*. In "Frege's Attack on Husserl and Cantor",[85] I endeavored to show the extent to which Frege directly incorporated into that review several criticisms he had already made of Cantor's work.

To begin with, Frege repeatedly criticized Husserl's use of a theory of number abstraction. He was perfectly correct to write that Husserl would have one "abstract from the peculiar constitution of the individual contents that make up the multiplicity and retain each one only in so far as it is a something or a one." He was also correct in writing that Husserl's "process of abstracting the number goes hand in hand with a process of emptying all content."[86] But, unwilling just to condemn Husserl for holding views he really had advocated, Frege unfairly went on to charge that Husserl's procedure would "cleanse things of their peculiarities ... in the wash-tub of the mind" where things "assume a quite peculiar pliancy. . . . " There, Frege said, "we can easily change objects by directing our attention towards them or away from them. . . . We attend less to a property, and it disappears. By thus making one characteristic mark after another disappear, we obtain more and more abstract concepts."[87] And he went on to provide this caricature of Husserl's procedure[88]:

> Suppose, e.g., that there are a black and white cat sitting side by side before us. We do not attend to their colour, and they become colourless—but they still sit side by side. We do not attend to their posture, and they cease to sit ... but each of them is still in its place. We no longer attend to the place and they cease to occupy one—but they continue presumably to be separate. We have thus perhaps obtained from each of them a general concept of cat. By continued application of this procedure, each object is transformed into a more and more bloodless phantom.

Frege maintained that it must surely be assumed that the process of abstraction effects some change in the objects and that they become different from the original objects which are either transformed or actually created by the abstraction process.[89] Nonetheless, in spite of the force of the charges he directed against Husserl in his review, Frege was honest enough to admit that Husserl himself did not hold that

---

[84]Cantor *Briefbücher I (1884–1888), II (1890–1895), III (1895–1896)*; Hill 2008, 2009 and 2012.

[85]Hill 1994.

[86]Frege 1894, 196.

[87]*Op. cit.*, 197.

[88]*Op. cit.*, 197–198.

[89]*Op. cit.*, 204.

the mind creates new objects or changes old ones, and actually acknowledged that Husserl "disputes this in the most vehement terms (p. 139)."[90]

When Frege went on to charge that Husserl had taken "the road of magic rather than of science",[91] we have a good clue as to whom else he wished to attack, for in reviewing the "Mitteilungen", he had accused Cantor of the very same thing. He called the verb 'abstract' a psychological expression to be avoided in mathematics"[92]; in a posthumously published draft of this review, he likened mathematicians who like Cantor talk of abstraction to "negroes from the heart of Africa." For them, words like 'abstraction' are supposed to have 'the kind of magical effects' that enable them to abstract from any properties of things that bother them.[93] In the spirit of the cat example of his review of Husserl, Frege complained that mathematicians like Cantor find a whole host of things in mice which are unworthy to form a part of the number. He writes[94]:

> Nothing simpler, one abstracts from the whole lot. Indeed when you get down to it everything in the mice is out of place: the beadiness of their eyes no less than the length of their tails and the sharpness of their teeth. So one abstracts from the nature of the mice ... one abstracts presumably from all their properties, even from those in virtue of which we call them animals, three-dimensional beings....

Frege also alludes to someone whom he suspects is one of Cantor's pupils, who when asked what general concept he arrives at when given a pencil exerts himself "to the utmost in abstracting from the nature of the pencil and the order in which its elements are given", to answer 'the cardinal number one.'[95]

In his review, Frege further characterized Husserl's endeavor as "an attempt to justify a naïve conception of number." He called naïve "any view on which a statement of number is not a statement about a concept or about the extension of a concept." If Husserl had used 'extension of a concept' "in the same sense as I", Frege declared, "we should hardly differ in opinion about the sense of a number statement."[96] He also complained that, for Husserl multiplicities, sets, are more indeterminate and more general than numbers.[97] In addition, he wrote: "A concept under which only one object falls has a determinate extension, as does a concept under which no object falls, or a concept under which infinitely many objects fall."[98]

As it happens, Cantor had written in his review of Frege's *Foundations of Arithmetic* that it was unfortunate that Frege had used extensions of concepts as the foundation of the number concept, that that was "a reversal of the proper

[90] *Op. cit.*, 205; Hill 1997c.
[91] Frege 1894, 205.
[92] Frege 1984, 180, 181.
[93] Frege 1979, 69.
[94] *Op. cit.*, 70.
[95] *Op. cit.*, 71.
[96] Frege 1894, 201–202.
[97] *Op. cit.*, 195.
[98] *Op. cit.*, 202.

order", because the extension of a concept was generally something quantitatively completely undetermined, and that for such quantitative determination, the concept of number would have to have been given from somewhere else.[99] So Frege's criticism of Husserl's views on the relationship between numbers and extensions of concepts echoed criticisms Cantor had made of Frege.

Cantor's presence is again felt when Frege wrote that according to Husserl "numbers are supposed to be *Vorstellungen*, the results of mental processes or activities" and charged Husserl with actually removing elements. Frege had also charged Cantor with "creating" numbers,[100] and with engaging in the "psychological and hence empirical" activity of psychically removing elements.[101] In addition, when criticizing Husserl for writing that when the number of items to be counted is beyond our capacity for presentation we are then to "idealize" our capacity for presentation,[102] Frege was surely also thinking of Cantor's statement in the *Mannigfaltigkeitslehre* that whenever "one comes to no greater number, one imagines a new one."[103]

8.11 Husserl's Doubts About Sets

In *On the Concept of Number* and *Philosophy of Arithmetic*, Husserl strove to make set theory the basis of mathematics. In *Formal and Transcendental Logic*, he explicitly described the *Philosophy of Arithmetic* as an initial attempt "to obtain clarity regarding the original genuine meaning of the fundamental concepts of the theory of sets and cardinal numbers."[104]

Stating that he would use the terms 'multiplicity' and 'set' interchangeably to neutralize any differences in meaning among the terms, and citing Euclid's classical definition of the concept of number as "a multiplicity of units", Husserl began his analyses by affirming that "the analysis of the concept of number presupposes the concept of multiplicity."[105] Just as cardinal numbers relate to sets, so ordinals relate to series that are themselves ordered sets, he maintained in the introduction to *Philosophy of Arithmetic*.[106] The most primitive concepts involved, he explained in the work, are the general concepts of set and number which are grounded in the concrete sets of specific objects of any kind whatsoever and to which particular numbers are assigned. There cannot be any doubt, Husserl went on to affirm there,

---

[99]Cantor 1885, 728–729.

[100]Frege 1884, §96 note.

[101]Frege 1984, 180–181; Dauben 1979, 220–228.

[102]Frege 1894 ('Review of Dr. E. Husserl's *Philosophy of Arithmetic*'), 207; *PdA* 219; *PoA* 231.

[103]Cantor 1883, 195. See Dauben 1979, 206.

[104]Husserl 1929, §27a; also §24 and note.

[105]*PdA* 14–15; *PoA* 15–16; also *PdA* 297; *PoA* 313–314.

[106]*PdA* 11; *PoA* 12.

as in *On the Concept of Number*, that the concrete phenomena that form the basis for the abstraction of the concepts in question are aggregates (*Inbegriffe*), multiplicities of determinate objects. Everyone, he stressed, knew what was meant by this, for despite difficulties experienced in analyzing it, the concept of multiplicity itself was perfectly precise and the range of its extension exactly delimited. It might therefore be considered as a given, he maintained, even though we might still be in the dark about the essence and origination of the concept itself.[107]

However, Husserl quickly came to judge his first attempts to clarify the true meaning of the fundamental concepts of the theory of sets and cardinal numbers to have been a failure. He not only began to have doubts about psychological analyses of sets, but he expressed doubts about set theory itself. He confessed to having been disturbed, and even tormented, by doubts about sets right from the very beginning.[108] He specifically put Cantorian sets, "the *Mannigfaltigkeitslehre* in the broadest sense", into the category of pure logic that was a source of torment to him.[109] While working on the logic of mathematical thought and mathematical calculation, he explained in introductions to the *Logical Investigations*, he had encountered disturbing problems while studying the logic of formal arithmetic and the theory of *Mannigfaltigkeiten*.[110]

In the early 1890s he was already expressing grave doubts about extensional logic, by which he meant the calculus of classes.[111] His chief target then was Ernst Schröder.[112] But his antipathy is evident in several articles of the period.[113] In those texts he sought to show "that the *total* formal basis upon which the class calculus rests is valid for the relationships between conceptual objects", and that one could solve logical problems without "the detour through classes."[114]

In Chapter XI of *Philosophy of Arithmetic*, Husserl warned against ascribing anything more to the concept of *infinite* sets than is actually logically permissible, and above all not the absurd idea of constructing the actual set. Treating "All S" as a closed set creates what he called a kind of "imaginary" concept whose anti-logical nature was harmless in everyday contexts precisely, because its inherent contradictoriness was never obvious in life. However, he warned, the situation changes when such an imaginary construct is actually carried over into reasoning and influences judgments.[115] This warning came at a time when

[107] *PdA* 14–15; *PoA* 15–16; *PdA* 298; *PoA* 314–315.

[108] Husserl 1913, 35.

[109] Husserl 1903a, "A Report on German Writings in Logic from the Years 1895–1899, Third Article" as translated in Husserl 1994, 250; Husserl 1913, 28.

[110] *LI* 41–43.

[111] Husserl 1893, 'A. Voigt's "Elemental Logic," in Relation to my Statements on the Logic of the Logical Calculus', as translated in Husserl 1994, ex. 121.

[112] *Op. cit.*, 52–91, 421–441.

[113] Ex. *op. cit.*, 92–114, 115–120, 121–130, 135–138, 443–451.

[114] *Op. cit.*, 109, 123.

[115] *PdA* 218–222; *PoA* 230–234.

Cantor was busy exploring, mapping and inventing what he himself called the strange world of transfinite sets.

Indeed, it is not surprising to find that Husserl had doubts about sets when one considers that he was on hand as Cantor began discovering the antinomies of set theory[116] and so had a foretaste of the crisis in foundations that broke out once his proof by diagonal argument that there is no greatest cardinal number opened Bertrand Russell's eyes to the famous contradiction of the set of all sets that are not members of themselves.[117] David Hilbert once described the reaction to that finding as having been dramatic and violent and having had a "downright catastrophic effect in the world of mathematics."[118]

In a November 7, 1903 postcard, Hilbert informed Frege that Ernst Zermelo had discovered "Russell's" paradox 3 or 4 years earlier after he himself had communicated his examples to him.[119] A record of Zermelo's discovery is to be found known in a note of April 16, 1902 conveyed to Husserl about the comments about the contradictions involving sets of sets that he had made in his review of Schröder.[120]

Volker Peckhaus and Reinhard Kahle have tried to elucidate Hilbert's remark. They consider that it can be assumed that Hilbert formulated his paradox during exchanges with Cantor documented in their 1897–1900 correspondence.[121] That correspondence shows Cantor coping with the antinomies of set theory during Husserl's years in Halle. As Dauben has commented, it is "not a little ironic" that the first mathematician to discover the antinomies of set theory was Cantor himself, who had anticipated the problem and by 1895 was already "trying to remedy the paradoxes with a minimum of damage to his system of transfinite numbers."[122] Using his diagonalization proof of 1891, Cantor "could argue that the set of all sets had to give rise to a set of larger cardinality; the set of all its subsets. But since this set had to be a member of the set of all sets, the paradoxical conclusion was inevitable that a set of lower cardinality actually contained a set of higher cardinality."[123]

It is to be noted that Husserl did not ultimately conclude that set theory itself was false. In later writings, he viewed it as a legitimate mathematical discipline of the second level of the purely logical sphere. It was a matter of a rigorously

[116]Dauben 1979, 240–270.

[117]Russell 1903, §§100, 344, 500; Russell 1959, 58–61. Hereto cp. Grattan-Guinness 1978, 1980, 2000.

[118]Hilbert 1925, 375.

[119]Frege 1980, 51.

[120]Husserl 1902, 'Memorandum of a Verbal Communication from Zermelo to Husserl', as translated in Husserl 1994, 442. Cp. Rang & Thomas 1981.

[121]Peckhaus & Kahle 2000/2001, 3–4. Also see Cantor 1991, "6. Die Phase des Gedankenaustausches mit Hilbert–Anerkennung der Mengenlehre und die Antinomien", 387–485.

[122]Dauben 1979, 241.

[123]*Op. cit.*, 165–168, 242.

scientific, a priori theory proceeding from purely logical concepts and axioms grounded in purely logical categories.[124] He concluded that it was faulty reasoning about a faulty concept of set had led to the set-theoretical paradoxes. He advocated making a fresh start and deriving set theory from non-contradictory concept of set and element, or more universally of whole and part without resorting to an axiom of extensionality.[125] He was lucid enough to see that mathematics would not crumble if it did not have "a single object to represent an extension", as Russell believed it would.[126] All the rigmarole that Russell went through to evade the contradictions derivable from Frege's system with its axiom of extensionality[127] serves to illustrate what Husserl meant in *Formal and Transcendental Logic* when he said that extensions generate contradictions requiring every kind of artful device to make them safe for use in mathematical reasoning.[128]

8.12 Arithmetization

Much of the initial intellectual kinship between Husserl and Cantor can be explained by the influence the great mathematician Karl Weierstrass exercised on both of them. Cantor had studied in Berlin from 1863 to 1869, where he had come under the influence of Weierstrass. In Halle, Husserl found him immersed himself in a project aimed at demonstrating that the positive whole numbers formed the basis of all other mathematical conceptual formations inspired by Weierstrass' famous theory to that effect. Any further progress of the work on set theory, Cantor explained in the beginning of his *Mannigfaltigkeitslehre* was absolutely dependent upon the expansion of the concept of real whole numbers beyond the present boundaries and in a direction which, as far as he knew, no one had yet searched. He had, he claimed, burst the confines of the conceptual formation of real whole numbers and broken through into a new realm of transfinite numbers. Initially he had not been clearly conscious of the fact that these new numbers possessed the same concrete reality the whole numbers did. He had, however, become persuaded that they did. As strange and daring as his ideas might now seem, he was convinced that they would one day be deemed completely simple, appropriate and natural.[129]

[124]Husserl 1906/07, 18d.
[125]Husserl Ms. A 1 35.
[126]Russell 1903, §489.
[127]Hill 1997b.
[128]Husserl 1929, §§ 23, 26.
[129]Cantor 1883, 165–166.

As for Husserl, he would say that it was from Weierstrass that he acquired the ethos of his intellectual endeavors.[130] Of his "great teacher", Husserl once wrote[131] that he

> was the one who ... awakened my interest in seeking radical foundations for mathematics. I came to understand the pains he was taking to transform analysis from the mixture of reason, and irrational instincts and knowhow it was at the time into a pure rational theory. His aim was to expose its original roots, its elementary concepts and axioms on the basis of which the whole system of analysis might be deduced in a completely rigorous, perspicuous way.

Late in his career, Husserl would say that he had sought to do for philosophy what Weierstrass had done for mathematics.[132]

Impressed by Weierstrass' work to arithmetize analysis and intent upon rigorously deducing all of mathematics from the least number of self-evident principles by first analyzing the concepts and relations that are in themselves simpler and logically prior, then the more complicated, more derivative ones, Husserl embarked upon a project to supply radical foundations for mathematics by submitting the concept of number itself to closer scrutiny. The natural and necessary starting point of any philosophy of mathematics, Husserl initially believed, was the analysis of the concept of whole number.[133] For the still faithful disciple of Weierstrass still believed the "domain of 'positive whole numbers' to be the first and most underivative domain, the sole foundation of all remaining domains of numbers."[134]

However, although in *On the Concept of Number*, Husserl had maintained that all of the more complicated and artificial forms of numbers had their origin and basis in the concept of positive whole numbers and their interrelations and were derivable from them in a strictly logical way,[135] in the *Philosophy of Arithmetic*, he never endorsed Weierstrass' thesis in that confident way. Rather, we find him explaining that he would provisionally use it as a springboard for his own analyses and warning readers that, although cardinal numbers in a certain way seem to be the basic numbers involved in arithmetic because the signs for them figure in expressions for positive, negative, rational, irrational, real, imaginary, alternative, ideal numbers, quaternions etc., the analyses of the second volume would perhaps show that thesis to be untenable.[136] In the April 1891 preface to the book, he went further to state that the analyses of the second volume would actually show that in no way does a single kind of concept, whether that of cardinal or ordinal numbers, form the basis of general arithmetic.[137] And before the *Philosophy of Arithmetic* had even

---

130 Schuhmann 1977, 7.

131 *Loc. cit.*

132 Becker 1930, 40–42; Schuhmann 1977, 34.

133 *PdA* 294–95; *PoA* 310–11.

134 Husserl 1890a, "The Concept of General Arithmetic", as translated in Husserl 1994, 2.

135 *PdA* 294; *PoA* 310.

136 *PdA* 12; *PoA* 13.

137 *PdA* 7; *PoA* 7.

made its way into print, he wrote to Carl Stumpf that the opinion by which had been guided in writing *On the Concept of Number* that the concept of cardinal number formed the foundation of general arithmetic had soon proved to be false, that through no manner of cunning could negative, rational, irrational and the various kinds of complex numbers be derived from the concept of the cardinal number.[138]

8.13 Husserl's Mannigfaltigkeitslehre

In the foreword to the *Logical Investigations*, Husserl specifically alluded to having been troubled by the theory of manifolds, the *Mannigfaltigkeitslehre*, with its expansion into special forms of numbers and extensions.[139] He always said that it was particularly difficulties he had experienced in trying to answers questions raised by "imaginary" numbers which arose while trying to complete the *Philosophy of Arithmetic*, that had marked the turning point in his thinking. He used the term "imaginary" in a very broad sense to cover negative numbers, negative square roots, fractions and irrational numbers, and so on.[140] He called infinite sets imaginary concepts.[141]

Now, the work on numbers Cantor was doing in the late 1880s makes Husserl's quandaries completely understandable. Those were years during which Cantor was particularly engaged in what Grattan-Guinness has called "rather strange work on theory of numbers",[142] producing what Dauben has called "dinosaurs of his mental creation, fantastic creatures whose design was interesting, overwhelming, but impractical to the demands of mathematicians in general."[143]

Husserl ultimately concluded that "the key to the only possible solution of the problem" as to "how in the field of numbers impossible (essenceless) concepts can be methodically treated like real ones" was to be found in the theory of complete manifolds, *definite Mannigfaltigkeiten* that he developed during his time in Halle.[144] According to his theory, logic defying creations like those flowing from Cantor's pen could be shown to have redeeming scientific value when integrated into a whole which was greater than the sum of its parts, where the anti-logical character of imaginary constructions would become neutralized within the context of a consistent, complete deductive system. He concluded that formal constraints banning meaningless expressions, meaningless imaginary concepts, reference to

[138] Husserl 1891, "Letter from Edmund Husserl to Carl Stumpf", as translated in Husserl 1994, 13.

[139] *LI* 41–42; Husserl 1913, 35.

[140] Husserl 1891, "Letter from Edmund Husserl to Carl Stumpf", as translated in Husserl 1994, 13–16; Husserl 1913, 33; Husserl 1929, §31. Hereto cp. Hill 1991, 81–86.

[141] *PdA* 221; *PoA* 233.

[142] Grattan-Guinness 1971, 369.

[143] Dauben, 158–59.

[144] *PRe* §70.

non-existent and impossible objects restrict theoretical, deductive work, but that resorting to the infinity of pure forms and transformations of forms frees us from such conditions and explains why having used imaginaries, what is meaningless, must lead, not to meaningless, but to true results. According to his theory of manifolds, one can operate freely within a manifold with imaginary concepts and be sure that what one deduced is correct when the axiomatic system completely and unequivocally determined the body of all the configurations possible in a domain by a purely analytical procedure.[145]

That solution was compatible with Cantor's conviction that mathematical concepts need only be both non-self-contradictory and stand in systematically determined relations established through definition from the previously formed, proven concepts and that mathematicians are only obliged to provide definitions of the new numbers determinate in this way and, if need be, to establish this relationship to the older numbers.[146] However, no one who has studied Cantor's *Mannigfaltigkeitslehre* could possibly confuse it with Husserl's theory of the same name. Cantor's *Mannigfaltigkeiten* were sets. Husserl's *Mannigfaltigkeiten* were not sets, which he placed on the second level of pure logic, whereas *Mannigfaltigkeiten* occupied the third and highest level of pure logic.[147] Husserl's *Mannigfaltigkeitslehre* was surely an attempt to arrive at a true understanding of the true intention of theories of numbers and manifolds as pure categorial forms of theory that would "banish all metaphysical fog and all mysticism" from mathematical investigations into sets and number like those of Cantor and others.[148]

8.14 Conclusion

I have examined several respects in which Husserl and Cantor might be said to be alike while pointing to dissimilarities between them. Husserl wrote of his time in Halle that it was a "decade of solitary, arduous labor" during which he saw all around him only ambiguously defined problems and profoundly unclear theories. Sick of the confusion and afraid of sinking into an "ocean of endless criticism", for "the sake of philosophical self-preservation", he had felt compelled to set aside his philosophical mathematical investigations and to risk striking out on his own. The course of his development, he wrote in the Foreword to the *Logical Investigations*, had led to his drawing apart from people and writings to whom he owed most of his

145 Husserl 1906/07, §19; Husserl 1917/18, §§56–57; Husserl 1929, §31. Hereto cp. Centrone 2010.

146 Cantor 1883, 182.

147 Husserl 1906/07, §§18–19, 434–35; Husserl 1996, Chap. 11; Husserl 1929, §33.

148 *PRe* 242. Hereto cp. Hill 2000 and 2002b.

intellectual training and to his drawing closer to a group of thinkers whose writings he had not been able to evaluate properly and had consulted all too little in the course of his labors.[149]

Initially, Husserl's ideas about mathematics and philosophy, sets, abstraction, and the arithmetization fit in with Cantor's ideas. That may be attributable to Cantor's position on his *Habilitationskommittee* and to the influence Weierstrass had exercised over both men.

In 1890, Husserl wrote that with "respect to the starting point and the germinal core of our developments toward the construction of a general arithmetic, we are in agreement with mathematicians that are among the most important and progressive ones of our times: above all with *Weierstrass*, but not less with *Dedekind*, Georg *Cantor* and many others."[150]

In *Concept of Number*, Husserl had expressed his concern that mathematicians had failed to examine the logic of the concepts and methods they introduced and used and called for logical clarification, precise analyses.[151] And, at that point, the naive epistemological theorizing in which Cantor was so earnestly engaging while Husserl was grappling with analogous questions could well have seemed to Husserl to be amenable to clarification through Brentano's teachings: He might have thought that Cantor's naïve talk of intuitions and presentations could grow in sophistication through the application of Brentano's ideas about presentation and intentionality; Brentano's collective unification might be what Cantor meant by the "special relationship" binding elements of a set; the objects of thought and intuition of Cantor's sets might be Brentano's intentional objects; Cantor's technique for extracting numbers from reality through abstraction might be a psychological process. Cantor's appeals to inner intuition and talk of things like the fingers of his right hand helping produce or awaken concepts in his mind do sound unabashedly empiricistic or psychologistic. Frege certainly saw them as such as he decried the "psychological and hence empirical turn" in the "Mitteilungen."[152] So Husserl might have initially hoped that Brentanian analyses could "banish all metaphysical fog and all mysticism" from mathematical investigations like Cantor's.[153]

At a later stage, Husserl renounced the psychologism, empiricism, and naturalism that Cantor was renouncing and drew near to the Platonic Idealism he was endorsing. But, the arithmetization that Cantor espoused failed Husserl too, and he confessed to having been troubled the theory of manifolds, the *Mannigfaltigkeitslehre*, with its expansion into special forms of numbers and extensions.[154] He also said he had been tormented by troubling puzzles about the interrelationship and intrinsic unity of what he called the "incredibly strange" worlds of the purely

[149] *LI* 43; Husserl 1913, 16–17.

[150] Husserl 1890a, 'The Concept of General Arithmetic', as translated in Husserl 1994, 1.

[151] *PdA* 294; *PoA* 310.

[152] Frege 1984, 180; 181.

[153] *PRe* 242.

[154] *LI* 41–42; Husserl 1913, 35.

logical and of actual consciousness that he saw opening up all around him while he was trying to understand the logic of mathematical thought and calculation and to achieve clarity regarding the true meaning of the concepts of set theory and the theory of cardinal numbers in the *Philosophy of Arithmetic*.

However, while it is tempting to think that Husserl's turnabout was conditioned, if not actually induced by Cantor's bold experiments with mathematics and epistemology, and although Cantor's ideas probably did have a hand in unseating Husserl from his earliest convictions by raising a number of unsettling questions, he said that it was Lotze's work that was responsible for his conscious, radical rejection of psychologism, his espousal of Platonism and newfound comprehension of Bolzano's work on pure logic[155] and, therefore, for his adoption of metaphysical and epistemological views that Brentano had taught him to consider odious and despicable.[156] That rejection of the "blinding prejudices of psychologism", he said, compelled him to recognize the purely logical ideal and the puzzles involved in the essential relationships between the ideal and actual consciousness at the heart of the dynamic that gave birth to phenomenology.

Brentano clearly figured among the thinkers to whom Husserl owed most of his intellectual training, but with whom he parted ways. But Husserl also abandoned Weierstrass' teaching on arithmetization. And, it is fair to say that although his ideas overlapped and crisscrossed with Cantor's in the various ways discussed here, Husserl drew apart from him too and that Lotze and Bolzano, not Cantor, were among those whose ideas he had not at first appreciated and to whom he drew closer. Husserl did say that he had felt intellectually isolated in Halle, that he had encountered confusion, ambiguously defined problems and profoundly unclear theories on all side and had felt obliged to dare to set out on his own. I know of no evidence that he remained in contact with Cantor once he had left Halle (Hill 1997a).

References

O. Becker, The Philosophy of Edmund Husserl (1930), in *The Phenomenology of Husserl, Selected Critical Readings* ed. by R. O. Elveton, (Quadrangle Books, Chicago, 1970), pp. 40–72

G. Cantor, Grundlagen einer allgemeinen Mannigfaltigkeitslehre. Ein mathematisch-philosophischer Versuch in der Lehre des Unendlichen (1883) (Teubner, Leipzig. in Cantor, 1932), pp. 165–246

G. Cantor, Rezension von Freges *Grundlagen der Arithmetik, Deutsche Literaturzeitung* VI (20) (1885), pp. 728–729

G. Cantor, Mitteilungen zur Lehre vom Transfiniten, *Zeitschrift für Philosophie und philosophische Kritik* 91 (1887), 81–125; 92 (1888), 240–265. in Cantor 1932, pp. 378–439

G. Cantor, in *Gesammelte Abhandlungen*, ed. by E. Zermelo (Springer, Berlin, 1932)

[155]Husserl 1903b, Review of Melchior Palágyi's *Der Streit der Psychologisten und Formalisten in der Modernen Logik*', as translated in Husserl 1994, 201–202; Husserl 1913, 36–38, 46–49.

[156]Cp. Hill 1998.

G. Cantor, in Briefe, ed. by H. Meschkowski, W. Nilson (Springer, Berlin, 1991)
G. Cantor, *Briefbücher I (1884–1888), II (1890–1895), III (1895–1896)* at the Niedersächsische Staats-und Universitätsbibliothek Göttingen, Abteilung Handschriften und Seltene Drucke (Cod. Ms. 18)
S. Centrone, *Logic and Philosophy of Mathematics in the Early Husserl* (Springer, Dordrecht, 2010)
J. Dauben, *Georg Cantor: His Mathematics and Philosophy of the Infinite* (Princeton University Press, Princeton, 1979)
G. Frege, *The Foundations of Arithmetic* (1884) (Blackwell, Oxford, 1986)
G. Frege, Review of Dr. E. Husserl's *Philosophy of Arithmetic* (1894), in Frege 1984, pp. 195–209
G. Frege, in *Philosophical and Mathematical Correspondence*, ed. by B. McGuinness, G. Gabriel, et al. (Blackwell, Oxford, 1980)
G. Frege, in *Posthumous Writings*, ed. by H. Hermes, et al. (Blackwell, Oxford, 1979)
G. Frege, in *Collected Papers on Mathematics, Logic und Philosophy*, ed. by B. McGuinness (Blackwell, Oxford, 1984)
H. M. Gerlach, H. R. Sepp (eds.), Es ist keine Seligkeit 13 Jahre lang Privadocent und Tit. 'prof'. zu sein. Husserls hallesche Jahre 1887 bis 1901, *Husserl in Halle*, (Peter Lang, Bern, 1994), pp. 15–39
I. Grattan-Guinness, Towards a biography of Georg Cantor. Ann. Sci. **27**(4), 345–391 (1971)
I. Grattan-Guinness, How Russell discovered his paradox. Hist. Math. **5**, 127–137 (1978)
I. Grattan-Guinness, Georg Cantor's influence on Bertrand Russell. Hist. Philos. Logic **1**, 61–93 (1980)
I. Grattan-Guinness, *The Search for Mathematical Roots, Logics, Set Theories and the Foundations of Mathematics from Cantor through Gödel* (Princeton University Press, Princeton, 2000)
M. Hallett, *Cantorian Set Theory and Limitation of Size* (Clarendon, Oxford, 1984)
C.O. Hill, *Word and Object in Husserl, Frege and Russell* (Ohio University Press, Athens, 1991)
C.O. Hill, Frege attacks Husserl and Cantor. Monist **77**(3), 347–357 (1994). Also in Hill & Rosado Haddock 2000
C.O. Hill, Did Georg Cantor influence Edmund Husserl? Synthese **113**, 145–170 (1997a). Also in Hill & Rosado Haddock 2000
C.O. Hill, *Rethinking Identity and Metaphysics. On the Foundations of Analytic Philosophy* (Yale University Press, New Haven, 1997b)
C.O. Hill, The varied sorrows of logical abstraction. Axiomathes **8**(1–3), 53–82 (1997c). Also in Hill & Rosado Haddock 2000
C. O. Hill, From empirical psychology to phenomenology: Husserl on the Brentano puzzle in *The Brentano Puzzle*, ed. by R. Poli (Ashgate, Aldershot, 1998), pp. 151–168
C. O. Hill, Abstraction and idealization in Georg Cantor and Edmund Husserl, in *Abstraction and Idealization. Historical and Systematic Studies, Poznan studies in the philosophy of the sciences and the humanities,* ed. by F. Coniglione et al. (Rodopi, Amsterdam, 1999). Also in Hill & Rosado Haddock 2000
C. O. Hill, Husserl's *Mannigfaltigkeitslehre*, in Hill & Rosado Haddock 2000, pp. 161–177
C. O. Hill, *On Husserl's mathematical apprenticeship and philosophy of mathematics*, in *Phenomenology World Wide*, ed. by Anna-Teresa Tymieniecka (Kluwer, Dordrecht, 2002a), pp. 76–92. Also in Hill & da Silva 2013
C.O. Hill, Tackling three of Frege's problems: Edmund Husserl on sets and manifolds. Axiomathes **13**, 79–104 (2002b). Also in Hill & da Silva 2013
C. O. Hill, Phenomenology from the Metaphysical Standpoint, *Diálogos* XLIII, 91, January (2008), pp. 19–35
C.O. Hill, Husserl and phenomenology, experience and essence, in *Phenomenology and Existentialism*, (Springer, Dordrecht, 2009), pp. 9–22
C. O. Hill, Cantor's paradise, metaphysics and Husserlian logic, in *Categories of Being, Essays on Metaphysics and Logic* ed. by H. Koskinen (Oxford University Press, Oxford 2012), pp. 217–240. Also in Hill & da Silva 2013

C.O. Hill, G.E. Rosado Haddock, *Husserl or Frege? Meaning, Objectivity, and Mathematics* (Open Court, La Salle, 2000)
C.O. Hill, J.J. da Silva, *The Road Not Taken, On Husserl's Philosophy of Logic and Mathematics* (College Publications, London, 2013)
E. Husserl, [1890a] Begriff der allgemeinen Arithmetik, in HGW XII, 375–379. English translation: The Concept of General Arithmetic, in HCW 1994, pp. 1–6
E. Husserl, [1890b] Zur Logik der Zeichen, in HGW XII, 340–373. English translation: On the Logic of Signs (Semiotic), in HCW 1994, pp. 20–51
E. Husserl, *Philosophie der Arithmetik*, in HGW XII, pp. 1–283
E. Husserl, *Philosophy of Arithmetic*, in HCW X, pp. 1–299
E. Husserl, [1891] Brief an Stumpf, in HGW XXI, 244–251. English translation: Letter from Edmund Husserl to Carl Stumpf, in HCW 1994, pp. 12–19
E. Husserl, [1893] A. Voigts "Elementare Logik", in HGW XXII, 73–82. English translation: A. Voigt's "Elemental Logic", in HCW 1994, pp. 121–130
E. Husserl, [1894] Psychologische Studien zur elementaren Logik, in HGW XXII, 92–123. English translation: Psychological Studies in the Elements of Logic, in HCW 1994, pp. 139–170
E. Husserl, [1896] *Logik Vorlesung 1896, Husserliana, Materialienbände* vol. I, ed. by E. Schuhmann (Kluwer, Dordrecht, 2001)
E. Husserl, [1898/99] Aus der Einleitung der Vorlesung Erkenntnistheorie und Hauptpunkte der Metaphysik 1898/99, in *Husserliana, Materialienbände* Vol III, ed. by E. Schuhmann (Kluwer, Dordrecht, 2001)
E. Husserl, [HGW] *Husserliana, Gesammelte Werke*, I–XXVI: Martinus Nijhoff, The Hague, 1950–; XXVII–XXXVII: Kluwer, Dordrecht, 1989–; XXXVIII–: Springer, New York, 2005–
E. Husserl, vol. XII: [PdA] *Philosophie der Arithmetik. Mit ergänzenden Texten (1890–1901)*, ed. by L. Eley, 1970a
E. Husserl, [Pre] *Prolegomena to Pure Logic*, translated by J. N. Findlay (Routledge & Kegan Paul, New York 1970b) (translation of HGW XVIII)
E. Husserl, [LI] *Logical Investigations*, translated by J. N. Findlay (Routledge & Kegan Paul, New York 1970c) (translation of HGW XVIII-XIX)
E. Husserl, vol. XVII: *Formale und transzendentale Logik. Versuch einer Kritik der logischen Vernunft*, ed. by P. Jannsen, 1974
E. Husserl, vol. XVIII: *Logische Untersuchungen, Erster Band: Prolegomena zur reinen Logik*, ed. by E. Holenstein, 1975
E. Husserl, vol. XXII: *Aufsätze und Rezensionen (1890–1910)*, ed. by B. Rang, 1979
E. Husserl, [HCW] *Husserliana, Edmund Husserl Collected Works*, I–XIII–: Kluwer/Springer, Dordrecht, 1980–
E. Husserl, vol. XXIV: *Einleitung in die Logik und Erkenntnistheorie, Vorlesungen 1906/07*, ed. by U. Melle, 1984
M. Husserl, Skizze eines Lebensbildes von E. Husserl. Husserl Stud. **5**, 105–125 (1988)
E. Husserl, vol. V: [1994] *Early Writings in the Philosophy of Logic and Mathematics*, translated by D. Willard (Kluwer, Dordrecht, 1994a)
E. Husserl, [1902] Notiz einer mündlichen Mitteilung Zermelos an Husserl, in HGW XXII, 399. English translation: Memorandum of a Verbal Communication from Zermelo to Husserl, HCW 1994b, p. 442
E. Husserl, vol. XXX: [1917/18] *Logik und allgemeine Wissenschaftstheorie, Vorlesungen 1917/18, mit ergänzenden Texten aus der ersten Fassung 1910/11*, ed. by U. Panzer, 1996
E. Husserl, [1902/03] in *Logik Vorlesung 1902/03, Husserliana, Materialienbände* vol. II, ed. by E. Schuhmann, (Kluwer, Dordrecht, 2001)
E. Husserl, vol. X: [PoA] *Philosophy of Arithmetic, Psychological and Logical Investigations with Supplementary Texts from 1887–1901*, translated by D. Willard (Kluwer, Dordrecht, 2003)
E. Husserl, vol. XIII: [1906/07] *Introduction to Logic and Theory of Knowledge, Lectures 1906/1907*, translated by C. O. Hill (Springer, Dordrecht, 2008)

E. Husserl, [1903a] Bericht über deutsche Schriften zur Logik in den Jahren 1895–99. Dritter Artikel, in HGW XXII, 201–15. English translation: A Report on German Writings in Logic from the Years 1895–1899, Third Article, in HCW 1994, pp. 246–259
E. Husserl, [1903b] Rezension von Palaygi, in HGW XXII, 152–61. English translation: Review of Melchior *Palágyi's Der Streit der Psychologisten und Formalisten in der Modernen Logik*, in HCW 1994, pp. 197–206
E. Husserl, [1905] Husserl an Brentano, 27. III. 1905, in *Briefwechsel, Die Brentanoschule* I (Kluwer, Dordrecht, 1994)
E. Husserl, [1906] Persönliche Aufzeichnungen, in HGW XXIV, 442–449. English translation: Personal Notes, in HCW 1994, pp. 490–500
E. Husserl, [1913] in *Introduction to the Logical Investigations, A Draft of a Preface to the Logical Investigations*, ed. by E. Fink. Translated by P. Bossert and C. Peters (Martinus Nijhoff, The Hague, 1975)
E. Husserl, [1919] Recollections of Franz Brentano, in *Husserl: Shorter Works* eds. by P. McCormick and F. Elliston (University of Notre Dame Press, Notre Dame, 1981), pp. 342–349
E. Husserl, [1929] *Formal and Transcendental Logic*, translated by D. Cairns (Martinus Nijhoff, The Hague, 1969)
E. Husserl, *[Ms A 1 35]* Manuscript on set theory available in the Husserl Archives in Leuven, Cologne and Paris
C. Ierna, La notion husserlienne de multiplicité: au-delà de Cantor et Riemann, Methodos, Savoirs et Textes 12, n. 23 (2012), http://methodos.revues.org/2943?lang=en#bibliography
V. Peckhaus, R. Kahle, Hilbert's Paradox, Report No. 38 (2000/2001), Institut Mittag-Leffler, The Royal Swedish Academy of Sciences, https://www.mittag-leffler.se/preprints/files/IML-0001-38.pdf
B. Rang, W. Thomas, Zermelo's discovery of Russell's paradox. Hist. Math. **8**, 16–22 (1981)
B. Russell, *Principles of Mathematics* (Norton, New York, 1903)
B. Russell, My Philosophical Development (1959) (Allen and Unwin, London 1975)
K. Schuhmann, *Husserl-Chronik* (M. Nijhoff, The Hague, 1977)

Chapter 9
Husserl and Frege on Sense

Christian Beyer

Abstract This article presents and compares Frege's and Husserl's conceptions of sense, also taking into account their 1891 and 1906 correspondence. It is argued that while the similarities between their views speak in favour of a Fregean interpretation of Husserl's notion of noematic sense, there are also important differences. With regard to the latter, it is argued that Husserl's view yields a more general criterion of propositional difference and also provides a more detailed conception of the use of indexicals and non-descriptive singular terms, and of (what determines) their reference. In this context, Husserl's conceptions of constitution and genetic constitution analysis, respectively, are invoked and interpreted in terms of the epistemic notion of processing mental files or individual "concepts" (as Husserl calls them).

Keywords Frege • Husserl • Sense • Meaning • Reference • Thought identity • Intentional content • Perception • Constitution • Individual concepts • Mental files • States of affairs • Indexicals • Singular terms

9.1 Introduction

In the last decade of the ninenteenth and at the beginning of the twentieth century, there was a debate between Edmund Husserl and Gottlob Frege on what Frege calls sense. I shall start by briefly describing some aspects of the first (1891) phase of this debate (Sect. 9.2). Following this, I present Frege's main arguments for the distinction between sense and reference and some of his most important theses regarding sense (Sect. 9.3), in order to set the stage for both the presentation of the second (1906) and last phase of the debate (Sect. 9.4) and for further comparison between Frege and Husserl in the course of my presentation of some important features of Husserl's conception of sense (Sect. 9.5). It is obvious that a comparison such as this is also highly relevant for the evaluation of the so-called Fregean

C. Beyer (✉)
Philosophisches Seminar, Universität Göttingen, Göttingen, Germany
e-mail: Christian.Beyer@phil.uni-goettingen.de

S. Centrone (ed.), *Essays on Husserl's Logic and Philosophy of Mathematics*,
Synthese Library 384, DOI 10.1007/978-94-024-1132-4_9

interpretation of Husserl's notion of noematic sense, and it will become clear along the way that this interpretation is basically correct, despite a number of crucial differences between Frege's and Husserl's views. Furthermore, it would seem to be a promising line of research to compare this notion to more recent conceptions of sense and content and to discuss its merits as compared to them. Following this line would however go beyond the scope of this contribution.[1]

9.2 The 1891 Exchange Between Husserl and Frege

In his letter to Husserl from May 24, 1891, Frege says that he did not yet have the distinction between sense and "meaning (*Bedeutung*)" when writing his *Grundlagen der Arithmetik* (*GA*) in 1884.[2] However, he at least uses the *term* "sense" in *GA* when characterizing his approach to the concept of number as follows:[3]

> [a] It is only in the context of a sentence that words mean something. [b] So it will be decisive to explain the sense of a sentence in which a given number word occurs. [. . .] [c] If the sign a is to designate an object for us [*Wenn uns das Zeichen a einen Gegenstand bezeichnen soll*], we must have a criterion [*Kennzeichen*] which decides in each case whether b is the same as a, even though it may be beyond our power to apply this criterion. [d] In our case we have to explain the sense [*Sinn*] of the sentence 'The number belonging to the concept F is the same as the number belonging to the concept G;' that is, we have to render [*wiedergeben*] the content [*Inhalt*] of this sentence in a different way, without employing the expression 'the number belonging to the concept F.'[4]

If (following the early Husserl) we interpret "meaning something" as "expressing a sense," then the *context principle* which Frege formulates in the first sentence of this quotation says that it is only in the context of a sentence that a "word" (i.e., a sub-sentential expression) expresses a sense.[5] In order to specify the sense of a

---

[1]The interested reader is referred to Beyer 2000, which relates the notion of noematic sense to neo-Russellian and neo-Fregean approaches and highlights some of its merits, and Beyer 2015a, which relates it to the debate about radical contextualism vs. semantic minimalism, among other things.

[2]Cf. Frege 1976, 96.

[3]Note, however, that in the letter in question Frege says that he would now replace the term '*Sinn*' by '*Bedeutung*' in many places in *GA* (Frege 1976, 96).

[4]Frege 1987, 94; my translation.

[5]Also cf. Frege 1987, 23. However, Linnebo (forthcoming) argues that the principle should be interpreted as concerning "meaning (*Bedeutung*)" rather than sense; also cf. Haaparanta 2006, 292. Note that Frege does not regard it as a *necessary* condition for a word to mean what it means that it be part of a particular sentence in whose context it means that; words even have meaning "outside the context" of a particular sentence. (This is also required by Frege's principle of *compositionality*, which Linnebo shows to be compatible with the context principle. This principle is explained in Sect. 9.2 below.) As Linnebo convincingly argues, Frege merely regards it as a sufficient condition for a word to mean something that it be part of a meaningful sentence (such as an identity sentence), and the context principle is probably intended as an "explanatory principle," as indicated by the following formulation from the introduction of *GA*: "One must ask for the meaning of words in the

word (in this case: a number term like 'the number belonging to the concept of a winged horse'), one therefore has to "explain the sense of a sentence" in which that word occurs, as Frege puts it in [b]. Which kind of sentence? Frege's answer is to be found in sentence [d]: One has to explain the sense of a sentence in which the identity sign (or rather one of its counterparts in the object language) is flanked by both that word and *another* word. Why choose *identity* sentences? Frege's answer to this question is found in [c]: In order to be able to use a sign *a* to designate an object, one must have an associated criterion ("*Kennzeichen*") of identity for that object in which the referent a of *a* as well the referent b (if any) of another sign *b* are mentioned – one must know in principle how to decide the question whether a and b are the same (Frege speaks of "recognition (*Wiedererkennen*)"). One plausible way to interpret this is to invoke Gareth Evans' notion of a *fundamental ground of difference* (FGD) and require that one must know the FGD of a in order to be able to use *a* to refer to a:

> For any object whatever [. . .], there is what may be called the *fundamental ground of difference* of that object (at a time). This will be a specific answer to the question 'What differentiates that object from others?' of the kind appropriate to objects of that sort. For example, the fundamental ground of difference of the number three is being the third number in the series of numbers [. . .].[6]

In *GA* Frege presents a different FGD of numbers than Evans ("being the [n^{th}] number in the series of numbers"), though. The corresponding Fregean definition of numerical identity is discussed by Husserl in his *Philosophie der Arithmetik* (*PA*). Before briefly turning to that discussion, let us register that Frege can be interpreted in such a way that he associates an *epistemic* requirement with the ability to refer to an object by a term. As we will see below, he also has an epistemic conception of (grasping) *sense*; and Husserl too sees a connection between sense and knowledge or cognition. This is already evident from *PA*, where he criticizes Frege's criterion, which refers to a one-to-one correspondence between the extensions of two "equinumerous" concepts, for missing the sense of number terms that correponds to our common use of those terms and our associated epistemic interests (while yielding the right extension):

> [According to Frege's definition,] [t]o ascribe a particular number to [a concrete] multitude [*Menge*] would mean to classify it as belonging to a particular group of multitudes all of which are equivalent to each other; but this is not at all the sense of a numerical statement. Just consider any specific example. Do we call a multitude of nuts lying before us four because it belongs to a certain class of infinitely many multitudes that can be mutually put in a one-to-one correspondence? No one probably ever has any such thought in this connection, and it is doubtful whether we would find any practical occasion whatever to become interested in something like this.[7]

---

context of a sentence, rather than in isolation." (Frege 1987, 23; my translation) Also cf. Haaparanta 2006, 293.

[6]Evans 1982, 107.

[7]Hua XII, 116; my translation. For a thorough interpretation and discussion, see Centrone 2010, 13–25. Also see Centrone 2013, 68–82.

So instead of referring to knowledge of one-to-one correspendence as the FGD of number, and the corresponding Fregean definition of numerical identity, Husserl makes recourse, in his conception of the conceptual content of number words, to the "thoughts" speakers have when using such words in order to ask or state how many unities a given multitude (such as a number of nuts) contains. Frege would reject this proposed alternative approach as psychologistic, i.e. as committed to the view that "the foundations of logic" and mathematics "can be clarified with the help of psychology" (Mohanty 1982, 1). Thus, in *GA* he writes:

> [A] description of the inner processes preceding a numerical judgment can never replace a proper conceptual analysis [*Begriffsbestimmung*], even if it is more correct [than the one given by Lipschitz in his *Lehrbuch der Analysis*; CB].[8]

However, in his first major work, *Logische Untersuchungen* (*LU*), Husserl would develop a conception of "meaning-bestowing acts" and their intentional contents which shows that the conceptual content of an expression can indeed be fruitfully studied with recourse to "thoughts" or "inner processes," without committing oneself to psychologism.[9] Part of this conception is already found in Husserl's manuscripts from the early 1890s, e.g. in the first (Bolzano-influenced) part of his treatise on "Intentional Objects."[10] (I will return to that conception in Sect. 9.5) Furthermore, in his later work *Erfahrung und Urteil* (*EU*) he would investigate into what he there calls (using the terminology of *PA*) the "origins" of (traditional) logical concepts (including *number*) in a way that makes recourse to "inner processes" (i.e., what he refers to as predicative and pre-predicative experiences), yet avoids psychologism by taking the form (which is characteristic of transcendental phenomenology)[11] of a phenomenological meta-justification – a justification of the claim that our use of these concepts in thought is justified (see Sect. 9.5 below).[12] This would represent his non-psychologistic alternative to Frege's attempt at an

---

[8]Frege 1987, 56; my translation. Also see Frege 1894.

[9]Dummett's "Humpty Dumpty" accusation notwithstanding; see Dummett 1992, 45 ff, and the discussion in Beyer 2000, 78 f.

[10]See Mohanty 1982, 10 ff. (Mohanty does not mention Bolzano in this context.) The citations Mohanty gives in ch. 1 of his book certainly show that Husserl "arrived at the Vorstellung-Sinn[—]reference distinction independently" of Frege (Mohanty 1982, 10), but both this distinction and Husserl's view, criticized by Frege in his "*Ausführungen über Sinn und Bedeutung*" (Frege 1983), that logic is concerned with senses or "conceptual contents" (Mohanty 1982, 4) rather than extensions (or Fregean concepts) seem to me to be compatible with psychologism as defined by Mohanty (1982, 1). So they do not speak against Føllesdal's (1958) hypothesis that Frege's 1894 review of *PA* (Frege 1894) was a decisive causal factor with respect to Husserl's turn against psychologism as manifested in his 1900 *Prolegomena* (Hua XVIII). Notwithstanding the aforesaid, I agree with Tieszen that Frege's review largely misses the point of Husserl's analyses in *PA* and that at least some of these analyses "represent an early and rather primitive attempt to provide what he would later refer to as a 'genetic' analysis of the concept." (Tieszen 1994, 98).

[11]See Beyer 2013.

[12]See Haaparanta 1988, 85–93. Haaparanta does not explicitly refer to the idea of meta-justification but stresses that Husserl's aim in *EU* is "to lay the epistemological foundation of the old logic" (Haaparanta 1988, 86).

analysis of concepts like *number*, which Husserl regards as indefinable.[13] I will return to that alternative approach in Sect. 9.5. For now, let us register that in *PA* Husserl employs some notion of sense, as opposed to extension, in his critique of Frege, arguing that his method does not enable us to define the *sense* of the relevant concepts.[14] Husserl would later withdraw this criticism, but it drew Frege's attention to Husserl's distinction between concept word, sense (what he also calls "*Inhalt*" in *PA*) and extension, as his letter to Husserl from May 24, 1891 demonstrates,[15] and the proper conception of sense (regarding declarative sentences) would play a central role in their later 1906 correspondence.[16]

Unfortunately, only one of Husserl's letters to Frege has been preserved; but at least all of Frege's letters survived. In the first of these, Frege (among other things) explains his conception of the "meaning (*Bedeutung*)" (reference), as opposed to the sense, of a concept word, stressing that (unlike Husserl) he wants to distinguish it from the objects, if any, falling under the concept, precisely because there are concept words whose extension is empty but which are still useful for scientific purposes ("*wissenschaftlich verwendbar*").[17] (Consider imaginary concepts whose emptiness can be established by indirect proof, for example.)[18] As a consequence, he points out that singular terms ("proper names") have quite a different semantics than concept words. On Frege's view concept words (which may contain general names) do not name objects but "mean (*bedeuten*)" functions mapping the referents of singular terms to truth-values – they are "unsaturated (*ungesättigt*)" in that

---

[13]Cf. Hua XII, 119.

[14]Cf. Hua XII, 122. There is a tension between this objection and Husserl's claim that numerical concepts are indefinable, for Husserl seems to suggest here that his own method does enable us to define (i.e., analyse) the required sense. But compare and contrast Hua XII, 20 f.

[15]Cf. Frege 1976, 94–98.

[16]In the first of his *LI* (1901) Husserl implicitly acknowledges that his notion of (respective) meaning roughly coincides with Frege's notion of sense: He there criticizes Frege's terminological decision to use "*Sinn*" but not "*Bedeutung*" to refer to what he (Husserl) calls "*Bedeutung*" (Hua XIX/1, 58).

[17]Frege 1976, 96. Another (related) reason one may give in this connection is the nature of predication; see Künne 2010, 251. Thus, to quote Künne's example, it is one thing to (plurally) refer to all Portuguesean islands in the Mediterranian Sea by 'the Portuguesean islands in the Mediterranian Sea' (compare the non-empty plural term 'the Balearics') and quite a different thing to apply the predicate 'is a Portuguesean island in the Mediterranian Sea' to an island. While the plural term is empty, the predicate is not – it denotes a property no object possesses, and it can be (falsely) ascribed to an object. – It might be replied, on Husserl's behalf, that he talks about "general names," and that a predicate like 'is an island,' which contains the "general name" 'an island,' can be conceived of as meaning the same as 'is among the islands.' However, this view does not fit in with the view of predication he develops in *EU*, according to which basic empirical predicates (to which 'is an island' does not belong, though) contain "adjectivity," i.e. the form of 'being an independent part of (a substrate),' as their "syntactic stuff (*syntaktischer Stoff*)" or "core-form (*Kernform*)" (see Husserl 1999, 247). Nor does it fit in with his general view of states of affairs (involving properties) as intentional objects of predicative judgments.

[18]See Frege 1983, 132 f, 135, where Frege makes it clear that only *vague* terms are scientifically useless and fail to express (*bedeuten*) concepts.

they contain an "empty place (*Leerstelle*)" which needs input by a non-empty singular term in order to yield an expression (notably, a "declarative sentence (*Behauptungssatz*)") whose sense (called "Thought (*Gedanke*)") bears a truth-value (the True, the False). Correspondingly, they also express a different kind of *sense* than singular terms (notably, an "unsaturated" or "dependent," as opposed to a "saturated" or "independent," sense). Husserl nowhere explicitly accepts Frege's criticism, but he would later agree with "the perspicacious Frege"[19] that:

> [...] all judgments fall into two great classes. Their matters [i.e., intentional contents; CB] are either quantified functions [such as instances of *That an A is b, holds generally, holds in every case*; CB] or instances of unquantified functions, deriving from corresponding functions by filling in the empty places, the arguments, by certain values. In this way, fixed judgments [*die festen Urteile*] are generated, which do not contain 'arguments' but 'fixed termini,' full termini or ones that are here functioning in a parallel way [*oder ihnen hier parallel stehende*], containing 'this' or 'certain.' Every fixed terminus contains a positing, while the quantified functions cannot contain partial positings at the corresponding argument places.[20]

This sounds like an implicit acknowledgement of Frege's point that so-called general names (like 'a bachelor' as it is used in 'A bachelor is unmarried') function quite differently than singular terms (like 'this bachelor'), whose counterparts at the level of thought Husserl here refers to as "fixed termini." Furthermore, he agrees with Frege's conception of the relationship between the use of a singular term and the acceptance of the existence of a referent, notably in terms of what Strawson[21] would refer to as an *existential presupposition* underlying that use of the term (see Sect. 9.3 below, thesis 9). Thus, in a recently published 1899 manuscript he writes:

> If I say 'the Kaiser,' then I do not assert anything, i.e. I do not predicate (in the expression *per se*), but it is 'implied' [*es 'liegt darin'*] that we are dealing with a real person [*dass es sich um eine wirkliche Persönlichkeit handelt*]. In this form, i.e., with such a subject, one can only predicate if one acknowledges that there is a Kaiser.[22]

Husserl also calls this acknowledgement "positing (*Setzung*)." He even gives the same linguistic argument as Strawson[23] for the claim that the existence of an *F* is not asserted but presupposed in the case of an assertion of the form 'The *F* is *G*:'[24]

> If someone says 'the Kaiser of France,' we will object: 'You believe that there is a Kaiser of France' etc.[25]

---

[19]Hua XXX, 180; my translation.

[20]Hua XXX, 181; my translation.

[21]Cf. Strawson 1950.

[22]Hua XL, 139; my translation.

[23]Cf. Strawson 1950, 330.

[24]Compare Husserl's claim, quoted in Frege's letter to Husserl from Dec. 12, 1906, that "The form with 'All' is normally understood in such a way that the existence of objects [...] is presupposed as conceded" (Frege 1976, 106), a view also shared by Strawson (but rejected by Frege).

[25]Hua XL, 139; my translation.

Note, however, that unlike Strawson Husserl also acknowledges what Donnellan[26] would call the "referential use" of a description like 'the Kaiser' (see Sect. 9.5 below).

Before turning to Frege's and Husserl's 1906 correspondence, I am going to explain Frege's conception of sense, which is presupposed in their correspondence. I shall mainly focus on his arguments for the claim that singular terms and declarative sentences express sense, as manifested in his 1892 article "*Über Sinn und Bedeutung*."

9.3 Frege's Main Arguments for the Distinction Between Sense and "Meaning"

By a "Thought," Frege means an "objective" judgeable "content,"[27] to be shared "by many,"[28] which gets acknowledged as true if someone answers a corresponding question in the affirmative. Thus, Thoughts can be expressed by interrogative and declarative sentences and be the content of questions, in which a Thought is merely "grasped" or "thought," and judgments, which acknowledge the truth of a Thought;[29] where "judgment" and "thinking" denote inner thoughts, so that it suggests itself to equate their intentional contents with Thoughts.[30] (In what follows, I shall use the term 'sentence' to refer to *declarative* sentences.) Frege starts from the observation that language "expresses an enormous number of Thoughts, some of which have not even been grasped and expressed before by any human being."[31] He postulates "Thought components" or "building blocks (*Gedankenbausteine*)" in order to explain this immense expressive power of language (and our corresponding interpretive ability), claiming that sub-sentential expressions ocurring in a given sentence are used to express such building blocks of Thought even in the context of many *other* sentences in which they may occur:

> Consider the sentence 'Aetna is higher than Vesuvius.' Here we have the name 'Aetna,' which also occurs in other sentences, such as the sentence 'Aetna is in Sicily.' Our ability [*Die Möglichkeit für uns*] to understand sentences that we have never heard before is obviously based upon the fact that we build the sense of the sentence out of parts

---

26 Cf. Donnellan 1966.

27 Of course, the notion of Thought is strictly speaking a successor concept of the notion of judgeable content; see Frege 1893, X. For a critical discussion of Frege's claim, in the foreword to his *Grundgesetze der Arithmetik*, that the notion of judgeable content "has now split up for me into what I call 'Thought' and 'truth value'" (Frege 1893, X), see Carl 1994, 76 ff.

28 First-person Thoughts are an exception; see Sect. 9.4 below.

29 Cf. Frege 1918, 62.

30 See McIntyre 1987, 532. The distinction between non-committal vs. judgmental propositional thought is also to be found in Husserl (who examines it in greater detail than Frege). Cf., e.g., Hua XIX/1, 462f; Hua XXX, 55.

31 Frege 1976, 243; my translation.

> corresponding to the words. If we thus find the same word, e.g. 'Aetna,' in two sentences, then we recognize a common element in the according Thoughts as well, which [common element] corresponds to that word.[32]

Frege here talks about the way we build the sense of the sentence out of parts, but the same consideration would go through if "sense" were replaced by "reference" or "*Bedeutung.*" So why are we to conceive of these Thought components as *senses*? This question is answered in "*Über Sinn und Bedeutung,*" to which I now turn.

Frege's argument for the sense/reference distinction with singular terms ("proper names") might aptly be called the *cognitive value argument.* It runs as follows.[33]

(P1) Sentences of the form 'a=a' and 'a=b' in which 'a' and 'b' "mean (*bedeuten*)" (i.e., designate, refer to)[34] the same object sometimes differ in their cognitive value ("*Erkentniswert*"). (For, the judgeable content of true sentences of the form 'a=a' is "analytic" and recognizable *a priori*, whilst true sentences of the form 'a=b' would sometimes contain "valuable extensions of our knowledge" and "cannot always be justified *a priori.*")

(P2) If the parts of the Thoughts expressed by sentences of the form 'a=a' and 'a=b' in which 'a' and 'b' "mean" the same object are always identical, then such sentences always share the same cognitive value. (Note: The Thought expressed is the content of the sentence that can be recognized as true, provided the Thought is true.)

(C1) Hence (by P1 and P2) the Thoughts expressed by sentences of the form 'a=a' and 'a=b' in which 'a' and 'b' "mean" the same object are not always identical.

(P3) If the names 'a' and 'b' contribute the object they "mean" to the Thoughts expressed by sentences of the form 'a=a' and 'a=b' in which 'a' and 'b' "mean" the same object, then the Thoughts expressed by such sentences are always identical.

(C2) Hence (by C1 and P3) the names 'a' and 'b' contribute something else than the object they "mean" to the Thoughts expressed by sentences of the form 'a=a' and 'a=b' in which 'a' and 'b' "mean" the same object. (Note: We have seen above, in connection with the expressive power of language and our corresponding interpretive capacity, that they must contribute some Thought component.)

(P4) If the names 'a' and 'b' contribute the semiotic relation ("*Bezeichungsweise*") they bear to the object they "mean" (i.e., roughly, the fact that they designate that object) to the Thoughts expressed by sentences of the form 'a=a' and 'a=b' in which 'a' and 'b' "mean" the same object, then the Thoughts expressed by such sentences always have the same cognitive value. (For, a difference regarding cognitive value can never be due to an arbitrary stipulation, but a semiotic difference can.)

(C3) Hence (by P1 and P4) the names 'a' and 'b' contribute something else than the semiotic relation they bear to the object they "mean" to the Thoughts expressed by sentences of the form 'a=a' and 'a=b' in which 'a' and 'b' "mean" the same object.

(C4) Hence (by C2 and C3) the names 'a' and 'b' contribute neither the object they "mean" nor their semiotic relation to it to the Thoughts expressed by sentences of the form 'a=a' and 'a=b' in which 'a' and 'b' "mean" the same object, but something else. This "something else" Frege calls "the *mode of presentation* of the object designated (*die Art des Gegebenseins des Bezeichneten*)." (Frege gives both a geometrical example – 'the intersection point of lines a and b' vs. 'the intersection point of lines b and c' – and his famous 'Morning Star'/'Evening Star' example to illustrate this notion.)

---

[32]Frege 1976, 127; my translation.

[33]Cf. Frege 1892a, 25.

[34]In *LI* Husserl criticizes Frege's use of the term "*Bedeutung*" as potentially misleading because the term is usually regarded as synonymous with 'sense;' cf. Hua XIX/1, 58.

Frege goes on to make a number of assertions with regard to the sense of a singular term.

1. The sense of a name contains the respective "mode of presentation of the object designated (*die Art des Gegebenseins des Bezeichneten*)."[35]
2. The sense of a name "is grasped by anyone knowing the language [. . .] it belongs to sufficiently well."[36]
3. The sense of a name "eludidates" its "meaning (*Bedeutung*)" (i.e., its reference) only "one-sidedly," thus merely allowing for one-sided "knowledge" or "cognition (*Erkenntnis*)" regarding its referent.[37] (This is one of the theses demonstrating that Frege has an epistemic conception of sense.)
4. In natural languages ("*Volkssprachen*") proper proper names (like 'Aristotle') sometimes express different senses in the mouth of different speakers, which can be expressed by different definite descriptions (such as 'the pupil of Plato and teacher of Alexander the Great' vs. 'the Stagira-born teacher of Alexander the Great').[38]
5. The sense of a name uniquely determines its "meaning" or reference, in the sense that "a particular meaning" is assigned to it,[39] "provided there is such a [meaning] (*falls eine solche vorhanden ist*)."[40] (The last addition takes account of the possibility of empty names such as 'Odysseus,' with regard to which Frege argues, at least in "*Über Sinn und Bedeutung*," that they do express a sense.)
6. Names are generally such that a particular sense and "meaning" (reference) are assigned to them. (Exceptions are empty names and names occuring in *oratio recta*, or in *oratio obliqua* and other cases concerning which Frege would speak of a "meaning shift;" see below)[41]
7. Unlike the idea ("*Vorstellung*") someone may associate with a name, its sense is objective; it can be "the common property of many," all of whom "grasp" it.[42]
8. The correct "translation" of a name (or another expression) must preserve the sense of the expression. (By contrast, the "colourings and shadings which

[35]Cf. Frege 1892a, 26.

[36]Cf. Frege 1892a, 27.

[37]Cf. Frege 1892a, 27; also cf. the telescope-analogy on p. 30.

[38]Cf. Frege 1892a, 27, fn. 2. This footnote may either be taken to formulate an exception to thesis 2 or to argue that different speakers of one and the same natural language sometimes speak different idiolects of that language. Note that either way the footnote seems to support the ascription of a descriptivist conception of the sense of proper proper names to Frege; also cf. Frege 1918, 65 (headword: "Dr. Gustav Lauben"). In Sect. 9.4 we will see that Husserl has a different view in this regard, arguing that (proper) proper names, demonstratives and referentially used description are more or less on a par as far as their sense is concerned.

[39]Cf. Frege 1892a, 27. Also cf. Frege 1983, 276: "Sentences and sentence parts which differ in meaning also have different senses."

[40]Cf. Frege 1892a, 32.

[41]Cf. Frege 1892a, 28.

[42]Cf. Frege 1892a, 29.

poetic arts and rhetoric try to give to the sense" are "not objective," and the corresponding intentions of the poet are "never to be exactly determined.")[43]
9. Whoever uses a name in order to talk about an object (such as the Moon) "presupposes" a corresponding "meaning."[44]

What about the *Thoughts* expressed by sentences? Are they to be regarded as senses as well? Frege's answer is affirmative. His argument makes recourse to the above-motivated idea of Thought components (see premises P4 and P5 below) as well as to a condition he regards as necessary for Thought identity which he would take up again in his 1906 correspondence with Husserl (see P3).[45]

Df.: The objective (judgeable) content of a sentence *s* is called the *Thought* expressed by *s*.[46]

(P1) The Thought expressed by a sentence *s* is either the sense or the "meaning" (reference) of *s*.

(C1) Let S1 be the sentence 'The Morning Star is illuminated by the Sun,' and let S2 be the sentence that results if we substitute the name 'the Morning Star' in S1 by the name 'the Evening Star,' which shares the "meaning" but (by the cognitive value argument) does not share the sense of 'the Morning Star;' then (by P2 and P3) S1 and S2 express different Thoughts.

(P2) For it is possible for a rational being who understands the sentences S1 and S2 as such to evaluate the Thoughts expressed by S1 and S2 differently with regard to their truth or falsity.

(P3) And if two sentences A and B express the same Thought, then it is impossible for a rational being who understands A and B as such[47] to evaluate the Thoughts expressed by A and B differently with regard to their truth or falsity. (This could be called *Frege's criterion of Thought difference*. It may be regarded as following, more or less immediately, from *Df.*)

(C2) S1 has the same "meaning" as S2 (by P4).

(P4) For, if one substitutes "a word by another one with the same meaning but a different sense" in a sentence, "then this cannot have a effect upon the meaning" of that sentence. (Note: This is an immediate consequence of the *principle of compositionality of "meaning"*, which says that the "meaning" of a sentence is a function of the meaning of the "words," i.e. sub-sentential expressions, it consists of. It follows that the "meaning" of a word is its complete contribution to the "meaning" of a sentence in which it occurs, what Dummett calls the word's "semantic role" and Tugendhat its "truth-value potential."[48])

(C3) By contrast, S1 does not have the same sense as S2 (by P5).

(P5) For, the sense of a sentence is a function of (is composed of) the senses of its sub-sentential components (*principle of the compositionality of sense*), and the names 'the Morning Star' and 'the Evening Star' express different senses (as established by the cognitive value argument; see above).

[43]Cf. Frege 1892a, 31.

[44]Cf. Frege 1892a, 31.

[45]Cf. Frege 1892a, 32.

[46]Cf. Frege 1892a, 32, fn. 5.

[47]For the point of the addition "as such" see Dummett 1992, 79 ff.

[48]Cf. Dummett 1973, 199.

(C4) Hence the Thought contained by S1 is the sense expressed by S1 (by P1, C1, C2 and C3).
(C5) The same holds true for all sentences: The Thought they contain is the sense they express (generalization of C4).

If one follows Frege in identifying the "meaning" of a sentence with its truth-value, then P4 entails that if one substitutes a name by another name with the same referent in a sentence, then the truth-value of the resulting sentence is the same as that of the original sentence – this thesis is often referred to as the *principle of substitutivity*. Alternatively, one could conceive of the "meaning" (reference) of the sentence as its truth-condition, i.e. with the (as Husserl would put it; see Sect. 9.5 below) *state of affairs* that obtains just in case the (sense of the) sentence is true. However, the notion of a state of affairs is not to be found in Frege.

The principle of substitutivity seems to be threatened by a number of alleged counter-examples that Frege discusses in detail in "*Über Sinn und Bedeutung*." One such class of examples is represented by *propositional attitude ascriptions*, such as 'Ed believes that the Morning Star is illuminated by the Sun.' If we substitute 'The Morning Star' by the co-referential name 'The Evening Star,' the resulting sentence may differ in truth-value from the original one. Frege's reply to this objection runs that in the context of such an ascription the words occuring in the subordinate clause specifying the content ('that the Morning Star is illuminated by the Sun') refer to their ordinary sense rather than to their ordinary "meaning." A "meaning shift (*Bedeutungsverschiebung*)" occurs in such contexts, and one may only substitute the words by expressions sharing the same ordinary *sense* if one wants to make sure that the truth-value is preserved. (Note that this consideration seems to rely upon both the context principle and the principles of compositionality.)

Husserl does not share this view.[49] He holds that the content clause refers to a state of affairs involving the ordinary referents of the words, which state of affairs is the intentional object of the ascribed attitude (given that the ascription is correct). We may here adapt an idea of Dummett's Frege interpretation and say that for Husserl, when we correctly ascribe a propositional attitude, say an act of judgment, we *say* what the intentional *object*, i.e. the judged state of affairs, is and thereby *show* what its intentional *content* (its "interpretive sense") is, notably by choosing a content clause that reflects the mode of presentation (the propositional content) under which the state of affairs is judged.[50] However, the relevant notion of state of affairs needs explanation. I will return to that notion in Sect. 9.5 below (headword: singular state of affairs).

---

[49]This disagreement has already been pointed out by Føllesdal in his 1969 paper on "Husserl's notion of noema." Cf. Føllesdal 1969, 686.

[50]Cf. Dummett 1973, 227. Also see Beyer 2000, 124 ff, where this idea is used to explain Husserl's notion of "phenomenological" or "noematic analysis of sense." This analysis is to be performed from the first-person point of view. However, it may be argued, very much in line with Husserl, that propositional attitude ascriptions are based (at least implicitly) on *empathy*, and thus have to simulate that point of view. For a defence of this view, see Beyer 2006, ch. 4.

9.4 The 1906 Correspondence

In his first 1906 letter to Husserl,[51] Frege begins by highlighting that Thoughts are objective, such that the Thought known as the Pythagorean Theorem, for instance, can be shared by "all men." (This would be one of the central themes of his later article "*Der Gedanke.*")[52] He adds that Thoughts are subject to logical laws, and that therefore they are to be individuated in such a way that the corresponding distinctions are "relevant to logical laws," where all distinctions going beyond this, in particular those that may refer to the "colouring" a particular language may lend to a given Thought, are to be ignored: They are irrelevant for logical purposes and lack a "criterion" of identity that allows for "an objective answer;" hence, they are useless for scientific purposes. (For Frege, science aims at the True; and Thoughts are what represents the True – or the False, as the case may be.)

On this basis, Frege criticizes Husserl's recourse to the notion of "congruence." Since the relevant letter by Husserl has been lost, we do not know what notion he has in mind here, but it is plausible to assume that he refers to a principle to be found in *LU* that he took over from Bolzano. (Note that in Husserl's lost letter answering the present one he refers to Bolzano.) (see., e.g., Bolzano 1837, vol. 1, 244).[53] I have elsewhere called the counterpart of this principle for sub-propositional thoughts ("*subjective Vorstellungen*," ideas – i.e. inner thoughts whose contents are Thought components rather than full Thoughts) and their contents the *principle of structural isomorphy (between subjective ideas and their stuff)*. It says that the manner in which a given complex (subjective) thought is built out of the thoughts composing it corresponds to the manner in which the (objective) intentional content (what Bolzano calls "stuff (*Stoff*)" and Husserl would refer to as "ideal matter (*Materie*)" in *LU*) of that thought is composed. Thus, to give a Bolzanian example, the thought you have while reading 'an educated son of an uneducated father' will have a different sub-propositional content than the one you have when reading 'an uneducated son of an educated father,' because it is composed out of its constituent thoughts in a different manner (order).[54] So there are manners in which thoughts are composed of their intentional elements, and the structure of sub-propositional content (its *mode of composition*, for short) can be read off from these. (This idea underlies Bolzano's notion of conceptual definition.) (see Bolzano 1837, vol. 1, 122). Likewise for (subjective) judgments and their propositional contents – Bolzano and Husserl both subscribe to *the principle of structural isomorphy between judgments and their stuff*, according to which the propositional content (what Bolzano calls the "sentence in itself (*Satz an sich*)") of the judgment expressed by '2 to the power of 3 is identical to 2 to the power

[51] Frege 1976, 101–105.

[52] Frege 1918.

[53] Cf. Frege 1976, 105.

[54] Beyer 1996, 107 f.

of 3' is different from the one of the judgment expressed by '2 to the power of 3 is identical to 3 to the power of 2,' for example.[55] Now if we apply this principle to the sentences presented in Frege's first 1906 letter to Husserl as examples of "equipollence" (see below) and Thought identity – (1.1) 'If *p*, then *q*' vs. (1.2) 'Not (*p* and not *q*)' and (2.1) 'If anything is an *F*, then it is a *G*' vs. (2.2) 'Not (there is an *F* which is not a *G*)' – the result seems to be that, *pace* Frege, the respective first element of these sentence pairs expresses a different propositional content than the second one, at least if we follow Bolzano and assume that the negation sign expresses a sub-propositional content. Since this latter assumption may appear controversial (especially in the context of a discussion of Frege), (see Bolzano 1837, vol. 1, 419),[56] let us consider a different sentence pair of which Frege says that its elements express the same Thought whilst Husserl would say that they express different propositional contents.

> [...] Singularity does not attach to a Thought as such but only with respect to a manner of possible analysis (*Zerlegung*). It is possible that the very same Thought can appear as particular (*partikulär*) with respect to another analysis (Christ won some people over to his doctrine).[57]

By the principle of structural isomorphy about judgments, the judgments expressed by 'Christ won some people over to his doctrine' and 'Some people have been won over by Christ to his doctrine' express different propositional contents, at least if we assume that the thought expressed by 'to win over' refers to a different relation (and thus displays a different sub-propositional content) than the one expressed by 'to be won over by.'[58] This applies all the more since the manners in which the first and the second judgment are composed of their component thoughts differ. (So

---

[55] Beyer 1996, 127 f.

[56] I am alluding to his view that a judgment that acknowledges a Thought *p* as true is "the same act (*dieselbe Tat*)" as a judgment that rejects its contradictory opposite non-*p* is false, which seems to imply that a sentence expresses the same Thought as its double negation (cf. Frege 1983, 201, 214), so that the two occurences of the negation sign to be found in the latter do not seem to contribute anything to the Thought expressed. Note, however, that this view also seems to imply that in the context of a *simple* negation the negation sign *does* contribute something to the Thought expressed, in turning a Thought into what Frege calls its "negation (*Verneinung*)."

[57] Frege 1983, 203; my translation. Also see Frege 1983, 218; Frege 1892a, 45 f.; Frege 1892b, 199 f. The latter passage shows that the idea of multiple *Zerlegbarkeit* of Thought plays an important role in Frege's semantics of "subordination" statements.

[58] Compare what Husserl says about the "a priori equivalence a > b and b < a": "Here we do not have a case in which the same is predicated [*ausgesagt*] of the same thing on both sides, as the subjects and predicates differ. [...] Rather, *the relation is now reversed.* [...] *A priori* equivalence, but not formal-logical equivalence, obtains in the case of reversed forms [*Umkehrungen*] like 'a is more intensive than b', 'b is less intensive than a.' [Provided, that is,] the situation of affairs [*Sachlage*] is the same on both sides." (Hua XVI, 98; translation and first emphasis mine) (Note that – somewhat misleadingly – Husserl sometimes refers to situation of affairs as "*Verhältnisse*," which could be translated as 'relations' as well. On this use of the term, both 'a > b' and 'b < a' are made true by the same relation.) For the thesis that if the intended relations differ, then the meaning or (sub-) propositional content differs, also see Hua XVI, 44 f. – Compare and contrast Frege's remarks, in §9 of his *Begriffsschrift*, about the – in his view identical – "conceptual content" expressed by the

it is not really colouring, in the sense of more or less conventional implicatures or "secondary thoughts (*Nebengedanken*)," as Frege would put it, but mode of composition and intentional object whose difference leads Husserl to regard the expressed propositions as diverging.) Notice also that Frege seems to hold that *logical form* (singularity, particularity) attaches to a Thought only relative to a (manner of) analysis, whereas Husserl would ascribe it to the propositional content as such. Frege nevertheless believes that logical laws concern Thoughts as such. So when do two sentences express the same Thought in his view? In his second 1906 letter to Husserl (Dec. 12, 1906) Frege tries to answer this question in terms of an "objective criterion" of identity for (appropriate) Thoughts. It runs as follows:

> Now in order to decide whether the sentence *A* expresses the same Thought as the sentence *B*, the following means seems to me to be the only possible one, where I assume that none of the two sentences contains a logically evident sense-component. Notably, if both the assumption that the content of *A* is false while the content of *B* is true and the assumption that the content of *A* is true while the content of *B* is false lead to a logical contradiction, such that in order to determine this contradiction one does not have to know whether the content of *A* or the content of *B* is true or false and also does not have to apply laws other than purely logical ones, then nothing which belongs to the content of *A*, so far as it is capable of being judged as true or false, can fail to belong to the content of *B*; for, there would be no justification whatsoever, within the content of *B*, for such a surplus, which also, by assumption, would fail to be logically evident. Likewise, by that very assumption of ours, nothing that belongs to the content of *B*, as far as it is capable of being judged as true or false, can fail to belong to the content of *A*. So everything about the contents of *A* or *B* that can be judged as true or false coincides, and this alone is relevant for logic, and that I call the Thought expressed by both *A* and *B*.[59]

So let us assume, with Frege, that (1) a given speaker/hearer S understands two sentences A and B (as such) and (2) the contents of A and B do not contain senses which are "logically evident," i.e. such that you cannot (as a rational being) grasp them without acknowledging them as true. (Clause (2) is supposed to prevent all immediately evident sentences, such as 'Hesperus is Hesperus' and 'If Hesperus is Phosphorus, then Hesperus is Phosphorus,' from expressing the same sense.) Then Frege's criterion runs: A and B express the same Thought (for S) iff A and B are logically equivalent. This is a very coarse-grained criterion of identity, and (as Künne has pointed out) it implies, first, that all formulations of logical laws which are not logically evident express the same Thought (for someone who understands them), and, secondly, that the sentences 'Beyer believes that Hesperus is Phosphorus' and 'Beyer believes that if Hesperus is not Phosphorus, then it is not the case that if Hesperus is Phosphorus, then Gottlob Frege is not the inventor of modern logic' express the same Thought (for someone who understands them).[60]

---

sentences 'Hydrogen gas is lighter than carbon dioxide gas' and 'Carbon dioxide gas is heavier than hydrogen gas,' respectively (Frege 1879, 15–18).

[59]Frege 1976, 105 f; my translation.

[60]See Künne 2010, 648 f.

While it may be debatable (in terms of condition (1)) whether these consequences are indeed unwelcome, it is clear that in his "*Ausführungen über Sinn und Bedeutung*," which contain at least some discussion of Husserl, Frege presents a different identity criterion, which he explicitly refers to as the criterion of "equipollence."[61] Assume, again, that the conditions (1) and (2) described in the foregoing paragraph are met. Then sentences A and B are equipollent, and (on Frege's view) express the same Thought (for S), iff (a) it is impossible for S (as a rational being who understands A and B) to acknowledge the content of A and B (as such), respectively, as true (or, we should add, to dismiss it as false, as the case may be)[62] without immediately also acknowledging the content of B and A, respectively, as true (or also dismissing it as false). This criterion fits in well both with the semantics of belief-ascriptions developed in "*Über Sinn und Bedeutung,*" the criterion of difference of Thoughts Frege there uses and, connectedly, with his identification of Thoughts with "judgeable contents (*beurteilbare Inhalte*)" (see Sect. 9.3 above). It also allows him to say, as he does in "*Über Sinn und Bedeutung,*" that sentences of the form 'If *p*, then *q*' express the same Thought as sentences of the form 'not-*p*, or *q*,'[63] and likewise for '*p*' and 'not-not-*p*' as well as other pairs of examples whose logical equivalence can be immediately derived by a logical law.

These identifications seem to cover all "that is relevant to logic," as Frege puts it in the foregoing quotation. So why should one follow Husserl and still individuate propositional contents in a more fine-grained way? There seem to me to be at least two reasons.

First, as Künne points out, Frege does not offer us a *general* criterion of identity for Thoughts. For instance, he does not say anything that could be used to justify the claim that the logically evident sentences 'It is not always Tuesday' and 'Bebel is Bebel' express different Thoughts. Künne comments on this as follows: We do not need a criterion to realize that one can think the first propositional content without thinking the latter.[64] Yet *why* is it (so obvious) that their propositional contents differ? I regard this as a legitimate question to be answered by an adequate theory of content. Husserl's theory enables us to meet this requirement, thanks to his principle of isomorphy, which he embeds in his conception of the "clarification" of "meaning"[65] in terms of the "things themselves," i.e. in terms of the "intuitive fulfillment" of a "meaning-bestowing act." This conception is sketched in *LU* and further developed in later writings (such as *EU*) in the context of Husserl's "transcendental" project of a phenomenological meta-justification of our natural understanding of ourselves and the world.

Secondly, Husserl's conception of meaning and content relies on a distinction between, roughly, (sub-)propositional content and conventional linguistic meaning

[61] Frege 1983, 213 f.

[62] See the thorough discussion of the clause about logical evidence in Künne 2010, 649 ff.

[63] Frege 1892a, 45 f. (fn. 13).

[64] Künne 2010, 652.

[65] Hua XIX/1. 10 f.

that is closer to more recent approaches to indexicality and may be preferable to Frege's approach, especially against the background of Husserl's detailed analysis of the intentionality of empirical thought and speech. I will explain these two points in the following section.

9.5 Husserl's Conception of Sense

For Frege, the use of an expression to refer to something (under a mode of presentation) involves an according referential intention on the part of the speaker (see Sect. 9.3 above, thesis 9). Husserl's semantic approach is also an intentionalist one.[66] Unlike Frege, however, he would study the (inner) *thoughts* (thoughts with a small 't,' i.e. intentional experiences or "acts") that the respective speaker deliberately presents herself as having – that she "intimates" or "gives voice to (*kundgibt*)" – when expressing the sense in question on a given occasion. For instance, if you claim 'One of my goals is to defend Husserl against Frege,' you give voice to a judgment, or occurrent belief, of yours, to the effect that defending Husserl against Frege is among your goals. This judgment is intentional, in that it represents a state of affairs, namely your having a particular goal; it is "about" that state of affairs, even if the latter does not exist (i.e., obtain) because you do not have that goal. Now it is the intentional content of this judgment (which may be intuitively empty or unfulfilled, i.e. made in the absence of a corresponding intuition, such as a corresponding perception) that a hearer has to know in order to understand your utterance, i.e. to grasp its literal meaning or sense. Thus, the (unfulfilled) judgment[67] functions as the "meaning-bestowing" or "meaning conferring act" regarding the sentence uttered.[68] This act, whose content has to be made out in order to grasp the intended literal meaning, is referred to as *the act given voice to*, or intimated, "*in the narrow sense.*"[69] However, in the present example ('One of my goals is to defend Husserl against Frege') the speaker also deliberately presents herself as someone who wants to defend Husserl against Frege; after all, she explicitly ascribes that intention to herself. This latter act (the intention in question) is given voice to "in the broader sense"[70] only – Frege would refer to

[66]For the following presentation of Husserl's theory of meaning cf. Beyer and Weichold 2011, 406 ff, as well as Beyer 1996, 44 ff.

[67]For the sake of simplicity, Husserl usually assumes that the speaker is sincere.

[68]Hua XIX/1, 44. Note that an act does not have to be given voice to in order to function as a meaning-bestowing act – even in the case of a soliloquy Husserl would speak of meaning-bestowing acts. Thus, consider the difference between your reciting to yourself a Greek verse whose meaning you have forgotten and your reciting that very verse to yourself while fully understanding it. In the second case, but not in the first one, you are experiencing a meaning-bestowing act.

[69]Hua XIX/1, 40.

[70]Hua XIX/1, 40.

it as a mere "secondary Thought" expressed by the speaker. On Husserl's view, the speaker intentionally presents herself as performing or undergoing that act, but if the hearer did not ascribe that intention to the speaker he would not thereby fail to grasp the literal meaning or sense of the utterance. Again, if you assert 'This is a blooming tree,' you give voice, in the narrow sense, to a demonstrative judgment; but you also present yourself as perceiving (or having perceived) something as a blooming tree, where the act of perception is given voice to in the broader sense. This perceptual act verifies the unfulfilled judgment (the meaning-bestowing act) by intuitively "fulfilling" it.[71] Since the meaning-bestowing act finds its aim, so to speak, in this intuitive fulfillment, Husserl also refers to it as the corresponding "meaning intention."[72] Since any meaning-intention aims at its intuitive fulfillment, every meaningful utterance can in principle be made to give voice (in the broader sense) to such an act of fulfillment, provided its literal meaning is not inconsistent as a matter of logical evidence. Meaning-bestowing acts belong to the broader class of what Husserl calls "*Intentions in need of fulfillment* (*erfüllungsbedürftige Intentionen*)," which also includes wishes, volitions, hopes, fears, doubts and conjectures.[73] (Arguably, this class coincides with the class of acts which have a "direction of fit" (Searle).)[74] The intuitive fulfillments of meaning-bestowing acts are either provided by corresponding perceptions or by corresponding acts of imagination, in which their intentional objects are presented (to a certain degree) just as they are *re*-presented (*vergegenwärtigt*) in the respective meaning-bestowing act (also called "meaning-intention") they fulfill. Husserl calls the intentional contents of these fulfilling acts "fulfilling senses (*erfüllende Sinne*)," and he formulates a *phenomenological principle of meaning*, which states that *a meaning-bestowing act can successfully refer to an (existing) object only if there is a fulfilling sense corresponding to it.*[75] Thus, inconsistent meanings like the sense of 'a round square' cannot successfully refer, because they have no corresponding fulfilling sense; rather, they inevitably lead to "intuitive frustrations (*Enttäuschungen*)," which are the negative counterparts of acts of fulfillment and show (when they are reflected upon) that there can be no such fulfillment, and no corresponding object, as the constituents of the intentional object (such as 'its' round and 'its' square form) are experienced, in the act of frustration, as conflicting (*widerstreiten*) with each other.[76] By contrast, any consistent sense can in principle be subjectively fulfilled, more or less adequately, by a unified intuition, where the structure and other essential features of the sense in question can be read off from the respective mode of intuitive fulfillment.

[71]Hua XIX/1, 44.

[72]Hua XIX/1, 44.

[73]Hua XIX/2, 572 f.

[74]See Beyer 1996, 45.

[75]Cf. Hua XIX/2, 625; also see Beyer 1996, 47.

[76]Cf. Hua XIX/2, 643; also see Beyer 1996, 48 f.

Now of course (to return to Künne's examples), you do not need perception to realize that Bebel is Bebel, or that it's not always Tuesday (see end of Sect. 9.4). However, the principle of structural isomorphy about judgments, when combined with Husserl's conception of giving-voice-to, implies that the intentional contents of these judgments contain different components, representing the senses of the expressions 'Bebel' and 'Tuesday' (among other things), as used in the sentences 'Bebel is Bebel' and 'It's not always Tuesday' that can be used to give voice to these judgments. (This consideration can be generalized, so that Frege's principle of compositionality of sense can be derived on this basis; which principle Husserl indeed explicitly endorses.[77]) These sub-propositional contents correspond to different fulfilling senses, as the meaning-bestowing act given voice to by 'Bebel' aims at a different kind of intuitive fulfillment than the one given voice to by 'Tuesday.' (Days, even days as types, are presented to us differently than people, even if both days and people are name-bearers. Their constitution is closer to that of numbers.) So Husserl's conception of sense yields a criterion of sense-difference, in terms of structural isomorphy (or mode of composition) and fulfilling sense, that can be used to explain *why* the relevant propositional contents differ. To be sure, if one simply makes recourse to Frege's notion of mode of presentation (and his compositionality principle about sense) one can thereby explain the difference in question as well. However, as we have just seen, Husserl's conception goes further in telling us how *modes of presentation* can in turn be distinguished from one another in a principled manner. By contrast, Frege "says practically nothing whatever directly about what the senses of expressions of different types consist in."[78] So much for the first assertion made at the end of Sect. 9.4; now to the second, concerning indexicality and different levels of (literal) meaning.

On Husserl's view, the "original function" of linguistic expressions is their communicative use to give voice to meaning-bestowing acts, or meaning intentions.[79] However, this "indicating (*anzeigende*)" function is not essential to their functioning as meaningful units, as they can also be employed "in solitary life (*im einsamen Seelenleben*)," thanks to meaning-bestowing acts not actually given voice to but had all the same.[80] But these acts and the meanings they bear are constrained by semantic factors concerning the linguistic expressions employed, factors determined

---

[77]Cf. Hua XXVI, 44, where Husserl approvingly states that "one says of the whole expression that it has a unitary meaning, to which all expression-parts make their contribution." (my translation)

[78]Dummett 1973, 227. Dummett continues: "Indeed, even when Frege is purporting to give the sense of a word or symbol, what he actually *states* is what the reference is. [...] In a case in which we are concerned to convey [...] the sense of the expression, we shall choose that means of stating what the referent is which displays the sense [...]." For a comparison of this idea of (Dummett's) Frege's to Husserl's closely related notion of a phenomenologically adequate ("noematic") specification of (sub-)propositional content, as manifested in his 1908 *Vorlesungen über Bedeutungslehre* (Hua XXVI), see Beyer 2000, 120–126.

[79]Cf. Hua XIX/1, 39.

[80]Cf. Hua XIX/1, 41 ff.

by linguistic conventions regarding the relationship between their meaning and the features of non-linguistic reality they serve to represent.[81]

To bring out the relevant semantic factors, consider what Husserl calls "essentially occasional expressions," i.e. systematically context-sensitive, or indexical, expressions like 'I,' 'here,' 'now,' 'I am here now.' In his pioneering discussion of these expressions in the first of *LU*, §26, he introduces the semantic distinction between, on the one hand, an expression's *general meaning-function* (which corresponds to what Kaplan calls "character," roughly: the linguistic meaning of the expression used) and, on the other hand, the propositional, or sub-propositional, content – the "*respective meaning*" – expressed in a given context of utterance.[82] If, for example, you and I both say 'I,' then our two utterances share the same general-meaning function but express different respective meanings, with different referents. Again, if you and I assert 'I have blood type A,' our utterances share the same general meaning-function but express different respective meanings, with different truth conditions (or represented states of affairs). These respective meanings are the (literal) propositional, or truth-conditional, contents expressed in the utterance of a sentence. If the truth-conditions of two sentences or sentential utterances differ, it follows that the expressed respective meanings differ, too.[83] Likewise, if the references of two uses or utterances of a sub-sentential expression (such as 'I') differ, then the sub-propositional contents expressed differ as well (compare Sect. 9.3 above, thesis 5).[84]

The level of respective meaning, alias (sub-)propositional content, is subject to what Husserl calls "pure grammar," which is the theory of what distinguishes sense (i.e. respective meaning) from nonsense. On this view, propositional content displays something like formal, "syntactic" or "categorical" structure (corresponding to what Husserl, in the fourth of his *LU*, refers to as "meaning categories"). This idea helps to explain the compositionality of sense, which of course in turn explains how speakers and hearers, or interpreters, are able to grasp the meaning of an infinite number of sentences, many of which they have never heard before, on the basis of a finite vocabulary and a finite set of linguistic rules or conventions (see Sect. 9.3 above, headword: expressive power of language). So far, so Fregean. But the view that propositional content *per se* displays logical form[85] is incompatible with Frege's view that logical form attaches to propositional contents (Thoughts) only relative to a (manner of) analysis (see Sect. 9.4 above). Note, however, that – like Strawson – Husserl construes the logical form of propositional contents more along the lines of *traditional* logic, as becomes particularly clear in *EU*. This is because he thinks that the corresponding forms reflect the structure of our actual thoughts,

[81]Cf. Hua XIX/1, 91 f.

[82]Cf. Hua XIX/1, 86 ff.

[83]Cf. Hua XIX/1, 84 (headword: "equivalence").

[84]Cf. Hua XIX/1, 88.

[85]Cf., e.g., Hua XXX, 180 (above).

as they occur in the ("active") pursuit of our epistemic interests, and their modes of intuitive fulfillment more accurately.

Husserl regards the *general meaning-function* as fixed by common usage.[86] Frege does not have the distinction between general meaning-function and respective meaning, but in some places he describes *sense* as being such that linguistic competence is sufficient to grasp it (see Sect. 9.3 above, thesis 2). The respective meaning determines the expression's reference, or truth condition (the represented states of affairs), in the sense that two expressions sharing that meaning are thus bound to refer to the same object(s), or to represent the same state of affairs, if any – just like Fregean sense is supposed to determine reference (see Sect. 9.3 above, thesis 5). Unlike Frege, however, Husserl construes sense *qua* "respective meaning" as two-factored, with the general meaning function plus the relevant context of utterancc (if any) determining the sense in question. Thus we have two levels of literal meaning, being expressed when a meaning intention is given voice to. (A third level of meaning acknowledged by Husserl would be the level of the (sub-)propositional contents of acts given voice to "in the broader sense," which corresponds to what the speaker *non*-literally means or suggests, what Grice calls "implicature.")[87] In the case of indexical expressions, the respective meaning is a function of both the context of utterance and the general meaning-function of the expression used, which differs from the respective meaning; in all other cases, the two levels can be said to coincide.

Let us take a look at Husserl's views regarding two (from a Fregean viewpoint) particularly difficult examples of indexical expressions and briefly compare them to Frege's, namely 'I' and 'this.'[88]

The meaning of 'I.' In "*Der Gedanke*" Frege contends that only the speaker herself can grasp the Thought expressed by 'I have been wounded,' as this sentence is used in a soliloquy (because no one can be presented to someone else under the mode of presentation under which she is presented to herself in a situation like this), and that the hearer has to grasp a different Thought, provided by the utterance context, in order to understand a corresponding sentential utterance, such as the Thought expressed by the expression consisting of the sentence 'She who

[86]Cf. Hua XIX/1, 91 f.

[87]The corresponding idea of different levels (*Stufen*) of understanding, which include the grasping of both character, content and implicitures, is borrowed from Künne, who is also to be credited for pointing out the close similarity between Kaplans character/content distinction and Husserl's distinction beween general-meaning function and respective meaning; cf. Künne 1982; also cf. Smith 1982, 184. In Beyer 2000 I have worked out the consequences of this distinction for Husserl's semantics and theory of intentional content ("noematic sense") in detail, arguing that the latter is to be rationally reconstructed as a moderate version of externalism and can be fruitfully compared to Evans' (radically externalist) neo-Fregean conception of sense, among others. That Husserl's view can be read this way lends further support to Føllesdal's so-called Fregean interpretation of Husserl's notion of noematic sense (cf. Føllesdal 1969).

[88]Here I will not go into Frege's idea of dynamic Thoughts; I have done so elsewhere (Beyer 2000, §§ 5, 7–10).

is speaking to you at this moment has been wounded'[89] and certain elements of the circumstances of utterance (here: addressee and time of utterance). Following Künne, one could refer to such *mixta composita*, "consisting of a linguistic sign and an element of non-linguistic reality," as *hybrid* expressions.[90] By contrast, the above-sketched Husserlian conception of linguistic communication requires that, in a case of successful communication, the hearer ascribes to the speaker of 'I have been wounded' what might be called an act of self-identification,[91] whose intentional content is nothing but that special 'I'-mode of presentation, which Husserl explains in a way that shows that (thanks to the general meaning-function of 'I') the hearer can properly ascribe (a meaning-bestowing act corresponding to) it to the speaker after all, so that *pace* Frege the speaker does not need to employ a different sense of 'I' in communication than in a soliloquy:

> Clearly, if someone says 'I,' he does not only refer to himself, but he is also aware of this referring to himself as such, and this awareness builds an essential part of what constitutes the meaning of the word 'I.' The current act of meaning oneself is functioning [. . .] in such a way that in the course of it its [intentional] object is [. . .] being meant *as* the object of an act of meaning oneself. [. . .] The hearer understands it, if he takes it as an indication for the whole structure of consciousness just described, that is to say, if the speaker is regarded by him as someone who refers to himself precisely as 'I,' i.e., as someone who refers to himself as the object of his recognition of himself recognized as a recognition of himself.[92]

Thus, if the speaker asserts 'I have been wounded,' she presents herself as someone who refers to herself *as* referring to herself (or as meaning herself/having herself in mind/thinking/speaking of herself), in order to state about herself that she has been wounded; and the hearer understands this assertion if he takes the speaker to refer to herself *as* referring to herself and to assert about herself that she has been wounded. Accordingly, the general-meaning function of 'I' requires this term to refer to the speaker under a (pre-predicative) aspect that is apt to motivate the (predicative) judgment given voice to by 'I am referring to (/meaning/thinking of/speaking of) *myself*.' I regard this metarepresentational view of the meaning-bestowing acts underlying the assertive use of 'I'-sentences as quite plausible. After all, if someone claims, say, 'I have a broken leg,' then she *eo ipso* knows that she *refers to herself* by 'I;' she could instantly add: 'I am speaking of *myself*.' (Contrast the case where the speaker unknowingly looks at herself in a mirror and exclaims 'She has a broken leg.')[93] Notice, however, that Husserl agrees with Frege about the unshareability of the sense of 'I' as used in a soliloquy. (Husserl would add that this sense is also unshareable in ordinary speech situations.) Unlike Frege, though, he does not maintain that successful communication therefore requires the speaker to express a different, shareable sense in public speech. Rather, the general meaning-function of

[89] Cf. Frege 1918, 66.

[90] Künne 2010, 456 f; my translation.

[91] Cf. Beyer 2015b, 2.

[92] Hua XIX/2, 813; my translation.

[93] See Beyer 2006, 33 f.

'I,' along with the utterance context, enables the hearer to ascribe the right meaning-bestowing act to the speaker and thus to identify the respective meaning expressed by 'I' in this context (without being able to *grasp* that respective meaning, in the sense of entertaining the same sense in thought as the speaker).

The meaning of 'this.' On Frege's view, demonstratives normally express sense in the context of hybrid expressions, also containing deictic gestures or other appropriate non-linguistic elements of the circumstances of utterance.[94] On Husserl's view, they express sense in virtue of meaning-bestowing acts whose content can be correctly ascribed by the hearer thanks in part to their general meaning-function, which refers to *the object the speaker wants to draw the hearer's attention to* in the context of utterance (the intended demonstratum).[95] This may sound like a view similar to Frege's.[96] However, Husserl has a lot to say about the content and intentional reference of the relevant meaning-bestowing act, which makes his approach more interesting for a theory of demonstrative reference than Frege's, who does not say much about how the sense of hybrid proper names involving demonstratives manage to determine their referent.

According to Husserl, these acts are (like all empirical experiences) embedded in a holistic structure, what Husserl calls their "intentional horizon," whose future elements are predelineated (in part) by the intentional content of the respective act. For example, if you consciously see something whose front side you are visually confronted by *as a house*, then you will "anticipate"[97] visual appearances of a back side and an inside, respectively, as future experiences you would or could undergo if you walked inside or walked around the object while observing it. As Husserl explains, the anticipations in question concern the way the represented object would present itself in the framework of further courses of experience compatible with what is currently experienced, and they also concern the way this object relates to other objects in the world, thus constituting the core of your current *world* horizon, what Husserl calls the "external horizon (*Außenhorizont*)"[98] of the experience (see below). When these anticipations are intuitively fulfilled, in the sense that an according pre-predicative experience of (what seems like) *verification* occurs, displaying a corresponding fulfilling sense, then our *individual concept* associated with the object in question, what Husserl sometimes calls "*Begriff*," will change – provided we make, at the level of *predicative* experience, an accordingly motivated judgment:

[94]See Künne 2010, 460 ff, and the quotations cited there.

[95]Cf. Hua XIX/1, 89.

[96]See Føllesdal 2006.

[97]For the close connection between anticipation and (internal) horizon, cf. Husserl 1999, 26–36. For an insigthful interpretation of Husserl's notion of horizon, cf. Smith and McIntyre 1982, 227–265.

[98]Husserl 1999, 28 f.

> The judgment-substrate in its logical sense, as it [sc. the judgment-substrate; CB] has acquired it [sc. that logical sense; CB] by the activity of of predicative determination, constitutes one *concept of concept* [. . .].[99]

These individual concepts, by recourse to which the determination of demonstrative reference can be explained (see below), may be looked upon as variable or dynamic systems of belief about a particular object, what Perry refers to as "mental files,"[100] and they (i.e., their respective current version) influence the further course of predicative experience:

> We speak of a *sedimentation of sense regarding the object* [*Sinnesniederschlag am Gegenstand*]. That is to say: Just like any step of receptive [i.e. pre-predicative; CB] experience, every step of predicative judging has its lasting result. It generates [*stiftet*] habitualities which influence the further course of actual judging in a great variety of ways.[101]

The identification of a "concept" in the present sense of the term with a mental file à la Perry is supported by the fact that Husserl characterizes such individual concepts as being infinitely "open" and "in flux."[102] Note that the elements of these concepts or files involve a *sense of identity through time*, which holds them together:

> I see an *object without a 'historic' horizon* [*footnote:* without a horizon of acquaintance and knowledge], and now it gets one. I have experienced the object multifariously, I have made 'multifarious' judgements about it and have gained multifarious [pieces of] knowledge about it, at various times, all of which I have connected. Thanks to this connection I now possess a 'concept' of the object, an individual concept[. . .] [W]hat is posited in memory under a certain sense [*mit einem gewissen Sinn*] gains an epistemic enrichment of sense, i.e. the x of the sense is determined further in an empirical way [*erfahrungsmäßig*].[103]

The "historic" horizon and the objects of the anticipations referred to in the preceding quotation constitute the "internal horizon (*Innenhorizont*)"[104] of the experience. They all belong to the same "*x* of the sense," also referred to by Husserl as the "*determinable X*," i.e. they share a sense of identity (of represented object) through time. This sense, along with the sense arising from the individual concepts associated with the respective singular terms, will under appropriate epistemic circumstances[105] give rise to non-trivial identity judgments,[106] such as the one given voice to, in the example used by Frege in his cognitive value argument (see Sect. 9.3 above), by 'The Morning Star is the Evening Star,' or, in one of Husserl's examples, the one given voice to by 'Napoleon is the conqueror of Waterloo:'

---

99 Husserl 1999, 277; my translation.

100 Cf. Perry 1980.

101 Husserl 1999, 250; my translation.

102 Hua XX/2, 359. For further textual evidence supporting this identification see the following quotation.

103 Hua XX/2, 358; my translation.

104 Husserl 1999, 28.

105 See Husserl 1999, 280 f.

106 See Beyer 2000, 171 ff.

> Tautological expressions [such as 'the victor of Jena' and 'the conqueror of Jena;' CB] mean one and the same thing [*besagen ein und dasselbe*], non-tautological expressions always mean different things. The object is the same, irrespective of whether I say 'Napoleon' or 'the conqueror of Jena,' but the two expressions mean something different. [...] The second expression [talks about] Jena and being a conqueror. Names with different meanings name the same person or thing. Thus, we may obviously distinguish between the thing itself which is named and the same thing, taken in the way in which it is meant here. [...] Expressions with different meanings can refer to the same object, provided we understand by 'meaning' that meant as such.[107]

In his second major work, "*Ideas*," Husserl would refer to the "meant as such" or "the object taken in the way it is meant" under the label "noematic sense (*noematischer Sinn*)," where the noematic sense contains the determinable X. As can be seen from both the preceding and the following quotation, Husserl introduces this notion in an argumentative context (headword: "tautological" vs. "non-tautological expressions") quite similar to the one in which Frege introduces his notion of sense in "*Über Sinn und Bedeutung*" (see Sect. 9.3 above, headword: cognitive value argument), a fact which lends support to the so-called Fregean interpretation of Husserl's notion of noema.[108]

> If [...] we wanted to say that it makes no difference [*es sei einerlei*] whether we say that the conqueror of Jena and the loser of Waterloo are the same person or that the conqueror of Jena and the conqueror of Jena are the same person, then one would protest. [...] After all, we do not want to assert a tautology here.[109]

Other past and anticipated experiences also belonging to what Husserl calls the "historic horizon" (or "experiential horizon [*Erfahrungshorizont*]")[110] bring it about that your "'concept' of the object" is *networked* with other concepts of (other) objects. They constitute the external horizon of the experience[111] and will (again, under appropriate epistemic circumstances) give rise to (i.e., motivate) relational judgments.

Unlike Frege, Husserl has a story to tell, in terms of what he calls "constitution" and the "determinable X," about how sense determines reference, at least in the case of singular empirical thought and speech. By "constitution," he means what he calls an "activity (*Tätigkeit*)" (not to be confused with an intentional action)[112] of the experiencing subject in a process of experience aiming at knowledge (where knowledge, or truth, defines the relevant *interest* of the subject leading it to pursue

[107]Hua XXVI, 28; my translation.

[108]Cf. Føllesdal 1969.

[109]Hua XXVI, 100 f.; my translation. See Beyer 2000, 67 ff, for further interpretation and discussion.

[110]Husserl 1999, 27.

[111]See Husserl 1999, 27 ff.

[112]See Husserl 1999, 89 ff. – Note that constitution *can* take the form of intentional action, but in the sphere of perception and empirical thought this is the exception rather than the rule. So clearly, "constitution" does not mean *creation*.

this activity).[113] So it occurs in a context of epistemic justification, which is why the phenomenological *analysis* of constitution (*Konstitutionsanalyse*), or "elucidation of the origin (*Ursprungsklärung*)," takes the form of a *meta*-justification.[114] On my interpretation, this activity consists, at the level of *predicative* experience (i.e., empirical judgment), in the processing of mental files or individual concepts (see above); at the *pre*-predicative level, it consists in the experiences *motivating* such (actual or potential) file processings.

Now Husserl's story about reference determination goes roughly as follows. The determinable X holding together the elements of an individual concept is apt to lead us back through time towards the original situation where the reference of the relevant concept was fixed, like for instance the occasion of the subject's first perceptual encounter with a particular object, which Husserl (in *EU*) describes as the constitution of an "original (*ursprüngliches*)" or "ultimate substrate (*letztes Substrat*)" (or as the "original doxa (*Urdoxa*)") for potential empirical judgments about this "individual object," which object would then be "categorically formed" by (attributes deriving from) subsequent judgments regarding it,[115] or by their pre-predicative foundations,[116] thus gaining an "epistemic enrichment of sense."[117] The original perceptual experience in which the relevant individual object "constitutes itself" will belong to the same determinable X as all of the experiences motivating these subsequent judgments, which in turn contribute to the processing of the mental file that has been opened on that original occasion. Now it is the original individual object, whose constitution initiated this file opening, that will normally count as the common object (to be subsequently "enriched by sense") of the experiences motivating the various file entries, all of which belong to the same determinable X. (In cases where the 'referent' of a mental file changes across time – i.e., is unnoticedly replaced by another object – the situation becomes more complicated, to be sure. The same goes for cases of perceptual judgments leading to, or taken by the respective subject to be confirming, entries into an already existent file.)[118] So in the case of empirical thought and speech there is a link between intentional content, or sense, and reference, such that the content can be said both to *determine* reference and to be *determined by* the referent as it "constitutes itself" empirically. In the final analysis, which also takes into account the *intersubjective* dimension of the constitution of empirical objects (which Husserl chooses to bracket in *EU*),[119] this referent turns out to be constituted as an element of the objective spatio-temporal

---

[113]To be more precise, it is only at the level of *predicative* experience that the subject's epistemic interests take the form "of the *will to knowledge* (*des Willens zur Erkenntnis*);" cf. Husserl 1999, 92.

[114]See Husserl 1999, 44; headword: "justification of the doxa (*Rechtfertigung der Doxa*)."

[115]See Husserl 1999, 18 ff, 60.

[116]See Husserl 1999, 132 f.

[117]Hua XX/2, 358.

[118]See Beyer 2000, §7.

[119]Cf. Husserl 1999, 58.

surrounding world (*Umwelt*). Thus, Husserl can be read as a non-naive externalist about sense and intentional content.[120]

Therefore, in order to determine the referent of a demonstrative, as used by a speaker, the hearer has to determine the object the speaker wants to turn his attention to, which object "constitutes itself" in the processing of the file or individual concept the speaker associates with that use of the term. Thus, when an interpretation problem occurs (in cases of mistaken identity, for example), the hearer should take this process of object-constitution on the part of the speaker into account (by means of empathy). However, he does not have to make out all *predicative entries* of the relevant file for this purpose. For, the predicative content of the speaker's individual concept does not enter into the respective meaning of her use of the demonstrative, not even as sediments in the categorically formed "object as meant." It is important to note that Husserl is not a descriptivist either about demonstratives or about (proper) proper names or referentially used descriptions, for that matter. Thus, regarding the referential use of a definite description like 'the Kaiser' he writes:

> [. . .] all that matters to us is that the person in question (the identical [sc. logical; CB] subject) [...] has such-and-such an identical predicate. [...] Note also that [sc. in such cases; CB] expressions such as 'the Kaiser' function quite like proper names, just as [']Wilhelm II[']. [. . .] An assertion is made, and all that matters is that the [sc. logical; CB] subject has such-and-such a property, this is all that is meant to be asserted, and the nominal representation of the [sc. logical; CB] subject merely serves as a tool for calling attention to the object in question, for naming it, by whatever means. [. . .] Wherever the nominal subject-representation has not only such naming but also 'assertive' function, wherever what it asserts matters, we will not be prepared to say that the state of affairs [asserted] remains invariably the same if the subject-representation gets replaced by any other, equivalent representation.[121]

A few lines later, Husserl refers to the "state of affairs [asserted]" described at the end of this quotation as the propositional content ("*das Propositionale*"), or the "asserted state of affairs as meant," and proposes to distinguish it from what might be called the *singular* state of affairs represented both by 'The present Kaiser of Germany has travelled to Paris,' where the definite description is used attributively to refer to (a person who happens to be) Wilhelm II, 'The Kaiser has travelled to Paris,' where 'the Kaiser' is used referentially to refer to Wilhelm II (who may just be in the focus of perceptual attention), and by 'Wilhelm II has travelled to Paris.' On his view, only the propositional content of the judgment given voice to by the first sentence (where the description is used attributively) involves an attribute like being Kaiser of Germany. The propositional contents of the judgments given voice to by the second and the third sentence (in the described situations) do *not* involve

[120]See Beyer 2000; cf. also Husserl's discussion of Twin Earth and twin examples in Hua XXVI, 211–214 (see Beyer 1996, 172–182, where Husserl's Twin Earth considerations are still interpreted in an internalist way, though).

[121]Hua XXVI, 170 f.; my translation. See Beyer 2000, 52–70, for a more extensive interpretation and discussion of this passage.

such descriptive content.[122] According to Husserl, the same holds true for judgments given voice to by demonstrative sentences (as ordinarily used). Apart from their general meaning-function, the meaning of demonstratives closely resembles the meaning of (proper) proper names and descriptions used referentially[123]:

> The proper name is a 'direct name of the object,' just like the [expression] 'this,' which is however ambiguous [sc. regarding its respective meaning; CB] and gets determined only by the occasion of utterance.[124]
>
> Already many years ago I have said that the proper name is in fact similar to a deictic meaning. The proper name [. . .] names the indeterminate identical [object] of a manifold of possible representations [. . .] of one and the same object. Only the form of [deictic] indication [*die Form des Hinweises*] is missing.[125]

By contrast, and somewhat paradoxically, Husserl can be said to be a 'descriptivist' about (many of) the perceptual contents which can serve as *fulfilling senses* for a given demonstrative identification (or referential intention, or act of proper naming, for that matter). For, he thinks that perceptual content is often formed by the results of pre-predicative experiences undergone in the process of detailed observation, or "explication *(Explikation)*," of an object and initiating mental dispositions or (as Husserl prefers to say)[126] *habitualities* of the experiencing subject, which could manifest themselves in predicative entries into a mental file about that object:

> With each stage of the process of explication, a sediment [*Niederschlag*] of habitual pieces of knowledge [*habitueller Kenntnisse*] forms itself with regard to the object of explication [*an dem . . . Gegenstand der Erfassung*] that has been indeterminate [. . .] before. When the process of explication in the mode of originality has run its course, the object is persistently constituted as determined by the respective determinations, even if it has sunk into passivity. It has acquired the forms of sense originally constituted in the acts of explication as *habitual knowledge*. Thus every explicating [*hineingehende*] examination of an object has a *lasting result* regarding that object [*an ihm*]. [. . .] From now on, it will be looked upon by the respective subject [. . .] as an object already known through these determinations, which have been ascribed [*zuerteilt*] to it by explicating cognition. That is to say: the new cognition, even if it [. . .] again takes the form of an original perceptual presentation of the object, will [. . .] display an essentially different sense-content than the former perceptions. [. . .] [T]he sediment of the [. . .] former ascription of a determination is now a component of the interpretive sense [*des Auffassungssinnes*] of the perception. [. . .] This becoming habitual of the results of original intuitive cognition occurs [. . .] even without any voluntary efforts on our part [...] However, it may also be deliberately intended. One can then say that

[122] Another argument for the distinction between propositional content and singular state of affairs (and the corresponding situation of affairs, for that matter) is given by Husserl in his 1917/18 lectures on "Logic and general theory of science:" "[. . .] [I]t is only when the judgment 'A is b' is true that we can say that the state of affairs 'A is b' corresponds to it. We cannot speak of the state of affairs that Sirius is a diamond, given that we know that such a judgment would be false." (Hua XXX, 58)

[123] See Beyer 2000, 85–100, for further interpretation and discussion.

[124] Hua XXVI, 174; my translation.

[125] Hua XXX, 357; my translation.

[126] Husserl reserves the term "disposition" for forces, tendencies etc. of substances as they are postulated in the framework of the "naturalistic attitude" of science and empirical psychology.

> the interest is directed towards a [...] memorization [*Einprägen*] of the perceptual image. [...] The individual features distinctly showing themselves [*sich abhebenden Eigenheiten*] become distinctive marks [*Merkmale*] [of the object]. [...] In most cases this will already go hand in hand with a predication [...].[127]

While the resulting rich perceptual contents or "interpretive senses (*Auffassungssinne*)" can *in principle* be shared, they will in fact most likely differ from subject to subject, and even intra-subjectively as time passes, due both to the dynamic nature of the individual concepts, and their pre-predicative foundations, in which perceptual objects "constitute themselves," and to their individual differences.

So perceptual content is generally unsuitable to play the role of the respective meaning of a demonstrative, provided that successful communication (in the case of demonstratives) requires sharing of respective meaning. Hence, these expressions must express a much less fine-grained respective meaning. On Husserl's view, this meaning, which he explicitly equates with a noematic sense (see the following quotation), derives from the meaning-bestowing act given voice to (in the narrow sense) by the use of a demonstrative on a particular occasion, which act he refers to as the "act of this-meaning (*Akt des Dies-Meinens*):"

> Noematically speaking: the wording as *signum* has an intentional (noematic) feature belonging to and founded in it, a feature in virtue of which the word 'means' something [...] Thus significance here amounts to this feature that is attached to the wording in conscious 'understanding' (or conscious assertion), whilst the meaning itself is the meant as such [*das Vermeinte als solches*; i.e., the noematic Sinn; CB] [...] Let us begin by examining a primitive case, where a mere 'that' or 'this' is taking the place of the subject, as in the case of 'That is a blackbird.' 'That' is an occasional expression, an expression whose meaning depends on the respective context [...] The meaning of 'this' [*sic*] differs according to circumstances, but in such a way that a common element is preserved across all those changes, an element that distinguishes this type of ambiguity from contingent equivocation. In all of these cases we are dealing with deictic indication [*Es ist immer auf etwas hingewiesen*]. [...] The proper meaning does not reside in the mere perception; rather, *a new act builds itself on the basis of it, an act which orients itself by the perception: the act of this-meaning*. Obviously, it is this latter act that serves the meaning-bestowing function proper. [...] The act of demonstrative reference remains essentially the same, regardless of which perception from the manifold of perceptions belonging together by presenting what is – recognizably – one and the same object throughout happens to underlie it.[128]

So the act of this-meaning merely *"orients itself by"* (one of) the elements of a "manifold of perceptions," such as a continuous observation of an object, which perceptions all belong to the same determinable X. In *LU* Husserl spells this out as follows:

> The perception is responsible for the relation of the word ['this'] to *this* object. However, the meaning does not reside in this very perception.[129]

[127] Husserl 1999, 137 ff; my translation.

[128] Hua XX/1, 74–78; my translation.

[129] Hua XIX/2, 553 f; my translation.

The above considerations on how sense determines reference explain this in greater detail, in terms of constitution and file processing. By contrast, we do not find much about the determination of demonstrative reference in Frege; he simply includes the elements of the circumstances of utterance which enable the hearer to make out the intended referent in the (hybrid) expression of demonstrative sense (see above); and he says nothing about the (to my mind, obvious) connection between perception and demonstrative sense and reference. To be sure, Husserl's conception of perception as the noetic-noematic shaping of ("hyletic") sensory content, such that a transcendent perceptual object "constitutes itself," is compatible with Frege's general view of perception. In fact, it can be used as an explication of Frege's idea that (in the case of visual perception):

> The having of visual impressions is necessary but not sufficient for seeing things. What needs to be added is nothing sensory. And yet it is precisely this additional element that opens the external world to us; for without this non-sensory element, everyone would remain encapsulated in his inner world. Since the decision thus lies in the non-sensory, something non-sensory could as well lead us outside the inner world and allow us to grasp Thoughts even [in cases] where no sensory impressions are involved [*mitwirken*].[130]

I hasten to add that Husserl does not conceive of the intentional content of perception as propositional; perception is *pre*-predicative experience (see above) and its content is *sub*-propositional, containing a determinable X it may share with the sub-propositional content of an "act of this-meaning" regarding which it can function as a fulfilling sense. Here and elsewhere Husserl goes more into detail than Frege when it comes to explaining the nature and structure of content and its relation to the objects referred to or represented.[131]

References

C. Beyer, *Von Bolzano zu Husserl. Eine Untersuchung über den Ursprung der phänomenologische Bedeutungslehre* (Kluwer, Dordrecht, 1996)

C. Beyer, *Intentionalität und Referenz. Eine sprachanalytische Studie zu Husserls transzendentaler Phänomenologie* (Mentis, Paderborn, 2000)

C. Beyer, *Subjektivität, Intersubjektivität, Personalität. Ein Beitrag zur Philosophie der Person* (De Gruyter, Berlin/New York, 2006)

C. Beyer, Husserl's transcendental phenomenology considered in the light of (recent) epistemology, in *Constructive Engagement of Analytic and Continental Approaches in Philosophy. From the Vantage Point of Comparative Philosophy*, ed. by Bo Mou, Richard Tieszen, (Brill Academic Publishing, Leiden, 2013)

---

[130] Frege 1918, 75; my translation.

[131] This article was written at the Centre for Advanced Study (Oslo) in the framework of the research group "Disclosing the Fabric of Reality – The Possibility of Metaphysics in the Age of Science" led by Frode Kjosavik and Camilla Serck-Hansen. I gratefully acknowledge the support. For helpful comments I would also like to thank Adriana Pavic.

C. Beyer, Meaning, context, and background, in *OpenMIND*, ed. by Thomas Metzinger, Jennifer Windt (MIND Group, Frankfurt/M., 2015a)
C. Beyer, Self-identification, intersubjectivity, and the background of intentionality. Reply to Anita Pacholik-Żuromska, in *OpenMIND*, ed. by Thomas Metzinger, Jennifer Windt (MIND Group, Frankfurt/M, 2015b)
C. Beyer, M. Weichold, Philosophy of language, in: *The Routledge Companion to Phenomenology*, ed. by Sebastian Luft, Søren Overgaard (Routledge, London/New York, 2011), pp. 406–416
B. Bolzano, *Wissenschaftslehre. Versuch einer ausführlichen und größtentheils neuen Darstellung der Logik*, 4 volumes, (Seidelsche Buchhandlung, Sulzbach, 1837)
W. Carl, *Frege's Theory of Sense and Reference. Its Origins and Scope* (Cambridge University Press, Cambridge, 1994)
S. Centrone, *Logic and Philosophy of Mathematics in the Early Husserl* (Springer, Berlin, 2010)
S. Centrone, Aspekte des Psychologismus-Streits. Husserl und Frege über Anzahlen und logische Gesetze, in *Versuche über Husserl*, ed. by S. Centrone (Meiner, Hamburg, 2013)
K. Donnellan, Reference and definite descriptions. Philos. Rev. **75**(3), 281–304 (1966)
M. Dummett, *Frege. Philosophy of language* (Duckworth, London, 1973)
M. Dummett, *Ursprünge der analytischen Philosophie* (Suhrkamp, Frankfurt/M, 1992)
G. Evans, *The Varieties of Reference* (Clarendon, Oxford, 1982)
D. Føllesdal, *Husserl und Frege* (Aschehoug, Oslo, 1958)
D. Føllesdal, Husserl's notion of noema. J. Philos. **66**, 680–688 (1969)
D. Føllesdal, Indikatoren und der Geist, in *Phänomenologie und Sprachanalyse*, ed. by Geert Keil, Udo Tietz (Mentis, Paderborn, 2006)
G. Frege, *Begriffsschrift. Eine der arithmetischen nachgebildete Formelsprache des reinen Denkens* (Verlag von Louis Nebert, Halle, 1879)
G. Frege, Über Sinn und Bedeutung. Zeitschrift für Philosophie und philosophische Kritik NF **100**, 25–50 (1892a)
G. Frege, Über Begriff und Gegenstand. Vierteljahresschrift für wissenschaftliche Philosophie **16**, 192–205 (1892b)
G. Frege, *Grundgesetze der Arithmetik. Begriffsschriftlich abgeleitet. I. Band* (Verlag von Hermann Pohle, Jena, 1893)
G. Frege, Rezension von Dr. E. G. Husserl: Philosophie der Arithmetik. Zeitschrift für Philosophie und philosophische Kritik NF **103**, 313–332 (1894)
G. Frege, *Der Gedanke, Beiträge zur Philosophie des deutschen Idealismus I.* Heft 2 (1918), 58–77
G. Frege, in *Wissenschaftlicher Briefwechsel*, ed. by G. Gabriel et al. (Meiner, Hamburg, 1976)
G. Frege, in *Nachgelassene Schriften*, 2nd edn., ed. by H. Hermes et al. (Meiner, Hamburg, 1983)
G. Frege, *Die Grundlagen der Arithmetik. Eine logisch mathematische Untersuchung über den Begriff der Zahl* (Reclam, Stuttgart, 1987)
L. Haaparanta, Analysis as the method of logical discovery. Some remarks on Frege and Husserl. Synthese **77**, 73–97 (1988)
L. Haaparanta, Compositionality, contextuality and the philosophical method. Acta Philos. Fenn. **78**, 289–301 (2006)
E. Husserl, *Husserliana, Gesammelte Werke* [Hua], I–XXVI: Martinus Nijhoff, Den Haag 1950–, XXVII–XXXVII: Kluwer, Dordrecht 1989–; XXXVIII–: Springer, New York 2005–
E. Husserl, *Erfahrung und Urteil*, 7th edn. (Meiner, Hamburg, 1999)
W. Künne, Indexikalität, Sinn und propositionaler Gehalt. Grazer Philosophische Studien **18**, 41–74 (1982)
W. Künne, *Die Philosophische Logik Gottlob Freges. Ein Kommentar* (Klostermann, Frankfurt, 2010)
Ø. Linnebo, The context principle in Frege's *Grundgesetze*, in *A Companion to Frege's Grundgesetze*, ed. by P. Ebert, M. Rossberg (Eds), (Oxford University Press, Oxford, forthcoming)
R. McIntyre, Husserl and Frege. J. Philos. **84**, 528–535 (1987)
J.N. Mohanty, *Husserl & Frege* (Indiana University Press, Bloomington, 1982)
J. Perry, A problem about continued belief. Pac. Philos. Q. **61**(4), 317–322 (1980)
D.W. Smith, What is the meaning of 'This'? Noûs **16**(2), 181–208 (1982)

D.W. Smith, R. McIntyre, *Husserl and Intentionality. A Study of Mind, Meaning and Language* (Reidel, Dordrecht, 1982)
P.F. Strawson, On referring. Mind. New Ser. **59**(235), 320–344 (1950)
R. Tieszen, The philosophy of arithmetic: Frege and Husserl. in *Mind, Meaning and Mathematics. Essays on the philosophical views of Husserl and Frege*, ed. by L. Haaparanta (Kluwer, Dordrecht, 1994), pp. 85–112

Chapter 10
Husserl and Riemann

Guillermo E. Rosado Haddock

Abstract After briefly showing that, contrary to the received view in analytic circles, Frege's influence on the evolution of Husserl's views on logic and mathematics, as well as on the distinction between sense and referent are either insignificant or, as in the last case, totally inexistent, and that Husserl's account of Leibniz and Bolzano's influence in *Logische Untersuchungen* is correct, we turn to Riemann, whose influence is certainly non-negligible. Husserl's conception of mathematics as a theory of manifolds (or structures) is a generalization of Riemann's notion of manifold – in fact, a sort of bridge between Riemann and the Bourbaki group. Moreover, Husserl's conception – since 1892 – of physical geometry as empirical is also strongly influenced by Riemann

Keywords Frege • Theory of manifolds • Geometry of physical space • Bourbaki

10.1 Introduction: Husserl's Place in the History of Philosophy[1]

It is said that Kurt Gödel considered Edmund Husserl as the greatest philosopher since Leibniz. Though such a statement could be questioned in at least two senses, e.g., whether Husserl was a greater philosopher than Kant, and whether Leibniz was a greater philosopher than Husserl, there should be little doubt that Husserl not only is one of the most influential recent philosophers, but also that he is one of the really great philosophers. His influence extends not only to the so-called phenomenological school, second in importance only to analytic philosophy during the twentieth century, to analytic philosophy itself, especially through Carnap, to logic and linguistics through Lesniewski's mereology and his

[1] All translations in this paper are mine.

G.E. Rosado Haddock (✉)
University of Puerto Rico, San Juan, PR, USA
e-mail: gerosadohaddock@gmail.com

S. Centrone (ed.), *Essays on Husserl's Logic and Philosophy of Mathematics*,
Synthese Library 384, DOI 10.1007/978-94-024-1132-4_10

and Ajdukiewicz' theory of semantic categories, and to existential philosophy, even though Husserl would probably not feel comfortable with the latter influence. Moreover, as pointed out by Stefania Centrone, Husserl already in his youth work *Philosophie der Arithmetik*[2] seems to have anticipated the theory of general recursive functions, and as pointed out by the present author, he anticipated the Bourbakian conception of mathematics,[3] and a possible at least indirect influence on the Bourbaki school should not be excluded. Husserl's *Logische Untersuchungen*[4] is one of the seminal works in the history of philosophy, but at least *Ideen I*,[5] *Formale und transzendentale Logik*[6] and *Erfahrung und Urteil*[7] could have been the *opus magnum* of many other prominent philosophers. In fact, Husserl's greatness as a philosopher can still not be completely measured, since about half of his manuscripts have still not been published. But in any case, in his *opus magnum Logische Untersuchungen* one finds, among other things: (1) the most thorough and decisive refutation of psychologism in logic[8]; (2) the distinction between nonsense and countersense and the corresponding introduction of what later Carnap will call 'rules of formation'[9] in contrast to the rules of logic in the strict sense or, what Carnap later called, 'rules of transformation'[10]; (3) a conception of mathematics as a theory of structures, in which each mathematical theory is either a fundamental one –like set theory or the formal theory of parts and wholes-, a specialization of a fundamental structure, a combination of at least two structures, or a combination of specializations[11]; (4) a very modern conception of physical theories as explanatory theories, including the anticipation of the deductive-nomological scheme both for statements of facts and for laws; (5) related with this last case, an analysis of the role of what Husserl called *hypotheses cum fundamento in re*, of the empirical equivalence of different alternative such hypotheses of higher level and the acknowledgement of the ensuing subdetermination of physical theories by its

[2]*Philosophie der Arithmetik* 1891, Hua XII, M. Nijhoff, Den Haag 1970. See Centrone's paper 'Husserl on the 'totality of all conceivable operations', *History and Philosophy of Logic 27 (3)*, 2006, pp. 211–228.

[3]See my 'Husserl's Philosophy of Mathematics: its Origin and Relevance' 2006, reprinted in *Against the Current*, Ontos Verlag, Frankfurt 2012, pp. 145–181.

[4]*Logische Untersuchungen* (2 vols.) 1900–1901, Hua XVIII & XIX, M. Nijhoff, Den Haag 1975 & 1984.

[5]*Ideen zu einer reinen Phänomenologie und phänomenologischen Philosophie I* 1913, Hua III, M. Nijhoff, Den Haag 1950, revised edition 1976.

[6]*Formale und transzendentale Logik* 1929, Hua XVII, M. Nijhoff, Den Haag 1974.

[7]*Erfahrung und Urteil* 1939, sixth edition, F. Meiner, Hamburg 1985.

[8]*Logische Untersuchungen I*, Chapters I–VIII.

[9]Ibid. Chapter XI. Compare with Carnap's *Die logische Syntax der* Sprache 1934, revised English edition 1937, where the distinction between the two sorts of rules is introduced in p. 4 and then used throughout the whole book.

[10]*Logische Untersuchungen I*, Chapter XI. For Carnap, see preceding footnote.

[11]Ibid.

empirical content[12]; (6) a theory of the ideality of meanings and a thorough analysis of so-called 'indexicals'[13]; (7) a theory of general objects[14]; (8) a theory of wholes and parts in general, and the ensuing characterizations of analytic statements and synthetic *a priori* statements[15]; (9) a theory of formal universal grammar[16]; (10) the beginning of what was going to be the most thorough analysis of all intentional acts[17]; and (11) the most serious attempt –at least since Plato- of explaining our categorial and, particularly, mathematical knowledge.[18]

10.2 The Myth of the Fregean Influence

What probably has been more difficult to assess are the influences Husserl suffered from earlier or contemporary authors. In analytic circles-that are not very distinguished in their historic studies of philosophy- it has been at least tacitly accepted that the mature Husserl of *Logische Untersuchungen* was influenced by Frege's unfriendly review of Husserl's *Philosophie der Arithmetik*[19] and by his now so popular distinction between sense and referent. In particular, the logician and historian of logic Evert W. Beth[20] and the analytic philosopher Dagfinn Føllesdal[21] have been the most influential proponents of that interpretation. And certainly Husserl in his youth work, *Philosophie der Arithmetik*, held a mild form of Brentanian psychologism. Nonetheless, that work, though published in 1891, was an expansion of his professorship's thesis *Über den Begriff der Zahl* of 1887, and reflects Husserl's views probably up to 1889 or the first half of 1890. In fact, the published *Philosophie der Arithmetik* was supposed to be the first volume of a two volume work, the first one concerned with psychological foundations, whereas the never published second volume was to be concerned exclusively with logical foundations. But Husserl never wrote the second volume, what certainly attests to the evolution of Husserl's thought during those years. By the way, that constant

---

[12]Ibid.

[13]*Logische Untersuchungen II*, U. I.

[14]Ibid., U. II.

[15]Ibid., U. III.

[16]Ibid., U. IV.

[17]Ibid., U. V.

[18]Ibid., U. VI.

[19]'Rezension von E. G. Husserl: *Philosophie der Arithmetik I*, 1894, reprinted in G. Frege, *Kleine Schriften* 1967, revised edition, Georg Olms 1990, pp. 179–192.

[20]See Evert W. Beth, *The Foundations of Mathematics*, North Holland, Amsterdam 1965, p. 353.

[21]See, e.g. Dagfinn Føllesdal's Master thesis *Husserl und Frege, ein Beitrag zur Beleuchtung der phänomenologischen Philosophie* 1964, translation in L. Haaparanta (ed.), *Mind, Meaning and Mathematics, Kluwer*, Dordrecht 1994, pp. 3–47, as well as his 'Husserl's Concept of Noema', *Journal of Philosophy 66 (20)*, 1969, pp. 680–697.

evolution is also attested by Husserl's *Studien zur Arithmetik und Geometrie*,[22] a posthumously published collection of manuscripts from 1886 to 1901 with which we will be concerned later in this paper.

In any case, it is in his paper 'Funktion und Begriff',[23] published in January 1891 that Frege presented for the first time his distinction between what he called 'Sinn' and 'Bedeutung', and that we will render as 'sense and referent'. But not only was Husserl's review of Ernst Schröder's book on the algebra of logic,[24] in which Husserl makes the same distinction, already written and in press, since it appeared, together with the *Philosophie der Arithmetik* in March of the same 1891, but there is a posthumously published paper by Husserl titled 'Zur Logik der Zeichen (Semiotik)'[25] written in 1890, in which Husserl clearly makes the distinction between sense and referent. Immediately after publishing both works Husserl sent copies to Frege, who in return sent Husserl a copy of 'Funktion und Begriff'. It seems pertinent to include here some passages of both of Husserl's papers, beginning with the review of Schröder's book.[26]

> Meanwhile . . . the author identifies the meaning of a name with the representation of the objects named by the name . . . .Moreover, he uses the term meaning equivocally, and [does] this in an already unacceptable level. In the above quotation, the incompatible and confusing explanations aside, what is pointed out to is the usual sense. But in another occasion what is really meant is the object named by the name, [Besprechung von E. Schröders *Vorlesungen über die Algebra de Logik I*, p. 11]
>
> With the unclearness about the concept of meaning is, moreover, connected [the fact] that Schröder puts names of the sort of "round square" as "senseless" side by side with those with one or more meanings. Obviously, he confuses here two different questions, namely, 1) whether a name has a meaning (a "sense"); and 2) whether an object corresponding to a name exists or does not exist. Senseless names in a strict sense are names without meaning, pseudonames, like Abracadabra. "Round square", however, is a univocal general name to which, nonetheless, nothing really corresponds. [Besprechung von E. Schröders *Vorlesungen über die Algebra der Logik I*, p. 12]
>
> That is why Schröder's distinction between "logical" and "psychological" content of a judgement, or more exactly, of a statement, is unacceptable. The truly logical content of a statement is the judgement content, hence, that what it means. [Besprechung von E. Schröders Vorlesungen *über die Algebra der Logik I*, p. 25]

Those three quotations from a paper published at the beginning of 1891, but most surely written in 1890, are decisive evidence that Husserl obtained the

---

[22] *Studien zur Arithmetik und Geometrie*, Hua XXII, M. Nijhoff, Den Haag 1983

[23] 'Funktion und Begriff' 1891, reprinted in Gottlob Frege, *Kleine Schriften*, pp. 125–142. The distinction is the central theme of his famous 'Über Sinn und Bedeutung' 1892, reprinted also in *Kleine Schriften*, pp. 143–162. See also Frege's paper written in the early 1890s, though only posthumously published, 'Ausführungen über Sinn und Bedeutung' in *Nachgelassene Schriften*, F. Meiner, Hamburg 1969, revised edition 1983, pp. 128–136.

[24] 'Besprechung von E. Schröders *Vorlesungen über die Algebra der Logik I*' 1891, reprinted in *Aufsätze und Rezensionen (1890–1910)*, Hua XXII, pp. 3–43.

[25] 'Zur Logik der Zeichen', written in 1890, but published for the first time as Appendix B.I to *Philosophie der Arithmetik*, Hua XII, pp. 340–373.

[26] See footnote 24.

distinction between sense and referent independently of Frege. Moreover, the last of the three quotations also serves to show that already at the time he wrote the review of Schröder's work Husserl had clearly distinguished the logical content of statements from the so-called psychological content, and acknowledged that only the logical content (or meaning) is genuinely the content of a statement, thus, that he had already put aside any sort of psychologism present in his *Philosophie der Arithmetik*. Hence, it seems pertinent to say that when Husserl published *Philosophie der Arithmetik* he had already surpassed some of the views present in that book, thus, that Husserl's first book was a dead born child. Concerning the sense-referent distinction, it seems appropriate to quote also some passages from Husserl's posthumous 'Zur Logik der Zeichen (Semiotik)'.

> In the case of indirect signs it is necessary to distinguish that which the sign means from that which it designates. In the case of a direct sign the two are the same. The meaning of a proper name, for example, consists precisely in that it names this determinate object. In the case of indirect signs, however, there are intermediaries between sign and thing, and the sign designates exactly by means of these intermediaries, and precisely because of this [fact] they constitute the meaning. ['Zur Logik der Zeichen (Semiotik)', p. 343]
>
> That is why, for example, the meaning of a general name consists in designating an object whatsoever on the basis and by means of some conceptual traits that it possesses. ['Zur Logik der Zeichen (Semiotik)', pp. 343–344]
>
> Two signs are identical when they designate in the same manner the same object or objects of one and the same contour. One of them is the mere repetition of the other, for example, horse and horse, five and five. ['Zur Logik der Zeichen (Semiotik)', p. 344]
>
> Two signs are equivalent in case they designate in different manner the same objects of the same contour, be it by different external or conceptual means, for example, a pair of names with the same meaning, like king and rex; Whilhelm II = present German emperor; $2 + 3 = 5 = 7 - 2 = \sqrt{25}$. ['Zur Logik der Zeichen (Semiotik)', p. 344]

Finally, it should be mentioned that Frege himself in his letter to Husserl of 24 May 1891,[27] in which he acknowledged the receipt of both *Philosophie der Arithmetik* and the review of Schröder's book, clearly accepts the fact that Husserl had also made the distinction between sense and referent and points to a difference between the two with respect to the referent of conceptual words. Thus, Frege says:

> I just want to point out here that there seems to exist a difference of opinion between us with respect to how the conceptual word (general name) relates to the objects. [*Wissenschaftlicher Briefwechsel*, p. 96]

After sketching his now famous sense-referent scheme, Frege goes on to sketch what he rightly believes is Husserl's sense-referent distinction for conceptual words. Of course, he is clearly acknowledging that Husserl had also made the distinction. Frege says:

Now, it seems to me that in your case the scheme would look like this:

[27] *Wissenschatlicher Briefwechsel*, F. Meiner, Hamburg 1976, pp. 94–98.

Conceptual Word
↓
Sense of the Conceptual Word
(Concept)
↓
Object that falls under the concept,

so that in your case there would be as many steps from the proper name as from the conceptual word to the object. [*Wissenschaftlicher Briefwechsel*, p. 98]

Moreover, in 'Ausführungen über Sinn und Bedeutung Frege also acknowledges Husserl's independent discovery of the famous distinction. Thus, Frege says[28]:

> Husserl says (p. 250) "Evidently he confuses here two very different questions, namely, 1) whether a name has a meaning (a sense); and 2) whether there exists or not some object corresponding to the name".

Hence, it is clear that Frege and Husserl arrived at the sense-referent distinction independently of each other and at the same time, though differed in some concrete details of the scheme. In fact, when Husserl finally completed the details of his scheme, there is a still a more important difference, namely, though the sense of a statement is as in Frege a thought (or proposition), the referent of a statement is a state of affairs, which is based on what Husserl called a "situation of affairs". Therefore, Husserl's scheme for statements has two additional steps between the statement and the truth-value.

Thus, there should not exist the least doubt that Husserl obtained the distinction between sense and referent with complete independence of Frege. Hence, Føllesdal's and others' contention that Husserl obtained that distinction from Frege is plainly false, and should rest in peace for all eternity. Moreover, as pointed out above, Husserl arrived at the clear separation of the genuine logical content of statements from their so-called psychological content at least 4 years before Frege's review of *Philosophie der Arithmetik*. Furthermore, Husserl's mature views on the nature of logic and mathematics, to which he seems to have arrived precisely in 1894, that is, in the year of the publication of Frege's review, are certainly different from Frege's at least in the following points: (1) Husserl rejected logicism, set-theoreticism and any such reduction of mathematics to a single fundamental area, rejecting also the current view that there is a fundamental mathematical notion, be it that of set or whatsoever; (2) hence, because of his rejection of logicism, Husserl did not need to postulate the existence of logical objects and, thus, logic was not ontologically committed, like mathematics, being a sort of syntactic (and in *Formale und transzendentale Logik* also semantic) discipline parallel to mathematics; (3) mathematics was ontologically committed and, thus, Husserl defended mathematical Platonism, though Husserl's mathematical Platonism was

[28] 'Ausführungen über Sinn und Bedeutung', p. 135. See also p. 134.

not one of objects, like Frege's, but one of structures based on different equally fundamental mathematical notions.[29]

In fact, with respect to logic there is no ground to doubt what Husserl wrote in Chapter X of the first volume of *Logische Untersuchungen*[30] and in his *Introduction to the Logical Investigations*[31] about his affinities with and influences from the views of Leibniz, Bolzano and Lotze, as well as the influence of Hume. It should also be clear that Husserl's conception of a fusion of the syntactical-semantic discipline of pure logic with the ontologically fat sister discipline of mathematics in a sort of *mathesis universalis* is a Leibnizian heritage. In fact, one can very well say that with respect to the relation between logic and mathematics Leibniz was the intellectual grand father of Husserl, as he also was of Frege and of Hilbert. As Leibniz, the three intellectual cousins considered logic and mathematics intimately related, but differed with respect to the sort of relation.

As already mentioned, Husserl conceived mathematics as a doctrine of structures, some more general, others more concrete obtained from the former, some simpler, some more complex, obtained by combining different simpler structures. To refer to mathematics Husserl did not use the word 'structure' (*Struktur* in German), but the word manifold (*Mannigfaltigkeit* in German), thus, mathematics was for Husserl essentially a doctrine of manifolds (*Mannigfaltigkeitslehre*). His friend Cantor had also used the word 'Mannigfaltigkeit', but to refer to what is now currently called 'set'. Husserl, however, meant by 'Mannigfaltigkeit' not a set but a structure and, thus, the originator of that usage is not Cantor. In fact, it is Riemann.

10.3 Riemann's Notion of Manifold

It is well known that in the nineteenth century mathematics probably experienced the most radical changes in its whole history, among them the rigourization and arithmetization of analysis, the origin of abstract algebraic systems, the origin of algebraic and set-theoretical topology, as well as that of set theory and, what is more relevant here, the birth of non-Euclidean geometries. As is also very well known, around 1830 János Bolyai and Nikolai Lovachevky discovered the first sort of Non-Euclidean geometry, namely, one in which in a plane, given a line l and a point P not on l, there exist more than one line through P, in fact, infinitely many, not intersecting l. In current geometric parlance, one says that Bolyai-Lovachevskyan geometry has negative curvature. In fact, the great Carl Friedrich Gauss had anticipated by about a quarter of a century the results of Bolyai and Lovaschevsky, but fearing

[29]On Husserl's philosophy of mathematics see *Logische Untersuchungen I*, Chapter XI, §§69–70, *Formale und transzendentale Logik*, 1929, Husserliana XVII, 1974, especially Chapters 2 and 3, and *Einleitung in die Logik und Erkenntnistheorie*, Husserliana XXIV, 1984, especially Chapter 2.

[30]See *Logische Untersuchungen I*, Chapter X, §§59–61 and Appendix.

[31]*Introduction to the Logical Investigations*, M. Nijhoff, Den Haag 1975, pp. 36–38

the power of the Kantians in German academic life, opted not to publish his results. In his extraordinary monograph *Über die Hypothesen, welche der Geometrie zugrunde liegen*,[32] in which, by the way, another sort of Non-Euclidean geometry was considered, namely, one in which, given a line *l* and a point *P* not in *l*, there are no lines through *P* parallel to *l*, briefly a geometry with positive curvature, one of Gauss' last students, Bernhard Riemann considered physical space and the different geometries as special cases of a much more general concept. He first introduced the notion of an n-fold extended magnitude, for n any natural number. Euclidean three-dimensional geometry, as well as its two three-dimensional rivals are simply particular cases, for n = 3, of the general concept of an n-fold extended magnitude. Physical space is also a special case in which n = 3, but, contrary to what the old Kant had taught, Riemann thought that it cannot be decided *a priori*, but only through experience, whether physical space has zero curvature (is Euclidean), negative curvature or positive curvature. Moreover, Riemann subsumed n-extended magnitudes under a still more general concept of a manifold. Riemann divided manifolds in two different classes, namely, discrete and continuous manifolds. In discrete manifolds one compares quantities by counting, whereas in continuous manifolds one compares them by means of a unit of measure. Whereas the manifold of natural numbers would be a good example of a discrete manifold, all sorts of geometric manifolds are examples of continuous ones. As one can see, in all this construction there is no reference to any sort of intuition, something Kant had considered an essential component of our mathematical knowledge. Thus, in particular, Riemann rejected Kant's *a priori* conception of physical space and, in general, any essential role of intuition –be it *a priori* or *a posteriori*- in our mathematical knowledge.

10.4 Brief Notes on Husserl's Conception of Mathematics

Husserl conceived mathematics as a formal ontology, that is, as a sort of theory of formal structures running parallel to its sister discipline logic. Husserl conceived both disciplines as based on fundamental concepts. In the case of mathematics, notions like those of set, cardinal number, relation, and whole and part gave rise to a variety of fundamental mathematical theories, whereas the whole edifice of mathematical theories was obtained either by specialization of those, in Bourbakian terminology, sorts of mother structures, combination of them or combination of their specializations. And the development of mathematics after 1900 –when Husserl made public his views on logic and mathematics in the last chapter of the *Prolegomena*-, has proven that Husserl was, in principle, right. Thus, for example, topological groups are mathematical structures that combine topological

[32] 'Über die Hypothesen, welche der Geometrie zugrunde liegen' 1867, third edition, Berlin 1923, reprint, Chelsea, New York 1973.

and algebraic structures, whereas more concrete mathematical structures, like that of real numbers or Banach spaces are much richer combinations of specializations of abstract structures.

Husserl called abstract mathematical theories 'manifolds',[33] clearly referring to Riemann's conception of manifold and his generalization of geometrical theory, and considered his notion of manifold as a generalization of that of Riemann. Moreover, Husserl considered the possibility of a sort of meta-mathematical treatment of mathematics in which one is not concerned with concrete mathematical theories, that is, with concrete manifolds, but with their abstract mathematical forms and their *a priori* relations, thus, mathematical theories are now the object of study, not the more concrete mathematical objects that constitute the manifold. On this very abstract level, Husserl discusses the possibility of different relatively more concrete mathematical theories having essentially the same form. In *Einleitung in die Logik und Erkenntnistheorie* Husserl makes this point more explicit by considering three different interpretations of an abstract commutative law, '$a + b = b + a$', where the symbol '+' can be interpreted either (i) in the familiar arithmetical way of addition of numbers, or (ii) as set-theoretical union, or (iii) as the juxtaposition of straight lines in a geometrical system. Thus, one can obtain a result for a mathematical form at the most abstract level and the result is immediately true also in any specialization of that manifold.

But Riemann's influence on Husserl is not limited to the notion of mathematical manifold. In fact, as attested by a posthumously published book of Husserl containing mostly material from the transition period of 1889–1893, namely his already mentioned *Studien zur Arithmetik und Geometrie*,[34] in which Riemann is referred to in more than twenty different pages, Riemann's influence was not limited to his views on pure mathematics. In fact, it also concerns the relation between geometry and physical space, an issue very little known both to Husserlian scholars and to analytic philosophers.

10.5 Husserl's Views on the Geometric Structure of Physical Space

From 1886 to 1893 Husserl worked intensively both on the philosophy of arithmetic and on the philosophy of geometry, as attested by the already mentioned *Studien zur Arithmetik und Geometrie*, and his views were in continuous evolution. In particular, around 1892 Husserl finally accepted Riemann's views on the nature of geometry, and he remained firmly convinced of the correctness of those views. In fact, in the manuscripts on space up to 1892 Husserl seems to have some discrepancies with

[33]See *Logische Untersuchungen I*, p. 252, and *Formale und transzendentale Logik*, p. 97 for references to Riemann in this context.

[34]*Studien zur Arithmetik und Geometrie*, Husserliana XXI, M. Nijhoff, Den Haag 1983.

Riemann's and Helmholtz' views. But, as attested by his correspondence, already at the end of that year Husserl had embraced both the Riemannian conception of the nature of physical space and the latter's conception of the abstract analytic treatment of geometrical manifolds. Nonetheless, before quoting from his letters it should be mentioned that already in a course of 1889–1890 titled "Geschichtlicher Überblick über die Grundlagen der Geometrie",[35] after discussing the impact of both Bolyai-Lovachevskyan non-Euclidean geometry and Riemann's more general analytic approach to the nature of geometry, Husserl stressed that the parallel Axiom distinctive of Euclidean geometry is not a logical necessity. Such an isolated point should not be given too much weight. In fact, though Frege was a staunch defender of Euclidean geometry and of the non-intuitiveness of the non-Euclidean ones, he had already acknowledged in *Die Grundlagen der Arithmetik*[36] that the geometrical axioms are not logical necessities. Nonetheless, Frege was so convinced of the validity of Euclidean geometry that he never seriously discussed, as Husserl did in that course, the impact both of Bolyai-Lobachevskyan geometry and of Riemann's more general and profound views.

In any case, there is clear evidence that at the end of 1892 Husserl had embraced Riemann's conception of both physical space and of geometric manifolds. Thus, in a letter to his teacher Franz Brentano dated 29 December 1892 Husserl lets Brentano know his basic commitment to Riemann's views on geometry, and in letters to the Neo-Kantian philosopher Paul Natorp dated 29 March 1897 and 7 September 1901 Husserl reiterates that commitment. Thus, Husserl accepted that there exists a diversity of geometric manifolds of n dimensions, for any natural number n, of a purely mathematical nature, and, on the other hand, physical space, whose structure –basically, curvature and dimensionality- should be empirically established, contrary to Kant's views. Hence, Husserl was far more receptive to a revision of traditional conceptions of geometry than other philosophers of that period, including the great mathematician and also important philosopher and physicist Henri Poincaré, and certainly by far much more receptive than his rival Frege, who in a posthumous paper at the turn of the twentieth century compared non-Euclidean geometries with alchemy and astrology.[37] Before referring to some of Husserl's manuscripts of that period, it should be pointed out that, contrary to Riemann, Husserl acknowledged a sort of intermediate area of a non-Kantian synthetic *a priori*. We are not going to discuss that still very obscure point here. It suffices to say that Carnap, in his dissertation, *Der Raum*, in which he was very strongly influenced by Husserl, assigns this synthetic apriorism to the topological level of space. There are, however, passages of Husserl that attest to a sort of not only topological but also even metric synthetic *a priori*. But let us now return to our main point.

[35] Ibid., pp. 312–347.

[36] *Die Grundlagen der Arithmetik*, §14.

[37] 'Über Euklidische Geometrie', in *Nachgelassene Schriften*, pp. 182–184.

In his letter to Brentano of 29 December 1892 Husserl states on p. 10[38]:

> Recently I have been working on the philosophical-geometrical problems. Some [things] that I formerly considered secure have now become very doubtful for me. I formerly believed that a continuum in which any 2 points are connected by a straight line, should *eo ipso* be characterized as plane (Euclidean). This was not correct. One cannot prove that with respect to any straight line one can trace through each point only one [straight line] not intersecting [it]; or that parallel segments between parallels are equal; that, thus, a rigid segment can be parallelwise displaced (i.e. without changing its length). Briefly, the parallel axiom is not true.
>
> Also my judgement about the Riemann-Helmholtz theories of space has changed.... the general theories that they build, following a strike of genius, conceal a valuable content that, philosophically elucidated, would be interesting even for a theory of geometrical knowledge.

In the letters of 29 March 1897 and 7 September 1901 to Natorp Husserl is even more explicit[39]:

> Without doubt, as you say, Euclidean space is an "unfounded", i.e. a by the natural scientists unfounded hypothesis. That has its roots in the origin of science in natural thinking.... The missing foundation can only be... an empirical one; it will, thus, be of an inductive nature, though could have an enormous probabilistic foundation.
>
> But when proceeding purely scientifically it is not allowed to take anything whatever from "intuition" except in the worst of cases what is already established in the Axioms.... The enormous significance of the intuitive procedure is thoroughly a methodical one; but the pure system of mathematics has to disown intuition completely.
>
> I would in general not expect that no matter what beneficial preference will be obtained for the 3-dimensional as against the n-dimensional Euclidean multiplicity. Thus, my conception about aprioricity in geometry is different from yours. The pure aprioristic in space is the system of fixed relation forms, which can be obtained by means of idealization of the empirical space intuitions and by going back to their categorial forms (abstracting from everything sensible...), i.e. reaching the categorial forms of their primitive (idealized) fundamental relations. What is pure in space is the Euclidean manifold of 3rd level. I acknowledge (against my former views) the possibility of other space intuitions, which give rise to different idealized geometric spaces and whose logical structuring could be exhibited in other pure manifolds. But what is completely certain for me, is that all in general existing possibilities are rigidly demarcated by aprioristic laws: ideal possibilities, Platonic ideas. Within this frame operates the mathematical "arbitrariness", with its "conventions", by means of which determined sorts of manifolds are selected from the totality of valid [ones], are "defined", but, of course, not created.... *A priori* = pure categorial lawfulness, in virtue of its extension[,] = the mathesis in its most general sense.

The last passage is especially important as a synthesis of Husserl's views on geometry. First of all, he reiterates his view that three-dimensional Euclidean geometry does not deserve any privileged status, because of two different reasons. There are no *a priori* grounds to prefer three-dimensionality instead of n-dimensionality, for n any positive integer. On the other hand, we can very well imagine other sorts of spatial intuition besides the three-dimensional Euclidean one, for example, n-dimensional or non-Euclidean. Moreover, each of those possibilities, be it three-

[38] *Briefwechsel I*, 1994, p. 10.

[39] *Briefwechsel V*, pp. 62, 63 and 83–84, respectively.

dimensional or n-dimensional, Euclidean or non-Euclidean, is grounded on *a priori* laws. The mathematician simply selects some of those structures for study, though from a purely mathematical standpoint there is no ground to attribute any of them a preferred status. And Husserl stresses the non-constructive character of mathematics by emphasising that the structures selected for study by the mathematician are in no way his creation or construction.

In *Studien zur Arithmetik und Geometrie* there are many passages that expand the views expressed in the three already quoted letters. We will quote only two passages from 1893 already quoted in our paper 'Husserl's Conception of Physical Theories and Physical Geometry in the Time of the *Prolegomena*: A Comparison with Duhem's and Poincaré's Views'. The first one is concerned with Hermann Lotze's argument on the logical necessity of Euclidean geometry. Husserl says:

> Lotze's and others' doctrine that there exists a logical necessity for the three-dimensionality and for the plane nature [Euclidicity GERH], then excludes the possibility of a non-three-dimensional and of a curved space. But that doctrine is false. The three-dimensionality, etc. is an empirical factuality, though a general fact (a law) like the law of gravitation. There is only an enormous improbability for space not being Euclidean, though we have to leave open its possibility. The probability [that space be Euclidean GERH] is not infinite.... Infinite is only the probability that space harmonizes with the Euclidean continuum in the limits traced by our capabilities of observation.[40] [*Studien zur Arithmetik und Geometrie*, p. 269]

After that brilliant refutation, clearly based on Riemann's views, of Lotze's and others' argument, Husserl goes on to question the requirement of intuitiveness of geometry and the ensuing attempt to disqualify non-Euclidean geometries on the basis of their lack of intuitiveness. But Husserl also criticizes Helmholtz for giving too much emphasis to a defence of the intuitiveness of non-Euclidean geometries. On this point it is clear that Husserl, the mathematician turned philosopher, feels rightly more comfortable with the mathematician Riemann's genial insights than with the natural scientist Helmholtz' observations. Thus, Husserl says:

> Here lies a complete misunderstanding of the significance of the theory of curvature.... Whether intuition fails or not is irrelevant. Intuitiveness is something completely inessential. Is it then what is decisive that we can intuitively represent other spaces besides Euclid's? Does a representation have to be carried out in order for it to be epistemologically useful? Big numbers. It is a fundamental misunderstanding of those opposing metageometrical research when they believe that by showing the non-intuitiveness of the metageometrical concepts they would also have refuted their epistemological contents. In any case, the representatives of metageometry, especially Helmholtz have committed the great error of giving too much weight to the intuitive representability that they presumably establish. Riemann, however, did not do that. [*Studien zur Arithmetik und Geometrie*, p. 411].

Many other passages of *Studien zur Arithmetik und Geometrie* also attest to the influence of Riemann on Husserl's views on geometric manifolds and physical

[40]In more modern parlance, one would say that the probability of space been Euclidean is less than 1, since the probability of any hypothesis or theory ranges from 0 to 1. See also the acknowledgement of the Riemannian restriction of local Euclidicity.

space. For example, in a passage of the same year 1893 Husserl stresses that, contrary to the geometry of physical space: “Pure geometry is a pure aprioristic science”.

10.6 Husserl and Riemann: A Brief Retrospect

By means of quotations from two writings of Husserl presumably written in 1890, we have shown the falsity of Føllesdal’s contention, accepted by the whole analytic family, that Husserl borrowed the distinction between sense and referent from Frege. Moreover, in a letter of Frege to Husserl of May 1891 and in his ‘Ausführungen über Sinn und Bedeutung’ Frege acknowledges that Husserl had made the distinction independently of him. In fact, after Bolzano, the sense-referent distinction was there to be taken, as happens with a ripe fruit. Only the obsession of some analytic philosophers trying either to ignore or to diminish Husserl’s stature as a philosopher can explain their philosophical-historical aberration of giving credit to Frege alone.[41] Moreover, there is no reason to doubt Husserl’s recognition of the importance for the evolution of his views on logic of Leibniz, Bolzano, Lotze and Hume. If Frege had played a major role in the evolution of Husserl’s views, the latter would have stated it. But in any case, as a matter of fact, Husserl’s mature views on logic and its relation to mathematics are clearly different from Frege’s.

That does not mean that Frege did not exert any, even indirect, influence on Husserl, or Husserl or Frege. Certainly, Frege’s level of rigour most surely elevated the level of discourse and, thus, indirectly influenced the most capable of his younger contemporaries, like Husserl and Hilbert.[42] Most probably, without Frege’s pioneering work, Husserl would not have arrived in his refutation of psychologism in logic in the first eight chapters of the Prolegomena to such a level of philosophical analysis that clearly surpassed Frege’s previous refutation of psychologism. But Frege also would probably had not felt compelled to discuss the problem of the indexicals in ‘Der Gedanke’,[43] were it not for the fact that Husserl had discussed it in more detail in the First Logical Investigation[44] and arrived at a similar, but once more finer conception.

[41]The above remark does not include Føllesdal, who certainly has tried to make Husserl known in analytic circles. In fact, it should be pointed out that when he wrote his MA thesis and the paper referred to above, neither Husserl’s ‘Zur Logik der Zeichen’ nor Frege’s *Briefwechsel* had been published.

[42]In this more general and sounder sense, Hume exerted an important influence on Kant, and Quine has exerted an influence on many recent rigorous philosophers, even those like the present author, who disagree with Quine in almost any concrete philosophical issue. But that is precisely the way in which philosophy is made.

[43]‘Der Gedanke’ 1918, reprinted in *Kleine Schriften*, pp. 342–362.

[44]*Logische Untersuchungen II*, U. I, Chapter 3, §§24–26.

Furthermore, Husserl's views on mathematics are not only different from Frege's, but much more modern, resembling the views of the Bourbaki group of French mathematicians. Husserl used the term 'manifold', as did his friend Cantor. But for Cantor 'manifold' meant essentially what we now call 'set', that is, a finite or infinite collection of objects, without taking into account any relations between those objects. For Husserl, manifolds are structures, and mathematics as a theory of manifolds is simply a theory of formal structures. On this point, one can clearly see the affinities with Riemann. Husserl's conception of mathematics as a theory of manifolds is a generalization and development of Riemann's views.

But Husserl not only obtained from Riemann the germ of his view of mathematics as a whole, but also had accepted by the end of 1892, thus, earlier than other important philosophers of his time, Riemann's conception of physical space as being such that both its curvature and dimensionality should be determined by experience.

References

E.W. Beth, *The Foundations of Mathematics* (North-Holland Publishing Company, Amsterdam, 1965)

R. Carnap, *Der Raum, Kant-Studien* (Ergänzungsheft 56) 1922, reprint (Vaduz, Lichtenstein 1991)

R. Carnap, *Die logische Syntax der Sprache* 1934. Expanded English edition, *The Logical Syntax of Language* (Routledge, London, 1937); reprint, Chicago et al. (2003)

S. Centrone, Husserl on the 'Totality of all conceivable arithmetical operations. *Hist. Philos. Logic* **27**(3), 211–228 (2006)

D. Føllesdal, Husserl's Concept of Noema. *J. Philos* **66**(20), 680–697 (1969)

D. Føllesdal, *Husserl und Frege, ein Beitrag zur Beleuchtung der phänomenologischen Philosophie*, translation in L. Haaparanta (ed.), *Mind, Meaning and Mathematics* (Kluwer, Dordrecht, 1994), pp. 3–47

G. Frege, *Die Grundlagen der Arithmetik* 1884. Centenary Edition. ed. by C. Thiel (Meiner, Hamburg, 1986)

G. Frege, 'Funktion und Begriff' (1891), reprinted in *Kleine Schriften*, 125–142

G. Frege, 'Über Sinn und Bedeutung' (1892), reprinted in *Kleine Schriften*, 143–162

G. Frege, 'Rezension von E. G. Husserl: *Philosophie der Arithmetik I*' (1894), reprinted in *Kleine Schriften*, 179–192

G. Frege, 'Der Gedanke' (1918), reprinted in *Kleine Schriften*, 342–362

G. Frege, *Kleine Schriften*, 1967. Revised edition. (Olms, Hildesheim, 1990)

G. Frege, *Nachgelassene Schriften*, 1969. Revised edn (Meiner, Hamburg, 1983)

G. Frege, *Wissenschaftlicher Briefwechsel* (Meiner, Hamburg, 1974)

G. Frege, 'Ausführungen über Sinn und Bedeutung', in *Nachgelassene Schriften*, 128–136

G. Frege, 'Über Euklidische Geometrie', in *Nachgelassene Schriften*, 182–184

E. Husserl, 'Besprechung von E. Schröders *Vorlesungen über die Algebra der Logik I*' (1891). in *Aufsätze und Rezensionen* (1890–1910), 3–43

E. HUSSERL, *Philosophie der Arithmetik*, in Hua XII, (1970a), 1–283

E. Husserl, Zur Logik der Zeichen (Semiotik), in Hua XII, (1970b), 340–373

E. Husserl, *Formale und transzendentale Logik*, 1929, Hua XVII, (1974)

E. Husserl, *Logische Untersuchungen* (2 vols), 1900–1901, Hua XVIII & XIX, (1975a & 1984)

E. Husserl, *Introduction to the Logical Investigations* (Martinus Nijhoff, Den Haag, 1975b)

E. Husserl, *Ideen zu einer reinen Phänomenologie und phänomenologischen Philosophie I* 1913, Hua III, 1950. Revised edn (Martinus Nijhoff, Den Haag, 1976)
E. Husserl, *Aufsätze und Rezensionen*, Hua XXII, (1979)
E. Husserl, *Studien zur Arithmetik und Geometrie*, Hua XXI, (1983)
E. Husserl, *Einleitung in die Logik und Erkenntnistheorie*, Hua XXIV, (1984)
E. Husserl, *Erfahrung und Urteil*, 1939. 6th edn. (Meiner, Hamburg, 1985)
E. Husserl, *Briefwechsel*, 10 vols. (Kluwer, Dordrecht, 1994)
E. Husserl, [Hua] *Husserliana, Gesammelte Werke*, I–XXVI: Martinus Nijhoff, Den Haag, 1950; XXVII–XXXVII: Kluwer, Dordrecht, 1989–; XXXVIII–: Springer, New York, 2005–
B. Riemann, *Über die Hypothesen, welche der Geometrie zugrunde liegen, 1867*. 3rd edn., Berlin 1923; reprint (Chelsea, New York, 1973)
G.E. Rosado Haddock, Husserl's philosophy of mathematics: its origin and relevance. *Husserl Studies 22*, (2006). Reprinted as Chapter 5 of *Against the Current*, 145–181
G.E. Rosado Haddock, Husserl's Conception of physical theories and physical geometry in the time of the prolegomena: a comparison with Duhem's and Poincaré's views. *Axiomathes 22*, (2012a). Reprinted as Chapter 6 of *Against the Current*, 183–214
G.E. Rosado Haddock, *Against the Current* (Ontos Verlag, Frankfurt, 2012b)

Chapter 11
Husserl and Hilbert

Mirja Hartimo

Abstract The paper examines Husserl's (1859–1938) phenomenology and Hilbert's (1862–1943) view of the foundations of mathematics against the backdrop of their lifelong friendship. After a brief account of the complementary nature of their early approaches, the paper focuses on Husserl's *Formale und transzendentale Logik* (1929) viewed as a response to Hilbert's "new foundations" developed in the 1920s. While both Husserl and Hilbert share a "mathematics first," nonrevisionist approach toward mathematics, they disagree about the way in which the access to it should be construed: Hilbert wanted to reach it and show it consistent by his formalism on the basis of sensuous signs, Husserl held that there should be a reduction to elementary judgements about individuals. Husserl's reduction does not establish the consistency of mathematics but he claims it is important for the considerations of truth.

Keywords Edmund Husserl • Foundations of mathematics • Hilbert's program • Dietrich Mahnke • Formalism

11.1 Husserl's and Hilbert's Friendship

Since 1901, from his arrival to Göttingen, until he left for Freiburg in 1916, Husserl's closest colleague and the main defender at the university seems to have been the mathematician David Hilbert. While Husserl befriended the mathematicians Felix Klein and especially Hilbert, the philosophers' attitude towards Husserl was sultry if not altogether hostile throughout his stay in Göttingen. The philosophical faculty had opposed Husserl's appointment initially in 1900 when the Prussian Ministry of Culture first expressed such an intention. The faculty had hoped

---

I wish to thank Øystein Linnebo, Matti Eklund, Besim Karakadilar, Mitsuhiro Okada, Volker Peckhaus, and Sören Stenlund for valuable feedback on earlier versions of this paper.

M. Hartimo (✉)
Norwegian University of Life Sciences, P.O. Box 5003, 1432 Aas, Norway,
e-mail: mirjahartimo@gmail.com

S. Centrone (ed.), *Essays on Husserl's Logic and Philosophy of Mathematics*, Synthese Library 384, DOI 10.1007/978-94-024-1132-4_11

for expertize in philologico-historical problems, but instead in 1901 they received Husserl whose background in mathematics and notorious anti-psychologism did not make him particularly popular among the psychologists of the department.[1] Husserl's problems continued in 1905 when ministry tried to promote Husserl as an ordinary professor. While the ordinary professors in philosophy, Julius Baumann and Georg Elias Müller objected to promoting Husserl, Hilbert defended Husserl and wrote seven reviews about his scientific achievements to the effect that he was eventually promoted as an "Ordinarius für Philosophie" in 1906. Hilbert promoted Husserl also in 1908 when a position as a "zusätzliches Philosophie-Ordinariat" was opened in Göttingen claiming that it is utmost important to try to keep Husserl in Göttingen. This time the position however was eventually given to Heinrich Maier.[2] The reportedly deep and respectful friendship between Husserl and Hilbert continued throughout their lives.[3] Given their common history, it is no wonder that in a letter for Hilbert's 60th birthday Husserl expressed gratitude for Hilbert's constant interest and trust in him in his philosophical struggles. In his letter Husserl also highlights the importance of Hilbert's mathematical creativity for philosophy.[4]

In this paper my aim is to clarify Husserl's view in respect of Hilbert's achievements in the foundations of mathematics. I will first discuss their views right after the turn of the twentieth century, but my main focus will be on their views in the 1920s. One motivation for this is Husserl's student's Friedrich Mahnke's insistence of the phenomenological nature of Hilbert's program. Accordingly, I will discuss Husserl's *Formale und transzendentale Logik* (1929)[5] as a response to Hilbert's program as developed in the early 1920s. Husserl thinks that a careful phenomenological analysis shows that the consistency of mathematical theories has to be approached semantically, by taking into account "the matters in the synthetic unity of the experience."[6] I will argue that Husserl's approach is model theoretical and for him *in mathematics* there is no need for direct consistency proof sought for by Hilbert. He nevertheless formulated a reduction of mathematics to primitive judgments, which for him shows the relationship of formal mathematics to truth and is important for *philosophical* reasons.

---

[1]Peckhaus 1990, 2016–208; Husserl Archive Mitteilungsblatt 34, 12, see also Hill & Da Silva 2013.

[2]Peckhaus 1990, 208–210.

[3]The most recent details of Husserl's and Hilbert's friendship and how it also extended to their families have been documented in the Husserl Archive Leuven *Mitteilungsblatt* 36, 2013.

[4]Husserl 1994, 119.

[5]Husserl 1974. Henceforth cited as *FTL*. English translations refer to (Husserl 1969) unless otherwise indicated.

[6]*FTL*, §89b.

11.2 The More General Background in the Development of Mathematics

The change of the century witnessed an emergence of modern abstract view of mathematics. The development towards abstract conceptual viewpoint centered in Göttingen of Felix Klein and David Hilbert where Husserl also moved in 1901.[7] Klein shared Hilbert's understanding of mathematics as a multifaceted but fundamentally unified body of knowledge.[8] While Klein's aim was to place mathematics on a healthy footing by means of his massive *Enzyklopädie der mathematischen Wissenschaften*, a survey in six volumes covering all mathematical branches, including applied mathematics, Hilbert developed the axiomatic approach. He held that[9]

> [m]athematics is an indivisible whole, an organism whose livelihood depends on the interconnections between its parts... The further a mathematical theory is developed, the more harmonious and unified its structure unfolds, leading to the discovery of relationships between previously distinct branches of knowledge. Thus it happens that with the expansion of mathematical knowledge, the holistic quality becomes enhanced rather than lost.

Like Klein and Hilbert Husserl also entertained a picture of a unifying theoretical framework for different mathematical theories. Husserl envisioned "a definite, ordered procedure which will enable us to construct the possible forms of theories, to survey their law-governed connections, and to pass from one to another by varying their basic determining factors, etc".[10] Husserl's theory of theories is a mathematical theory within which individual theories could be examined formally. As Husserl's student Dietrich Mahnke puts it, Husserl emphasized the importance of the "eternally valid systematic structure of formal mathematics as the ideal storeroom of the forms of theories of all exact science that it places with clear consciousness, for the first time since Leibniz, at the entrance to the unified theory of science."[11]

However, the theory of theories was not something a philosopher could impose on mathematicians. "No one can debar mathematicians from staking claims to all that can be treated in terms of mathematical form and method."[12] In his

[7]cf. Ferreirós 2007, 31. Soon after the move to Göttingen, Husserl's wife, Malvine, reported that in Göttingen a ‚ganz anderer Zug im geistigen Leben der Universität als in Halle <herrsche>, u. besonders sind es die Mathematiker (Klein u. Hilbert), die Edmund in ihren Kreis ziehen u. ihn <...> anregen." David Hilbert and Edmund Husserl developed a ‚tiefe achtungsvolle Freundschaft" which, according to Husserl's wife, was a consequence of the ‚gleichen Ethos einer restlosen Hingabe an sein Werk" (Husserl Archive Mitteilungsblatt 2013, 15).

[8]Rowe 1989, 198.

[9]Cited from Rowe 1989, 212.

[10]Husserl 1975, henceforth cited as *Prolegomena*, §69, translation modified.

[11]Mahnke 1977 [1923], 75.

[12]*Prolegomena*, §71.

Prolegomena Husserl advocated a division of labor, according to which mathematicians' task was to be ingenious technicians constructing theories and solving problems, while philosophers' task was to provide ultimate insight to the essence of these mathematical theories.[13] The division of labor reflects Husserl's general approach towards mathematics: mathematics comes first, only afterwards its theories should be examined by a philosopher. Similar, "mathematics-first" approach to the foundations of mathematics can also be found in Hilbert's explanation of how, contrary to construction of buildings, in the house of knowledge the foundations are given only after the comfortable rooms have been built. Eventually, however, for both, Husserl and Hilbert, the examination of the foundations became more seriously motivated by the paradoxes found around the turn of the century. What distinguishes the two is that in accordance to the division of labor, Hilbert (and other mathematicians) focused on freely constructing theories, Husserl saw as his own task to provide understanding for their essence and clarifying fundamental concepts of the constructed theories. He thus held his own approach, not as abandoning or sidestepping mathematics, but as importantly complementing it.

11.3 Hilbert on Foundations

Hilbert's interest in the foundations came in two periods: first in 1899–1905 and later again in 1917 onwards. The first period marks the beginning of Husserl's career in Göttingen until his discovery of phenomenological reduction. During the second period, Hilbert developed his 'Program.' At this point Husserl had already moved to Freiburg. However, he still worked on the foundations of mathematics as is witnessed by his *Formal and Transcendental Logic* (1929).

The purpose of Hilbert's axiomatics was to provide mathematics and physical sciences secure foundations. Hilbert compared his axiomatic method to the genetic method used by Weierstrass, Kronecker, and Dedekind among others. In the genetic method one typically started from the number 1 and then extended the number domain by adding the results of calculation into it. In the axiomatic method, on the contrary, one begins by assuming the existence of all the elements, and then one brings these elements into relationship with one another by means of certain axioms:

> The necessary task then arises of showing the consistency and the completeness of these axioms, i.e. it must be proved that the application of the given axioms can never lead to contradictions, and, further, that the system of axioms is adequate to prove all geometrical propositions. We shall call this procedure of investigation the axiomatic method.

Hilbert continues: "Despite the high pedagogic and heuristic value of the genetic method, for the final presentation and the complete logical grounding [Sicherung] of our knowledge the axiomatic method deserves the first rank."[14] Contrary to the

[13] *Loc. cit.*

[14] Hilbert 1900a, 1092–1093.

earlier genetic approach, with the axiomatic approach it is possible to compress what we know in some field of knowledge into a logical axiom system and thus it enables obtaining an overview of this entire field.

In 1899 Hilbert's axiomatization of geometry, *Grundlagen der Geometrie* (*Foundations of Geometry*) appeared in print. In it Hilbert presented the axioms of Euclidean geometry divided into five groups: axioms of incidence, axioms of order, axioms of congruence, axiom of parallels, and axioms of continuity. Hilbert established the consistency of the axiom system by constructing a model of real numbers that satisfies the axioms. The consistency of geometry is thus established assuming the consistency of real number system, and thus he managed to prove only a relative consistency of geometry. But, similarly, by model-theoretical means, he also established the independence of various axioms from each other.

To ensure that the axiomatic method indeed provides the complete logical grounding of our knowledge Hilbert soon added the so-called axiom of completeness (*Vollständigkeitsaxiom*) to his axiomatization. The axiom appeared in print already in the French translation of the *Grundlagen die Geometrie* as well as in his axiomatization of arithmetic *Über den Zahlbegriff* (1900a). According to Hilbert's Completeness axiom[15]:

> To a system of points, straight lines, and planes, it is impossible to add other elements in such a manner that the system thus generalized shall form a new geometry obeying all of the five groups of axioms. In other words, the elements of geometry form a system which is not susceptible of extension, if we regard the five groups of axioms as valid.

The number theoretical formulation of the same axiom is the following[16]:

> It is not possible to add to the system of numbers another system of things so that the axioms I, II, III, and IV are also all satisfied in the combined system; in short, the numbers form a system of things which is incapable of being extended while continuing to satisfy all the axioms.

With this axiom Hilbert made his system of axioms categorical. A set of axioms is categorical if it has a unique model up to isomorphism. The completeness axiom is thus an axiom about models of axioms. Hilbert however did not have a general notion of isomorphism, nor did he have a precise notion of formal deduction at use, so it is not entirely clear what exactly he meant by "completeness".[17]

The consistency of the axiomatization remained a problem for Hilbert. In "On the Concept of Number" Hilbert thought that "one needs only a suitable modification of familiar methods of inference".[18] This proved to be more difficult than what he first anticipated. Already in the same year Hilbert listed the consistency of the arithmetical axioms as the second problem in his list of 23 unsolved problems in his famous address to the International Congress of Mathematicians in Paris. Hilbert's

---

[15]Hilbert 1950, 15.

[16]Hilbert 1900a, 1094.

[17]Awodey & Reck 2001, 11–20.

[18]Hilbert 1900a, 1095.

formulation is now more careful: "I am convinced that it must be possible to find a direct proof for the consistency of the arithmetical axioms, by means of a careful study and suitable modification of the known methods of reasoning in the theory of irrational numbers."[19]

From about 1917 onwards Hilbert, in collaboration with Bernays, gave a series of lectures, in which the modern mathematical logic and proof theory were conceived.[20] The development of the so called 'Hilbert Program' was first made public in the lectures given in the winter term 1921/1922.[21] Incidentally, Dietrich Mahnke held that the presentation of Hilbert's axiomatics in lectures in 1921 comes closest to Husserl's phenomenological foundation.[22] The basic idea in Hilbert's 'Program' is to turn the propositions that constitute mathematics into formulas. Certain formulas are called axioms, by my means of which further formulas can be derived. "The axioms and provable propositions, that is, the formulas resulting from this procedure, are copies (*Abbilder*) of the thoughts constituting customary mathematics as it has developed till now".[23] With such formal deductive systems he sought to capture the customary mathematics, his intention was not to restrict it. By means of such approach Hilbert thought he could cover all the infinitistic mathematics and finally prove the consistency of arithmetic by finitary methods.

Hilbert motivates his project philosophically. As a presupposition of his program, Hilbert mentions that[24]

> something must already be given to us in our faculty of representation [in der Vorstellung], certain extralogical concrete objects that are intuitively [anschaulich] present as immediate experience prior to all thought. If logical inference is to be reliable, it must be possible to survey these objects completely in all their parts, and the fact that they occur, that they differ from one another, and that they follow each other, or are concatenated, is immediately given intuitively, together with the objects, as something that neither can be reduced to anything else nor requires reduction.

This, according to Hilbert, is the basic philosophical requirement for mathematics and all scientific thinking. In mathematics, the objects in question are concrete signs themselves, "whose shape ... is immediately clear and recognizable".[25] The aim of Hilbert's theory is that, in Hilbert's own words, "contentual inference is replaced by manipulation of signs (*äußeres Handeln*) according to rules; in this way the axiomatic method attains that reliability and perfection that it can and must reach if it is to become the basic instrument of all theoretical research".[26]

---

[19]Hilbert 1900b, 1104.

[20]Sieg 1999, 12.

[21]*Op. cit.*, 23, Sieg 2013, 115.

[22]Mahnke 1977, 76–77.

[23]*Op. cit.*, 465.

[24]*Op. cit.*, 464–465.

[25]*Op. cit.*, 465.

[26]*Op. cit.*, 467.

Hilbert's words may suggest an interpretation that to him mathematics *is* "manipulation of signs according to rules," and that Hilbert thus is a dogmatic formalist about mathematics. This however is not true and accordingly in the secondary literature, the view has been repeatedly rejected. David E. Rowe, for example, characterizes Hilbert as the leading proponent of the 'structuralist' tradition in modern mathematics rooted in the axiomatics, Cantor's set theory and the algebraic tradition. According to Rowe, it is a mistake to identify Hilbert's philosophy of mathematics with his formalist program, the purpose of which was to legitimate the entire corpus of mathematical knowledge.[27] Rowe suggests further that, as Herbert Mehrtens has put it, Hilbert's position in the *Grundlagenkrise* was that "Mathematicians can philosophize about mathematics, but they are not permitted to draw normative consequences that would constrain scientific production".[28] Similar view is expressed also by W. Sieg who points out that Hilbert and Bernays always present their theories together with a structure, calling it the 'existential aspect' of the axiomatic method.[29] According to him, Hilbert's program seeks a uniform structural reduction: "intended structures are projected through their assumed complete formalizations into the properly mathematical domain ..., i.e., finitist mathematics."[30] Furthermore, in the spirit of his "mathematics-first" approach, Hilbert's view of the rival approaches to the foundations is that they, too, enhance our understanding of the mathematical content of theories. Thus he viewed Brouwer's and Weyl's, work as part of axiomatic investigations.[31] I will argue soon that something like this is also Husserl's view, namely that Husserl, too, wants to show how abstract mathematics is accessible from the meaningful concrete basis.

11.4 Husserl on the Foundations

As already mentioned above, Hilbert's view of mathematics as a unified but multifaceted axiomatic science seems to be close to what Husserl means by a theory of theories in the *Prolegomena*. The theory of the possible forms of theories deals a priori with the essential forms of theories and the relevant laws of relation. Husserl certainly wants the theories to be axiomatic, or "nomological," that is the term used by Husserl (§64). In 1901 upon arrival at Göttingen Husserl was invited to give two lectures on the notion of definiteness of axiom systems. In those lectures Husserl defines the concepts 'relatively' and 'absolutely definite' axiom systems. The relatively definite axiom system is one to which one cannot add any new axioms without defining a new domain. It thus defines its objects so that they are

[27] Rowe 1989, 199–200.

[28] *Op. cit.*, 211.

[29] Sieg 2013, 106.

[30] *Op. cit.*, 316.

[31] Sieg 2009, 450.

completely unambiguous (i.e., in modern terminology, they do not admit any non-standard interpretations). The absolutely definite axiom system is the one that has the maximal set of axioms.[32] Both are categorical at least in some general sense. The relatively definite axiom system has a unique model, and thus it defines its objects completely unambiguously. An absolutely definite theory is one that cannot be extended consistently (for more detail, see Hartimo 2016). Indeed, in it "the numbers form a system of things which is incapable of being extended while continuing to satisfy all the axioms"[33] as Hilbert defined his completeness axiom.

Husserl himself held that absolute definiteness captures completeness in Hilbert's sense.[34] But Husserl disagreed with Hilbert in holding that completeness should not be viewed as an axiom but as a theorem.[35] Later, in the *Ideen I* Husserl writes that "the close relationship of the concept of definiteness to the 'axiom of completeness' introduced by Hilbert for the foundation of arithmetic will be immediately obvious to every mathematician".[36] In *Formal and Transcendental Logic* (§31) Husserl points out that

> Hilbert arrived at his concept of completeness (naturally quite independently of my [Husserl's] still-unpublished investigations); he attempts, in particular, to complete a system of axioms by adding a separate 'axiom of completeness.' The above-given analyses should make it evident that, even if the inmost motives that guided him mathematically were inexplicit, they tended essentially in the same direction as those that determined the concept of the definite multiplicity.

Husserl thus, also retrospectively, held that like Hilbert, he was advocating categoricity of the axiom systems with his notion of definiteness. Similar interpretation is supported by Dietrich Mahnke's 1923 explanation of Husserlian "theory-forms" as the common deductive scaffolding of all 'logically isomorphic' or 'formally equivalent' disciplines.[37] Obviously, as in Hilbert's case, without the precise notions

---

[32]In Husserl's words: "Eine axiomatisch definierte Mannigfaltigkeit kann die Eigenschaft haben, daβ jedes ihrer Objekte operativ bestimmbar ist, und zwar eindeutig. D. h. jedes Objekt, das für sie als existierend definiert ist (in die Sphäre der Existenz gehört, welche die Axiome umschreiben), ist durch die zugrunde liegenden oder eine endliche Zahl willkürlich anzunehmender bestimmter Existenzen unmittelbar oder mittelbar zu bestimmen, und zwar eindeutig. Eine solche Mannigfaltigkeit ist eine mathematische und ist definit (d.h. ihr Axiomensystem ist definit). [...] Relativ definit ist ein Axiomensystem, wenn es zwar für sein Existential gebiet keine Axiome mehr zuläβt, aber es zuläβt, daβ weiteres Gebiet dieselben und dann natürlich auch neue Axiome gelten. Neue Axiome, denn die bloβ alten Axiome bestimmen ja nur das alte Gebiet. Relativ definit ist die Sphäre der ganzen, der gebrochenen Zahlen, der rationalen Zahlen, ebenso der diskreten Doppelreihenzahlen (komplexen Zahlen). Absolut definit nenne ich eine Mannigfaltigkeit, wenn es keine andere Mannigfaltigkeit gibt, welche dieselben Axiome hat wie sie (alle zusammen). Kontinuierliche Zahlenreihe, kontinuierliche Doppelzahlenreihe" (Schuhmann & Schuhmann 2001, 101–102).

[33]Hilbert 1900, 1094.

[34]*Op. cit.*, 103.

[35]*Op. cit.*, 102.

[36]Husserl 1950, §72.

[37]Mahnke 1977, 80.

of isomorphism and formal deduction the exact nature of Husserl's notion of completeness is not clear. Further complications arise from the fact that the exact content of Husserl's lectures is not known either.[38, 39]

11.5 Dietrich Mahnke on Husserl and Hilbert

An interesting angle to Husserl's and Hilbert's views can be obtained from the writings of Dietrich Mahnke who had studied with both Hilbert and Husserl in Göttingen in 1902–1906. He wrote his dissertation on Leibniz under Husserl's supervision and defended it in 1922. Husserl and Mahnke had a long correspondence, to the effect that in 1921 Husserl wrote to Mahnke that he, Mahnke, was closer to him (Husserl) than any of his other students. This makes it particularly curious that Mahnke, in 1923, wrote that it is not Hilbert's early work but Hilbert's later axiomatics "in its latest exposition (first presented in lectures in Copenhagen and Hamburg 1921)" that is the most complete 'foundation' of mathematics "which comes closest to Husserl's phenomenological foundation. . . . ".[40]

Hilbert's reliance on sensuous signs in his 1922 lectures resembles Husserl's early approach to calculations in the *Philosophie der Arithmetik* (1891). Husserl however gave up the view by the time of the turn of the century. In the beginning of the century Husserl was well aware of Hilbert's views about mathematics and it would be easy to see Husserl's views of mathematics to be compared to Hilbert's axiomatics rather than to his proof-theory.[41] Thus Mahnke's claim may sound surprising, for in the lectures of 1921 Hilbert first publicly discussed what was to become his 'Program.' Indeed, Mahnke was so inspired by the

[38]Hence, it has given a rise to several competing interpretations. See for example Lohmar (1989), Da Silva (2000), Centrone (2010), Hartimo (2007), Recently Mitsuhiro Okada has defended a computational view of Husserl's completeness (Okada 2013).

[39]This paper was originally written in 2013 and represents my thoughts about the matter then. More papers on the topic have been published since, most notably, Da Silva (2015) and Hartimo and Okada (2015), and most recently in Hartimo (2016).

[40]Mahnke 1977, 77.

[41]Within a few months after Husserl's Definitheit lectures Hilbert showed Husserl his so called Memoir, the second foundations to geometry, on which Husserl took detailed notes (cf. Hartimo 2008). Husserl's interest in it, like Hilbert's, shows his unprejudiced interest in different kinds of axiomatic systems. Husserl was also well aware about the set theoretical paradoxes that plagued Hilbert's school. Zermelo's version of 'Russell's paradox' has been found written down by Husserl (Husserl 1979, 399). Hilbert also showed Husserl his correspondence with Frege about the nature of the axioms in geometry. Husserl's comment to the exchange is that Frege does not understand Hilbert's axiomatic foundations of geometry (Husserl 1970, 447–451). Husserl was also aware of the contents of Hilbert's 1905 lectures thanks to Dietrich Mahnke, who sent the lecture notes for him. In that connection Husserl expressed the wish, „recht viel aus Hilberts Darstellungen zu lernen, wie es ja eigentlich selbstverständlich ist" (Husserl Archive Leuven Mitteilungsblatt 2013, 15).

phenomenological nature of Hilbert's approach that he thought that phenomenology could be introduced by discussing Hilbert's view of formal mathematics. The paper in which he sought to do that, "Von Hilbert zu Husserl: Erste Einführung in die Phänomenologie, besonders der formalen Mathematik," appeared in 1923.

Mahnke views Hilbert's new approach as a continuation of the earlier task of axiomatics to achieve 'existence proofs' through the establishment of the consistency of systems of axioms. Mahnke describes Hilbert's proof-theoretical approach and explains how Hilbert now mounts a step higher and formalizes the statements about numbers and the deductive connections between them. He explains Hilbert's aim correctly to be to bring the "non-existence of a contradiction ... available to 'inspection' in the true sense of the word."[42] Mahnke goes on[43]:

> The definitive foundation of mathematics is thus attained, according to Hilbert, by way of a 'proof critique' or 'metamathematical' theory of the mathematical process of inference, analogous to the philosophical 'critique of reason' or epistemology—I would almost say by way of a 'Copernican revolution', equivalent to the Kantian one from the object to the subject of mathematical knowledge or, in Husserl's more precise terminology by way of going back from noema to noesis, from the intended objects to the intending acts of reason.

For Mahnke, Hilbert's metamathematics is thus a shift from a focus on mathematical theories to the acts of constructing them. But, as Mahnke writes,[44]

> there always still remains for philosophers a task which lies in another dimension: that of elucidating epistemologically the inner nature and the true sense of the axiomatic method and its objects. For example, what really are those 'things' of which Hilbert speaks? Not, of course, physical realities; but neither are they ideal concepts like those of intuitive geometry; nor are they individual thought-entities (a 'particular' triangle), nor generic concepts of such individuals ('the' triangle). Rather, they are, mere conceptual skeletons without the covering of sensory material: emptied thing-forms, similar to the 'variable functions with empty places f(*)' which Hilbert introduces. They are interchangeable carriers of relations, mere supports which one can take to be of any arbitrary constitution, provided relations of the same formal-logical character can be 'fastened onto' them.

Mahnke thus holds with Husserl's view expressed in *Prolegomena* that philosopher's task is to examine the essence of mathematical theories in "another dimension". As an example of the phenomenologist's task Mahnke takes up the question of what are the things Hilbert speaks of. According to him, they are abstract objects in structures. To Mahnke, the phenomenological analysis shows that Hilbert's attempt to bring existence proofs "under inspection" is for this reason fundamentally misguided. Mahnke explains that "[t]he system of formal arithmetic is thus not really a definite scientific discipline, but only the logical form of a theory which is employed in all 'logically isomorphic' or 'formally equivalent' disciplines as their common deductive scaffolding, despite their intuitive

---

[42] Mahnke 1977, 78.

[43] *Op. cit.*, 79.

[44] *Loc. cit.*

incomparability."[45] Mahnke then goes on to point out that "this elucidation of Hilbert's 'things' makes it immediately clear that it is philosophically wrongly expressed when Hilbert calls the intuitive numerals the true objects of arithmetic."[46] In other words, Hilbert's formalist analysis of the objects of mathematics as constructed from sensuous signs is fundamentally misguided. However, for him, Hilbert is phenomenologically correct when he demands the givenness of 'extra-logical discrete objects' as the foundation of the arithmetically evident. According to Husserl's *Logical Investigations*, in Mahnke's view, only singular real individuals are sensorily perceived on the basis of which "higher acts of reason" are constructed:

> But on the basis of these sense perceptions higher acts of reason are constructed, such as ... colligation or counting; and it is precisely in these logical, but sensorily 'founded' experiences that aggregates and numerical totals are 'manifestly given' us, and that what is actually meant by the utterances 'and' and 'two' attains to a fully accomplished 'categorial intuition.'

Hence, Hilbert should not say that mathematics is about sensuous signs. Instead, numerals "'express' formal-logically 'isomorphically' and yet at the same time intuitively-clearly all the purely arithmetical relations, such as ordering and connection, among all arbitrary countable things."[47] Thus in phenomenology these signs have a meaning: "True objects of arithmetic are simply these logical relations."[48] In other words, according to Mahnke, Husserl in the *Logical Investigations* holds that the arithmetical theory, built upon categorial intuition of sensible objects, defines numbers, so to say, up to isomorphism. Numerals then refer to these pure numbers defined by the categorical theory. Furthermore he holds that Hilbert's problem is that he held the extra-logical sensory objects to be signs, which is a wrong philosophical analysis of the essence of numbers.

11.6 Formal and Transcendental Logic 1929

Let us now return to Husserl's view. In the *Formal and Transcendental Logic* (1929) Husserl claims to give "definitive clarification of the sense of pure formal mathematics ... , according to the prevailing intention of mathematicians: its sense, namely, as a pure analytics of non-contradiction, in which the concept of truth remains outside the theme".[49] Presumably 'the prevailing intention,' to at least some degree, means Hilbert's intention, possibly also that of Zermelo who also was at the time in Freiburg. Husserl was still in friendly terms with Hilbert, having visited

[45] *Op. cit.*, 80.
[46] *Op. cit.*, 80–81.
[47] *Op. cit.*, 81.
[48] *Op. cit.*, 82.
[49] *FTL*, 11.

Hilbert at least once in 1928.[50] It is rather likely that on that occasion he and Hilbert discussed the views that came out in print as the *Formal and Transcendental Logic* (1929).

The "other dimension" mentioned by Mahnke above obviously amounts to transcendental phenomenological clarification. Accordingly, elucidating the inner nature and the true sense of the axiomatic method and its objects epistemologically, and for example, what really are those 'things' of which Hilbert speaks of, are roughly among the tasks for what Husserl called transcendental logic. Its task is to ask transcendental questions about logic, the task that Kant, according to Husserl, failed to address. Such reflection examines logicians' intentions, their aimings and fulfillment – "the activity that is hidden ... throughout the naive doing and only now becomes a theme in its own right – we examine that activity after the fact".[51] Husserl's motivation is very similar to the way in which Hilbert characterizes his motivation to formulate his proof theory: "The fundamental idea of my proof theory is none other than to describe the activity of our understanding, to make a protocol of the rules according to which our thinking actually proceeds."[52] However, Husserl wants to examine the mathematicians' implicit, hidden presuppositions transcendentally, not proof-theoretically. According to Husserl, the transcendental logic helps to clarify the "internal shiftings of intentionality" that lead to equivocations.

Such examination has, for example, led Husserl to distinguish between three strata of logic and the corresponding three different evidences with their corresponding modes of empty expectation and fulfillment.[53] With such clarification, Husserl claims, the fundamental concepts of the sciences are clarified and criticized so that there should be no paradoxes.[54] The three levels of logic are grammar, formal mathematics (logic of non-contradiction) and applied mathematics (truth-logic). The first level defines the grammar of pure logic, what he also calls a theory of the pure forms of significations. It is a theory of how from certain fundamental forms a system of all conceivable judgment-forms emerges.[55] The second level is the logic of non-contradiction. Theories of this level are mathematical theories that are non-contrary. There is no need for them to be applicable to the world. The third level, the logic of truth, is the level that requires empirical applicability.[56]

Husserl exemplifies the different layers by means of Euclidean geometry that can be regarded either purely mathematically or else as something that aims at true

[50]During his Freiburg years Husserl considered himself as an old friend of Hilbert's household. Husserl visited Hilbert about which he reported to Heidegger that Hilbert's reception had been very friendly, „Sehr freundschaftlich kam uns Hilbert entgegen" (Husserl to Heidegger 9.5.1928, Husserl-Archive Mitteilungsblatt 26, 2013, 16).

[51]*FTL*, §69.

[52]Hilbert 1927, 475.

[53]*FTL*, §70a.

[54]*Op. cit.*, §71.

[55]*Op. cit.*, §13.

[56]For a nice exposition of Husserl's conception of logic in *FTL*, see Cavaillès 1970, 386–409.

judgments about the world (even though still certain idealizations are required). In the latter case, in the logic of truth, "one must draw fullness of adequation, not from evidence of the judgment-senses, but instead from evidence of the 'matters' or 'affairs' corresponding to them," Husserl writes.[57] In addition, much later in the text, Husserl also defines a supplement by means of which the complex judgments will be reduced to judgments about individuals. Such a supplement is located between the logic of non-contradiction and the logic of truth. I will next discuss in more detail Husserl's logic of non-contradiction and Husserl's reduction. My intention is to show that the former, the logic of non-contradiction consists of theories with a structure, whereas the syntactic reduction is analogous to Hilbert's proof-theory in that it shows how the former is accessible from our concrete meaningful experiences. I will not discuss Husserl's logic of truth in this connection. It would however offer another interesting connection between Husserl and Hilbert, for it shows how for Husserl mathematics is applied to the empirical world.

11.7 The Logic of Non-contradiction: Formal Mathematics

Like in Hilbert's axiomatics, in the mathematics of non-contradiction as Husserl conceives it, mathematical 'existence' derives from consistency. Husserl holds that the mathematician does not need to think that the multiplicities for example, exist concretely.[58] For him, for a "*'pure' formal mathematics, there can be no cognitional considerations other than those of 'non-contradiction'*, of immediate or mediate *analytic consequence or inconsistency*, which manifestly include all questions of mathematical *'existence'* ".[59]

The evidence related to the formal mathematics is called "distinctness" [Deutlichkeit]. The distinct judgements may or may not be true of the world. In the formal mathematics the fundamental question is not whether the judgment is true or not, but: "When, and in what relations, are any judgments – as judgments, and so far as mere form is concerned – possible within the unity of one judgment."[60] Later, as a result of a more detailed analysis, Husserl explains that the distinct judgment is a proper judgment in a sense that it has a unitary, non-contradictory sense (in addition to the purely grammatical sensefulness). The source of this unity is in the ideal "existence" of the judgment-content.[61] It is rooted not only in syntactical forms but also syntactical stuffs [Stoffe].[62] At this point Husserl remarks that this is "easily overlooked by the formal logician, with his interest directed one-sidedly to the

[57] *Op. cit.*, §82.
[58] *Op. cit.*, §51.
[59] *Op. cit.*, §52.
[60] *Op. cit.*, §18.
[61] *Op. cit.*, §89a.
[62] *Op. cit.*, §89a.

syntactical—the manifold forms of which are all that enters into logical theory—and with his algebraizing of the cores as theoretical irrelevancies, as empty somethings that need only be kept identical."[63]

Husserl thus wants to analyze the contentful mathematics as something that contains syntactical stuff in addition to pure form. The above quote appears to be directed against Hilbert, who in Husserl's view seems to be one-sided in his focus on the syntactical side of mathematics alone. In other words, for Husserl formal logic is not mere syntax, but the content of the judgments should be taken into account as well.

Husserl's discussion of syntactical forms and stuffs is notoriously difficult to understand. One way of trying to make sense out of them is to understand them and thus the unity of judgment-content proto-model theoretically: According to Husserl, all judging presupposes a harmonious unity of possible experience. In this harmony, everything has 'to do' materially with everything else. Husserl writes, "in respect of its content, every original judging and every judging that proceeds coherently, has coherence by virtue of the coherence of the matters in the synthetic unity of the experience, which is the basis on which the judging stands".[64] These matters in the synthetic unity of the experience are presumably the syntactical stuffs. The synthetic unity of experience is further something one experiences when provided with a structure, and the evidence related to this experience is then evidence of distinctness.

In Husserl's words, then a consistent theory does not merely have a grammatical unity, but "*coherence by virtue of the coherence of the matters in the synthetic unity of experience*," which Husserl also calls a universe of possible experience [ein Universum möglicher Erfahrung].[65] Its "cores are congruous in respect of sense—that is: all judgments that fulfill the conditions for unitary sensefulness."[66] Interestingly, Husserl adds that

> [w]e do not intend to say in advance that there can be only one universe of possible experience as the basis for judgment, and that therefore every intuitive judgment has the same basis and all judgments belong to a single materially coherent whole. To reach a decision about that would require a separate investigation.[67]

Husserl thus leaves open the possibility of several "universes of possible experiences."

The evidence obtained from the structure is evidence of distinctness. It has the same expectation and fulfillment structure as the other evidences.[68] Also the law of the excluded middle holds of the distinct judgments. This means that they can be brought to either adequation, in which case the judgment explicates

[63] *Op. cit.*, §89b.

[64] *Loc. cit.*

[65] *FTL*, §89 a–b.

[66] *Op. cit.*, §90.

[67] *Op. cit.*, §89b.

[68] *Op. cit.*, §70a.

and apprehends categorially what is given in harmonious experience, or to the negative of adequation, in which case what it predicates conflicts with something experienced.

These universes of possible experiences, or harmonious unities of possible experience (or a unitary material province, ein einheitliches sachliches Gebiet, as he also calls it in §92) function like models in the contemporary model theoretic semantics. Husserl's harmonious unity of possible experience thus realizes or satisfies the non-contradictory theory. The experience of distinctness derives from the existence of a model in which the judgment can be brought to adequation, i.e., "satisfied" by means of categorial intuition.[69]

Husserl's conception of formal mathematics is thus in agreement with Hilbert's view in so far as the latter viewed mathematics as a study of structures. Husserl's view seems also close to Zermelo's view. Zermelo held that the models, what he called 'substrates,' are presupposed in mathematics.[70] But while Hilbert wanted to analyze the inferences syntactically with his proof theory, Husserl held that the formal content of mathematics cannot be analyzed away and that the models are presupposed in mathematics. In the *Formal and Transcendental Logic* (§33) Husserl also discusses a possibility of building a

> discipline comprising deductive games with symbols, which does not become an actual theory of multiplicities itself, one builds only a discipline comprising deductive games with symbols, which does not become an actual theory of multiplicities until one regards the game-symbols as signs for actual Objects of thinking—units, sets, multiplicities—and bestows on the rules of the game the significance of law-forms applying to these multiplicities.

This is easy to read as a critique of Hilbert. Husserl, like Mahnke before, holds that the numerals should not be regarded as objects of mathematics. Husserl explicitly points out that

> we must not define merely in terms of signs and calculational operations—for example: 'It shall be allowed to manipulate the given signs in such a manner that the sign b + a can always be substituted for a +b'. Rather we must say: 'There shall obtain among the *objects* belonging to the multiplicity (conceived at first as only empty Somethings, 'Objects of thinking') a certain *combination-form* with the law-form a+b=b+a'—where equality has precisely the sense of actual *equality* such as belongs to the categorial logical forms.[71]

Nevertheless, using symbols as a technique facilitating the calculations is not a problem for Husserl. But whereas for Hilbert such technique expresses our thinking

[69]'Categorial intuition' is Husserl's term for perception of formal structures, typically states of affairs. From Husserl's remarks it is difficult to say what everything could be an object of categorial intuition. Dietrich Mahnke and Oskar Becker discussed this matter and disagreed about it: Becker held that categorial intuition is restricted to human consciousness and that we cannot intuit transfinite elements. Mahnke thought that this is not the case, and that the consciousness in question is an ideal consciousness (Mancosu & Ryckman 2010, 350–355). Given Husserl's overall non-revisionist attitude Mahnke's view about the matter seems to be closer to Husserl's intentions.

[70]Ebbinghaus 2007, 156.

[71]*FTL*, §34.

irreducibly that is not the case for Husserl. Thus when Hilbert explains that "[t]his formula game is carried out according to certain definite rules, in which the technique of our thinking is expressed," Husserl would agree. But when Hilbert continues that "[t]he fundamental idea of my proof theory is none other than to describe the activity of our understanding, to make a protocol of the rules according to which our thinking actually proceeds,"[72] Husserl would point out that when so doing, Hilbert ignores the formal content of mathematics and the evidence of *Deutlichkeit* related to it. Transcendental examination however makes it explicit that such formal content cannot be ignored. Without the transcendental clarification of the evidences, there will be equivocations "which we cannot remove while confining out attention to the language itself and simply examining it with respect to the significations to which it points associationally."[73] Husserl thus seems to suggest that Hilbert's way of founding mathematics with concretely intuitable signs will not solve the equivocations and paradoxes that have their sources in the contents of the judgments, not only in their syntactic formulation.

11.8 Syntactic Reduction

Despite of the above criticism Husserl seems to agree with Hilbert that something like Hilbert's proof-theoretical reduction is needed. Indeed, in the *Formal and Transcendental Logic*, too, something similar can be found. Husserl establishes a "reduction" of formal mathematics to logic of truth and then to judgments about individuals. According to Husserl, "*any actual or possible judgment leads back to ultimate cores* when we follow up its syntaxes; accordingly that it is a syntactical structure built ultimately, though perhaps far from immediately, out of *elementary cores, which no longer contain any syntaxes*".[74] By means of this kind of reduction the ultimate substrates, i.e., absolute subjects, ultimate predicates, ultimate universalities, and ultimate relations, are reached. According to Husserl, these are of no interest for general mathematics, but if we are interested in truth, then this is important: "because ultimate substrate-objects are *individuals*, about which very much can be said in formal truth, and *back to which all truth ultimately relates*".[75] Similar reduction takes place among truths, so that every true judgment ultimately relates to individual objects. These individual objects are then objects of our experiences about which we can predicate something as in "The man is pale", "The paper is white", etc. The constitution of such objectivities and the further structures related to such experiences is then the topic for phenomenological

---

[72]Hilbert, 1927, 475.

[73]*FTL*, §70.

[74]*FTL*, §82.

[75]*FTL*, §82.

analyses. That is what Husserl in the above quote refers to as formal truth as he gives the "time-form" as an example.

Husserl thus agrees with Hilbert that mathematics presupposes intuition of extra-mathematical objects, and hence a reduction to primitive judgments about individual objects, although not numerals, is needed. This takes place by means of a *transitional link* between mathematics of non-contradiction, i.e., formal mathematics, and truth-logic. It is an analytic reduction of mathematics to ultimate substrate-objects that are individuals.

What is curious is that Husserl claims that for formal mathematics such a reduction is of no particular interest.[76] For him mathematics as such is not in need of a direct proof-theoretical consistency proof, but that the above described model theoretical approach is quite sufficient for the formal mathematicians. However, he claims that the reduction to the ultimate cores is important for the question of truth. In the logic of truth the judgments are viewed as striving for truth, anticipating the possible fulfillment by means of intuition. The focus of judging is in the states of affairs about which something is judged.[77] The logic of truth is thus about the world, and thus the reduction to judgments about individuals show how mathematics is related to our judgments about the world. The motivation for it is primarily philosophical: to understand the nature and essence of mathematical theories so that it is all as if "available to inspection", as for Hilbert. Such reduction thus starts to resemble in its motivation very much what Wilfried Sieg has termed Hilbert's 'structural reduction,' in which the abstract structures are projected into the finitist domain, "ever more encompassing part of mathematics to a fixed, elementary, and meaningful part of itself".[78] Such a reduction thus shows how the abstract mathematical structures are accessed from a concrete meaningful experience.

11.9 Conclusion

Husserl shares Hilbert's approach to mathematics as "mathematics first," approach to formal mathematics that deals with axiomatically characterized abstract structures. Thus in accordance to the division of labor between mathematicians and philosophers, Husserl left the task of constructing the theories to Hilbert and his colleagues, while Husserl saw as his task to examine their essence and their presuppositions. Husserl developed the view further in his *Formal and Transcendental Logic,* which appropriates many of Hilbert's insights but is also critical about his view. In particular, Hilbert should not conceive his reduction purely syntactically and base it on sensuous signs. Rather mathematics is about structural objects and thus the reduction should take into account the "formal content" of the judgments

---

[76] *FTL*, §82.

[77] *FTL*, §19.

[78] Sieg 2013, 17.

as well. Husserl seems to think that for mathematics model theoretical consistency proofs are entirely adequate. Yet Husserl offers us an analogous syntactic-semantic reduction to objects as a transitional link between formal mathematics and truth-logic. This however is not Husserl's view of how the direct consistency proof should be established, but rather a philosophically motivated analysis of how mathematical theories relate to our experiences.

References

Husserl Archive, Mitteilungsblatt 34. Leuven 2011. http://hiw.kuleuven.be/hua/mitteilungsblatt/mitteilungsblatt34.pdf. Accessed 19 Dec 2013

Husserl Archive, Mitteilungsblatt 36, Leuven 2013. http://hiw.kuleuven.be/hua/mitteilungsblatt/mitteilungsblatt36.pdf. Accessed 19 Dec 2013

J. Cavaillès, On logic and the theory of science, in *Phenomenology and the Natural Sciences*, ed. by J.J. Kockelmans, T.J. Kisiel (Northwestern University Press, Evanston, 1970), pp. 353–409

S. Centrone, *Logic and Philosophy of Mathematics in the Early Husserl* Springer, Dordrecht, 2010

J. Da Silva, Husserl's two notions of completeness. Synthese **125**, 417–438 (2000)

J. Da Silva, Husserl and Hilbert on completenes, still. Synthese (2015). doi:10.1007/s11229-015-0821-2

H.D. Ebbinghaus in cooperation with V. Peckhaus, *Ernst Zermelo, An Approach to His Life and Work* Springer, Berlin/Heidelberg/New York, 2007

W. Ewald, *From Kant to Hilbert: A Source Book in the Foundations of Mathematics*, vol II (Clarendon Press, Oxford, 1996)

J. Ferreirós. *Labyrinth of Thought, A History of Set Theory and Its Role in Modern Mathematics* (Birkhäuser Verlag AG, Basel/Boston/Berlin, 2007)

M. Hartimo, Towards completeness: Husserl on theories of manifolds 1890–1901. *Synthese* **156**, 281–310 (2007)

M. Hartimo, From geometry to phenomenology. *Synthese* **162**, 225–233 (2008)

M. Hartimo, M. Okada, Syntactic reduction in Husserl's early phenomenology of arithmetic. Synthese (2015). doi:10.1007/s11229-015-0779-0

M. Hartimo, Husserl on completeness, definitely. Synthese (2016). doi:10.1007/s11229-016-1278-7

E. Hellinger, Logische Prinzipien des Mathematischen Denkens. Vorlesungen von Professor Dr. Hilbert im Sommer-Semester 1905, unpublished lecture notes (1905)

D. Hilbert, 1900a, On the concept of number. In: Ewald (1996), pp. 1092–1095

D. Hilbert, 1900b, From mathematical problems. In: Ewald (1996), pp. 1096–1105

D. Hilbert, 1927, The foundations of mathematics. In: van Heijenoort (1967), pp. 464–479

D. Hilbert, *Foundations of Geometry*. Translated by Leo Unger. Open Court, Illinois [1971], 1990. (Open Court, La Salle, 1950)

D. Hilbert, Über die Grundlagen der Geometrie, Mathematische Annalen **56**, 381–422 (1902). English translation in Hilbert (1990), 150–190

C.O. Hill, J.J. da Silva, *The Road Not Taken, On Husserl's Philosophy of Logic and Mathematics* (College Publications, London, 2013)

E. Husserl, *Ideen zu einer reinen Phänomenologie und phäneomenologischen Philosophie. Erstes Buch. Allgemeine Einführung in die reine Phänomenologie.* Herausgegeben von Walter Biemel. Husserliana Band III. (Martinus Nijhoff, Haag, 1950). English translation: *Ideas Pertaining to a Pure Phenomenology and to a Phenomenological Philosophy. First book. General Introduction to a Pure Phenomenology* (Martinus Nijhoff, The Hague/Boston/Lancaster, 1983)

E. Husserl, *Philosophie der Arithmetik.* Husserliana Band 12, ed. by L. Eley (Martinus Nijhoff, Den Haag, 1970)

E. Husserl, *Formale and transzendentale Logik. Versuch einer Kritik der logischen Vernunft.* Husserliana Band 17. ed. by Paul Janssen (Martinus Nijhoff, The Hague, 1974). English translation: *Formal and Transcendental Logic,* transl. by Dorion Cairns (Martinus Nijhoff, The Hague, 1969)

E. Husserl, *Logische Untersuchungen. Erster Teil. Prolegomena zur reinen Logik.* Text der 1. und der 2. Auflage. Halle 1900, rev. ed. 1913. Husserliana Band 8. ed. by Elmar Holenstein (Martinus Nijhoff, The Hague, 1975). English translation: *Logical Investigations. Prolegomena to pure logic*, transl. by J.N. Findlay (Routledge, London/New York, [1970], 2001), pp. 1–161

E. Husserl, *Aufsätze und Rezensionen (1890–1910).* Husserliana Band 22. ed. by Bernhard Rang (Martinus Nijhoff, The Hage/Boston/London, 1979)

E. Husserl, *Briefwechsel Band VII. Wissenschaftlerkorrespondenz,* Kluwer, Dordrecht/Boston/London, 1994

D. Lohmar, *Phänomenologie der Mathematik* Kluwer, Dordrecht/Boston/London, 1989

D. Mahnke, From Hilbert to Husserl: First Introduction to Phenomenology, especially that of Formal Mathematics (1923). *Studies in the History and Philosophy of Science* **8**, 71–84 (1977)

P. Mancosu, *The Adventure of Reason. Interplay between Philosophy of Mathematics and Mathematical Logic 1900–1940* Oxford University Press, Oxford, 2010

M. Okada, Husserl and Hilbert on Completeness and Husserl's Term Rewrite-based Theory of Multiplicity, in *24th international conference on rewriting techniques and applications (RTA'13)*, ed. by F. van Raamsdok (LIPIcs, Eindhoven, 2013), pp. 4–19

V. Peckhaus, *Hilbertprogramm und Kritische Philosophie. Das Göttinger Modell interdisziplinärer Zusammenarbeit zwischen Mathematik und Philosophie* (Vandenhoeck & Ruprecht, Göttingen, 1990)

D.E. Rowe, 'Klein, Hilbert, and the Göttingen Mathematical Tradition', Osiris, 2nd Series, vol V. Science in Germany: The Intersection of Institutional and Intellectual Issues, 186–213 (1989)

W. Sieg, Hilbert's Programs: 1917–1922. *Bulletin of Symbolic Logic* **5**, 1–44 (1999)

W. Sieg, Beyond Hilbert's Reach? in *Logicism, Intuitionism, and Formalism*, ed. by S. Lindström et al. (Eds), (Springer, 2009), pp. 449–483

W. Sieg, *Hilbert's Programs and Beyond* Oxford University Press, Oxford, 2013

A. Steve, E.H. Reck, *Completeness and Categoricity: 19th Century Axiomatics to 21st Century Semantics*. Technical Report no. CMU-PHIL-118, Carnegie Mellon, Pittsburgh, Pennsylvania (2001). http://www.hss.cmu.edu/philosophy/techreports/118_Awodey.pdf. Accessed 9 Jan 2014

J. van Heijenoort, *From Frege to Gödel. A Source Book in Mathematical Logic, 1979–1931* Harvard University Press, Cambridge, MA/London, 1967

Chapter 12
Construction and Constitution in Mathematics

Mark van Atten

Abstract I argue that Brouwer's notion of the construction of purely mathematical objects and Husserl's notion of their constitution by the transcendental subject coincide. Various objections to Brouwer's intuitionism that have been raised in recent phenomenological literature (by Hill, Rosado Haddock, and Tieszen) are addressed. Then I present objections to Gödel's project of founding classical mathematics on transcendental phenomenology. The problem for that project lies not so much in Husserl's insistence on the spontaneous character of the constitution of mathematical objects, or in his refusal to allow an appeal to higher minds, as in the combination of these two attitudes.

Keywords Phenomenology • Intuitionism • Brouwer • Husserl • Gödel • Construction • Constitution • Mathematics

12.1 Introduction

In the following, I argue that L.E.J. Brouwer's notion of the construction of purely mathematical objects and Edmund Husserl's notion of their constitution coincide. That conclusion will be the combined result of a defence of the following two claims:

Originally published as Mark van Atten. 2010. "Construction and constitution in mathematics". The New Yearbook for Phenomenology and Phenomenological Philosophy 10:43–90. Corrections and the occasional additional note in this reprint (marked as such) are those of the reprint in van Atten (2015a).

M. van Atten (✉)
SND (CNRS/Paris IV), 1 rue Victor Cousin, 75005, Paris, France
e-mail: vanattenmark@gmail.com

S. Centrone (ed.), *Essays on Husserl's Logic and Philosophy of Mathematics*, Synthese Library 384, DOI 10.1007/978-94-024-1132-4_12

1. From a systematical point of view, Brouwer's intuitionistic mathematics should be considered part of Husserl's transcendental-phenomenological foundations of pure mathematics;
2. Transcendental phenomenology cannot provide a foundation for a pure mathematics that would go beyond intuitionism.

Thus, according to the first claim, Husserl's notion of constitution subsumes Brouwer's notion of construction, but it is, according to the second claim, not a wider one. Already Oskar Becker, in his *Mathematical Existence*, readily assimilated Brouwer's mathematical constructions to Husserl's purely categorial formations; he comfortably writes:

> To the intuitionist, this cognition [that arithmetic cannot be obtained from formal logic] is no surprise, but just a consequence of his view that pure mathematics in general is based on a number of acts and categorial objects that have to be recognised intuitively, such as collecting, ordering, assigning, exchanging, etc., which even allow for an endless expansion.[1]

But he made no attempt at an explicit justification of that assimilation. The second claim, also made by Becker, is worth confronting with (later) efforts, notably by Gödel, to provide a phenomenological foundation of classical mathematics.

Obviously, a defence of these claims requires finding answers to the following questions:

1. Do Husserl and Brouwer deal with consciousness at the same level? Husserl's later phenomenology deals with consciousness at the level of a transcendental subjectivity, but Brouwer's intuitionism is generally seen as a form of (mundane) psychologism.
2. Can any part of phenomenology be mathematics? One might think that phenomenology, which exercises epoché and is purely descriptive of consciousness, has nothing to say on the ontology of mathematics at all.
3. Can Brouwer's notion of the construction of mathematical objects indeed be understood as a form of Husserl's notion of their constitution?

The following discussion will propose answers to these questions, and also address various objections to mathematical intuitionism that have been raised in the recent phenomenological literature, in particular by Claire Ortiz Hill (2010), Guillermo Rosado Haddock (2006, 2010), and Richard Tieszen (1989, 2010), as well as suggestions by Kurt Gödel.

With respect to the physical world, Brouwer was, unlike Husserl, a phenomenalist. In the following, I shall be concerned only with his thoughts on mathematics.

[1] 'Für den Intuitionisten ist diese Erkenntnis [daß die Arithmetik nicht aus der formalen Logik herzuleiten ist] nicht verwunderlich, sondern nur eine Folge seiner Auffassung, daß die reine Mathematik überhaupt auf eine Reihe von intuitiv zu erkennenden Akten und kategorialen Gegenständlichkeiten beruht, wie Kolligieren, Ordnen, Zuordnen, Vertauschen usw., die sogar eine Ausdehnung ins Endlose gestatten.' (Becker 1927, 23; see also 69 and 196)

12.2 Intuitionistic Mathematics is Part of Transcendental Phenomenology

12.2.1 Husserl: Pure Mathematics as Formal Ontology

Husserl's introduction of purely categorial objects in the *Logical Investigations* opened the way to a conception of pure mathematics that he never abandoned. Pure mathematics, as a science that aims at knowledge and true judgements, is the theory of the purely categorial formations,[2] which, in turn, he came to identify with formal ontology[3]: 'The whole mathesis universalis is therefore the analytics of possible categorialia, the theory of their essential forms and eidetic laws'[4] And:

> The formal ontology [can] also directly be set as a task from the outset, without taking the idea of a theory of science as the starting point. Its question is then: what can one assert within the empty region 'any object whatever'? Purely a priori, in this formal generality are available the syntactical formations, by means of which from any objects of thought pre-given as possible (anything whatever), always new categorialia can be thought of as

---

[2]In Sect. 31 of *Formal and Transcendental Logic*, Husserl describes the ideal of the definite manifold, and attempts to characterise it precisely by saying that such a manifold indeed is described by a syntactically complete axiom system. The actual claim that pure mathematics is a definite manifold is implied in a text from 1920–1921: 'The ideal total extension of the purely rational objects and the extension of the objective objects [*objektive Gegenstände*] cognizable to every possible rational I, coincide. It turns out that this extension is a definite manifold, and infinitely constructible system' (Husserl 1974, 388). But Gödel's first Incompleteness Theorem, established shortly after *Formal and Transcendental Logic* was published, shows that in any consistent theory containing arithmetic there are undecidable sentences. A fortiori, the same holds for set theory and the whole of pure mathematics. See also Lohmar (1989, 197n27). Clearly, for Gödel, this fact was not an essential obstacle to embracing Husserl's transcendental-phenomenological approach to the foundations of mathematics, and, more generally, in Gödel's foundational project there hardly seems to have been a role for Husserl's theory of formal systems; given Gödel's realist conception of mathematical objects, this is not surprising. Note that Dietrich Mahnke had raised the question whether pure mathematics is one definite manifold in his *Neue Monadologie* of 1917, and conjectured the answer is yes, while realising that an argument is needed (Husserl marked this passage in his own copy): 'That not all, indeed even rather few, manifolds in the actual world have this property of being definite, is obvious. But also in formal mathematics it is still a big question whether its totality is a heap of infinitely many different and unrelated theories of manifolds, or rather can be organised into one big, definite system. The concept of mathematics seems to demand that the latter is the case. Yet a proof is still to be found' (Mahnke 1917, 32). 'Dass nicht alle, ja sogar nur herzlich wenige Mannigfaltigkeiten der wirklichen Welt diese Eigenschaft der Definitheit haben, liegt auf der Hand. Aber auch in der formalen Mathematik ist noch eine grosse Frage, ob ihre Gesamtheit ein beziehungsloses Nebeneinander von unendlich vielen verschiedenen Mannigfaltigkeitslehren ist oder vielmehr selbst in ein einziges, grosses definites System geordnet werden kann. Der Begriff der Mathematik scheint zu erfordern, dass das letztere der Fall ist. Doch steht der Nachweis dafür noch aus.'

[3]See also Husserl (1974, 82), Husserl (1975, 245, 247), Husserl (1976a, 26–27), Husserl (1979, 166), a text from 1903, Husserl (1985a, 52, 55, 61, 78, 167), and Husserl (2002b, 266).

[4]'Die gesamte mathesis universalis ist also Analytik möglicher Kategorialien, Theorie ihrer Wesensformen und Wesensgesetze.' (Husserl 1976a, 143)

> produced. One will thereby also come to distinguish possible productions that yield merely distinct meanings, but, being contradictory, can never lead to possible objects themselves, etc. *Obviously, the whole formal mathesis then arises.* (Husserl 1973d, 148, emphasis mine)[5]

The insight that the pure categorialia are available in formal ontology, even though these are not limited to 'any object whatever' (*Etwas-überhaupt*) but are for the most part internally structured objects composed from it, marked an important step for Husserl. In 1910 he wrote:

> A cross to bear for me is the relation between mathematics and ontology as formal ontology. But once I have placed the idea of the 'material' (*des 'sachhaltigen'*), of 'reality' in the sphere of meaning, I obtain the idea of the non-material (*des Nicht-Sachhaltigen*), and that is the ontological in the formal sense.[6]

What changed over the years, however, is that Husserl came to see the difference between the logic of non-contradiction[7] and the logic of truth, and that it is the

---

[5] 'Die formale Ontologie [kann] auch von vornherein direkt als Aufgabe gestellt werden, ohne von der Idee einer Wissenschaftslehre auszugehen. Ihre Frage ist dann: was kann man innerhalb der Leerregion Gegenstand-überhaupt aussagen? Rein apriori in dieser formalen Allgemeinheit stehen die syntaktischen Gestaltungen zur Verfügung, durch die aus irgendwelchen als möglich vorgegeben gedachten Gegenständen (Etwas-überhaupt) immer neue Kategorialien erzeugt gedacht werden können. Man wird dabei auch auf den Unterschied möglicher Erzeugungen kommen, die bloß deutliche Meinungen liefern, aber als widerspruchsvolle nicht zu möglichen Gegenständen selbst führen können usw. *Offenbar erwächst dann die ganze formale Mathesis.*' (Husserl 1974, 153–154, emphasis mine)

[6] 'Ein Kreuz ist für mich das Verhältnis von Mathematik und Ontologie als formaler Ontologie. Nachdem ich aber die Idee des 'sachhaltigen', der 'Realität' in der Bedeutungssphäre untergebracht habe, so gewinne ich ja die Idee des Nicht-Sachhaltigen, und das ist das Ont<ologische> im formalen Sinn' (Husserl 1995a, 343). Note the contrast with Alexander Pfänder, who, working in the period 1928–1935, saw mathematics not as part of formal ontology, but as a material region (Pfänder 1973, 42). In his lectures on 'first philosophy', Husserl speaks of both the 'formal region' and the 'formal-ontological quasi-region'; (Husserl 1956a, 187n1).

[7] Husserl's description of the mathematics of non-contradiction as 'the mathematics of the mathematicians' (*die Mathematik der Mathematiker*, Husserl 1974, 146) is problematic. On the one hand, it is true that only a mathematician will engage in the study of formal systems for their own sake; on the other hand, after criticism by Henri Poincaré and Brouwer, David Hilbert acknowledged in 1922 that study requires accepting at least part of pure mathematics as contentual (Hilbert 1922, 165, 174). (Husserl knew this paper; see Lohmar 1989, 216n10.) Roughly, that part is arithmetic including the principle of induction; for elaborate discussion see, e.g., Kreisel (1965), Tait (1981), and Parsons (1998). In other words, pure mathematics cannot be limited to the logic of consequence. James Dodd (in an article that I in many respects agree with) speaks of 'Brouwer's polemical accusation, invited one must say by Hilbert himself, that finitism amounts to a mere empty game with symbols devoid of all sense, and which in the end has no genuine connection to mathematical objectivity at all' (Dodd 2007, 268). (The reference given in Dodd's footnote to this sentence is not, as one would have expected, to a passage where Brouwer makes the alleged accusation but to Husserl 1974, Sects. 33–34.) But Brouwer never did, and never would, equate finitism with empty symbolism devoid of all sense: finitism's characteristic acceptance of the mathematics of natural numbers as contentual is precisely due to Brouwer's influence on Hilbert. (For a discussion of this influence, with full references, see Brouwer (1928A2). Husserl owned a copy of the latter; see van Atten 2007, 128n7).

latter, not the former, that governs formal ontology; being entirely clear about this distinction is one of the ways in which *Formal and Transcendental Logic* makes significant progress over the *Prolegomena* (Husserl 1974, 15, 76). The mathematics of non-contradiction is not yet formal ontology (Husserl 1974, 150), and not yet formal mathesis in the sense relevant to an interpretation of Brouwer, who recognised the existence of the mathematics of non-contradiction, but was himself exclusively concerned with mathematical truth.[8]

When seen as a positive science, formal ontology is exercised in the natural attitude and, as such, no part of transcendental phenomenology. Yet Husserl came to say that, from the transcendental perspective, it is a part, together with all other a priori ontologies. The rationale of this claim is, briefly, as follows. For the transcendental Husserl, all being, including mathematical being, is constituted being; moreover, every type of existing objects is, as a matter of essence, correlated to the type of act in which they are constituted by the transcendental subject. This is an essential correlation (*Wesensbeziehung*; Husserl 1976a, 159) between transcendental being and transcendent being. It allowed Husserl to hold that transcendental phenomenology is not limited to an eidetic, descriptive study of consciousness as such, but in a sense, namely, through that correlation, comprises all a priori ontologies: 'any clarifying ontological insight obtained within axiomatic clarity, which is not directly phenomenological, becomes so by a mere *change of attitude* [*Blickwendung*], just as, conversely, among all phenomenological insights there must occur ones that by a mere change of attitude become ontological.'[9]

[8]In 1908 Brouwer both showed the consistency of the principle of the excluded middle (as its double negation is true, the principle itself cannot be false) and justified his doubts about its truth. 'Consequently the theorems which are usually considered as proved in mathematics, ought to be divided into those that are true and those that are non-contradictory' (Brouwer 1975, 110n2) ('Men behoort dus in de wiskunde de gewoonlijk als bewezen geldende stellingen te onderscheiden in juiste en niet-contradictoire', Brouwer 1908C, 158n2). Brouwer characterises the difference between a true proposition and a merely non-contradictory one by the presence, in the former case, of a mathematical construction that proposition adequately describes. This corresponds to Husserl's characterisation of truth-logic as the one that, unlike consequence-logic is concerned with the existence of the objects. Note that, before Gödel proved his Incompleteness Theorem, Brouwer was quite optimistic about Hilbert's program to establish the consistency of classical mathematics; in his first Vienna lecture of 1928, he said that 'An appropriate mechanisation of the language of this intuitionistically non-contradictory mathematics should therefore deliver exactly what the formalist school has set as its goal' (Brouwer 1929A, 164). ('Eine geeignete Mechanisierung der Sprache dieser intuitionistisch-nichtkontradiktorischen Mathematik müßte also gerade das liefern, was die formalistische Schule sich zum Ziel gesetzt hat.') Of course, he at the same time insisted that would have no value for mathematics as such, given its concern, in his view, with constructions.

[9]'alle klärende und im Rahmen der axiomatischen Klarheit vollzogene ontologische Einsicht, die nicht direkt phänomenologisch ist, wird dazu durch eine bloße *Blickwendung*, wie umgekehrt im All der phänomenologischen Einsichten solche auftreten müssen, die durch bloße Blickwendung zu ontologischen werden.' (Husserl 1952, 105) Husserl also comments on the phenomenological clarification of geometry by tracing it back to its constitution in nexuses of consciousness: 'That is an application of phenomenology, not phenomenology itself. The predicate 'phenomenological' carries over, of course, to the applications' 'Das ist Anwendung der Phänomenologie, nicht Phänomenologie selbst. Das Prädikat "phänomenologisch" überträgt sich natürlich auf die An-

In the 1920s he makes the point by saying that formal logic as formal ontology and formal apophantics is 'a stratum necessarily belonging' to transcendental phenomenology (*eine ihr notwendig zugehörige Schichte*, Husserl 1974, 277). More generally, all possible a priori disciplines are branches (*wesensnotwendige Verzweigungen*, Husserl 1962, 298) of transcendental phenomenology,[10] which is 'as opposed to the only seemingly universal ontology in positivity, the truly universal'.[11] He is explicit about the aptness of transcendental phenomenology to provide the foundations of mathematics: 'For the a priori disciplines that are grounded within phenomenology (for example as mathematical sciences) [there can] be no "paradoxes", no "foundational crises".'[12]

Hill, on the other hand, has recently claimed[13] that

> Transcendental phenomenology has no dealings with a priori ontology, none with formal logic and formal mathematics ... Transcendental phenomenology is phenomenology of the constituting consciousness, and consequently not a single objective axiom, meaning one relating to objects that are not consciousness, belongs in it, no a priori proposition as truth for objects, as something belonging in the objective science of these objects, or of objects in general in formal universality. (Hill 2010, 62)

In light of the passages from Husserl that we have just seen, as a claim about Husserl's conception of transcendental phenomenology this cannot be right. Admittedly, Husserl does in the same period also say that the universal epoché is exercised 'with respect to all objectivity that is valid for me'[14] which certainly includes mathematics. But there is no contradiction: the point of the passages above, when applied to mathematics, is that the mathematical objectivity known in the natural attitude can be reconstituted, to the extent that this can be done with full evidence, from within the transcendental attitude. An a priori ontology delineates the pure possibilities for objects of a certain type, but in general does not by itself suffice to bring an individual object of that type to intuitive givenness, as in general

---

wendungen', Husserl (1952, 83). As we will see in the main text, later Husserl expressed an even stronger view.

[10]See also Husserl (1988, 18–19), and Husserl (2002c, 300–301).

[11]'gegenüber der nur scheinbar universalen Ontologie in der Positivität die wahrhaft universale' (Husserl 1962, 297).

[12]'Für die apriorischen Disziplinen, die innerhalb der Phänomenologie zur Begründung kommen (z.B. als mathematische Wissenschaften) [kann es] keine "Paradoxien", keine "Grundlagenkrisen" geben.' (Husserl 1962, 297)

[13]Noted added in this reprint: Even though it does not affect the content of my argument, here I committed a howler. All the lines I quote from Hill's article here are actually translations of sentences written by Husserl. Hill presents these without quotation marks or indentation, but at the end of her paragraph there is a reference to the corresponding pages in Husserliana XXIV (411, 422, 423; and 425 must be added). I thank Dr. Hill for bringing this to my attention, and apologise for this mistaken attribution to her. The reason my argument is not affected is that the sentences Hill quotes come from a group of texts of 1908, and what I document here is precisely how, as Husserl developed his transcendental phenomenology over the following years, he came to oppose the view that a priori ontology is not part of transcendental phenomenology.

[14]'hinsichtlich aller mir geltenden Objektivität' (Husserl 1959, 445).

that depends on the availability of specific material content (*Sachhaltiges*). Purely formal objects are the exception. Their constitution, although ultimately founded on material content, does not depend on any specific material content or type thereof. But then there are no further conditions on the constitution of any particular formal object than those that exist a priori.[15] As Husserl had already written in the sixth *Logical Investigation*, 'The ideal conditions of the possibility of categorial intuition in general are, correlatively, the ideal conditions of the possibility of the objects of categorial intuition and of the possibility of categorial objects as such'.[16] For purely categorial objects, possibility and being coincide. As a consequence, in this unique case transcendental phenomenology provides the ontology in its whole extension.[17] One can turn immediately from concrete formal-mathematical insights (which are part of positive science) to purely phenomenological ones and vice versa by an appropriate shift of one's regard. As Husserl put it in a text that he dates at the end of the war or perhaps St. Märgen 1921:

> Over all sciences stands a mathesis universalissima, and not as a naive mathematics which, going much farther still than Leibniz' mathesis universalis, constructs the formal-ontological Apriori in a systematical order and develops it in theories, but as a mathematics of cognitive performances, the noetic study of which, carried out in pure subjectivity, comprises the mathematical as a noematic formation of reason and hence as a correlate of consciousness.[18]

That conclusion does not contradict the fact, elaborated on by Husserl in Sects. 72–75 of *Ideas I*, that transcendental phenomenology and mathematics are two different kinds of science, the former being descriptive, the latter deductive. Transcendental phenomenology can describe the constituting performances of the subject's consciousness when it is engaged in doing deductive science. In fact, according to Husserl it is only by giving such descriptions that the axioms and the mediate

---

15 In a text from 1923, Husserl writes of the formal Apriori and the purely formal objects (Husserl 1959, 225n): 'But here the matter is quite different from that of a geometrical, nature-ontological etc. Apriori; here, on the ontological side the specifications and correlations that have an influence on the sense are lacking, only the general constitutive relation remains' ('Aber hier ist die Sache doch anders als bei einem geometrischen, naturontologischen usw. Apriori; hier fehlen auf ontologischer Seite die sinn-mit-bestimmenden Besonderungen und Korrelationen, es bleibt nur die *allgemeine* konstitutive Korrelation').

16 'Die idealen Bedingungen der Möglichkeit kategorialer Anschauung überhaupt sind korrelativ die Bedingungen der Möglichkeit der Gegenstände kategorialer Anschauung und der Möglichkeit von kategorialen Gegenständen schlechthin.' (Husserl 1984b, 718–719)

17 This point I have argued for in greater detail in van Atten (2002).

18 'Über allen Wissenschaften steht eine Mathesis universalissima, und nicht als eine naive Mathematik, die noch weit über die Leibniz'sche Mathesis universalis hinaus das formalontologische Apriori systematisch geordnet konstruiert und in Theorien entfaltet, sondern als eine Mathematik von Erkenntnisleistungen, deren noetisches und in der reinen Subjektivität vollzogenes Studium das Mathematische als noematisches Gebilde der Vernunft und somit als Bewußtseinskorrelat begreift.' (Husserl 1959, 249)

results of such a science can be philosophically grounded (Husserl 1952, 83; Husserl 2002c, 301). The difference in kind is precisely a condition of possibility of such a grounding.[19]

12.2.2 Brouwer: Mathematics as Mental Constructions

Brouwer's intuitionism aims to provide a philosophical foundation for pure mathematics by seeing it as 'an autonomous interior constructional activity' (Brouwer 1981A, 92).[20] In this activity, the subject builds up a stock of mathematical objects. The material out of which these mental constructions are made is provided by the intuition of time, which Brouwer calls 'the basic intuition'. (As I argue below, this sounds more Kantian than it is.) The corresponding notion of existence of a mathematical object then should be defined in terms of mental construction: to say that a mathematical object exists is to say that, in principle, it can be constructed in the mind out of the basic intuition. We will discuss two of Brouwer's own, more detailed characterisations of intuitionism below.

It turns out that various parts of classical mathematics are not constructible according to the principles of intuitionism, and to that extent are, from the intuitionistic point of view, not mathematics proper. For example, intuitionism rejects the universal validity of the principle of the excluded middle, together with most of Cantorian set theory. Conversely, Brouwer introduces objects and principles of reasoning about them that are not acceptable in classical mathematics, such as choice sequences (*Wahlfolgen*), e.g. Brouwer (1918B, 3). It should be emphasised that this revisionism is an outcome, rather than a preset goal, of the intuitionistic reconstruction of mathematics. In principle, it is indifferent to the intuitionist whether the theorems of mathematics as founded on his notion of construction turn out to coincide with those of classical mathematics; what matters is the philosophical foundation.[21] As point of departure, I take two characteristic fragments from Brouwer, which I call A and B, and in which I label various parts for reference in the following discussion.

[19]For further discussion of the relation between phenomenology and mathematics, see Gödel (*1961/?) and Yoshimi (2007).

[20]For recent introductions to intuitionism and its history, see Troelstra and van Dalen (1988), Hesseling (2003), and van Atten (2004b). Also the following articles in the on-line *Stanford Encyclopedia of Philosophy*: van Atten (2008), Iemhoff (2009), Moschovakis (2008), van Atten (2009b), and Bridges (2009).

[21]Dodd writes that 'Brouwer's revolution [has] run aground on the insuperable technical difficulties in re-establishing classical analysis on exclusively intuitionist principles' (Dodd 2007, 300–301). For an intuitionist, the impossibility to re-establish classical analysis is of no particular philosophical importance. Moreover, a number of theorems of intuitionistic analysis formally contradict classical analysis. That fact also shows that, while previously established theorems of classical analysis may of course serve as a heuristic in the search for new intuitionistic theorems, this is only so up to a point.

Fragment A is from the lecture 'Consciousness, Philosophy, and Mathematics' (held in 1948):

A1. First of all an account should be rendered of the phases consciousness has to pass through in its transition from its deepest home to the exterior world in which we cooperate and seek mutual understanding. This account does not imply mutual understanding and in some way may remain a soliloquy. The same can be said of some other parts of this lecture too.

A2. *Consciousness* in its deepest home seems to oscillate slowly, will-lessly, and reversibly between stillness and sensation.

A3. And it seems that only the status of sensation allows the initial phenomenon of the said transition. This initial phenomenon is a *move of time*. By a move of time a present sensation gives way to another present sensation in such away that consciousness retains the former one as a past sensation, and moreover, through this distinction between present and past, recedes from both and from stillness, and becomes *mind.*

A4. As mind it takes the function of a subject experiencing the present as well as the past sensation as object. And by reiteration of this two-ity-phenomenon, the object can extend to a world of sensations of motley plurality ...

In the world of sensation experienced by the mind, the free-will-phenomenon of *causal attention* occurs. It performs identifications of different sensations and of different complexes of sensations, and in this way, in a dawning atmosphere of forethought, creates *iterative complexes of sensations* ...

On the other hand there are iterative complexes of sensations whose elements are permutable in point of time. Some of them are completely estranged from the subject. They are called *things*. Forinstance *individuals*, i.e. human bodies, the home body of the subject included, are things ...

The whole of things is called the *exterior world of the subject.* ...

A5. Mathematics comes into being, when the two-ity created by a move of time is divested of all quality by the subject,

A6. and when the remaining empty form of the common substratum of all two-ities, as basic intuition of mathematics, is left to an unlimited unfolding,

A7. creating new mathematical entities in the shape of *predeterminately or more or less freely proceeding infinite sequences* of mathematical entities previously acquired,

A8. and in the shape of *mathematical species*, i.e., properties supposable for mathematical entities previously acquired and satisfying the condition that if they are realized for a certain mathematical entity, they are also realized for all mathematical entities which have been defined equal to it. (Brouwer 1949, 1235 and 1237)

Fragment B is taken from the lecture 'Points and Spaces' (1954):

B1. *The first act of intuitionism* completely separates mathematics from mathematical language, in particular from the phenomena of language which are described by theoretical logic. It recognizes that mathematics is a languageless activity of the mind having its origin in the basic phenomenon of the perception of a *move of time*, which is the falling apart of a life moment into two distinct things, one of which gives way to the other, but is retained by memory.

B2. If the two-ity thus born is divested of all quality, there remains the common substratum of all two-ities, the mental creation of the *empty two-ity*.

B3. This empty two-ity and the two unities of which it is composed, constitute the *basic mathematical systems*. And the basic operation of mathematical construction is the *mental creation of the two-ity of two mathematical systems previously acquired*, and the consideration of this two-ity as a new mathematical system.

B4. It is introspectively realized how this basic operation, continually displaying *unaltered* retention by memory, successively generates each natural number, the infinitely proceeding sequence of the natural numbers, arbitrary finite sequences and infinitely proceeding sequences of mathematical systems previously acquired, finally a continually extending stock of mathematical systems corresponding to 'separable' systems of classical mathematics.

...

B5. In the edifice of mathematical thought based on the first and second act of intuitionism, language plays no other part than that of an efficient, but never infallible or exact, technique for memorizing mathematical constructions, and for suggesting them to others;

B6. so that the wording of a mathematical theorem has no sense unless it indicates the construction either of an actual mathematical entity or of an incompatibility (e.g., the identity of the empty two-ity with an empty unity) out of some constructional condition imposed on a hypothetical mathematical system.

B7. So that mathematical language, in particular logic, can never by itself create new mathematical entities, nor deduce a mathematical state of things. (Brouwer 1954A, 2–3)

12.2.3 A Systematic Comparison

Given Husserl's conception of pure mathematics as formal ontology, which in turn is a branch of transcendental phenomenology, the two questions whether Brouwer's intuitionism is part of transcendental phenomenology, and, if so, whether it is mathematics as understood in that framework, together amount to the question whether Brouwer's construction of mathematical objects is a form of Husserl's constitution of purely categorial objects by the transcendental subject. These questions can only be answered by taking as concrete an approach as possible: one has to see what the acts have in common in which, according to their writings,

Brouwer's constructions and Husserl's purely categorial objects are brought to givenness.[22]

With the two fragments A and B from Brouwer as a guiding thread, I shall now defend the claim that Brouwer's acts of mathematical construction have the following eleven properties in common with Husserl's acts of the constitution of purely categorial objects as described in, notably, the sixth *Logical Investigation*, *Formal and Transcendental Logic*, and *Experience and Judgment*:

C1. These acts result in non-sensuous intuitions of formal objects.
C2. In these acts, the objects are, in the ontic sense, produced.
C3. These acts are active and voluntary.
C4. These acts are prelinguistic.
C5. They are synthetic, objectifying acts that give form to 'matter'.
C6. These acts can be iterated, limitlessly.
C7. These acts are, however, ultimately founded on sensuous intuition.
C8. Although these acts are active and voluntary (C3), the fulfilment of the corresponding intentions is not arbitrary.
C9. The intuition obtained in these acts is closely related to the awareness of inner time.
C10. In the intuitions obtained in these acts, the intuited objects can be adequately given.
C11. The subject that is correlate to these acts is not a psychological, but a transcendental subject, considered in its essential properties; this motivates the introduction of the notion of a single, ideal subject.

Since Husserl's theory of constitution of categorial objects is the better known of the two, references to it in the discussion below sometimes go without further comment. Nor are they meant to be exhaustive.

C1. These acts result in non-sensuous intuitions of formal objects. (Husserl 1984b, Chap. 6)

For Brouwer the result of acts of mathematical construction is an object given in a non-sensuous intuition. The general point was already made in Brouwer's dissertation, where he said that 'in the intellect, mathematical sequences [are] not sensed, yet perceived' (Brouwer 1975, 53).[23] (Note how this shows that Brouwer, like Husserl (e.g., Husserl 1984b, 732) understands perception (*waarneming*) to include the givenness of non-sensuous objects.) In particular, the specification that

[22]On p. 14 of Eley (1969), there is an interesting reference to 'H.-R. Brennecke, Untersuchungen zur Genealogie der Menge. Das Vorprädikative und Prädikative in der Begründung der Mengenlehre bei E. Husserl und L.E.J. Brouwer (Köln 1968).' Unfortunately, it turns out that this was a projected dissertation by an assistant at the Husserl Archive in Cologne that was never finished; moreover, at the Archive, not even fragments of it are known. I thank Matthias Wille, Dirk Fonfara, and Dieter Lohmar (all in Cologne) for their clarification of this matter.

[23]'in het intellect . . . niet gevoelde, doch waargenomen wiskundige volgreeksen' (Brouwer 1907, 81).

infinitely proceeding sequences need not be determined by a law (A7) again reveals the intuitive character of these sequences, for it means that even in absence of a (finitely stated) law to go proxy for them, they can be given as themselves. (More on infinite sequences as categorial objects in C10 below.)

C2. In these acts, the objects are, in the ontic sense, produced.

The transcendental Husserl explicitly held that the objects of mathematics are produced by the subject; this means that (ontological) constructivism is not only 'perfectly compatible with transcendental phenomenology', as Rosado Haddock concedes (Rosado Haddock 2006, 218n144), but transcendental phenomenology leaves, pace Rosado Haddock, no other possibility. How this ontological constructivism of Husserl is related to Brouwer's, and whether it allows Husserl to think of classical mathematics as a constructivism for higher minds, are questions that will be discussed later in the present paper. This view of Husserl on the ontology of mathematical objects is mainly documented in the lecture course *Phenomenological Psychology* of 1925, in *Formal and Transcendental Logic*, and in *Experience and Judgment*.[24] For example, in Sect. 63 of the latter, Husserl draws the contrast between objects of sense perception and categorial objects as follows. Of the former he says: 'What the I in its doing produces is precisely only the representations of [the object of perception], but not the object itself'.[25] Of the latter: 'By contrast, in spontaneous production the state of affairs itself is produced and not a representation of it.'[26] It is clear that Husserl here names the state of affairs just as an example of a categorial formation. A footnote to the first of these two passages emphasises that what is at stake in this discussion is the ontic:

> The fact that from the transcendental point of view also the object [of perception] itself is a product of constitution, can be left out of consideration when drawing this contrast, which concerns an ontic difference.[27]

This stands in sharp contrast to Sects. 22 and 23 in the earlier *Ideas I*. In particular, in Sect. 22 Husserl says, 'I form my representations of numbers ... but ... a representation of a number is not the number itself',[28] and in Sect. 23, 'In this manner, also in spontaneous abstraction it is not the essence, but the awareness of

[24] Husserl (1962, 25–26), Husserl (1974, 86, 87, 176, 267, 270, 323), and Husserl (1985b, 301). Also, e.g., Husserl (1950a, 87–88), Husserl (1966, 291), and Husserl (2003b, 147).

[25] 'Was das Ich hier in seinem Tun erzeugt, sind eben nur die Darstellungen von ihm, nicht aber der Gegenstand selbst.' (Husserl 1985b, 301)

[26] 'Hingegen wird im spontanen Erzeugen der Sachverhalt selbst erzeugt und nicht eine Darstellung von ihm.' (Husserl 1985b, 302)

[27] 'Daß auch der [Wahrnemungs-Gegenstand selbst unter transzendentalem Gesichtspunkt Produkt der Konstitution ist, kann im Rahmen dieser Kontrastierung, die einen ontischen Unterschied betrifft, außer Betracht bleiben.' (Husserl 1985b, 301)

[28] 'Ich bilde meine Zahlvorstellungen ... aber ... Zahlvorstellung ist nicht Zahl selbst.' (Husserl 1976a, 48–49)

it that is something produced'.[29] On the basis of such a distinction between the object and our awareness of it, one might begin to develop a phenomenological justification of realism in mathematics.[30] However, Husserl's marginal notes in two of his personal copies of *Ideas I* question and criticise Sects. 22 and 23 on precisely this point (Schuhmann 1973, 161). Notably, regarding Sect. 23 Husserl marked 'wrong, improve!' (*falsch, bessern!*) and

> There are real and ideal products. The production of a 'centaur' in phantasy is production of something ideal and not production of a psychic act (which is confused with its noematic 'object'). Similarly, an essence is an ideal product.[31]

These comments date from the period 1914–1921 (Husserl 1976b, 478). In the 1922–1923 lectures, Husserl puts it thus: 'Every ideal object is what it is only as actus verus, as an everlasting potentiality that I can actualise anytime and perhaps now indeed actualise'.[32] Correspondingly, in *Formal and Transcendental Logic*, Husserl says that the question to be answered is 'how subjectivity *can produce within itself, purely from the sources of its spontaneity, formations* that can have the status of ideal objects of an ideal "world"'.[33]

The view, then, that it is not the mathematical object that is constructed but(only) our awareness of it (Tieszen 1992, 189), or, put differently, that it is in and through the activity of the mind itself that mathematical objects are given to us as mind-independent (Tieszen 2010, Sects. IV and V), finds no support in transcendental idealism as Husserl developed and refined it after *Ideas I*. For the same reason, I see no justification for Rosado Haddock's claims that in *Formal and Transcendental Logic* 'there is no sign of any constructivism' (Rosado Haddock 2006, 218), and that, more generally, not only before but also after the transcendental turn Husserl propounded an 'ontological Platonist mathematics' as opposed to a constructive one (Rosado Haddock 2006, 200n40; Rosado Haddock 2010, 26–27).[34] In the

---

29 'So ist auch im spontanen Abstrahieren nicht das Wesen, sondern das Bewußtsein von ihm ein erzeugtes.' (Husserl 1976a, 50)

30 E.g., Tieszen (1989, 178–179), Tieszen (1992, 189)l, and Tieszen (2010).

31 'Es gibt reale und ideale Erzeugnisse. Eine Erzeugung eines "Kentauren" in der Phantasie ist Erzeugung eines Idealen und nicht Erzeugung des psychischen Aktes (der verwechselt wird mit seinem noematischen "Gegenstand"). Ebenso ist Wesen ein ideales Erzeugnis.' (Husserl 1976b, 482–483)

32 'Jede ideale Gegenständlichkeit ist, was sie ist, nur als actus verus, als immerwährende Potenz, die ich jederzeit aktualisieren könnte une eventuell jetzt wirklich aktualisiere.' (Husserl 2002c, 280)

33 'wie die Subjektivität *in sich selbst rein aus Quellen ihrer Spontaneität Gebilde schaffen kann*, die als ideale Objekte einer idealen "Welt" gelten können' (Husserl 1974, 267, emphasis mine).

34 Note also that if the transcendental Husserl had been a Platonist in any sense that is not compatible with constructive mathematics – there is a compatible sense: one might, perhaps somewhat redundantly, hold that the objects that exist independently correspond exactly to what we can mentally construct – one would not have expected him to be as reserved about the validity of the principle of the excluded middle in truth-logic as he is in *Formal and Transcendental Logic* (Husserl 1974, Sects. 77, 79).

lectures *Phenomenological Psychology* of 1925, Husserl diagnoses psychologism as a misinterpretation of the exact sense of this production:

> Precisely this circumstance that if the occasion arises, we encounter ideal objects as subjective formations in our forming consciousness and doing, has been the source of the then almost universal psychologisation of the ideal objects. Even if it was made evident that ideal objects, in spite of the fact that they are formed in consciousness, have their own being, being-in-themselves, this posed a large task, never earnestly seen and tackled: namely, that of making this peculiar correlation between ideal objects of the purely logical sphere and subjective-psychic consciousness as a forming doing into a topic of research.[35]

One important element in the clarification of the *Ansich-sein* of ideal objects is that the way in which they are produced in the acts in which they come to givenness is such that the subject can at any time repeat it and produce identically the same object:

> Indeed, we find ourselves even urged to say: numbers are produced in counting, judgments are produced in judicative doing ... When I 'repeat' a similar producing, when I carry out once more similar actions of counting, predicating, inferring, then they are, to be sure, psychically a new fact, but I can cognise with evidence that what has come to be there is identically the same pure number, identically the same truth, etc.[36]

And in *Formal and Transcendental Logic*:

> [Ideal objects] are what they are, only 'as arising from' original production. But that does not at all mean that they are what they are only in and during the original production. Their being 'in' the original production means that in it, they are given in consciousness as a certain intentionality of the form of spontaneous activity, in fact in the mode of the original self. This manner of givenness arising from original activity is nothing but the type of 'perception' proper to them.[37]

Similarly, in Sect. 64c of *Experience and Judgment*, Husserl explains in what sense a proposition is omnitemporal, even though it is a 'unity of becoming'

---

[35] 'Eben dieser Umstand, daß ideale Gegenstände uns gegebenenfalls als subjektive Gebilde im bildenden Erleben und Tun entgegentreten, war die Quelle der damals fast allgemeinen Psychologisierung der idealen Gegenstände gewesen. Wenn nun auch evident gemacht wurde, daß ideale Gegenstände, trotzdem sie zur Bildung im Bewußtsein kommen, ihr eigenes Sein, Ansichsein haben, so bestand hier doch eine große und nie ernstlich gesehen und in Angriff genommene Aufgabe: nämlich die, diese eigentümliche Korrelation zwischen idealen Gegenständen der rein logischen Sphäre und subjektiv psychischem Erleben als bildendem Tun zum Forschungthema zu machen.' (Husserl 1962, 25–26) See also Husserl (1974, 177–178).

[36] 'Ja wir sehen uns sogar gedrängt zu sagen: erzeugt werden die Zahlen in Zählen, erzeugt werden die Urteilssätze im urteilenden Tun. ... 'Wiederhole' ich ein gleiches Erzeugen, übe ich noch einmal gleiche zählende, prädizierende, schließende Aktionen, so sind sie zwar seelisch ein neues Faktum, aber ich kann evident erkennen, daß, was da geworden ist, identisch dieselbe reine Zahl, identisch dieselbe Wahrheit usw. sei.' (Husserl 1962, 25–26).

[37] '[Ideale Gegenständlichkeiten] sind, was sie sind, nur "aus" ursprünglicher Erzeugung. Das sagt aber keineswegs, sie sind, was sie sind, nur in und während der ursprünglichen Erzeugung. Sind sie "in" der ursprünglichen Erzeugung, so sagt das, sie sind in ihr als einer gewissen Intentionalität von der Form spontaner Aktivität bewußt, und zwar im Modus des originalen selbst. Diese Gegebenheitsweise aus solcher ursprünglichen Aktivität ist nichts ander als die ihr eigene Art der "Wahrnehmung".' (Husserl 1974, 176)

(*Werdenseinheit*) and 'the becoming is here, seen from the subject, a being-created' (Husserl 1985b, 309):

> The irreality of the proposition as the idea of a synthetic unity of becoming is the idea of something that can occur at any point in time, occurs at each of them necessarily temporally and temporally becoming, and yet is 'at all times' the same.[38]

This view of the transcendental Husserl is in fact an adaptation of views he expressed much earlier. Already in *On the Concept of Number* of 1887, he had written:

> The numbers are mental creations inasmuch as they are the results of actions that we perform on concrete contents; but what these actions create are not absolute contents that we can find again somewhere in the 'outside world' but rather peculiar relational concepts that can time and again only be produced, but by no means somewhere be found pre-made.[39]

Correspondingly, in Sect. 27a of *Formal and Transcendental Logic*, Husserl reads his *Philosophy of Arithmetic*, in hindsight, as giving a constitution analysis of categorial objects.

Brouwer always considered mathematics to be a mental creation (A7, B2, B3). Already in his dissertation, he had stated: 'Existence in mathematics means: to be constructed in intuition ...Mathematics is a free creation, independent of experience'.[40] In his own copy of his dissertation, Brouwer later substituted 'constructible' for 'constructed',[41] which is more appropriate; see C11 below. The term 'free' here does not mean 'arbitrary', by C8 below.

In his first Vienna lecture in 1928, Brouwer criticised the tendency both in philosophy and mathematics to introduce 'fetish-like "concepts"' and 'ideal truths' (*fetischartige 'Begriffe', ideale Wahrheiten*) about them that do not originate in the activity of the mind (Brouwer 1929A, 159–160); the point being that it is through such activity that the genuine concepts and truths about them are given in intuition. Unlike Hill (2010, 55), I take this to indicate a broad agreement, not a disagreement, between Brouwer and Husserl.

C3. These acts are active and voluntary (Husserl 1984b, Sect. 62; Husserl 1974, Sect. 63; Husserl 1985b, Sects. 61 and 63).

---

[38] 'Die Irrealität des Satzes als Idee einer synthetischen Werdenseinheit ist Idee von Etwas, das an jeder Zeitstelle auftreten kann, an jeder notwendig zeitlich und zeitlich werdend auftritt, und doch "allzeit" dasselbe ist.' (Husserl 1985b, 311)

[39] 'Geistige Schöpfungen sind die Zahlen, sofern sie Resultate von Tätigkeiten bilden, die wir an konkreten Inhalten üben; aber was diese Tätigkeiten schaffen, das sind nicht absolute Inhalte, die wir irgendwo in der 'Aussenwelt' wiederfinden können, sondern es sind eigentümliche Relationsbegriffe, die immer wieder nur erzeugt, aber keineswegs irgendwo fertig vorgefunden werden können.' (Husserl 1970, 317)

[40] 'bestaan in de wiskunde beteekent: intuïtief zijn opgebouwd ...De wiskunde is een vrije schepping, onafhankelijk van de ervaring.' (Brouwer 1907, 177, 79)

[41] In the original, *op te bouwen* for *opgebouwd*; see van Dalen (2001a, 134n(g)).

This third common property is really an aspect of C2 that is worthy of particular emphasis.

A categorial object is a spontaneous accomplishment of the ego ('in the presence of the I' (*unter Dabeisein des Ich*', Husserl 1985b, 301.)). To each part of a purely categorial object that is given in intuition necessarily corresponds a preceding voluntary act (an 'operation') in which that part was explicitly intended and produced. The syntheses involved are active and carried out in freedom. A consequence that Husserl notes is that to break off the constitution of a particular categorial object at an arbitrary point implies the failure to constitute it at all (Husserl 1985b, 302). For Brouwer, having sensations is passive (A2). But mathematics, he wrote in his dissertation (Brouwer 1907, 179), is 'a free construction' (*een vrije schepping*). In 'Mathematik, Wissenschaft und Sprache', he writes that mathematics is an 'activity of mankind' which has its origin in 'the individual man's will to live'. The perception of a two-ity, which he there calls 'the intellectual ur-phenomenon', is not a 'passive attitude' but an 'act of the will' (Brouwer 1929A, 153–154).[42] A particular role of the will in Brouwer's foundations of mathematics is seen in the construction of infinitely proceeding sequences of elements that the subject freely chooses (A7).

C4. These acts are prelinguistic.

For Husserl, authentic acts of thought, hence in particular also authentic acts of purely categorial thought, are not of a linguistic nature. These intuitive acts can, within certain limits, be expressed in language, but these acts as such do not depend on anything signitive (Husserl 1984b, 672–3, 731; Husserl 1985b, Sect. 50b).

Rosado Haddock has argued that the view that for Husserl, linguistic expressions are not essential for meaningful thinking, is, in the case of mathematics, 'questionable, to say the very least': 'A study of Husserl's philosophy of mathematics ... shows very clearly that in his theory of deductive theories in which logic and mathematics build, respectively, the logico-linguistic and the ontological main components, the most basic part, presupposed by the rest of the edifice, is a theory of pure logical grammar, which is essentially a theory of the syntactic categories and the syntactic rules of formation' (Rosado Haddock 1991, 357–358). But as we saw in the quotation at the beginning of Sect. 12.2.1, Husserl explicitly noted the fact that formal ontology can be approached directly as the theory of purely categorial formations, leaving aside the theory of scientific theories. Scientific theories are linguistic objects, but purely categorial formations are not.[43]

Brouwer takes pure mathematics to be 'essentially languageless' (B1). The practical importance of language is recognised (B5), but nothing linguistic enters

[42] *Aktivität der Menschheit*; *der Wille zum Leben des einzelnen Menschen*; *das intellektuelle Urphänomen*; *passive Einstellung*; *Willesakt*.

[43] In this context, it is interesting that Husserl decided to use the term 'syntactic' in two distinct senses: with respect to linguistic form, and with respect to the categorial form of an object: Husserl (1976a, Sect. 11), Husserl (1974, Sect. 42b–d), Husserl (1974, Beilage I), and Husserl (1985b, 247n).

into the conditions of possibility of acts of mathematical construction as such: 'Formal language accompanies mathematics as a score accompanies a symphony by Bach or an oratorio by Handel.'[44] No linguistic act can bring a mathematical object into existence; that can only be done in a languageless act of construction (C2). 'I can conceive mathematical truth which can never be fixed down in any system of formulas ... But even when the formal system coincides with intuitive mathematics, or expressed more exactly, when they are parallel, then exactness lies in the intuition, never in the formula' (Brouwer 1975, 452).[46] Similarly: 'The question where mathematical exactness does exist, is answered differently by the two sides; the intuitionist says: in the human intellect, the formalist says: on paper' (Brouwer 1913C, 83).

For one specific type of mathematical object and one specific mathematical object Brouwer gave arguments to the effect that not only are they originally non-linguistic, but they cannot be properly represented in language. Of mathematical demonstrations understood as mental objects, he said:

> These mental mathematical demonstrations, which in general have infinitely many parts, should not be confused with their linguistic accompaniments, which are finite, necessarily inadequate, and hence not belonging to mathematics.[47]

And he goes on to call this 'my main argument against the aspirations of Hilbert's metamathematics'.[48] And of the continuum, he remarked that it is given in a priori intuition, yet cannot be exhausted by a language with denumerably many

---

[44] 'de formuletaal, die de wiskunde begeleidt als het notenschrift een symfonie van Bach of een oratorium van Händel' (Brouwer et al. 1937, 262). For a reason unknown to me, the (partial) translation 'Signific Dialogues' (Brouwer 1975, 447–52), which as far as I know is not by Brouwer, substitutes 'Formal language accompanies mathematics as the weather-map accompanies the atmospheric processes' (Brouwer 1975, 451); that simile is not inappropriate, but the original is better, because it is a richer analogy.[45]

[45] ⟦Note added in this reprint: Here I was mistaken. As is now evident to me, the translation in Brouwer (1975) was made from the publication of the 'Signifische Dialogen' in book form Brouwer et al. (1939), not from the prior publication in the journal *Synthese* (Brouwer et al. 1937); the bibliographical details as stated in Brouwer (1975), p. x, do not mention there are differences between the two. But already in the book, the musical image had been replaced by '... de formuletaal, die de wiskunde begeleidt als het weerkaartje het atmosferisch gebeuren' – which the English quoted above renders correctly. In the preface to the book, Mannoury says that this edition is 'more complete', and he thanks the editorial board of *Synthese*, but he does not say that changes have been made, let alone by whom or why. Be that as it may, I apologise to the translators.⟧

[46] 'kan ik mij wiskundige juistheid denken, die nooit in enig formulesysteem kan worden vastgelegd ... Maar ook áls de intuïtieve wiskunde en de formalistiek overeenstemmen, of juister uitgedrukt: parallel lopen, dan ligt het exakte in de intuïtie, maar nooit in de formule.' (Brouwer et al. 1937, 262–263)

[47] 'Diese gedanklichen, im allgemeinen unendlich viele Glieder aufweisenden mathematischen Beweisführungen dürfen mit ihren endlichen, notwendigerweise inadäquaten, mithin nicht zur Mathematik gehörenden sprachlichen Begleitungen nicht verwechselt werden.' (Brouwer 1927B, 64n8)

[48] 'mein Hauptargument gegen die Anspruche der Hilbertsche Metamathematik'.

expressions, as the real numbers are not denumerable (Brouwer 1930A, 3, 6). Thus, Husserl and Brouwer agree that there is prelinguistic thought, and that this is moreover the authentic mode of thinking, to which other subjects have only indirect access, through the mediation of language. This renders authentic thought (and hence ideal cognition) to a certain extent private (Bernet et al. 1989, 177), and this is the sense in which Brouwer argues that there is no 'plurality of mind' (Brouwer 1949, 1239–1240). For Brouwer, the construction of mathematics can, ideally, wholly take place in authentic thought. As argued below (C11), the constructions in question are based on essential properties of the mind, which are therefore shared by any other subject. There is no reason, then, to think that Brouwer's epistemological solipsism entails a denial of the intersubjective validity of mathematics, nor that Husserl, in so far as he too holds that mathematics can ideally wholly be done in authentic thought, would have an essentially different view on this matter.[49]

Logic, on Brouwer's understanding of the term, is an application of mathematics to the language of mathematicians. It is the study of patterns in linguistic recordings of our activities of mathematical construction, and in particular the patterns that characterise validity of inferences, defined as the preservation of constructibility. This understanding is, in one sense, much closer to modern mathematical logic than Husserl's, for whom it also included formal ontology. But the difference is only terminological: Husserl would call Brouwer's logic 'formal apophantics', whereas logic in Husserl's wide sense corresponds to Brouwer's logic and pure mathematics combined.

C5. They are synthetic, objectifying acts that give form to 'matter' (Husserl 1984b, Sect. 40). Husserl formulates this idea particularly clearly in his lectures of 1906–1907:

> But as a matter of principle, mathematics and pure logic in general ... are a fund of purely intellectual truths ... They are thoroughly and purely grounded in the pure 'forms of thought', in the essence of formal thoughts about meanings and objects, which resemble templates in which first a material must be poured in order that material thoughts relating to material objects can result.[50]

In Brouwer, creating the two-ity of two mathematical systems previously acquired is to synthesise the latter into a new object of which they are parts, and thereby to give a form to the matter which these systems are (B3). In A3 and A4, where the passage of sensations comes to found a new act (see Husserl 1984b, 679), one recognises the steps of preconstitution and subsequent objectivation that according to Husserl are

[49]For a different view, see Tieszen (1995, 453). For extensive discussion of intersubjectivity and intuitionism, see Placek (1999) and van Atten (2004b, Chap. 6).

[50]'Prinzipiell aber bildet die Mathematik und reine Logik überhaupt ... ein[en] Fonds reiner Verstandeswahrheiten ... Sie gründen vielmehr durchaus und rein in den bloßen 'Denkformen', in dem Wesen der formalen Bedeutungs- und Gegenstandsgedanken, die Schablonen gleichen, in die erst ein Stoff eingefüllt sein muß, damit sachhaltige Gedanken mit Beziehung auf sachhaltige Gegenstände resultieren können.' (Husserl 1985a, 61)

characteristic for the constitution of syntactical objects (Husserl 1985b, Sects. 58 and 61).

The same matter may admit of different categorial formations (Husserl 1984b, Sect. 62; Husserl 1985b, 303), in which case the subject is free to choose which one to make. Correspondingly, Brouwer writes that this is how the basic intuition gives rise to a number of different fundamental notions:

> There are elements of mathematical construction that in the system of definitions must remain irreducible, and which therefore, in communication, must be understood from a single word, sound or symbol; they are the elements of construction that are immediately read off from the basic intuition or intuition of the continuum; notions such as continuous, unity, once more, and so on are irreducible.[51]

In a handwritten note in his own copy of his dissertation, Brouwer said that these notions are each different 'polarisations' (*polarizeringen*) of the basic intuition (van Dalen 2001a, 136).

C6. These acts can be iterated, limitlessly (Husserl 1984b, Sect. 60; Husserl 1974, Beilage I 6b; Husserl 1985b, Sect. 61).

For Brouwer, the object that is the result of combining mathematical systems in a two-ity is itself again a mathematical system (B3). Mathematical objects, once constructed, can be used as parts in the construction of further mathematical objects (B4); this iteration is the 'unlimited unfolding' mentioned in A6. Note that Husserl, similarly, speaks of 'the ideally iterative production of forms in open infinity' (Husserl 1974, 305).

The status of infinitely proceeding sequences as mathematically constructible objects, and thereby given in intuition (A7, B4; also the quotation from Brouwer's dissertation in C1 above) depends on the givenness of this iterative form itself. In the early development of intuitionism, Brouwer had not yet recognised the possibility of choice sequences, but he later commented that 'the extension [of mathematics with choice sequences] is an immediate consequence of the self-unfolding' (Brouwer 1981A, 93n). Husserl recognises this iteration as a categorial form in *Experience and Judgment* (Husserl 1985b, 258–259); the matter is discussed further in (van Atten 2007, Sect. 6.2).

Rosado Haddock sees in the iterative character of purely categorial formation an affinity with classical set theory as opposed to constructivism:

> His epistemology of mathematics of the second part of the sixth *Logical Investigation* . . . offers an iterative constitution of mathematical objects in categorial intuition. Such a view is clearly related to the views of his friends Cantor and Zermelo on the iterative notion of set, which is not to be related with constructivisms of Kantian or Brouwerian, or any other sort. (Rosado Haddock 2006, 219)

[51] 'Er zijn elementen van wiskundige bouwing, die in het systeem der definities onherleidbaar moeten blijven, dus bij mededeeling door een enkel woord, klank of teeken, weerklank moeten vinden; het zijn de uit de oer-intuïtie of continuumintuïtie afgelezen bouwelementen; begrippen als *continu*, *eenheid*, *nog eens*, *enzoovoort* zijn onherleidbaar.' (Brouwer 1907, 180)

But we see that iteration is as prominently present in Brouwer's notion of construction. More generally, iteration is such a conspicuous phenomenon in both classical and constructive mathematics, that Husserl's emphasis on it can, as such, hardly be taken to indicate a proximity to the one rather than the other.

C7. These acts are, however, ultimately founded on sensuous intuition (Husserl 1984b, Sects. 48, 60).

The basic intuition of the empty two-ity is generated by emptying any concrete two-ity of its sensuous content, and hence depends on it for its existence (A5; B2). As any further mathematical objects are constructed from that basic intuition, by transitivity these objects are likewise founded on sensuous intuition.

C8. Although these acts are active and voluntary (C3), the fulfilment of the corresponding intentions is not arbitrary (Husserl 1984b, Sect. 62).

The fact that constructions are voluntary but not arbitrary, that there is an interplay of the active and the passive, is explicitly recognised in intuitionism:

> The only possible foundation of mathematics must be sought in this construction under the obligation carefully to watch which constructions intuition allows and which not. [52,53]

Similarly, note the 'limits' that Heyting invokes when he writes that

> [Brouwer's] construction of intuitionist mathematics is nothing more nor less than an investigation of the utmost limits which the intellect can attain in its self-unfolding (Heyting 1968, 314).

C9. The intuition obtained in these acts is closely related to the awareness of inner time.

---

[52] '... dat dus in dezen opbouw, onder de verplichting, zorgvuldig acht te geven, wat de intuïtie veroorlooft te stellen en wat niet, de eenig mogelijke grondvesting der wiskunde is te zoeken' (Brouwer 1907, 77).

[53] This also provides an answer to the following objection to intuitionism, formulated by Lohmar (1989, 212): 'Other doubts in turn are directed at the view that the objects of mathematics are produced in the mathematician's acts. However understandable this view is as a counter-reaction to Platonism, and in spite of its pointing to the contribution of actions to the constitution of mathematical objects, it conceals the fact that cognition and itself-givenness of mathematical connections are founded on something which, in the activity that leads up to them, occurs passively. In mathematics, too, all we can do is to bring ourselves to the point where cognition either takes place or not.' ('Andere Bedenken richten sich wiederum auf die Ansicht, daß die Gegenstände der Mathematik im Handeln des Mathematikers erzeugt werden. So verständlich dies als Gegenreaktion zum Platonismus ist und so klarsichtig hiermit auf den Anteil an Handlungsaktivität hingewiesen wird, der in der Konstitution mathematischer Gegenständlichkeiten enthalten ist, so wird damit doch überdeckt, daß Erkennen und Selbstgegebenheit mathematischer Zusammenhänge auf etwas beruht, das sich in der Aktivität des Heranführens passiv einstellt. Auch in der Mathematik gilt, daß wir nur an den Punkt heranführen können an dem sich Erkennen einstellt oder nicht.')

In transcendental phenomenology, there is an intimate relation between time and formal ontology, as the awareness of inner time constrains the form of all particular objects and multiplicities of objects:

> While time consciousness is the primal site of the constitution of the unity of identity or of objecthood [*Gegenständlichkeit*], and then of the combination-forms of coexistence and succession of all objects [*Gegenständlichkeiten*] that come to givenness in consciousness, it is nevertheless no more than the consciousness that produces a general form. Mere form is of course an abstraction, and hence the intentional analysis of time consciousness and its accomplishment is from the outset an abstractive one. It apprehends, is only interested in the necessary temporal form of all singular objects and pluralities of objects, or correlatively in the form of the manifolds that constitute the temporal [object].[54,55]

Anticipating on Sect. 12.3 below, I remark that, in particular, the form of time constrains the purely categorial formations, and thereby constrains the cardinality of mathematical multiplicities that can ideally be given in intuition. (Outside a strictly phenomenological context, the relation between the form that time has for a subject and the cardinality of the mathematical constructions that this subject can carry out has been remarked on by Charles Parsons, in a discussion of the iterative concept of set. More on that below.)

In Brouwer, the constraint of the form of time on the mathematical objects is immediate (A3–A5, B1–B2). Moreover, like transcendental phenomenology (Mensch 1996, 109), intuitionism recognises the fundamental role of time awareness in the genesis of intentionality itself (A3). In this context, Brouwer takes care not to understand time as mundane:

> And since in this intuition [i.e., the basic intuition of mathematics] we become conscious of time as change per se, we can state:
> The only a priori element of science is time. (Brouwer 1975, 61)[56]

to which he adds in a footnote:

> Of course we mean here intuitive time which must be distinguished from scientific time. By means of experience and very much a posteriori it appears that scientific time can suitably be introduced for the cataloguing of phenomena, as a one-dimensional coordinate having a one-parameter group. (Brouwer 1975, 61n2)[57]

---

[54] 'Ist nun das Zeitbewusstsein die Urstätte der Konstitution von Identitätseinheit oder Gegenständlichkeit, und dann der Verbindungsformen der Koexistenz und Sukzession aller bewusst werdenden Gegenständlichkeiten, so ist es doch nur das eine allgemeine Form herstellende Bewusstsein. Bloße Form ist freilich eine Abstraktion, und so ist die intentionale Analyse des Zeitbewusstseins und seiner Leistung von vornherein eine abstraktive. Sie erfasst, interessiert sich nur für die notwendige Zeitform aller einzelnen Gegenstände und Gegenstandsvielheiten, bzw. korrelativ für die Form der Zeitliches konstituierenden Mannigfaltigkeiten.' (Husserl 1966, 128) Also Husserl (1966, 312); Husserl (1974, Beilage II 2c, 318), Husserl (1950a, 99).

[55] ⟦Note added in this reprint: For a detailed discussion of the role of inner time awareness in the constitution of sets, see van Atten (2015b).⟧

[56] 'En daar deze samenvalt met de bewustwording van den tijd als verandering zonder meer, kunnen we ook zeggen:
Het eenige aprioristische element in de wetenschap is de tijd.' (Brouwer 1907, 99)

[57] 'Natuurlijk wordt hier bedoeld de intuïtieve tijd, wel te onderscheiden van de wetenschappelijke tijd, die, wel zeer a posteriori, eerst door de ervaring blijkt, als met een eenledige groep

C10. In the intuitions obtained in these acts, the intuited objects can be adequately given.

Husserl writes in the sixth *Logical Investigation*, 'The ideal conditions of the possibility of categorial intuition in general are, correlatively, the ideal conditions of the possibility of the objects of categorial intuition and of the possibility of categorial objects as such'.[58] The justification of this claim would seem to be that, because of the wholly spontaneous character of its constitution (C3), a purely categorial object can only have a certain part if an act of the will has been carried out in which that part arose by giving form to some matter at hand. This means that if we make the idealisation of a perfect memory, all parts of a purely categorial object are necessarily simultaneously and adequately perceivable. Husserl explicitly denies the possibility of ideal objects that cannot, in principle, be adequately given, and he therefore says that ideal objects are 'ideally immanent' (Husserl 1974, 389–90). Since for Brouwer, too, these acts are wholly voluntary, the same argument applies. The emphasis on 'unaltered retention by memory' (B4) reflects a point made in a lecture of 1932, 'Will, Knowledge, Speech':

> The languageless constructions which arise from the self-unfolding of the basic intuition are, on the sole basis of their presence in memory, exact and correct, but the human faculty of memory which must survey these constructions is, even when it seeks the support of linguistic signs, by its nature limited and fallible. (Brouwer 1975, 443, translation modified)[59]

With an eye on Brouwer's infinitely proceeding sequences (A7, B4), it should be noted that ideal, adequate givenness of a potentially infinite sequence does not consist in its being given as an actually infinite sequence, for that would contradict the essence of the object qua potentially infinite. (On this see also Dummett 2000b, 41–43.) Rather, it consists in the givenness of the whole finite initial segment generated so far, however large the number of its elements may be, together with the open horizon that adequately indicates the ever present possibility to construct additional elements of the sequence. The absence of such further elements from an intuition of the sequence at a given moment does not render that intuition inadequate, because they do not yet even exist. In contrast, the reason why our intuition of a physical object at a given moment is necessarily inadequate is precisely that, as a matter of three-dimensional geometry, any concrete view of the object hides parts that do at that moment exist. Such perspectival givenness is not a feature of purely mathematical objects. Like Brouwer, Husserl holds that an infinite

voorziene eendimensionale coördinaat geschikt te kunnen ingevoerd tot het katalogizeeren der verschijnselen.' (Brouwer 1907, 99n1)

[58] 'Die idealen Bedingungen der Möglichkeit kategorialer Anschauung überhaupt sind korrelativ die Bedingungen der Möglichkeit der Gegenstände kategorialer Anschauung und der Möglichkeit von kategorialen Gegenständen schlechthin.' (Husserl 1984b, 718–719)

[59] '... dat de door de zelfontvouwing der oerintuïtie ontstaande taallooze constructies, uit kracht van hun in de herinnering aanwezig zijn alleen, exact en juist zijn, dat echter het menschelijk herinneringsvermogen, dat deze constructies heeft te overzien, ook als het linguïstische teekens te hulp roept, uit den aard der zaak beperkt en feilbaar is' (Brouwer 1933A2, 58).

sequence can only be given as a potentially infinite sequence, for example in Sect. 143 of *Ideas I*, and again in *Nature and Spirit* of 1927:

> How can an infinity be experienced? Only in this way, that a finite stock at a time directly falls within experience, and is at the same time the carrier of a horizonal presumption, a reference to a subjectively possible progress to new experience, etc.'[60]

To the extent that Husserl is here describing consciousness eidetically, and therefore is not describing just human consciousness, it follows from this passage together with his observation that the ideal conditions of categorial intuition are, correlatively, the conditions of possibility of categorial objects (quoted above), that for him infinite sequences are possible only as potentially infinite sequences. But perhaps one considers the view Husserl here expresses on the experience of infinite sequences to be valid for beings like us, but not necessarily for higher minds, who might be able to experience actually infinite sequences. The question whether in Husserl's phenomenology there indeed is room for a consideration of this type is dealt with in Sect. 12.3 below, where it will be answered in the negative.

That potentially infinite sequences are indeed categorial objects is stated explicitly in *Formal and Transcendental Logic*:

> I mention here only the fundamental form, never stressed by the logicians, of the 'and so on', of the iterative 'infinity',[61] which has its subjective correlate in 'one can again and again' ... Mathematics is the realm of infinite constructions, a realm of ideal existences, not only in a 'finite' sense, but also of constructive infinities. Obviously the problem of the subjective constitutive origins repeats itself here as the hidden construction method which should be revealed and reconstructed as norm, the method in which the 'and so on' in various senses and the infinities as categorial formations of a new kind become evident.[62,63]

---

60 'Wie kann eine Unendlichkeit erfahren sein? Nur so, dass ein endlicher Bestand jeweils direkt in die Erfahrung fällt und zugleich Träger ist einer Horizontpräsumtion, einer Verweisung auf einen subjektiv möglichen Fortgang zu neuer Erfahrung usw.' (Husserl 2001b, 107)

61 Husserl's historical claim here is not quite correct: e.g., Brouwer in his dissertation (1907) and Ludwig Wittgenstein in his *Tractatus logico-philosophicus* (1921) had thematised the notion 'and so on'. Brouwer (Brouwer 1975, 80n): 'The expression "and *so* on" means the indefinite repetition of *one and the same* object or operation, even if that object or that operation is defined in a rather complex way' ('Waar men zegt "en *zoo* voort", bedoelt men het onbepaald herhalen van *eenzelfde* ding of operatie, ook al is dat ding of die operatie tamelijk complex gedefinieerd', Brouwer 1907, 143n); Wittgenstein: 'The concept of the successive application of an operation is equivalent to the concept "and so on".' ('Der Begriff der successiven Anwendung der Operation ist äquivalent mit dem Begriff "und so weiter".', Wittgenstein ((1921) 2013, 52 (5.2523))

62 'Ich erinnere nur noch an die von den Logikern nie herausgehobene Grundform des "Und so weiter", der iterativen "Unendlichkeit", die ihr subjektives Korrelat hat im "man kann immer wieder". ...Die Mathematik ist das Reich unendlicher Konstruktionen, ein Reich von idealen Existenzen, nicht nur "endlicher" Sinne, sondern auch von konstruktiven Unendlichkeiten. Offenbar wiederholt sich hier das Problem der subjektiven konstitutiven Ursprünge als der verborgenen, zu enthüllenden und als Norm neu zu gestaltenden Methode der Konstruktionen, der Methode, in der das "und so weiter" verschiedenen Sinnes und die Unendlichkeiten als neuartige kategoriale Gebilde ... evident werden.' (Husserl 1974, 196)

63 See also Husserl (1985b, 258–259).

(I think Husserl's references to construction and method here should indeed be interpreted as the constructive mathematics it suggests. But as Husserl does not quite make explicit how he understands the notions of construction and method, I prefer to let this reading be justified by the arguments in the rest of this paper, instead of using this quotation itself as evidence for Husserl's constructivism.)

C11. The subject that is correlate to these acts is not a psychological, but a transcendental subject, considered in its essential properties; this motivates the introduction of the notion of a single, ideal subject.

Like phenomenology, intuitionism studies essential, structural properties of consciousness, not those of any particular individual's consciousness. Brouwer characterised intuitionism as 'inner architecture' (Brouwer 1949, 1249), and was interested in the question what mathematical constructions this inner architecture in principle allows, given unlimited memory, time, and so on. Among the various indications of that fact is Brouwer's claim that the subject can construct infinitely proceeding sequences; any factual notion of subject would have ruled out that claim, as factual humans are limited to the construction of finite sequences before they pass away. To use Noam Chomsky's distinction: Intuitionism does not study the performance of human consciousness in making certain constructions, but its competence.[64] Its focus on essential properties of the mind also allows one to account for the intersubjective validity of intuitionistic mathematics. Brouwer's statement that, for the subject, language serves to suggest mathematical constructions to others (B5) implicitly contains the claim that other subjects exist and have the same constructional means at their disposal.

As Brouwer wrote, 'The stock of mathematical entities is a real thing, for each person, and for humanity' (Brouwer 1981A, 90).

Like phenomenology, intuitionism recognises that the fundamental notion of subject is not psychological but transcendental. Brouwer's notion, as described in fragment A, does not presuppose the construction of the world. As a comparison of A4 and A5 shows, the construction of the world, including the subject as a mundane subject, requires preserving the sensations that make up the concrete two-ities, while the construction of mathematics requires exactly the opposite, namely, abstracting from all qualitative aspects of the two-ities. The notion of subject that is correlate to Brouwer's acts of mathematical construction is therefore not that of a subject in the world, and cannot be psychological. 'Mathematics is certainly completely independent of the material world'[65]; moreover, for Brouwer, the constitution of the mundane presupposes the mathematical, because he conceives of the mundane as a mathematical construction out of sense data (Brouwer 1929A, 153; Brouwer 1949, 1235). Lohmar has objected to intuitionism that its notion of the subject is mundane

[64] For Brouwer's rejection of psychological interpretations of intuitionism in a letter to Van Dantzig from 1949, see van Atten (2004b, 75–76).

[65] 'De wiskunde is zeker geheel onafhankelijk van de materieele wereld.' (Brouwer 1907, 177)

(Lohmar 1989, 211; Lohmar 2004, 63); while this seems to be the case for Heyting, to whom Lohmar refers in this context, it is not the case for Brouwer.

Given that Brouwer and Husserl are interested in the essential possibilities the subject has, it is no surprise to see both introduce in their writings the notion of a possible ideal subject who suffers from no limitations on the practical realisation of these possibilities due to non-essential factors (e.g., limits of time, space, attention, memory), and to whom therefore any object that can ideally be given in these acts can indeed be given adequately. Husserl points to the correlation between, on the one hand, ideal objects, ideal possibilities, ideal necessities, and, on the other hand, ideally possible subjects ('ideal mögliche Subjekte', Husserl 2002c, 279) and speaks of the ideal cognising subject (*ideales Erkenntnissubjekt*, Husserl 1974, 383–387).[66] That it is a justifiable idealisation to hold that this ideal knowing subject can always continue the constitution of a potentially infinite sequence has its ground in a structural property of the transcendental subject's time awareness. As Husserl says, the transcendental subject is immortal, it is a subject for whom 'the future means infinite time', and who is 'an eternal being in becoming'.[67]

Brouwer first thematises the notion of an ideal subject in the form of 'hypothetical human beings with an unlimited memory' Brouwer (1975, 443),[68] and later calls it 'the Creating Subject' (Brouwer 1975, 478–479).[69] That such a hypothetical being is conceived as having an infinite future is implied by the fact that Brouwer's mathematics requires it to be able to extend, for example, the sequence of the natural numbers and also choice sequences arbitrarily far (the latter is necessary for a choice sequence of rationals to converge to an irrational value). For both Husserl and Brouwer, then, to say that a certain mathematical object exists is to say that it can be brought to givenness by their respective ideal subjects. Whether this also means that they arrive at the same mathematics is discussed in Sect. 12.3.

These common points C1–C11 express how constitution (of purely categorial objects) and construction function in Husserl and Brouwer, and hence how these notions are to be understood. They agree on the relevant aspects of the ego (C11), the cogito (C3–C7, C9), and the cogitatum (C1, C2, C8, C10). Such, then, are my grounds for the claim that Brouwer's intuitionism can be interpreted as belonging to Husserl's theory of purely categorial formation. Some objections to this claim will be discussed below, after the following elucidations.

Given Husserl's view on the relation between transcendental phenomenology and the a priori sciences, discussed in Sect. 12.2.1 above, intuitionism is therefore both mathematics and part of phenomenology. For this reason, I do not think that Lothar Eley's comment on the *Philosophy of Arithmetic* also holds for the transcendental

[66] See on this idealisation also Husserl (2001b, 200–201), Husserl (1975, 188–189), Husserl (1984b, Sect. 64), and Becker (1927, 285, 287, 292f, 304, 320f).

[67] *unsterblich, die Zukunft unendliche Zeit bedeutet, ein ewiges Sein im Werden* (Husserl 1966, 378, 379, 381).

[68] *hypothetische mensen met onbeperkt herinneringsvermogen* (Brouwer 1933A2, 59).

[69] *het scheppend subject* (Brouwer 1948A).

Husserl: 'Husserl's approach to philosophy recognises a thematising subjectivity, however not a constructive one in the intuitionists' sense.'[70]

The common points C1–C11 also show that Brouwer is, systematically speaking, much closer to Husserl than to Kant, who did not acknowledge categorial intuition,[71] and for whom the flow of time is not self-given to us in a mode of intuition proper to it. In Brouwer, it is essentially an application of categorial intuition to the intuition of the flow of time that allows him to accept potentially infinite sequences as objects given in intuition. There is no parallel possibility in Kant: For him the result of acts of mathematical construction is a sensuous intuition (an image) in which a mathematical concept is instantiated by actual or possible empirical objects, but in a determinate image there is no intuitive givenness of the open horizon and the categorial form 'and so on'. This seems to me to be the reason behind Kant's refusal, in his letter to Rehberg from Autumn 1790 (Kant 1900–, 11:207–210), to identify the square root of 2 with a potentially infinite sequence of ever closer rational approximations, an identification that Brouwer on the other hand did make. Although Brouwer in his writings never remarks on it, his acceptance of such sequences as objects constructed in intuition is the strongest indication that his notion of intuition is in fact different from Kant's.[72]

In his inaugural lecture 'Intuitionism and formalism' (1912), Brouwer referred to Kant to characterise his own position as 'abandoning Kant's apriority of space but adhering the more resolutely to the apriority of time' (Brouwer 1913C, 85).[73] Perhaps Brouwer here had Kant's full doctrine of time in mind, perhaps only the apriority it ascribes to time. Be that as it may, in his later writings this characterisation no longer occurs, and in particular it is absent from his later historical introductions to intuitionism (Brouwer 1952B; Brouwer 1954A).[74] Brouwer introduced choice sequences soon after his inaugural lecture (Troelstra 1982). For the reasons explained in the previous paragraph, from a systematical point of view it would be no coincidence that Kant disappears as choice sequences appear. That

[70]'Der Husserlsche Ansatz der Philosophie kennt eine thematisierende Subjektivität, hingegen nicht eine konstruktive im Sinne der Intuitionisten.' (Husserl 1970, xxvi) Note that Eley, in his *Metakritik der formalen Logik*, is sympathetic with Brouwer's thought (Eley 1969, 14, 15, 64, 264n1, 329n2, 332).

[71]'Admittedly, in Kant's thought the categorial (logical) functions play an important role; but he does not arrive at the fundamental extension of the concepts of perception and intuition over the categorial realm' ('In Kants Denken spielen zwar die kategorialen (logischen) Functionen eine große Rolle; aber er gelangt nicht zu der fundamentalen Erweiterung der Begriffe Wahrnehmung und Anschauung über das kategoriale Gebiet.', Husserl 1984b, 732).

[72]A detailed discussion of this matter can be found in van Atten (2012).

[73]'van de theorie van Kant de aprioriteit der ruimte prijs te geven, doch aan de aprioriteit van de tijd des te vastberadener vast te houden' (Brouwer 1912A, 11).

[74]In 1929, in his second Vienna lecture, Brouwer expresses his 'fundamental [*im wesentlichen*] agreement' with Kant and Schopenhauer on the specific point of taking the continuum to be given in a priori intuition (Brouwer 1930A, 1, 6). That formulation does not imply agreement on the details.

this was also historically Brouwer's motivation no longer to present his intuitionism as a direct modification of Kant's position can, at present, only be conjectured.

12.2.4 Discussion of Some Remaining Objections

But against the background of this agreement, perhaps one sees a difference between the temporal characteristics of Husserl's purely categorial objects and Brouwer's mental constructions. As Tieszen formulates his objection to identifying these notions:

> Constructions are given as having temporal duration, for example, but numbers and finite sets are not. As Husserl puts it in [*Experience and Judgment*], intuitions are temporal processes but objects like numbers and finite sets that are given in these acts are given as 'omnitemporal'.(Tieszen 1989, 178)

But for Husserl there is no contradiction here. As we saw in C2 above, Husserl in Experience and Judgment and other texts holds that objects like numbers ontically indeed are constructions.

This does not contradict their omnitemporality, because by that Husserl means that this ontical construction is of such a kind that the subject can, in principle, repeat it at any later time and could have made it at any earlier time. The intuitionist can readily acknowledge that many of his constructions also have this property, namely, those that do not depend on choices. When a mathematical construction depends on free choices, we have to reckon with the fact that, when at different times the subject is confronted with the same situation in which to choose, it can do so differently each time, and hence these objects cannot be omnitemporal. Indeed, Brouwer thematises acts that satisfy properties C1 to C11, yet result in the construction of such objects: these are the 'more or less freely proceeding infinite sequences' (A7) or 'arbitrary infinitely proceeding sequences' (B4), now better known as 'choice sequences'. They come into being when the subject begins making a series of choices, and they grow in time with each choice the subject makes.

It can be argued that, phenomenologically, a choice sequence is an object in the emphatic sense of a 'correlate of an identification to be accomplished in an open, endless, and free repetition' (Husserl 1985b, 64; see also van Atten 2007, Sect. 6.2). In each of these repeated identifications, which can take place once the choice sequence has been begun, this potentially infinite sequence is given originally (see C10 above). Moreover, it is a specifically mathematical object (van Atten 2007, Sect. 6.3). Besides the fact that the acts in which choice sequences are constructed share the characteristics C1–C11, they preserve the intersubjectivity[75] and monotonicity of mathematical truth. The burden of proof that in spite of

---

[75] It is true that in a context of multiple subjects, a choice sequence is owned by the subject that creates it. But all that this subject knows about the sequence can be communicated by it to other subjects.

having all these properties, choice sequences are not mathematical objects, clearly lies with those phenomenologists who insist that omnitemporality is essential to mathematical objects as such. It is in the theory of choice sequences that theorems are obtained that formally contradict classical mathematics. Husserl's attitude towards revisionism in logic and mathematics seems to have been negative. But, as I have argued elsewhere, that was more a matter of Husserl's psychology than an intrinsic characteristic of transcendental phenomenology (van Atten 2002, 14).

As another argument against the identification of purely categorial objects and mental constructions, Tieszen (1989, 178) cites the following passage from *Experience and Judgment* in order to support his claim that 'Husserl offers arguments in a number of places in his later writings to show that mathematical objects could be said to 'exist' independently of our minds, or of our constructions':

> 'there are' mathematical and other irreal objects which no one has yet constructed. Their existence, to be sure, is revealed only by their construction (their 'experience'), but the construction of those already known opens in advance a horizon of objects capable of being further discovered, although still unknown. As long as they are not discovered (by anyone), they are not actually in spatiotemporality; and as long as it is possible (how far this is possible, there is no need to decide here) that they never will be discovered, it may be that they will have no world-reality (Husserl 1973e, 260).[76]

However, this passage equally admits of an intuitionistic interpretation. Perhaps one already finds sufficient ground for such a reading in the infinite iterability of the operation of forming two-ities out of previously constructed objects. Actual human beings certainly cannot concretely overview the constructions implied in the possibility of that iteration. In Husserl's passage, it is clearly constructions that the horizon is opening up to, yet he also says the mathematical objects are 'capable of being discovered.' But this need not be taken in the sense of discovering an object that exists independently of our acts; there is also an epistemic sense in which a constructional possibility can appropriately be said to be discovered. A closely related argument depends on Brouwer's 'denumerably unfinished sets' (*aftelbaar onaffe verzamelingen*). These are sets such that 'only denumerable subsets of it can be indicated in a well-defined way, but from each such denumerable subset we can immediately obtain, following some previously defined mathematical process, new elements which are considered also to belong to the set in question' (Brouwer 1975, 82, modified).[77] Brouwer gave as examples the definable real numbers and

[76]That is the translation Tieszen uses. 'Mathematische und sonstige irreale Gegenstände "gibt es", die noch niemand konstruiert hat. Ihr Dasein erweist freilich erst ihre Konstruktion (ihre "Erfahrung"), aber die Konstruktion der schon bekannten eröffnet voraus einen Horizont weiter entdeckbarer, wenn auch noch unbekannter. Solange sie nicht entdeckt sind (von niemandem), sind sie nicht faktisch in der raum-Zeitlichkeit, und sofern es möglich ist (darüber wie weit dies möglich ist, braucht nicht entschieden zu werden), daß sie nie entdeckt worden wären, hätten sie überhaupt keine Weltwirklichkeit. (Husserl 1985b, 312).

[77]'waarvan niet anders dan een aftelbare groep welgedefinieerd is aan te geven, maar waar dan tevens dadelijk volgens een of ander vooraf gedefinieerd wiskundig proces uit elke zoodanige aftelbare groep nieuwe elementen zijn af te leiden, die gerekend worden eveneens tot de verzameling in kwestie te behooren' (Brouwer 1907, 148–149).

the ordinals of Georg Cantor's second number class (respectively, Brouwer 1907, 82, and Brouwer 1908A, 570). He then observed that, a fortiori, the totality of all intuitionistic constructions is denumerably unfinished. In effect, he thereby confirmed the existence of such a horizon as Husserl is speaking of here; note that this horizon exists for both actual humans and the ideal subject (see C11 above).

As for the second half of Husserl's passage ('As long as ...'), Brouwer defines mathematical existence in terms of constructibility by an ideal subject (C11 above). In that sense, there may well exist mathematical objects of which a given actual human or group of humans will, as a matter of fact, never discover their constructibility. But it would be contradictory to assume that a mathematical object can exist without it being in principle discoverable for humans, because the ideal subject, and hence mathematical existence, is defined by what is in principle constructible by the human mind. (The question whether the same would also be true according to Husserl will be addressed in Sect. 12.3 below.)

Hill (2010, 64–65) sees a contrast between Brouwer's view that mathematics has its origin in the perception of a move of time and Husserl's rejection of theories of number based on intuitions of time in *On the Concept of Number* and the *Philosophy of Arithmetic*. But there is no such contrast. What Husserl criticises are views according to which time enters into the content of the concept of number, but it is not Brouwer's view that it does.[78] For him, the concept of ordinal number is defined in terms of nested empty two-ities, which are purely formal (categorial) objects. The order of the ordinal numbers is certainly founded on the order that is intrinsic to time, but time as such does not enter into the content of the concept of ordinal number. The same holds for cardinal numbers, which are defined in terms of ordinal numbers and abstraction from ordering. It is rather in the genesis of the two-ities that the intuition of time plays its role; hence, Brouwer's reference to that intuition as the 'origin'. Brouwer's position actually agrees with (a transcendental reading of) Husserl's conclusion of his discussion of this matter in relation to cardinal numbers in the *Philosophy of Arithmetic*:

> Thus we see that time only plays the role of a psychological precondition for our concepts, and that in a two-fold manner:
>
> (1) It is essential that the partial representations united in the representation of the multiplicity or number be present in our consciousness simultaneously.

[78] Nor, for that matter, is it Kant's. E.g., in his letter to Johann Schultz of November 25, 1788, there is this well-known passage: 'Time has, as you very well remark, no influence on the properties of numbers (as pure determinations of magnitude), as it does for example on the property of any alteration (as of a quantum), which itself is possible only relative to a specific property of the inner sense and its form (time), and the science of number is, regardless of the succession that any construction of magnitude requires, a pure intellectual synthesis, which we represent to ourselves in thought' ('Die Zeit hat, wie Sie ganz wohl bemerken, keinen Einflus auf die Eigenschaften der Zahlen (als reiner Größenbestimmungen), so wie etwa auf die Eigenschaft einer jeden Veränderung (als eines Qvanti), die selbst nur relativ auf eine specifische Beschaffenheit des inneren Sinnes und dessen Form (die Zeit) möglich ist, und die Zahlwissenschaft ist, unerachtet der Succession, welche jede Construction der Größe erfodert, eine reine intellectuelle Synthesis, die wir uns in Gedanken vorstellen', Kant 1900–, 10:556–557). I take it that when in the *Critique of Pure Reason*, A142–43/B182, Kant speaks of number as a schema, he is speaking of number in so far as that concept is constructible by us.

> (2) Almost all representations of multiplicities – and, in any case, all representations of numbers – are results of processes, are wholes originated gradually out of their elements. Insofar as this is so, each element bears in itself a different temporal determination.
>
> But we found that neither simultaneity nor successiveness in time enters in any way into the (logical) content of the representation of the multiplicity; and so, likewise, into that of the representation of number. (Husserl 2003a, 33)[79]

One readily finds agreement on related points in Husserl's discussion: like Husserl, Brouwer does not hold that the mere noticing of a temporal sequence of contents suffices to mark out a determinate multiplicity; to mark those contents as a determinate multiplicity, they have to be combined, and this happens by repeatedly forming, in spontaneous acts, two-ities out of the present content and the preceding ones. (See also Brouwer 1907, 179n.) This operation is also central to the formation of 'complexes of sensations' (A4 above). For the same reason, Brouwer would agree with Husserl that to perceive temporally successive contents does not yet mean to perceive contents as temporally successive (Husserl 1970, 29). The latter requires framing the contents in a categorial structure.

Finally, in his concern with essential properties of the mind (C11), and his recognition of inner time as opposed to physical time as the time that is relevant for the constitution of pure mathematics, Brouwer agrees with Husserl that mathematics is not a chapter of psychology; Brouwer is not concerned, as Hill (2010, 64) thinks he is, with 'what happens in or to real temporal matters of fact that we call mental experiences of experiencing individuals'.

More generally, Brouwer can readily acknowledge that even though the construction of any mathematical object requires the perception of the move of time, this does not by itself mean that time enters into the content of mathematical concepts (the concept of number being an example). When the later Husserl (around 1917) abandoned the view that purely ideal objects are atemporal, and came to hold that they are omnitemporal (because their constitution presupposes that of time), he must have relied on a very similar observation (Lohmar 1993; van Atten 2007, Sect. 5.4.1). But it was a fundamental insight of Brouwer that the possibility that (some notion of) time does enter into some mathematical concept is not excluded either. So far, no other example than that of choice sequence (or structures based

---

[79]'Wir sehen also, die Zeit spielt für unsere Begriffe nur die Rolle einer psychologischen Vorbedingung und dies in doppelter Weise:

(1) Es ist unerläßlich, daß die in der Vorstellung der Vielheit bzw. Anzahl geeinigten Teilvorstellungen zugleich in unserem Bewußtsein vorhanden sind.
(2) Fast alle Vielheitsvorstellungen und jedenfalls alle Zahlvorstellungen sind Resultate von Prozessen, sind aus den Elementen sukzessive entstandene Ganze. Insofern trägt jedes Element eine andere zeitliche Bestimmtheit an sich.

Wir erkannten aber, daß weder die Gleichzeitigkeit noch die Aufeinanderfolge in der Zeit in den Inhalt der Vielheits- und somit auch der Zahlvorstellungen irgendwie eintreten.' (Husserl 1970, 32)

on it, such as the 'spread') has been found. Husserl, who knew about choice sequences through works by Becker and by Hermann Weyl that he read, never directly discussed the question whether choice sequences are mathematical objects or not.[80]

A real contrast between Husserl's account of number in the *Philosophy of Arithmetic* and Brouwer's lies elsewhere: Brouwer holds that cardinal numbers genetically depend on ordinal numbers, while Husserl thinks it is the other way around and that his analyses will confirm this (Husserl 1970, 11, 13). Elsewhere, I have argued at length that Brouwer's view here is the phenomenologically more accurate one, and I will not repeat that discussion here (van Atten 2004c).[81]

Rosado Haddock sees in Husserl's distance from Kant also a distance from Brouwer:

> Moreover, Husserl never seemed to have retracted of his classification of Kant's views in Chap. VII of LU I as a sort of specific relativism. Hence, one should not press too much the affinities between Husserl's and Kant's transcendental philosophies, and beware of assessing Husserl's views as a foundation of Brouwer's Fichtean mathematical subject. (Rosado Haddock 2006, 200n40)[82]

For the transcendental Husserl, the relevant contrast drawn in that chapter of the *Logical Investigations* is that between an individual or species on the one hand and the ideal cognising subject (see C11 above) on the other. Husserl rejects the necessity of a correspondence between the subjective capacities of the former and the laws of logic and mathematics; but he says such a correspondence does exist with the latter. As it is the latter that serves, as argued above, as a foundation of Brouwer's mathematical subject, and not an individual or species, the arguments in Chap. 7 of the *Prolegomena* are no obstacle to a transcendental-phenomenological justification of intuitionism.

12.3 Beyond Intuitionistic Mathematics?

So far, the fundamental constructive principles in intuitionism have been limited to the 'two acts' described in fragment B; and eventually, Brouwer came to see that the second act is a special case of the first (Brouwer 1981A, 93n). However, it is also a basic tenet of intuitionism to be prepared to extend these two acts whenever further descriptive analysis of consciousness warrants this; see the first quotation at C8 above.[83] In particular, had Becker's attempt in *Mathematische*

---

[80]See van Atten 2007, 72–74 for further discussion.

[81]A somewhat revised version has been published as an appendix in van Atten (2007).

[82]Note that the qualification 'Fichtean' is not Brouwer's. It is of course a separate question to what extent that qualification is applicable; I will not go into it here.

[83]Also Heyting: 'It is, as a matter of principle, impossible to devise a system of formulas that would be equivalent to intuitionistic mathematics, because the possibilities of thinking do not admit a

Existenz at a constructive, phenomenological foundation of Cantor's ordinals in terms of structures of self-reflection succeeded beyond the denumerable, this would have readily provided an intuitionistically acceptable extension of the two acts; but, in fact, it did not, as Becker acknowledges.[84] In his book, he admits that he can only give a clear phenomenological interpretation of ordinals 'up into the first epsilon numbers' (Becker 1927, 129),[85] which are still countable. Brouwer, in the notebooks for his dissertation from 1904–1907, had also briefly attempted to get the whole second and further number classes by appropriately differentiating the meaning of 'and so on' for each class (thus, like Becker, referring to differently structured horizons; Brouwer Archive, Notebook VI, 38). But he noticed that one cannot give a reasonable constructive sense to such differentiations, as witnessed by the explanation in his dissertation: 'The expression "and so on" means the indefinite repetition of one and the same object or operation, even if that object or that operation is defined in a rather complex way' Brouwer (1975, 80n).[86]

A more general question is whether Husserl's phenomenology can provide a foundation also for nonconstructive, classical mathematics, as notably Gödel hoped.[87] Is there a phenomenological basis for claiming that the mathesis universalis, understood as the theory of purely formal objects (see the citation from *Formal and Transcendental Logic* at the beginning of Sect. 12.2.1), is richer than Brouwer's intuitionism? In other words: are there purely categorial formations that go beyond the ones that Brouwer indicates or could have indicated?

Becker and Weyl held the view that constructive mathematics admits of a phenomenological foundation, but classical does not. The question has been raised to what extent their conceptions of phenomenology coincide with Husserl's, and to what extent Husserl (would have) agreed with them (Lohmar 1989, 195). The discussion below is meant to emphasise that in the transcendental Husserl one indeed finds all the elements needed to arrive at the same conclusion. Some

---

reduction to a finite number of rules that can be set up in advance.' ('Es ist prinzipiell unmöglich, ein System von Formeln aufzustellen, das mit der intuitionistischen Mathematik gleichwertig wäre, denn die Möglichkeiten des Denkens lassen sich nicht auf eine endliche Zahl von im voraus aufstellbaren Regeln zurückführen', Heyting 1930a, 3.)

[84]Note that Becker in his *Grundlagen der Mathematik in geschichtlicher Entwicklung* of 1954 describes the horizon phenomenon and the possibilities of reflection, but without reference to his work of 1927. He says: 'And thus one can endlessly proceed through the series of indices numbered by transfinite ordinal numbers – in so far as the ordinal numbers used can be defined constructively, and hence can be univocally and exactly named.' ('Und so kann man unbegrenzt fortfahren in der Reihe der durch transfinite Ordnungszahlen numerierten Indices – soweit sich die verwendeten Ordnungszahlen konstruktiv definieren und infolgedessen eindeutig und exakt bezeichnen lassen', Becker 1954, 386.) For a clear recent note on constructive transfinite ordinals, see Jervell (2006).

[85]These numbers are the infinite ordinals ε such that $\varepsilon = \omega^{\varepsilon}$. The smallest is $\varepsilon_0 = \omega^{\omega^{\omega^{\cdot^{\cdot^{\cdot}}}}}$

[86]'Waar men zegt "en *zoo* voort", bedoelt men het onbepaald herhalen van *eenzelfde* ding of operatie, ook al is dat ding of die operatie tamelijk complex gedefinieerd'. (Brouwer 1907, 143n)

[87]For Gödel's philosophical views, in particular in their relation to phenomenology, see, e.g., Tragesser (1977), Tieszen (1992), Føllesdal (1995), Parsons (1995), and van Atten and Kennedy (2003).

remarks on mathematics and minds that Gödel made after his turn to transcendental phenomenology will serve as a foil, and reveal a significant tension in his position.

The starting point is the voluntariness of categorial formation. Our experience shows that voluntary acts of human consciousness are, structurally, limited to the denumerable, in the sense that any series of them that we can carry out is either finite or potentially infinite (see C10 above); and Brouwer's two acts of intuitionism cover precisely that. There is in our experience no motivation whatsoever to think that, if only we could abstract from certain empirical limitations on our memory, lifetime, and so on, we could, on the basis of our intuition of time, make either the uncountable or the actual infinite intuitive.

In recognition of the limitation of our mental acts to the denumerable,[88] Gödel suggested that it is not necessary to insist that, ideally, mathematical objects can be completely given in our intuition. In his paper on Cantor (1964 version), he wrote:

> This negative attitude toward Cantor's set theory [is] only the result of a certain philosophical conception of the nature of mathematics, which admits mathematical objects only to the extent to which they are interpretable as our own constructions or, at least, can be completely given in mathematical intuition. For someone who considers mathematical objects to exist independently of our constructions and of our having an intuition of them individually, and who requires only that the general mathematical concepts must be sufficiently clear for us to be able to recognize their soundness and the truth of the axioms concerning them, there exists, I believe, a satisfactory foundation of Cantor's set theory in its whole original extent and meaning (Gödel 1964, 262).

Note that by the time Gödel wrote this passage, he had already made his turn to transcendental phenomenology (Wang 1987, 121–122); and although it is written in such a way as to admit different interpretations, evidence from Gödel's archive strongly suggests that he wrote it with the phenomenological perspective in mind. In a draft of Gödel's paper, there is an additional final paragraph that starts: 'Perhaps a further development of phenomenology will, some day, make it possible to decide questions regarding the soundness of primitive terms and their axioms in a completely convincing manner.'[89] Clearly, Gödel saw the Continuum Hypothesis[90] as a question about (or very close to) the primitive terms and their axioms: the problem shows (at least from a realist perspective) that we do not yet have a full grasp of the relation between the basic notions of exponentiation and cardinality. Therefore, I do not agree with Rosado Haddock (2010, 25) that the question is not about the very basis of classical set theory. Similarly, and again unlike Rosado Haddock, I

---

[88]'It has some plausibility that all things conceivable by us are denumerable' (Gödel 1946, 152). In his introduction to that paper, Parsons notes that this is ambiguous between 'For any x, if x is conceivable by us, then x is denumerable' and 'Only denumerably many things are conceivable by us', and says that the latter reading seems more likely (Gödel 1990, 148). I agree: at the point in the text where Gödel makes this remark, it is used to support an objection to accepting 'undenumerably many sets.'

[89]As quoted on p. 466 of van Atten and Kennedy (2003).

[90]This is Cantor's hypothesis that the power of the (classical) continuum, $2^{\aleph_0}$, is equal to the first uncountable cardinal, $\aleph_1$.

would say that questions about large cardinals lie at the heart of classical set theory because they show that we do not yet have a sufficient grasp of the concept of set and of the universe of sets to decide what basic objects there are in the universe.

A Husserlian objection to Gödel's suggestion in the passage just quoted is that, for a given mathematical concept to be admitted into truth-logic, we should be able to constitute at least one object falling under it with evidence. It is true that Husserl says that the starting point of eidetic variations may as well be an imagined object as one given in experience (Husserl 1985b, 411–412), which suggests that we can also constitute essences without having constituted with evidence an object that falls under it. But in the case of purely categorial objects, the distinction between constitution with evidence and imagination collapses: an object can only be clearly imagined to the extent that it is a possible object, and for purely categorial objects, possibility and existence coincide.[91]

Without having constituted a mathematical object that falls under the concept in question, the concept remains empty, a mere meaning but not an essence (Husserl 1952, Sect. 16). Remaining at the level of mere meanings, without intuitions of the objects meant, genuine knowledge of these objects cannot be obtained, and no eidetic variations can be performed on them. At best, we might be able to obtain purely analytic knowledge about the concepts involved. From an ontological perspective, this would not get us beyond if-thenism.

Although it has sometimes been said that a distancing from questions of truth and ontology is characteristic of the practice of modern pure mathematics,[92] that is certainly not the conception of anyone who, like Gödel and Brouwer, holds that mathematics describes a domain of objects (be they Platonic or mental).[93] If, on the other hand, a mathematical concept is not empty, and a purely categorial object can be given to us that falls under it, then, according to Husserl, that object can in principle be given adequately, it can, to borrow Gödel's expression, be 'completely given in mathematical intuition' (see C10 above). Gödel, in the passage quoted here, is concerned with the part of Cantorian set theory that is not constructive. But given that the human mind seems, as Becker's failed attempts emphasise, unable to constitute even a single uncountable ordinal, it seems that we cannot come to 'recognize the soundness and the truth of the axioms' concerning uncountable ordinals, for the reason given in the previous paragraph. A phenomenological foundation for Cantor's set theory will then not be forthcoming.

There seem to be two ways out of this difficulty that still respect in some sense the idealistic tenet in transcendental phenomenology: an appeal to ideas in the Kantian sense, and an appeal to higher minds.

---

[91]See Sect. 12.2.1 above, and van Atten (2002).

[92]In *Formal and Transcendental Logic*, Husserl says that the then prevailing understanding of purely formal mathematics among mathematicians is that of 'a pure analytics of consistency', and adds that on that understanding, the concept of truth remains unthematised (Husserl 1974, 15–16).

[93]With respect to mathematics, Gödel and Brouwer are both what may be called ontological descriptivists; see Sundholm and van Atten (2008, 71).

In Sects. 142–44 of *Ideas I*,[94] Husserl invokes ideas in the Kantian sense to resolve the following apparent contradiction. On the one hand, the fundamental principle of his idealism equates existence of an object with its being given adequately to a possible consciousness (Husserl 1976a, 329). On the other hand, there clearly exist objects (such as physical objects) that can never be given adequately. His solution is to say that in the latter case, adequate givenness of the object is nevertheless the correlate of an incompletable, infinite series of acts, in each of which a further partial determination of the same object obtained. Husserl says that such a series is lawlike or governed by a rule that depends on the type of the object. To the extent that we have evidence that we can actually begin such a series of experiences and then in principle can always continue it according to that rule, the positing of an object that is the correlate of the infinite series is rationally motivated. Husserl does not discuss whether ideas in the Kantian sense can play a role in the constitution of objects of pure mathematics, but Lohmar (1989, 194–195n15)[95] suggests that they can, and sees it as a means for Husserl to accept mathematical objects that do not satisfy the intuitionistic existence criteria. While Lohmar incorrectly holds that intuitionists accept only finite constructions, and correspondingly sees a use for Kantian ideas in attempts to go beyond the finite, a valid question is whether Kantian ideas provide a way to accept more infinite objects than the intuitionist can.

However, it seems to me that, if the suggestion to accept Kantian ideas in the account of mathematics works, it will not motivate positing anything beyond the denumerable. As soon as we try to be precise about the law or rule that governs the process of generating the infinite series, we run into the fact that humans can only follow it if it prescribes no more than a finite or potentially infinite number of steps. As noted above, it was because of this limitation that Becker's attempt at founding Cantor's ordinals failed for the non-denumerable. Moreover, it can be argued that this suggestion cannot work to begin with. For Husserl, purely categorial objects are immanent and can in principle be given adequately (C10 above), so the notion of a purely categorial object that as a matter of principle cannot be adequately given is contradictory. But only for objects that cannot be adequately given does it make sense to introduce adequate givenness as a Kantian idea (Husserl 1966, 21, Husserl 1976a, 332).[96] Conceptually at least, this would also explain the fact that Husserl does not give examples from pure mathematics (formal ontology) to illustrate the notion of a Kantian idea.[97] (It is important to recall here that the inapplicability

---

[94] See also Husserl (1974, 66–67n1), Husserl (1985b, 346), and Husserl (2009, 217).

[95] See also Lohmar (1989, 141–143).

[96] Note that my objection does not apply to Husserl's appeal to Kantian ideas in the context of geometry (which is not part of purely formal mathematics), e.g., in Sect. 74 of *Ideas I*. There, in an act of ideation an ideal geometrical notion is given through the elements of an incompletable infinite series of sensuous intuitions, as its ideal limit. But as that notion is not a (mereological) composition out of these elements, for our present purpose there is no relevant analogy with the constitution of higher-order categorial objects.

[97] ⟦Note added in this reprint: An exception would seem to be a passage in a manuscript probably from the early 1920s:

of Kantian ideas here does not mean that all purely categorial formations must be finite: as argued above, a potentially infinite sequence can also be adequately given.)

Turning now to the appeal to higher minds, the idea behind it is that, if a categorial object cannot be constituted by our minds, perhaps it can by minds of a higher type. So it is, perhaps, no coincidence that we find Gödel saying, to Hao Wang, 'For every set there is some mind which can overview it in the strict sense' (Wang 1996, 260). and

> What this idealization [to the integers as a totality, and also with arbitrary omissions] – realization of a possibility – means is that we conceive and realize the possibility of a mind which can do it. (Wang 1996, 220)

Indeed, Gödel believed that higher minds exist: in a private note (ca. 1960) headed 'Meine philosophische Ansicht[en?]', Gödel stated, 'There are other worlds and rational beings of a different and higher kind' (Wang 1996, 316).[98] In particular, Gödel appeals to a higher mind in an argument for the existence of the classical power set of ℕ (the set of the natural numbers), based on the idea that a subset of ℕ may be characterised by specifying what members of the number series are not elements of it:

> To arrive at the totality of integers involves a jump. Overviewing it presupposes an [idealized] infinite intuition. In the second jump we consider not only the integers as given but also the process of selecting integers as given in intuition. 'Given in intuition' here means [an idealization of] concrete intuition. Each selection gives a subset as an object. Taking all possible ways of leaving elements out [of the totality of integers] may be thought of as a method for producing these objects. What is given is a psychological analysis, the point is whether it produces objective conviction. This is the beginning of analysis [of the concept of set]. (Wang 1996, 220, emendations by Wang)

> Finally, there is the concept of a Kantian idea, which needs its own clarification. This comprises real and 'ideal', irreal objects, e.g., the number sequence '1, 2, 3 and so on' is a truly existing object, seeable as such, with this 'and-so-on'. A law of iteration and iterative construction is given to me with evidence, together with the indeterminate 'idea' of an open multiplicity. ('Endlich haben wir den Begriff der Kantischen Idee, der seiner eigenen Klärung bedarf. Umspannt sind hier reale und 'ideale', irreale Gegenstände, z.B. die Zahlenreihe '1, 2, 3 usw.' ist ein wahrhaft seiender, als das erschaubarer Gegenstand, mit diesem 'Und-so-Weiter'. Ich habe ein Gesetz der Iteration und iterativen Konstruktion einsichtig gegeben, neben der unbestimmten 'Idee' einer offenen Vielheit', Husserl 2012, 79.)

But I argue that the sequence as a potentially infinite one is given adequately here; it is only the sequence as an actually infinite one that might be given as a Kantian idea, inadequately. But in that case, it would be inconsistent with Husserl's ideas about categorial objects to think of this sequence as one of them. Be that as it may, it should be noted that, in its dependence on an iterative construction that is given to us with evidence, Husserl's use of the notion of Kantian idea is not one that would generalise to the non-denumerable.⟧

[98] 'Es gibt andere Welten und vernünftige Wesen einer anderen und höheren Art.' Transcription Robin Rollinger, according to whom the word 'einer' is actually hard to make out on the microfilm.

And Gödel elucidates: 'What this idealization [to the integers as a totality, and also with arbitrary omissions] – realization of a possibility – means is that we conceive and realize the possibility of a mind which can do it' (Wang 1996, 220). From other remarks to Wang, it is clear that Gödel not only accepts the psychological analysis but it also gives him the objective conviction: 'We can form the power set of a set, because we understand the selection process (of singling out any subset from the given set) intuitively, not blindly' (Wang 1996, 259), and 'From the idealized subjective view, we can get the power set' (Wang 1996, 260). This kind of reasoning leads to the idea of classical mathematics as constructive mathematics for higher minds,[99] with, perhaps, God as the highest mind.[100] The intuitionist, on the other hand, will see Gödel's psychological analysis as the analysis of a mistake: to speak of 'all possible ways of leaving elements out' of ℕ is to speak of something that is not denumerable, and hence not constructible in the intuitionistic sense.[101]

Objections to such quasi-constructive interpretations of axioms of classical set theory are of course well-known outside the phenomenological literature, from the discussions of the so-called iterative concept of set,[102] notably by Parsons ([1977] 1983) and by Hallett (1984, Chap. 6). Indeed, the point noted above that a higher being would need to have a time awareness with a far richer structure than ours was made in that context by Parsons ([1977] 1983, 273). When Wang in a letter drew Gödel's attention to Parsons' paper, Gödel wrote a shorthand remark, 'Often [he] continually confuses concepts and sets, and moreover he does not understand "idealization" broadly enough' (Gödel 2003a, 390, trans. Parsons)[103]; according to Parsons, it seems clear that that remark indeed refers to his paper. Gödel apparently thought that such idealisations as he made posed, or would be shown to pose, no obstacle to the phenomenological foundation of classical set theory that, as is known

[99]Compare Gödel's remark on Frank Ramsey's idea of using propositions of infinite length to provide a foundation of a classical theory of classes: 'Ramsey's viewpoint is, of course, everything but constructivistic, unless one means constructions of an infinite mind' (Gödel 1944, 145; see also 142).

[100]In the period of his intensive study of Leibniz in the 1940s, one idea of Leibniz that appealed to Gödel was that the objects of mathematics exist in God's mind; see van Atten (2009a, 7–8).

[101]Gödel's argument was first published in Wang (1974, 182); for criticism, see Parsons (1977) 1983 and Hallett (1984, 220).

[102]For expositions and defences of that conception see, e.g., Boolos 1971; Wang 1974, Chap. 6; Shoenfield 1977, 322–327.

[103]'Oft verwechselt fortwährend concepts und sets und ausserdem versteht er "Idealisierung" nicht genug weit'.

from other sources, he hoped to find (e.g., Gödel *1961/?, van Atten and Kennedy 2003, esp. 466, 69).[104]

But also on the basis of Husserl's texts an objection to Gödel's proposal can be developed, starting from the question: Can we say, with evidence, what categorial objects a certain mind higher than ours can constitute? It would not suffice if we know about this greater ability because it follows analytically from the concept of the specific higher being we are imagining, for then the evidence would still be lacking that the existence of such a mind is not a mere empty or a problematic possibility, but one that we could say with insight could indeed obtain, and hence one that we could appeal to in order to increase our knowledge about categorial objects.[107]

On the other hand, if our judgement on the higher mind's capacities is not analytic, but is arrived at by making its categorial formations intuitive to ourselves (perhaps after appropriate idealisations), then we are, in principle, able to do what that mind can, which is therefore, after all, not essentially higher than ours. In fact, in a text of 1909, Husserl emphatically denies that the notion of an essentially different knowing subject whose performance cannot even ideally be matched by ours makes any sense:

---

[104]It should be noted that, on another occasion, Gödel admitted that arguments that depend on idealising human mental (constructive) capacities are not always the most evident ones. In a letter to Paul Cohen of August 13, 1965, he wrote:

> As far as the axiom of the existence of inaccessibles[104] is concerned I think I slightly overstated my view. I would not say that its evidence is due solely to the analogy with the integers.[105]But I do believe that a clear analogy argument is much more convincing than the quasi-constructivistic argument in which we imagine ourselves to be able somehow to reach the inaccessible number. On the other hand, Levy's principle[106] might be considered more convincing than analogy. (Gödel 2003, 386)

[104]*Note MvA.* A cardinal number κ is (strongly) inaccessible if it is uncountable and neither the sum nor the product of κ numbers smaller than κ. Such an inaccessible, itself a set, can be used to build a model of the axioms of ZFC, and thereby establish the consistency of that theory. From Gödel's second Incompleteness Theorem it follows that, if ZFC is consistent, it cannot prove the existence of inaccessible numbers, because otherwise ZFC could prove its own consistency. The question of the existence of inaccessibles therefore concerns the acceptability of a new axiom.

[105]*Note MvA.* All numbers smaller than the countable infinite cardinal $\aleph_0$ are finite, but $\aleph_0$ is neither the sum nor the product of finitely many finite numbers. If one believes that $\aleph_0$ is not very special among the infinite cardinals, then there should exist an uncountable cardinal with the analogous property.

[106]*Note MvA.* This is a specific form of the general reflection principle. The latter (roughly) says that, if the universe of all sets has a certain property, then there is a set in the universe that also has it; the property of the universe is reflected in that set. Note that the reflection principle does not yield a construction of that set. The argument Gödel refers to uses reflection (roughly) as follows: If the cardinality of the universe is inaccessible, then by reflection the same is true of a set in the universe. For a discussion of Gödel's justification of the reflection principle by an analogy to Leibniz' monadology, see van Atten 2009a.

[107]Errett Bishop, who saw constructive mathematics as the mathematics that finite beings are capable of, said: 'If God has mathematics of his own that needs to be done, let him do it himself' (Bishop 1967, 2).

> One can entertain the thought that, just as man stands higher intellectually than minerals or [the] jellyfish, there might actually be beings that, compared to man, are more highly developed intellectually, in this sense, that they have fundamentally new ways of knowing at their disposal ... The common talk about possible cognitive natures that are not at all ours and have nothing to do with ours, is pointless, indeed nonsensical: for [then] there is nothing to sustain the unity of the concept of cognition. For there to be a point to talk of such possibilities, it should be about ways of knowing that are discernibly essentially identical to ours according to their generic character; and even when on factual grounds, on empirical psychological ones, such ways of knowing will never be present in our psyche and will never really be conceivable, a priori there should exist the possibility of an extension of our cognition as an ideal possibility, by which our cognition would itself become the cognition of that higher intellect that we are thinking of in an indirect and empty representation.[108]

Gödel, in the list referred to earlier, stated that 'The higher beings are connected to the others by analogy, not by composition' (Wang 1996, 316),[109] which strongly suggests that, correspondingly, at least part of our knowledge about these higher beings would be obtained by drawing analogies. But Husserl's insistence on the unity of knowledge rules out an appeal to analogy where idealisation falls short, and indeed should keep us from accepting the possibility of such beings altogether.[110] Thus we find a tension in Gödel's position: he wishes to use transcendental phenomenology to provide a foundation for classical mathematics, but the reference to higher beings that this seems to require has no phenomenological support. When I spoke to Robert Tragesser about these issues, it reminded him of Kierkegaard's *Fear and Trembling*: How unsatisfactory it is, when trying to make rational sense of the story of Abraham and Isaac, to be told, 'Surely no one was as great as Abraham. Who is able to understand him?' (Kierkegaard [1843] 2006, 11).

In various texts by the transcendental Husserl, we find this idea of the unity of knowledge applied to (knowledge of) categorial objects, and to mathematics in particular:

---

108 'Mann kann den Gedanken erwägen, dass wie der Mensch intellektuell höher steht als die Mineralien oder <die> Qualle, so es in Wirklichkeit Wesen geben mag, die dem Menschen gegenüber intellektuell höher entwickelt sind, und zwar so, dass sie über ganz neue, prinzipiell neue Erkenntnisarten verfügen. ... Die allgemeine Rede von möglichen Erkenntniswesen, die durchaus nicht unsere sind und mit unseren gar nichts zu tun haben, ist sinnlos, ist in der Tat widersinnig: da nichts vorhanden ist, was die Einheit des Begriffs der Erkenntnis aufrecht erhält. Soll von solchen Möglichkeiten sinnvoll die Rede sein, so muss es sich um Erlebnisarten handelen, die einsehbar wesensidentisch ihrem Gattungscharacter nach sind mit den unseren; und wenn auch aus faktischen Grunden, aus empirisch psychologischen, in unserer Seele nie solche Erkenntnisarten auftreten und wirklich vorstellbar sein können, so müsste es a priori die Möglichkeit einer Erweiterung unserer Erkenntnis bestehen als ideale Möglichkeit, durch die unsere Erkenntnis selbst zu der Erkenntnis jener in indirekt-leerer Vorstellung gedachten höheren Intellekte würde.' (Ms. K II 4, 109a/b, October 1909; quoted from Kern 1964, 129–130.)

109 The original reads: 'Die höheren Wesen sind durch Analogie, nicht durch Komposition mit den anderen verbunden.' Transcription Robin Rollinger.

110 For further discussion of Husserl's denial of essentially higher minds, see Kern (1964, 125–134); also of interest here is the correspondence between Becker and Mahnke (2005), together with Mancosu's introduction to it (Mancosu 2005).

> Speaking in eidetic generality, should every subject as such be capable of seeing every eidetic object [*Gegenständlichkeit*] (and likewise every object of sense, every ideal object in the widest sense)? The question should be answered affirmatively. This is a matter of an a priori [truth]. Of course, not every subject (not every actual or, in the attitude of the pure consideration of possibilities, every possible [subject], that is, posited as possible actuality) needs be thought of as actually cognising every eidetic object (or be thought of as cognising it).[111]

> The ideal total extension of the purely rational objects and the extension of the objective objects [*objektive Gegenstände*] cognizable to every possible rational I, coincide.[112]

> Any mathematical step that someone else makes, I should be able to repeat originaliter in myself ... It should be observed that a monosubjective [*einzelsubjektive*] mathematics is eo ipso intersubjective, and that, conversely, none is intersubjectively possible that is not already fully and completely founded monosubjectively.[113]

On Husserl's view, if one being is able actually to achieve less in mathematics than another, this is due only to contingent limitations:

> It should be realised that, what an I thinks (given the matter of thought), any I could think. This 'could' implies, however, that with every logical formation and with every grounding of truth, every possible inhibition is compatible, and that no essential law can reach into a logical one, as was shown in the *Logical Investigations*: just as every matter is freely variable, every inhibition is 'variable.' That means: the essential laws of freedom do presuppose an unfreedom ('lower' Psyche), but not one arising from positing and intervening.[114]

It seems to me that in Husserl's transcendental idealism, the fundamental 'unfreedom' is that imposed by the basic structure of inner time consciousness (C9 above); in Brouwer, this is wholly explicit. In the considerations so far, I have taken intuitive givenness as the criterion for accepting certain mathematical objects and

---

[111] 'Muss jedes Subjekt überhaupt, in eidetischer Allgemeinheit gesprochen, jede eidetische Gegenständlichkeit (und ebenso jede Sinnesgegenständlichkeit, jede ideale Gegenständlichkeit im weitesten Sinn) erschauen können? Die Frage ist zu bejahen. Es handelt sich hier um ein Apriori. Nicht jedes Subjekt braucht natürlich (nicht jedes wirkliche oder, in der Einstellung reiner Möglichkeitsbetrachtung, jedes mögliche, d.i. als mögliche Wirklichkeit angesetzte) jeden eidetischen Gegenstand wirklich zu erkennen (oder als ihn erkennend gedacht zu werden).' (Husserl 2003b, 147, from 1918 at the latest)

[112] 'Der ideale Gesamtumfang der rein rationalen Gegenstände und der Umfang der einem jeden möglichen Vernunft-Ich erkennbaren objektiven Gegenstände deckt sich.' (Husserl 1974, 388, from 1920–1921)

[113] 'Jeden mathematischen Schritt, den ein anderer macht, muß ich in mir selbst originaliter nachmachen können. ... Die Feststellung [ist] zu machen, dass eine einzelsubjektive Mathematik eo ipso intersubjektiv sei und umgekehrt keine intersubjektiv möglich ist, die nicht schon voll und ganz einzelsubjektiv begründet ist.' (Husserl 1974, 344, November 1926)

[114] 'Es muß eingesehen werden, daß, was ein Ich denkt (die Denkmaterie aber vorausgesetzt), jedes Ich denken könnte. Dieses Könnte besagt aber, daß mit jedem logischen Gebilde und jeder Wahrheitsbegründung jede mögliche Hemmung verträglich ist und daß kein Wesensgesetz in ein logisches hineingreifen kann, wie in den Logischen Untersuchungen gezeigt ist: Wie jede Materie frei variabel ist, so ist auch jede Hemmung "variabel". Das sagt: Die Wesensgesetze der Freiheit setzen zwar eine Unfreiheit ("niedere" Psyche) voraus, aber keine durch Setzen und Eingreifen.' (Husserl 1974, 386n1, from 1920–1921)

axioms about them; but that need not be the only rationally motivated criterion. Examples of additional criteria would be the inductive one of the success of an axiom in deciding a number of important open questions (Gödel 1964, 265), or the cognitive-aesthetic one of the extent to which the introduction of certain objects simplifies a theory. But in the context of Husserl's transcendental idealism, if we are not to break the tie to truth and evidence, decisions based on such additional criteria can only be provisional, to be replaced, ideally, by the insight that the objects and axioms thus introduced can be constituted by the ideal subject. Such additional criteria can therefore only play a heuristic role. Finally, it should be remarked that the transcendental Husserl not only saw a correlation between consciousness and being, according to which existence is equivalent to accessibility to a possible consciousness, but moreover held that the existence of consciousness is absolute, while that of all beings constituted by it is not (Husserl 1976a, 120). Without going into systematic issues concerning the meaning and strength of this second claim, I note that at least one occasion can be documented on which Gödel was critical of it.[115] The discussion of Gödel's view in the present section, however, has only been concerned with the correlation as such, and in particular with the notion of 'possible consciousness' that occurs in it; my argument that at this point there is an incompatibility between Gödel's view and Husserl's transcendental idealism does therefore not depend on the correctness of Husserl's further-going claim.

12.4 A Historical Note

The claim defended above that Brouwer's intuitionistic mathematics should be considered part of Husserl's transcendental phenomenology is not meant in a historical sense. In contrast to other mathematicians who reflected phenomenologically on mathematics, Brouwer did not take Husserl's (or any other phenomenologist's) writings as his point of departure.

He certainly could have done so. For example, by the time Brouwer began to develop intuitionism as a student, around 1904, Husserl had published two major works on the philosophy of mathematics and logic: *Philosophy of Arithmetic* in 1891 and *Logical Investigations* in 1900–1901. But Brouwer seems to have been quite unaware of them. In the nine notebooks in which he prepared his dissertation during the period 1904–1907 (Brouwer Archive, Notebooks I–IX) Husserl is not mentioned at all. (Incidentally, in his dissertation he did not refer to Gottlob Frege either; but in the notebooks there are some references to Frege's discussion with Hilbert on axiomatic geometry. Brouwer agrees with Frege's criticism (Brouwer Archive, Notebook VIII, 29, 32). The philosophers of logic and mathematics whom the young Brouwer did study extensively were Cantor, Louis Couturat, Richard Dedekind, David Hilbert, Kant, Felix Klein, Henri Poincaré, and Bertrand Russell.)

[115] For a brief discussion of this matter, see van Atten and Kennedy (2003, 454–455).

There is in Brouwer's archive and writings no evidence that he read Husserl's work at a later stage either; so we must suppose that, when the two met in person and had discussions, in Amsterdam in 1928, Brouwer's intuitionism had already become mature independently. Martin Heidegger, in his 1925 lectures entitled *History of the Concept of Time*, claims that intuitionism, of which he names Brouwer and Weyl as representatives, was 'essentially influenced by phenomenology' (*wesentlich von der Phänomenologie beeinflusst*, Heidegger 1979, 4); in Weyl's case, this is clear from his preface to *The Continuum* (Weyl 1918), but in Brouwer's case there is no evidence whatsoever for this. It is, on the other hand, known that Husserl owned a copy of Brouwer's paper 'Intuitionistische Betrachtungen über den Formalismus' (Brouwer 1928A2), which appeared in the year before Husserl's *Formal and Transcendental Logic*, and is highly relevant to the themes of that book (see van Atten 2007, 128n7).

12.5 Concluding Remark

Rather than a direct influence from Husserl (or other phenomenologists) on Brouwer, it seems there was a close intellectual kinship between them. The object and methods of Brouwer's intuitionism have sufficiently much in common with Husserl's phenomenology as to be interpretable in the latter. Compared to Brouwer's own explicitly formulated philosophy (Brouwer 1949), Husserl's is the broader and more detailed framework. Moreover, Brouwer's many philosophical passages about his own position tend to state conclusions rather than give analyses that lead to them. His original descriptions and explanations are not explicitly phenomenological. Phenomenology can clarify, explicate and deepen the intuitionistic position; and this allows us to see intuitionism as an example of a fruitful application of phenomenological analysis to mathematics. Philosophically, of the applications of transcendental phenomenology to mathematics so far, intuitionistic mathematics has been the most successful one. It has led to phenomenologically grounded foundations that are coherent, detailed and comprehensive: it includes not just arithmetic, but also analysis, measure theory, topology, and algebra (Heyting 1956; Troelstra and van Dalen 1988). To show that one can go beyond this and that classical mathematics is a true mathematics in the sense of *Formal and Transcendental Logic*, general considerations about intentionality, meanings, essences, idealisations, and, perhaps, non-revisionism, will not do: what is still wanting is a concrete and detailed phenomenological foundation of even just one characteristically classical alleged truth, such as the existence of the power set of $\mathbb{N}$. On the basis of the foregoing considerations, the prospects for such a foundation within transcendental phenomenology seem to me to be dim. The reason for this lies not so much in Husserl's insistence on the spontaneous character of the constitution of mathematical objects, or in his refusal to allow an appeal to higher minds, as in the combination of these two attitudes.

Acknowledgements This text (except for the appendix) grew out of my talk at the conference 'Phénoménologie discrète : Le parcours intellectuel de Gian-Carlo Rota entre mathématiques et philosophie', Lille, November 8 and 9, 2009, of which a later version was presented in the 'Ideals of Proof Seminar', Paris, March 24, 2010. I thank the organisers for their invitations, and the audiences for their questions and comments. I am indebted to the Institute for Advanced Study for permission to quote from the Kurt Gödel Papers, The Shelby White and Leon Levy Archives Center, Institute for Advanced Study, Princeton, on deposit at Princeton University. I am grateful to Robin Rollinger for his transcriptions from Gödel's Gabelsberger shorthand. Mirja Hartimo kindly provided photocopies of some of the items in the bibliography. Special thanks for discussion and comments are due to Robert Tragesser.

Appendix: Null on Choice Sequences

Gilbert Null's review of my *Brouwer Meets Husserl* is centred around his attribution to me of two claims (Null 2008, 119, 128):

1. choice sequences are objects without identity criterion;
2. choice sequences are real (*reell*) parts of inner time.

Not only are these claims made nowhere in the book, they are directly and explicitly contradicted in it.The passage that Null quotes from p. 36 of my book to support the attribution of claim (1) denies only that (non-lawlike) choice sequences have an identity criterion that is extensional. It does not at all say that choice sequences can have no identity criterion whatsoever. Indeed, as I argue extensively in Chap. 6, in particular on pp. 92–93 (which Null nowhere refers to), the principle of individuation (in terms of the moment of the beginning of a choice sequence) provides the identity criterion, and this criterion I identify explicitly as intensional (e.g., p. 90).[116]

Null further maintains that my book 'notes ... but leaves the need for a closer analysis of choice event protentions ... unfulfilled' (Null 2008, 129). Not only is their role made fully explicit in the analysis on pp. 92–93, but this role is discussed at many other places as well because, as is explicit on p. 92, protentions are essential to the constitution of choice sequences as open-ended (pp. 6, 15, 36, 92, 97, 98, 105). In this light, it is incomprehensible that Null makes it seem as if my book, and 'Brouwerians' in general, neglect 'any present (actual) choice event's inner horizon of openly possible contrary futures' (Null 2008, 129).[117]

[116] In the intuitionistic literature, 'intensional identity' (and the related 'intensional equivalence') is a standard notion, both in philosophical and mathematical discussions (e.g., Dummett 2000a, 16–17, and Troelstra 1977, 5, respectively). Applied to choice sequences, 'extensional' means 'in terms of the numbers chosen in the sequence', 'intensional' means 'in terms of other aspects of how the sequence is given to us'. In particular, the moment at which a sequence was begun is such an intensional aspect. So are any of the restrictions that we may have imposed on our choices.

[117] It is worth pointing out that the identity criterion for choice sequences that Null himself suggests ('CIP', Null 2008, 127), is a criterion in which protentions play no role whatsoever: for $n < \omega$, CIP operates only on initial segments (which are finite sequences and as such given to us without

Claim (2) is denied outright on p. 92 of my book, for reasons given on p. 91. Null's mistaken attribution of claim (2) to me may well stem from my expressing agreement with Guido Küng's thesis (Küng 1975) that the noema is a real (*reell*) moment (p. 70 and note 162). But whatever the merits of the thesis of the noema as a real moment, I do consider it a thesis about the noema in its fullest extent; and I state just as explicitly (p. 71 and pp. 89–90) that I consider the noematic essence and the noematic nucleus as ideal objects that are omnitemporal.[118] On p. 128 of his review, Null says that I am also committed to the thesis that a number can occur in a choice sequence only once. But since it is in virtue of the noematic nucleus that an intentional act has whatever object it has, also on my conception of the noema it is perfectly possible to choose the same number more than once in a choice sequence. One might think that, if the full noema is taken to be a real (*reell*) moment of the noesis, then the noematic nucleus cannot be omnitemporal, as the concrete noesis can, by definition, not be repeated in time. But it does not follow from this that the noematic nucleus cannot be omnitemporal. According to Sect. 64c of *Experience and Judgment*, Husserl calls an object 'omnitemporal' (*allzeitlich*) if in its constitution, which always takes place at a specific moment in time, this moment in time does not 'enter into' (*eingehen*) that object, that is, this temporal determination is not part of what makes the object the object it is. But then it is not at all excluded that some appropriate part of a given noesis can be constituted at a different time as identically the same: namely, if in the constitution of that part the moment in time does not 'enter into it', is not part of what makes it the object it is.[119] Not only is this not at all excluded, it is exactly parallel to Husserl's explanation how categorial objects are, ontically, productions, yet omnitemporal (Husserl 1985b, 311)[120]; see the discussion of point C2 above.

This also bears on page 124 of Null's review. He there says that a choice sequence contains either (i) the senses or (ii) the referents of choice events in an associated choice process.[121] The problem with (ii), according to Null, is that in that case 'the available Husserlian approach [sic] leaves Van Atten's characterization of choice sequences as intratemporal prima facie unsupported', because 'numbers are omnitemporal if anything is' (Null 2008, 124). If this argument is to work, then Null must hold that, if a higher-order object (here, a choice sequence) is founded on omnitemporal objects (here, numbers), then that higher-order object must also be omnitemporal. As Null does not supply any argument for this idea, he is begging the question against those who defend that choice sequences are intratemporal objects.

---

an open horizon (van Atten 2007, 90)), and for $n = \omega$, we are dealing with sequences of infinitely many given numbers, which are not given as open-ended either. So CIP is not suitable as an identity criterion for Brouwer's choice sequences; they are not even among the objects it operates on.

[118]On this difference see also the remarks in Bernet et al. (1989, 94–95).

[119]This is discussed in van Atten (2008, 70–71).

[120]Quoted in van Atten (2008, 11–12).

[121]Incidentally, characterisation (ii) is explicitly stated in my book to be the appropriate one (e.g., van Atten (2008, 1, 24)).

One may ask, of course, whether an argument such as Null fails to supply nevertheless exists. This is not the case, because there are counterexamples (independently from the one that, as I argue, choice sequences are). A specific act in which I judge, with full evidence, that the number 2 is an omnitemporal object, is founded on the number 2: for if the number 2 did not exist, neither could this specific act of judgement with full evidence. But this act, when objectified, is not an omnitemporal object, as it exists only for a certain stretch of time.

Husserl makes a closely related point: 'That a subject conceives a proposition with evidence, lends the proposition locality,[122] and, as the thought of this thinker etc. a unique one, but not to the proposition as such, which would be the same when thought at different times'.[123] In Husserl's example, the particular thinker's particular thought episode of the proposition as evident is founded on the proposition, but does not share the temporal characteristics of the latter.

More generally, Husserl had already remarked, speaking of the constitution of a higher-order object on the foundation of lower-order ones, 'And even when the time-constituting acts of the lower level also enter, they need not do this in such a way that the times enter, like the objects themselves, into the objects constituted at the higher level'.[124] (Recall that Husserl says that 'such an irreality has the temporal being of supertemporality, of omnitemporality, which however is a mode of temporality'.[125])

In view of the above, Null's ascription of claims (1) and (2) to me is entirely mistaken, and, since his review turns on them, his discussion is, in effect, not about my book.[126]

---

[122] *Note MvA*. Husserl here means spatio-temporal locality; see Husserl (1985b, 311).

[123] 'Daß ein Subjekt ein Satz evident denkt, das gibt dem Satz Lokalität, und als gedachtem dieses Denkers etc. eine einzige, aber nicht dem Satz schlechthin, der derselbe wäre als zu verschiedenen Zeiten etc. gedachter.' (Husserl 1985b, 312–313)

[124] 'Und gehen auch die zeitkonstituierenden Akte der Unterstufe mit ein, so brauchen sie es doch nicht so zu tun, dass die Zeiten wie der Gegenständlichkeiten selbst in die höher konstituierten Gegenständlichkeiten eingehen.' (Husserl 1985b, 310)

[125] 'eine solche Irrealität [hat] das zeitliche Sein der Überzeitlichkeit, der Allzeitlichkeit, die doch ein Modus der Zeitlichkeit ist.' (Husserl 1985b, 313)

[126] There are other infelicitous claims in Null's review, of which I here briefly mention the following two (which, moreover, are incompatible with each other). (1) Discussing the tree he set up on p. 126 ('Fig. 1a'), Null claims of the paths through it that 'each of these 2 to the $\aleph_0$ many sequences is an ideal object of the sort accepted by Husserl, Weyl, Becker, and Kaufmann' (Null 2008, 127). This is not correct. On the one hand, these sequences are all predeterminate (in each sequence the n-th element is fixed from the beginning for each n); on the other hand, by laws one can specify at most denumerably many sequences. As a consequence, most of these $2^{\aleph_0}$ many sequences are predeterminate but not given by a law. But that combination is possible in neither of the respective varieties of constructivism by which Weyl, Becker and Kaufmann defined their philosophical positions. Husserl is not explicit about the matter, but, as I argue in the present paper, accepting non-constructible objects in mathematics is not an option in his framework either. (2) Null claims that Husserl's formal objects should be 'countable', which, as defined by Null, implies decidability of equivalence. But although classical mathematics, intuitionistic mathematics, recursive analysis, Bishop's constructive mathematics all have different conceptions of real number, in none of them is equality of real numbers decidable. This would leave Husserl's phenomenology incapable of founding any of them. But I do not think that Husserl anywhere actually poses or implies Null's countability condition.

References

Except where required otherwise,

1. Brouwer's writings are referred to according to the scheme in Van Dalen's bibliography (van Dalen 2008);
2. Gödel's writings are referred to according to the scheme in the *Collected Works* (Gödel 1986, 1990, 1995, 2003,a);
3. Husserl's writings are referred to in the *Husserliana* edition (Husserl 1950a–).

M. van Atten, Why Husserl should have been a strong revisionist in mathematics. Husserl Stud. **18**(1), 1–18 (2002)

M. van Atten, Intuitionistic remarks on Husserl's analysis of finite number in the *philosophy of arithmetic*. Grad. Fac. Philos. J. **25**(2), 205–225 (2004b)

M. van Atten, Review of *Gnomes in the Fog. The reception of Brouwer's intuitionism in the 1920s*, by Dennis Hesseling. Bull. Symb. Log. **10**(3), 423–427 (2004c)

M. van Atten, *Brouwer Meets Husserl: On the Phenomenology of Choice Sequences* (Springer, Dordrecht, 2007)

M. van Atten, Luitzen Egbertus Jan Brouwer, in Zalta 1997 (Winter 2008). http://plato.stanford.edu/archives/win2008/entries/brouwer

M. van Atten, Monads and sets: On Gödel, Leibniz, and the reflection principle, in Primiero and Rahman 2009 (2009a), pp. 3–33

M. van Atten, The development of intuitionistic logic, in Zalta 1997 (Summer 2009b). http://plato.stanford.edu/archives/sum2009/entries/intuitionistic-logic-development

M. van Atten, Construction and constitution in mathematics, in *The New Yearbook for Phenomenology and Phenomenological Philosophy*, vol. 10 (2010), pp. 43–90

M. van Atten, Kant and real numbers, in Dybjer et al. (2012), pp. 3–23

M. van Atten, *Essays on Gödel's Reception of Leibniz, Husserl, and Brouwer* (Springer, Dordrecht, 2015a)

M. van Atten, On the fulfillment of certain categorial intentions, in *The New Yearbook for Phenomenology and Phenomenological Philosophy*, vol. 13 (2015b), pp. 173–185

M. van Atten, P. Boldini, M. Bourdeau, G. Heinzmann (eds.), *One Hundred Years of Intuitionism (1907–2007): The Cerisy Conference* (Birkhäuser, Basel, 2008)

M. van Atten, J. Kennedy, On the philosophical development of Kurt Gödel. Bull. Symb. Log. **9**(4), 425–476 (2003)

J. Barwise (ed.), *Handbook of Mathematical Logic* (North-Holland, Amsterdam, 1977)

O. Becker, Mathematische Existenz: Untersuchungen zur Logik und Ontologie mathematischer Phänomene. Jahrbuch für Philosophie und phänomenologische Forschung **8**, 439–809 (1927)

O. Becker, *Grundlagen der Mathematik in geschichtlicher Entwicklung* (Alber, Freiburg/München, 1954)

O. Becker, D. Mahnke, Briefwechsel mit Dietrich Mahnke. With an introduction by Bernd Aust and Jochen Sattler, in Peckhaus (2005), pp. 245–278

P. Benacerraf, H. Putnam (eds.), *Philosophy of Mathematics: Selected Readings*, 1st edn. (Cambridge University Press, Cambridge, 1964)

P. Benacerraf, H. Putnam (eds.), *Philosophy of Mathematics: Selected Readings*, 2nd edn. (Cambridge University Press, Cambridge, 1983)

R. Bernet, I. Kern, E. Marbach, *Edmund Husserl: Darstellung seines Denkens* (Meiner, Hamburg, 1989)

E. Beth, H. Pos, J. Hollak (eds.), *Proceedings of the 10th International Congress of Philosophy, Amsterdam 1948*, vol 2, bk 1 (1949)

E. Bishop, *Foundations of Constructive Analysis* (McGraw-Hill, New York, 1967)

L. Boi, P. Kerszberg, F. Patras (eds.), *Rediscovering Phenomenology* (Springer, Dordrecht, 2007)
G. Boolos, The iterative concept of set. J. Philos. **68**, 215–231 (1971)
P. Bossert (ed.), *Phenomenological Perspectives* (Martinus Nijhoff, Den Haag, 1975)
D. Bridges, Constructive mathematics, in Zalta 1997– (Summer 2009). http://plato.stanford.edu/archives/sum2009/entries/mathematics-constructive/
L.E.J. Brouwer, Archive (Noord-Hollands Archief, Haarlem, The Netherlands)
L.E.J. Brouwer, Over de grondslagen der wiskunde. PhD Dissertation, Universiteit van Amsterdam. English translation in Brouwer 1975, 1907, pp. 11–101
L.E.J. Brouwer, *Die mögliche Mächtigkeiten* (1908A). Lecture, published in Castelnuovo 1909, pp. 569–571. Facsimile reprint in Brouwer 1975, pp. 102–104
L.E.J. Brouwer, De onbetrouwbaarheid der logische principes. Tijdschrift voor Wijsbegeerte **2**, 152–158 (1908C). English translation in Brouwer 1975, pp. 107–111
L.E.J. Brouwer, *Intuïtionisme en formalisme* (Clausen, Amsterdam, 1912A). English translation in Benacerraf and Putnam 1983, pp. 77–89
L.E.J. Brouwer, Intuitionism and formalism. Bull. Am. Math. Soc. **20**, 81–96 (1913C). Facsimile reprint in Brouwer 1975, pp. 123–138
L.E.J. Brouwer, Begründung der Mengenlehre unabhängig vom logischen Satz vom ausgeschlossenen Dritten: Erster Teil: Allgemeine Mengenlehre. KNAW Verhandelingen **5**, 1–43 (1918B). Facsimile reprint in Brouwer 1975, pp. 150–190
L.E.J. Brouwer, Über Definitionsbereiche von Funktionen. Math. Ann. **97**, 60–75 (1927B). Facsimile reprint in Brouwer 1975, pp. 390–405. English translation of Sects. 1–3 in van Heijenoort 1967, pp. 457–463
L.E.J. Brouwer, Intuitionistische Betrachtungen über den Formalismus. KNAW Proc. **31**, 374–379 (1928A2). Facsimile reprint in Brouwer 1975, pp. 409–414. English translation in Mancosu 1998, pp. 40–44
L.E.J. Brouwer, Mathematik, Wissenschaft und Sprache. Monatshefte für Mathematik und Physik **36**, 153–164 (1929A). Facsimile reprint in Brouwer 1975, pp. 417–428. English translation in Mancosu 1998, pp. 45–53
L.E.J. Brouwer, *Die Struktur des Kontinuums* (Komitee zur Veranstaltung von Gastvorträgen ausländischer Gelehrter der exakten Wissenschaften, Wien, 1930A). Facsimile reprint in Brouwer 1975, pp. 429–440. English translation in Mancosu 1998, pp. 54–63
L.E.J. Brouwer, Willen, weten, spreken, in Brouwer et al. 1933 (1933A2), pp. 45–63. English translation of Sect. 3 in Brouwer 1975, pp. 443–446. Full English translation in van Stigt 1990, pp. 418–431
L.E.J. Brouwer, Essentieel negatieve eigenschappen. Indag. Math. **10**, 322–323 (1948A). English translation in Brouwer 1975, pp. 478–479
L.E.J. Brouwer, Consciousness, philosophy and mathematics, in Bethetal 1949, pp. 1235–1249. Facsimile reprint in Brouwer 1975, pp. 480–494
L.E.J. Brouwer, Historical background, principles and methods of intuitionism. S. Afr. J. Sci. **49**, 139–146 (1952B). Facsimile reprint in Brouwer 1975, pp. 508–515
L.E.J. Brouwer, Points and spaces. Can. J. Math. **6**, 1–17 (1954A). Facsimile reprint in Brouwer 1975, pp. 522–538
L.E.J. Brouwer, *Philosophy and Foundations of Mathematics*. Vol 1 of *Collected Works*, ed. by A. Heyting (North-Holland, Amsterdam, 1975)
L.E.J. Brouwer, *Brouwer's Cambridge Lectures on Intuitionism*. ed. by D. van Dalen (Cambridge University Press, Cambridge, 1981A)
L.E.J. Brouwer, J. Clay, A. de Hartog, G. Mannoury, H. Pos, G. Révész, J. Tinbergen, J. van der Waals Jr. (eds.), *De uitdrukkingswijze der wetenschap: Kennistheoretische openbare voordrachten gehouden aan de Universiteit van Amsterdam gedurende de kursus 1932–1933* (Noordhoff, Groningen, 1933)
L.E.J. Brouwer, F. van Eeden, J. van Ginneken, G. Mannoury, Signifische dialogen. Synthese **2**(5,7,8), 168–174, 261–268, 316–324 (1937)
L.E.J. Brouwer, F. van Eeden, J. van Ginneken, G. Mannoury, *Signifische dialogen* (Erven J. Bijleveld, Utrecht, 1939). English translation in Brouwer 1975, pp. 447–452

R. Butts, J. Hintikka (eds.), *Logic, Foundations of Mathematics and Computability Theory* (D. Reidel, Dordrecht, 1977)

G. Castelnuovo (ed.), *Atti del IV Congresso internazionale dei matematici, Roma, 6–11 aprile 1908: Comunicazioni delle sezioni III-A, III-B e IV* (Tipografia della Reale Accademia dei Lincei, Roma, 1909)

D. van Dalen, *L.E.J. Brouwer en de grondslagen van de wiskunde* (Epsilon, Utrecht, 2001a)

D. van Dalen, A bibliography of L.E.J. Brouwer, in van Atten 2008, pp. 343–390

M. Davis (ed.), *The Undecidable: Basic Papers on Undecidable Propositions, Unsolvable Problems and Computable Functions* (Raven Press, Hewlett, 1965)

J. Dodd, Husserl between formalism and intuitionism, in Boi et al. 2007, pp. 261—265

M. Dummett, Is time a continuum of instants? Philosophy **75**, 497–515 (2000a)

M. Dummett, *Elements of Intuitionism*, 2nd rev edn. (Clarendon Press, Oxford, 2000b)

P. Dybjer, S. Lindström, E. Palmgren, G. Sundholm (eds.), *Epistemology versus Ontology: Essays on the Philosophy and Foundations of Mathematics in Honour of Per Martin-Löf* (Springer, Dordrecht, 2012)

L. Eley, *Metakritik der Formalen Logik* (Martinus Nijhoff, Den Haag, 1969)

D. Føllesdal, Gödel and Husserl, in Hintikka 1995, pp. 427–446

K. Gödel, Russell's mathematical logic, in Schilpp 1944, pp. 123–153. Reprinted, with original page numbers in the margin, in Gödel 1990, pp. 119–141

K. Gödel, *Remarks before the Princeton bicentennial conference on problems in mathematics* (1946). Lecture, first published in Davis 1965, pp. 84–88. Page references are to the reprint in Gödel 1990, pp. 150–153

K. Gödel, "What is Cantor's continuum problem? Am. Math. Mon. **54**, 515–525 (1947). Reprinted, with original page numbers in the margin, in Gödel 1990, pp. 176–187

K. Gödel, *The modern development of the foundations of mathematics in the light of philosophy* (*1961/?). Lecture draft in German, published, with an English translation, in Gödel 1995, pp. 374–387. The English title is Gödel's

K. Gödel, What is Cantor's continuum problem? in Benacerraf and Putnam 1964, pp. 258–273. Revised and expanded version of Gödel 1947. Reprinted, with original page numbers in the margin, in Gödel 1990, pp. 254–270

K. Gödel, *Publications 1929–1936*. Vol 1 of *Collected Works*, eds. by S. Feferman, J. Dawson Jr., S. Kleene, G. Moore, R. Solovay, J. van Heijenoort (Oxford University Press, Oxford, 1986)

K. Gödel, *Publications 1938–1974*. Vol 2 of *Collected Works*, eds. by S. Feferman, J. Dawson Jr., S. Kleene, G. Moore, R. Solovay, J. van Heijenoort (Oxford University Press, Oxford, 1990)

K. Gödel, *Unpublished Essays and Lectures*. Vol 3 of *Collected Works*, eds. by S. Feferman, J. Dawson Jr., W. Goldfarb, C. Parsons, R. Solovay (Oxford: Oxford University Press, 1995)

K. Gödel, *Correspondence A–G*. Vol 4 of *Collected Works*, eds. by S. Feferman, J. Dawson Jr., W. Goldfarb, C. Parsons, W. Sieg (Oxford University Press, Oxford, 2003)

K. Gödel, *Correspondence H–Z*. Vol 5 of *Collected Works*, eds. by S. Feferman, J. Dawson Jr., W. Goldfarb, C. Parsons, W. Sieg (Oxford University Press, Oxford, 2003a)

M. Hallett, *Cantorian Set Theory and Limitation of Size* (Clarendon Press, Oxford, 1984)

M. Hartimo (ed.), *Phenomenology and Mathematics* (Springer, Dordrecht, 2010)

M. Heidegger, *Prolegomena zur Geschichte des Zeitbegriffs. Heidegger Gesamtausgabe*, vol 20, ed. by P. Jaeger (Vittorio Klostermann, Frankfurt am Main, 1979)

J. van Heijenoort (ed.), *From Frege to Gödel: A Sourcebook in Mathematical Logic, 1879–1931* (Harvard University Press, Cambridge, 1967)

D. Hesseling, *Gnomes in the Fog: The Reception of Brouwer's Intuitionism in the 1920s* (Birkhäuser, Basel, 2003)

A. Heyting, Die formalen Regeln der intuitionistischen Logik. Pt. 1. Sitzungsberichte der Preussischen Akademie der Wissenschaften 42–56 (1930a). English translation in Mancosu 1998, pp. 311–327

A. Heyting, *Intuitionism: An Introduction* (North-Holland, Amsterdam, 1956)

A. Heyting, L.E.J. Brouwer, in Klibansky 1968, pp. 308–315

D. Hilbert, Neubegründung der Mathematik (Erste Mitteilung). Abhandlungen aus dem Mathematischen Seminar der Hamburgischen Universität **1**, 157–177 (1922). English translation in Mancosu 1998, pp. 198–214

C. Hill, Husserl on axiomatization and arithmetic, in Hartimo 2010, pp. 47–71

J. Hintikka (ed.), *From Dedekind to Gödel: Essays on the Development of the Foundations of Mathematics* (Kluwer, Dordrecht, 1995)

E. Husserl, *Cartesianische Meditationen und Pariser Vorträge. Husserliana*, vol. 1, ed. by S. Strasser (Martinus Nijhoff, Den Haag, 1950a)

E. Husserl, *Ideen zu einer reinen Phänomenologie und phänomenologischen Philosophie: Drittes Buch: Die Phänomenologie und die Fundamente der Wissenschaften. Husserliana*, vol. 5, ed. by M. Biemel (Martinus Nijhoff, Den Haag, 1952)

E. Husserl, *Erste Philosophie (1923/1924): Erster Teil: Kritische Ideengeschichte. Husserliana*, vol. 7, ed. by R. Boehm (Martinus Nijhoff, Den Haag, 1956a)

E. Husserl, *Erste Philosophie (1923/1924): Zweiter Teil: Theorie der phänomenologischen Reduktion. Husserliana*, vol. 8, ed. by R. Boehm (Martinus Nijhoff, Den Haag, 1959)

E. Husserl, *Phänomenologische Psychologie. Husserliana*, vol. 9, ed. by W. Biemel (Martinus Nijhoff, Den Haag, 1962)

E. Husserl, *Analysen zur passiven Synthesis (1918–1926). Husserliana*, vol. 11, ed. by M. Fleischer (Martinus Nijhoff, Den Haag, 1966)

E. Husserl, *Philosophie der Arithmetik. Husserliana*, vol. 12, ed. by L. Eley (Martinus Nijhoff, Den Haag, 1970)

E. Husserl, *Cartesian Meditations*, trans. by D. Cairns (Martinus Nijhoff, Den Haag, 1973d)

E. Husserl, *Experience and Judgment*, trans. by J. Churchill, K. Ameriks (Routledge & Kegan Paul, London, 1973e)

E. Husserl, *Formale und transzendentale Logik. Husserliana*, vol. 17, ed. by P. Janssen (Martinus Nijhoff, Den Haag, 1974)

E. Husserl, *Logische Untersuchungen: Erster Band. Husserliana*, vol. 18, ed. by E. Holenstein (Martinus Nijhoff, Den Haag, 1975)

E. Husserl, *Ideen zu einer reinen Phänomenologie und phänomenologischen Philosophie: Erstes Buch. 1. Halbband: Text der 1.–3. Auflage. Husserliana*, vol. 3/1, ed. by K. Schuhmann (Martinus Nijhoff, Den Haag, 1976a)

E. Husserl, *Ideen zu einer reinen Phänomenologie und phänomenologischen Philosophie: Erstes Buch. 2. Halbband: Ergänzende Texte (1912—1929). Husserliana*, vol. 3/2, ed. by K. Schuhmann (Martinus Nijhoff, Den Haag, 1976b)

E. Husserl, *Aufsätze und Rezensionen (1890–1910). Husserliana*, vol. 22, ed. by B. Rang (Martinus Nijhoff, Den Haag, 1979)

E. Husserl, *Logische Untersuchungen: Zweiter Band, 2. Teil. Husserliana*, vol. 19/2, ed. by U. Panzer (Martinus Nijhoff, Den Haag, 1984b)

E. Husserl, *Einleitung in die Logik und Erkenntnistheorie: Vorlesungen 1906/07. Husserliana*, vol. 24, ed. by U. Melle (Martinus Nijhoff, Den Haag, 1985a)

E. Husserl, *Erfahrung und Urteil*, ed. by L. Landgrebe (Meiner, Hamburg, 1985b)

E. Husserl, *Aufsätze und Vorträge (1922–1937). Husserliana*, vol. 27, ed. by T. Nenon, H. Sepp (Martinus Nijhoff, Den Haag, 1988)

E. Husserl, *Logik und allgemeine Wissenschaftstheorie: Vorlesungen 1917/18. Mit ergänzenden Texten aus der ersten Fassung 1910/11. Husserliana*, vol. 30, ed. by U. Panzer (Kluwer, Den Haag, 1995a)

E. Husserl, *Natur und Geist: Vorlesungen Sommersemester 1927. Husserliana*, vol. 32, ed. by M. Weiler (Kluwer, Dordrecht, 2001b)

E. Husserl, *Zur phänomenologischen Reduktion: Texte aus dem Nachlass (1926–1935). Husserliana*, vol. 34, ed. by S. Luft (Kluwer, Dordrecht, 2002b)

E. Husserl, *Einleitung in die Philosophie: Vorlesungen 1922/23. Husserliana*, vol. 35, ed. by B. Goossens (Kluwer, Dordrecht, 2002c)

E. Husserl, *Philosophy of Arithmetic: Psychological and Logical Investigations with Supplementary Texts from 1887–1901. Edmund Husserl Collected Works*, vol. 10, ed. and trans. by D. Willard (Kluwer, Dordrecht, 2003a)

E. Husserl, *Transzendentaler Idealismus: Texte aus dem Nachlass (1908–1921). Husserliana*, vol. 36, ed. by R. Rollinger in collaboration with R. Sowa (Kluwer, Dordrecht, 2003b)

E. Husserl, *Untersuchungen zur Urteilstheorie: Texte aus dem Nachlass (1893–1918). Husserliana*, vol. 40, ed. by R. Rollinger (Springer, Dordrecht, 2009)

E. Husserl, *Zur Lehre vom Wesen und zur Methode der eidetischen Variation: Texte aus dem Nachlass (1893–1918). Husserliana*, vol. 41, ed. by D. Fonfara (Springer, Dordrecht, 2012)

R. Iemhoff, Intuitionism in the philosophy of mathematics, in Zalta 1997– (Winter 2009). http://plato.stanford.edu/archives/win2009/entries/intuitionism/

H. Jervell, Constructing ordinals, in *Cahier spécial 6: Constructivism: Mathematics, Logic, Philosophy and Linguistics. Philosophia Scientiae*, ed. by G. Heinzmann, G. Ronzitti (2006), pp. 5–20

I. Kant, *Gesammelte Schriften*, 29 vols., ed. by A. der Wissenschaften (Reimer, Berlin, 1900–). (from 1920 De Gruyter)

I. Kern, *Husserl und Kant* (Martinus Nijhoff, Den Haag, 1964)

S. Kierkegaard, *Fear and Trembling*, ed. by S. Evans, S. Walsh (Cambridge University Press, Cambridge, 2006)

R. Klibansky (ed.), *Logic and Foundations of Mathematics*. Vol. 1 of *Contemporary Philosophy. A Survey* (La Nuova Italia editrice, Firenze, [Johannes de Silentio, pseud.]. (1843) 1968)

G. Kreisel, Mathematical logic, in Saaty 1965, pp. 95–195

G. Küng, Das Noema als reelles Moment, in Bossert 1975, pp. 151–153

D. Lohmar, *Phänomenologie der Mathematik: Elemente einer phänomenologischen Aufklärung der mathematischen Erkenntnis nach Husserl* (Kluwer, Dordrecht, 1989)

D. Lohmar, On the relation of mathematical objects to time: are mathematical objects timeless, overtemporal or omnitemporal? J. Indian Counc. Philos. Res. **10**(3), 73–87 (1993)

D. Lohmar, The transition of the principle of excluded middle from a principle of logic to an axiom: Husserl's hesitant revisionism in the field of logic, in *The New Yearbook for Phenomenology and Phenomenological Philosophy*, vol. 4 (2004), pp. 53–68

D. Mahnke, *Eine neue Monadologie. Kantstudien Ergänzungsheft*, vol. 39 (1917)

P. Mancosu (ed.), *The Debate on the Foundations of Mathematics in the 1920s*, in *From Brouwer to Hilbert* (Oxford University Press, Oxford, 1998)

P. Mancosu, Das Abenteuer der Vernunft: O. Becker and D. Mahnke on the phenomenological foundations of the exact sciences, in Peckhaus 2005, pp. 229–243

J. Mensch, *After Modernity: Husserlian Reflections on a Philosophical Tradition*. Chap. Intersubjectivity and the constitution of time (State University of New York Press, Albany, 1996), pp. 57–66

J.R. Moschovakis, Intuitionistic logic, in Zalta 1997– (Fall 2008). http://plato.stanford.edu/archives/fall2008/entries/logic-intuitionistic/

G. Null, Entities without identities vs. temporal modalities of choice: review of Mark van Atten, *Brouwer Meets Husserl*. Husserl Stud. **24**, 119–130 (2008)

C. Parsons, (1977) What is the iterative conception of set? in Parsons 1983, pp. 268–297. Originally in Butts and Hintikka 1977, pp. 335–367

C. Parsons, *Mathematics in Philosophy: Selected Essays* (Cornell University Press, Ithaca, 1983)

C. Parsons, Platonism and mathematical intuition in Kurt Gödel's thought. Bull. Symb. Log. **1**(1), 44–74 (1995)

C. Parsons, Finitism and intuitive knowledge, in Schirn 1998 (Oxford, 1998), pp. 249–270

V. Peckhaus (ed.), *Oskar Becker und die Philosophie der Mathematik* (Wilhelm Fink Verlag, München, 2005)

A. Pfänder, *Philosophie auf phänomenologischer Grundlage: Einleitung in die Philosophie und Phänomenologie*, ed. by H. Spiegelberg (Fink, München, 1973)

T. Placek, *Mathematical Intuitionism and Intersubjectivity: A Critical Exposition of Arguments for Intuitionism* (Kluwer, Dordrecht, 1999)

G. Primiero, S. Rahman (eds.), *Judgement and Knowledge: Papers in honour of B.G. Sundholm* (College Publications, London, 2009)
G.E. Rosado Haddock Review of *mathematical intuition* by Richard Tieszen. J. Symb. Log. **56**(1), 356–360 (1991)
G.E. Rosado Haddock, Husserl's philosophy of mathematics: its origin and relevance. Husserl Stud. **22**(3), 193–222 (2006)
G.E. Rosado Haddock, Platonism, phenomenology and interderivability, in Hartimo 2010, pp. 23–46
T. Saaty (ed.), *Lectures on Modern Mathematics*, vol. 3 (Wiley, New York, 1965)
P.A. Schilpp (ed.), *The Philosophy of Bertrand Russell. The Library of Living Philosophers*, vol. 5 (Nortwestern University Press, Evanston, 1944). 3rd edn (Tudor, New York, 1951)
M. Schirn (ed.), *The Philosophy of Mathematics Today* (Oxford University Press, Oxford, 1998)
K. Schuhmann, *Reine Phänomenologie und phänomenologische Philosophie* (Martinus Nijhoff, Den Haag, 1973)
J. Shoenfield, Axioms of set theory, in Barwise 1977, pp. 321–344
B. Smith, D. Smith (eds.), *The Cambridge Companion to Husserl* (Cambridge University Press, Cambridge, 1995)
W. van Stigt, *Brouwer's Intuitionism* (North-Holland, Amsterdam, 1990)
G. Sundholm, M. van Atten, The proper interpretation of intuitionistic logic: on Brouwer's demonstration of the Bar theorem, in vanAtten 2008, pp. 60–77
W. Tait, Finitism. J. Philos. **78**(9), 524–546 (1981)
R. Tieszen, *Mathematical Intuition: Phenomenology and Mathematical Knowledge* (Kluwer, Dordrecht, 1989)
R. Tieszen, Kurt Gödel and phenomenology. Philos. Sci. **59**, 176–194 (1992)
R. Tieszen, Mathematics, in Smith and Smith 1995, pp. 438–462
R. Tieszen, Mathematical realism and transcendental phenomenological idealism, in Hartimo 2010, pp. 1–22
R. Tragesser, *Phenomenology and Logic* (Cornell University Press, Ithaca, 1977)
A. Troelstra, *Choice Sequences: A Chapter of Intuitionistic Mathematics* (Oxford University Press, Oxford, 1977)
A. Troelstra, On the origin and development of Brouwer's concept of choice sequence, in Troelstra and van Dalen 1982, pp. 465–486
A. Troelstra, D. van Dalen (eds.), *The L.E.J. Brouwer Centenary Symposium* (North-Holland, Amsterdam, 1982)
A. Troelstra, D. van Dalen, *Constructivism in Mathematics: An Introduction*, 2 vols. (North-Holland, Amsterdam, 1988)
H. Wang, *From Mathematics to Philosophy* (Routledge/Kegan Paul, London, 1974)
H. Wang, *Reflections on Kurt Gödel* (MIT Press, Cambridge, 1987)
H. Wang, *A Logical Journey: From Gödel to Philosophy* (MIT Press, Cambridge, 1996)
H. Weyl, *Das Kontinuum: Kritische Untersuchungen über die Grundlagen der Analysis* (Veit, Leipzig, 1918)
L. Wittgenstein, *Tractatus logico-philosophicus*, ed. by J. Schulte (Suhrkamp, Frankfurt am Main, (1921) 2013)
J. Yoshimi, Mathematizing phenomenology. Phenomenol. Cogn. Sci. **6**(3), 271–291 (2007)
E. Zalta (ed.), *The Stanford Encyclopedia of Philosophy* (The Metaphysics Research Lab, CSLI, Stanford University, Stanford, 1997–). http://plato.stanford.edu

Chapter 13
Husserl and Weyl

Jairo José Da Silva

Abstract In this paper, I carry out a comparative study of the philosophical views of Edmund Husserl and Hermann Weyl on issues such as mathematical existence and mathematical intuition, the validity of classical logic, the concept of logical definiteness, the nature of symbolic mathematics, the role of mathematics in empirical science, the relation of scientific theories with perception, space representation and the philosophy of geometry, and intentional constitution in general. My main goal is not simply to assess the extent of Husserl's influence on Weyl, although this is an ever present concern, but to clarify the views of one by contrasting them with those of the other.

Keywords Weyl • Philosophy of mathematics • Philosophy of logic • Philosophy of science

Like all original thinkers Weyl was not a faithful follower of those who influenced him. He preferred to 'digest' his intellectual masters, Husserl, Leibniz, Kant or Fichte, to make of theirs flesh of his own. Husserl's ideas in particular, regardless of the considerable influence they had on him, were far from being absorbed unaltered into Weyl's system of thought, despite his close ties with Husserl on both the intellectual and personal levels. His wife, the beautiful Friedericke Bertha Helene Weyl, née Joseph from a Jewish father and best known as a translator of Ortega y Gasset, had been a student of Husserl's and he himself followed, at the age of 19, the course on the phenomenology of time Husserl gave in Göttingen in the winter term 1904/05. Nonetheless, Weyl is definitely not a 'Husserlian' in the sense of someone who endorses Husserl's philosophical views unreservedly.

This doesn't mean that we cannot trace some of Weyl's most important philosophical ideas, particularly his intuition-based epistemology, back to Husserl, even though Husserl himself would certainly not endorse these ideas in the form they took in Weyl's hands. My aim here is twofold, to carry out an 'archeological' investigation in order to identify the Husserlian lineage of some of Weyl's main

J.J. Da Silva (✉)
Sao Paulo State University, Sao Paulo, Brazil
e-mail: dasilvajairo1@gmail.com

S. Centrone (ed.), *Essays on Husserl's Logic and Philosophy of Mathematics*,
Synthese Library 384, DOI 10.1007/978-94-024-1132-4_13

philosophical ideas, particularly those concerning the role of intuition and symbolization in mathematics and the empirical sciences, among a few others that are to this day relevant to the philosophy of logic, mathematics and the empirical sciences, but also to contrast Weyl's and Husserl's views on matters that interested both, even if no direct influence on either direction can be safely identified.

A general 'psychoanalytic' comment is, I believe, relevant. Weyl gradual allegiance to ever more radical forms of 'intuitionism,' a development that may have started under the influence of Husserl's emphasis on the epistemological primacy of the 'given,' but that assumed radical forms under the spell of the definitely non-Husserlian brand of intuitionism associated with Brouwer and his foundational 'school,' seems to me a form of 'killing' his formalist father Hilbert. A rebellious disposition that was not free of gilt, for Weyl could never, unlike the enfant terrible Brouwer, bring himself to give up non-intuitive forms of knowledge in mathematics and physics. In the end, he settled for a compromise, an epistemology in which intuition and 'symbolic knowledge,' including symbolic mathematics, were reconciled, most importantly, in science,[1] whose task, he thought, was to provide a 'symbolic reconstruction' of the world, which only as a whole could be confronted with experience[2] (as we'll see, Husserl couldn't accept without reservations such a mingling of the symbolic and the intuitive, particularly when empty, referentially void signs are involved).

Since I'll be mostly concerned with Weyl's views on mathematics, physics and to some extent also logic in their relation to Husserl's, it would be useful to briefly state what I think to be Husserl's views on these matters, beginning with logic.

For Husserl, logic was not only the science of concepts, propositions and theories insofar as distinctiveness (i.e. intrinsic consistency), truth content, and truth

[1]"The probing of the foundations of mathematics during the last decades seems to favor a realistic conception of mathematics: its ultimate justification lies in its being a part of the theoretical construction of the one real world" (Weyl 1946, 169). "For even from a purely philosophical standpoint, the conception that mathematics is essentially a part of the theoretical constructions of the one real world is in better accord with our probing of the foundations of mathematics than more idealistic views" (Weyl 1946, 171).

[2]"Thus we had better not commit ourselves to any definition and rather develop the theory as a symbolic construction with unexplained symbols and only at the end indicate in which way certain derived quantities may be checked by observation. The theory then becomes a connected system that only as a whole may be confronted with experience" (Weyl 1949, 183). "Only this entire connected theory, into the texture of which geometry also is interwoven, is capable of being checked by observation" (Weyl 1949, 184). "An individual law isolated from this theoretical structure simply hangs in the air. Ultimately all parts of physics including geometry coalesce into an individual unit" (Weyl 1949, 184). "We have a symbolic reconstruction, but nothing which we could seriously pretend to be the true real world" (Weyl 1949, 184). "In this manner a theory of nature emerges which only as a whole can be confronted with experience, while the individual laws of which it consists, when taken in isolation, have no verifiable content. This discords with the traditional idea of truth, which looks at the relation between Being and Knowing from the side of Being, and may perhaps be formulated as follows: 'A statement points to a fact, and it is true if the fact to which it points is so as it states.' The truth of physical theory is of a different brand" (Weyl 1954, 199).

preservation are concerned, but also the science of objects merely as objects of possible scientific interest. Husserl believed that logic, as the theory of science, should also be concerned with how objects, no matter which, can in principle be presented to consciousness, that is, we could say, with logical forms. Husserl's logic has, then, an apophantic and a formal ontological dimension that are in strict correspondence.[3]

For him, domains of objects of some determinate type or concepts having these domains as extensions, be they given adequately in intuition or not, can be objects of theories whose content of intuitiveness are equal to that of their subject matter (these I call *contentual* theories, they can be intuitive, partially intuitive or even purely symbolic). The *ideal* of science is the elaboration of *complete* theories relative to scientifically interesting domains.[4] A theory is complete relative to a domain if every proposition that is true in this domain is, if not itself an axiomatic presupposition, *provable* from those presuppositions by strictly logical means – a complete theory is the 'master' in its domain.[5] Formal ontology requires that we also consider theory-forms, that is, non-interpreted axiomatic systems. For Husserl, formal theories also have their domains, *formal* domains, constituted by *formal objects*, that is, empty forms of objects characterized only formally by their relations with other similar objects of the domain.[6] In a way that reminds one of the modern methods of constructing models from constants, Husserl believed that *only* the formal objects the theory *requires* to exist do exist satisfying the properties the theory requires them to satisfy. Formal theories are supposed to be (formally) true in their formal domains.

The formal ontological correspondent of the notion of completeness relative to a domain is that which Husserl called *relative definiteness*, that is, completeness relative to the formal domain of the formal theory. This condition requires that any sentence *referring* to the formal domain of a theory[7] must be logically decidable in the theory. The theory must not, in principle, leave without a decision any assertion that purports to describe a fact in its domain. Relative definiteness can, of course, be redefined as *syntactic* completeness of the theory with respect to a *subclass* of sentences of its language (those precisely 'referring' to its formal domain). Husserl's

---

[3]See da Silva 1999.

[4]By this I simply mean that the goal of scientific theories in general, no matter in which scientific field, is to find out everything that is true about their domains of investigation (in Husserl's words, "to master" their domains). That this is an *ideal* is hardly disputable.

[5]Completeness relative to a domain is equivalent to *syntactic* completeness, i.e. the property of a theory that can either prove or disprove any assertion expressed in its language. There are, of course, different notions of completeness of theories, but given that (interpreted) theories are logically articulated systems of *true* assertions *about their domains*, the notion of completeness that offers itself naturally in this context is, I believe, that of a theory that contains *everything* that is true about its domain, a *maximally* consistent set of assertions. The *idea* of a domain that can be completely mastered by a theory is, for Husserl, a regulative idea of science.

[6]See da Silva 2000a for details.

[7]Not an easily definable notion, see da Silva 2000a.

concept of *absolute* definiteness is just syntactic completeness with respect to *all* sentences of the language of the theory.[8] Since the formal domain of a relatively definite theory cannot be extended and still be the domain of *this* theory (since this domain has already all the elements that the theory requires it to have), the *domain* of a relatively definite theory is also complete, but in a different sense, it is non-extendible, which is the sense of completeness associated with Hilbert's *axioms* of completeness.[9]

Husserl epistemology is intuition-based, but not only objects, concrete or abstract, real or ideal, can be intuited; concepts and essences can also be presented in full clarity to consciousness. This is enough to neatly distinguish Husserl's from Brouwer's conception of intuition. For Husserl, there can be an intuitive theory of a domain of non-adequately intuitable *objects*, such as, for instance, the domain of natural numbers (or, one may argue, maybe stressing a bit the concept, the domain of real numbers, whose concept can be spelled out in a system of axioms that articulate our intuitive grasp of it) if this domain is unified under an intuitable *concept*. In general in mathematics, Husserl thinks, contentual theories are conceptual, and since concepts can be intuited, mathematical theories of inadequately intuited *domains* can very well be intuitive (and this, I believe, is why Husserl never advocated trimming mathematical domains of their non-intuitable elements – for him, *conceptual*, not *objectual* intuition was the relevant form of intuition in mathematics). But Husserl considered also non-contentual, purely formal mathematical theories, either formally abstracted from conceptual theories, and sometimes generalized, or created ex nihilo in an exercise of free mathematical imagination. These are not theories in the proper sense, theories of determinate domains of things, but forms of theories or, better, theories of forms, i.e. a priori theories of ways objectual domains can *in principle* be presented to consciousness. As theories of pure possibilities they are of concern to logic, and this is why, for Husserl, symbolic or formal mathematics belonged to that chapter of formal logic he called formal ontology. Incidentally, Weyl said something to the same effect, i.e. that a formal axiomatic system is "a logical mold (*Leerform*) *of possible sciences*."[10]

Despite Weyl's acceptance of intuition as the ground where knowledge is rooted, he manifests reservation concerning Husserl's notion of the intuition of essences (*Wesensshau*)[11]:

---

[8]There are other, competing interpretations of Husserl's notions of absolute and relative definiteness in the literature. Here, I offer mine. See Ortiz Hill 1995, Majer 1997, da Silva 2000a, b, Hartimo 2007, Centrone 2010, and Okada 2013.

[9]In *Ideas* I, §72, footnote, Husserl calls our attention to the connection between his and Hilbert's notions of completeness.

[10]Weyl 1963, 25.

[11]Weyl 1954, 202.

> At the basis of all knowledge there lies ... *intuition*, mind's act of 'seeing' what is given to him, limited in science to the *Aufweisbare* [*something to which we can point in concreto*[12]] but in fact extending far beyond these boundaries. How far one should include here the *Wesensschau* of Husserl's phenomenology, I prefer to leave in the dark.

Weyl made it quite clear that he doubted, for example, the possibility of intuiting the essence of space. This makes it doubtful that he could accept without reservation Husserl's germane notion of conceptual intuition. Weyl, however, believed that once being is revealed to consciousness, phenomenology's 'decisive question' can be raised: "to what extend does the limitation of those aspects of being which are finally revealed to consciousness express an innate structure of what is given, and to what extent is this a mere matter of convention?"[13] Weyl's dialectic of knowledge involves *cognition* and reflection, not only, as for Husserl, *intuition* and reflection (details below).

Husserl didn't give physics and the empirical sciences in general the amount of attention he dedicated to logic and mathematics. Nonetheless, we can extract from his last published book, *The Crisis of European Sciences and Transcendental Phenomenology* (*Crisis*, 1936), some elements of a philosophy of science. In this work, Husserl is basically concerned with the *historicity* of mathematical physics, i.e. the continuous process (the historical *flow*) of *transmutation* and *sedimentation* of sense that went into the constitution of the mathematical science of empirical reality that we know today. Husserl attributed *philosophical* relevance to genetic investigations of this sort. The philosopher's task, as he saw it, was to clarify the process of constitution by going through layers of sedimented sense, 'desedimenting' it in order to render explicit the original intentionality that went into its constitution. For the philosopher of science, the sense formation 'empirical reality' (*physis*) poses particular interest; his task is to identify the intentional acts that go into its constitution (and that may have been 'forgotten'), justify, based on the intrinsic sense of the formation 'empirical reality,' the methods which were developed for its investigation, delimiting moreover their range of *lawful* applicability and, eventually, identify at the basis of the series of intentional constitution the source from where sense originally flowed. For Husserl, the original source of sense is the life-world, where the scientist lives as man among men, but also where scientific theorizing must eventually be put to test. Clarification of sense constitution, with the identification of 'blind' formalization and the 'technization' that goes with it as the causes of the loss of sense that he thought was at the heart of the 'crisis' that menaced the European science (and humanity), constituted the core of Husserl's philosophical analyses of science.

For Husserl, the world *as perceived* is not simply what meets the senses. Rather, it is an *intentional* elaboration of hyletic *sensibilia* (and Weyl is in full agreement with him: "A full act of perception is always impregnated with certain mental interpretations which perform a kind of vitalizing integrating function on the mere

[12] *Op. cit.*, 200.

[13] Weyl 1955a, b, 211.

sense data."[14] "[I]t is not a real component of sensation itself but rather an entity pertaining to the intentional object, which arises before my consciousness in the perceiving acts."[15] The senses, on the other hand, do not provide inordinate masses of impressions, as empiricists or positivists believe, but *systems* of impressions that are already largely articulated. The world as perceived, however, is not *yet* the physical world, whose constitution requires further intentional action (formal abstraction and idealization, in particular, but also, and in different degrees, intuitively empty formalization[16]). Some aspects of empirical reality as science conceives it can be *directly* perceived, if only in rough sketch, but others will necessarily extrapolate the possibility of *direct* perception (the global aspects of physical space, for example).[17]

The formalization of empirical science, as Husserl reminded us, started with the Galilean geometrization of nature (which required, as Husserl also showed, that the sense of the geometry on which Galileo relied had already been altered vis-à-vis the sense of geometry as it was originally conceived),[18] but reached undreamed of forms in contemporary science. This poses a particularly difficult question concerning the role of *symbolic* (non-intuitive) mathematics in science. *How* symbolic mathematics came to play a role in science is not difficult to understand. In science, *physis* is *already* represented as a mathematical or, at least, proto-mathematical domain. That the 'Galilean' science of nature is not concerned with reality 'out there,' only with our conveniently idealized and intentionally elaborated *perception* of reality seems to have been clear to Husserl (this, incidentally, as he also clearly saw, opens the possibility for a *science*—phenomenology, of course—that deals *directly* with the perceived and the a priori structure of perception, besides acts of intentional positing in general, intuitive or not).

As Husserl argued, it is not the empirical world *itself* that is mathematical, but an *idealized* version of our *perception* of *abstract* aspects of the world. Mathematics, then, is naturally invited to contribute to science. The problem Husserl (and Weyl) had to face was the *extent* to which *purely symbolic* mathematics, whose terms do *not* have referents in our representation of experienceable reality, can also be allowed in. For Husserl, the role of purely symbolic mathematics had to be restricted to protect science from shifts or loss of meaning. Weyl, on the other hand, believed that instead of putting a danger, symbolization, even empty symbolization, offered the only way science could escape the *factuality*, *immediacy*, *subjectivity*, *superficiality* and *materiality* of perception. Although, for Weyl, science starts with intuition, it ends up with "a purely symbolic construction that uses as its material nothing but mind's free creations: symbols." However, he continues, "it is evident

[14] Weyl 1934, 102.

[15] *Loc. cit.*

[16] By *formalization* I mean the imposition of abstract form on material content.

[17] Although this is not a rule, perception can also *suggest* scientific concepts; for example, the geometric concept of congruence, suggested by the perceptual fact that bodies can be brought together in space and superposed, to cite a case on which both Weyl and Husserl agree.

[18] See Husserl's fundamental essay on this, "The Origins of Geometry."

that now the words 'in reality' must be put between quotation marks; who could seriously pretend that the symbolic construct is the real world?"[19]

Husserl was not so bold. For him, since formal mathematical theories are theories of *possible* realms of being, to use formal mathematics in dealing with domains that do *not* actualize the possibilities envisaged in the theories is tantamount to treating reality from the perspective of mere possibility, immerging the real into imaginary worlds so as to bring to light the intrinsic properties of reality (if only the purely formal ones). Husserl seemed cautious as to the validity of such a procedure, which Weyl unreservedly embraced: "[I]n the natural science we are in contact with a sphere that is impervious to intuitive evidence; here cognition necessarily becomes symbolical construction"[20]; "with reference to the formation of constructive notions and theories in the natural sciences, ... the field of the possible and free choice in such a field must necessarily be opposed to and placed ahead of the really occurring and the confinement to mere actuality."[21]

Husserl allowed a place for symbolization in science provided symbols and symbolic reasoning played only a *surrogating* role. As for the introduction of purely formal theories in the empirical sciences, however, he was cautious; they could produce, he believed, since 'empty' symbols are being used, undesirable shifts or maybe even complete loss of meaning. We, apparently, would not be able to trace our steps back from theory to perception. Husserl seems to have believed that scientific assertions must meet the test of experience *individually*, since each one corresponds individually, even though through a series of idealizations, to a content of *possible* experience. Weyl, on the other hand, admitting as he did purely formal theories as components of our scientific image and scientific explanation of reality, believed that only whole theories, including their mathematical components (which could be meaningless in isolation) can meet the tribunal of experience (I will call this view *scientific holism*).

Let's now proceed to a closer inspection of Weyl's philosophical views in their relation to Husserl's.

1. Let's first consider the extent to which creative definitions (of which definitions by abstraction are particular cases), seen as intentional acts, have, for Weyl, the right to grant objective existence to the objects they posit. This will show the extent to which Weyl was willing to accept the object-positing rights of non-intuitive intentional acts. Did Weyl see creative definitions as ways of introducing new entities? For example, geometrical forms; are they objects on their own or only aspects of bodies, equal but not identical with similar aspects of other bodies? The following quote seems to indicate that, for Weyl, geometrical forms do not enjoy objective existence: "But that which is common to all congruent triangles or to all bodies occupying the same spatial position, I find myself unable to represent by an objective feature ... but merely by the indication: congruent to this triangle,

[19] Weyl 1954, 198–99.

[20] Weyl 1932, 80.

[21] Weyl 1934, 84.

occupying this spatial position."[22] But, then, he continues, "the transformation of a common feature into an ideal object, e.g. of the property 'red' into an objectified 'red color' of which the red things 'partake,' is an essential step (Plato's μέθεξις)."[23] The mind, Weyl thinks, "is dissatisfied with equality, it seeks identity."[24] In the end, "if one wants to speak, all the same, of ... ideal objects one must at any rate refrain from giving them independent existence."[25]

Husserl would agree with this last sentence; ideal objects do not have, for him, *independent* existence. But he would not deny them *objective* existence; they are not, for him, as seems to be the case for Weyl, mere façons de parler. Weyl's conception of the nature of ideal objects is, or so it seems, closer to nominalism, which may betray the influence of Poincaré. This is particularly true, as we'll see below, in Weyl's predicative analysis as developed in *The Continuum.* In it, he refuses to grant objective existence to real numbers emptily posited by intentional acts embodied in arithmetical definitions.

2. A minor point, but worth mentioning, is Weyl agreement with Husserl's conception of analytic laws and analytic necessities as expressed in Logical Investigations (III Inv. §12), which Weyl quotes[26]:

> Analytic laws are unconditional universal propositions containing no concepts other than formal. As opposed to the analytic laws we have the particular instances, which arise through the introduction of material concepts or ideas positing individual existence; as particular cases of laws yield necessities, so particular cases of analytic laws yield analytic necessities

3. Weyl *explains* Husserl's concept of definiteness thus[27]: a scientific theory is *definite* in Husserl's sense if it provides all the concepts and basic laws from which all other concepts of this science can be defined and all other truths logically derived. Supposing that basic concepts and laws refer to the domain of the theory, and *truths* are *truths in this domain*, definite means complete with respect to the domain (or *syntactically* complete with respect to the class of propositions referring to the domain).[28] This, as we've already seen, is Husserl's notion of *relative* definiteness; that of *absolute* definiteness coincides with syntactic completeness. But then at the conclusion of a series of considerations Weyl *defines* an axiom system to

---

[22] Weyl 1963, 11.

[23] *Op. cit.*, 1.

[24] *Op. cit.*, 11.

[25] *Op. cit.*, 36.

[26] *Op. cit.*, 18.

[27] *Loc. cit.*

[28] By "domain" Husserl sometimes means an independently existing realm of objects, which a theory is designed to describe (completely, if possible, so that no meaningful assertion about the domain is left undecided). In this case the theory is an interpreted one and the domain of the theory is its intended interpretation. But by "domain" he also sometimes means the realm of objects the theory *requires* to exist. We can maintain the ambiguity for it does not affect the definitions substantially.

be *complete* if it is *categorical*.[29] If a system is syntactically complete all of its models are logically equivalent, i.e. any proposition expressed in the language of the theory that is true in one model is also true in any other model, but this does not imply that the theory is categorical, i.e. that all its models are *isomorphic*. On the other direction, a theory can be categorical (for example, second-order arithmetic) but not syntactically complete. Obviously, Weyl is clearly conscious that his conception of completeness as categoricity does *not* coincide with Husserl's notion of relative definiteness. It may be that he only redefines it thus in face of the theoretical impossibility of obtaining syntactically complete theories (due to limitations imposed by Gödel's theorem). Mathematics, then, must settle for the feasible, namely, categoricity, a condition moreover more attuned to the formal nature of mathematics (form, not matter, matters, and categoricity is the way of guaranteeing that a theory singles out a particular form). In short, Weyl, I think, is not offering an *interpretation* of Husserl's notion of definiteness as categoricity, but an *alternative* to it.

4. Another point of logic where Husserl and Weyl disagree is the interpretation of existential and universal propositions, particularly in mathematics, where the scope of quantification can be, and often is, infinite. For Weyl (in agreement with Brouwer)[30]:

> An existential statement, such as 'there exists an even number' is not considered a proposition in the proper sense that it asserts a fact.... '[T]here exists an even number' is nothing but a *propositional abstract* derived from [*a proper proposition, for example, '2 is an even number' JJS*] ... the propositional abstract is merely a document indicating the presence of a treasure without disclosing its location.... Whenever nothing but the *possibility* of a construction is being asserted, we have no meaningful proposition; only by virtue of an effective construction, an executed proof, does an existential statement acquire meaning. In any of the numerous existential theorems of mathematics, what is valuable in each case is not the theorem as such but the construction carried out in the proof; without it the theorem is an empty shadow.

Husserl's interpretation of existential assertions in mathematics, contentual or formal, is in complete disagreement with Weyl's (or Brouwer's). One quote is enough to show this[31]:

> All mathematical propositions of existence have this modified sense ... not simply a 'there is' but rather: it is possible a priori that there is.... All existential judgments of mathematics, as a priori existential judgments, are in truth judgments of existence about possibilities.[32]

[29] *Op. cit.*, 25.

[30] *Op. cit.*, 50–1.

[31] Husserl 1973, 370–71.

[32] This demands clarification. Obviously, Husserl cannot be saying that in mathematics the existential assertion $\exists x A(x)$ (there is ...) *means* $\diamond(\exists x A(x))$ (it is possible that there is ...), since this last assertion is consistent with $\diamond(\neg\exists x A(x))$ and he was aware that once mathematics proves that certain entities exist, they exist by *necessity*, and so it is *inconsistent* to claim that these things could not exist. What he is saying is that, in mathematics, to assert the existence of something is to assert the *purely intentional* existence of this thing, more or less like asserting the existence of

But for Weyl[33]:

> Universal propositions are incapable of negation. Even a universal statement does not refer to a fact, it is not to be interpreted as the logical product of infinitely many singular propositions but as a *hypothetical* statement: if applied to a single definite given number it yields a definite proposition. There is no occasion here for the application of a principle of *tertium non datur*.

This is how I understand him. For Weyl, $\forall x A(x)$ is a pseudo-assertion standing for the meta-assertion "*if* an object a of the domain of quantification is *given*, *then* $A(a)$." Now, how can the meta-assertion be justified? I believe that Weyl would say, with Brouwer, that only by displaying a schema that once some a of the domain is exhibited, by construction or definition, would provide a proof of $A(a)$. Husserl saw matters differently. For him, $\forall x A(x)$ is a proper assertion, which could be proved by showing that the *meaning* associated with the domain of quantification, i.e. the *type* of objects it contains, requires that the property A befits *by necessity* every one of them. The proof of this *fact* would not involve directly the elements of the domain of quantification U, only the intentional features of the meaning-formation 'element of U.' In other words, it would be a proof of the fact that the *conception* of U entails the property A for each and every object belonging to U. Since this proof involves only meaning analyses it can be a priori, independently of how many object belong to U, if any, or their intuitive 'accessibility.'

For Weyl, the negation of the meta-assertion translating $\forall x A(x)$ is "an a is given but there is no proof of $A(a)$." This, of course, does not entail that there is a proof of $\neg A(a)$ (i.e. not-$A(a)$), that is, the proposition $\exists x(\neg A(x))$. So, in Weyl's interpretation, $\exists x(\neg A(x))$ is *not* the negation of $\forall x A(x)$ and the disjunction $\forall x A(x)$ *or* $\exists x(\neg A(x))$ is not in general valid. Husserl, on the other hand, saw the matter from the perspective of classical mathematics. *Provided* U is *intentionally conceived* as an ontologically complete domain of being where every possibility is *in itself* either definitively the case or definitively not the case, if $\forall x A(x)$ is false, then *there must* exist some a for which A does not apply, i.e. $\exists x(\neg A(x))$ *must* be true.

An analogous situation occurs in literature. Even if the author of the novel we're reading does not say anything about what the Countess of G. was doing that fateful Friday immediately before the mischievous Duke of M. arrived for tea, we assume that she *must* had been doing *something*, even though we'll probably *never* know what. Because we presuppose that the novel depicts a *reality*, and ontological completeness belongs to the *sense* of reality, whatever reality, we assume that every possibility is decided in itself, i.e. it is either a fact or not a fact. 'The Countess was

a character in a work of fiction. He had a stronger notion of possibility in mind, which coincides with that of "classical" mathematical existence. The fact is that Husserl did not give existential claims in mathematics the "constructive" sense Weyl gave them. For Husserl, I believe, existential assertions can follow from clarifications of *intentional meaning*, not *only* as consequences of intuitive presentations (so that A(a) implies $\exists x A(x)$, but not the converse, if a denotes an object of *intuition*).

[33] Weyl 1963, 51.

writing a letter to Justine immediately before the Duke arrived for tea' expresses a *possibility*; *so*, it is either true or false. In fact, the worlds of mathematics are not so very different from the fictitious worlds of literature, and can be as useful as them in our representation and understanding of reality.

The difference between Weyl's interpretation of mathematical propositions, including general ones, and that of classical mathematics with which Husserl apparently agreed is that whereas for Weyl these propositions refer to mathematical *objects*, for the 'classical' mathematicians, among them Husserl, they express a priori relations between *meaning formations*. And it is precisely because of this that mathematics is closer to logic than to the ontological, material sciences, composing with it the group of the formal sciences.

On these matters, Weyl turned completely the back to Husserl and the teachings of phenomenology, embracing Brouwer ideas *cum gusto*. With regard to the scope of classical logic (the validity of tertium non datur in particular) Weyl's agreement with Brouwer was complete: "Classical logic was abstracted from the mathematics of the subsets of a definite finite set ... an *a priori* existence independent of mathematics was ascribed to this logic and ... on the basis of this suppositious apriority it was unjustifiably applied to the mathematics of infinite sets."[34]

The influence of Brouwer extended to the analysis of the continuum. For Weyl, the intuitive continuum is not analyzable in terms of element and set, but part and whole; it is, for him, an 'extensive whole' in the sense of Husserl (*LI*, IV Inv.). But, as often with Weyl, after endorsing intuitionism on epistemological grounds he backs off for practical reasons: the price to be paid is too high; he resented the awkwardness of intuitionist mathematics and the trimming it impinges on the tree of mathematics.[35]

5. Despite his 'intuitionism' (by which I mean the view according to which intuition of objects and facts is the privileged—although not unique—form of justification of beliefs) Husserl never imposed any restriction on the validity of classical logical principles of reasoning. Weyl, on the other hand, has; his interpretation of quantification is, as he himself noted, inconsistent with the universal validity of excluded-middle (a principle, however, the mathematician in him found difficult to abandon). Nonetheless, Weyl seems to have been aware of a possible way of justifying this principle; one that, as I have argued somewhere else,[36] can be traced back to Husserl himself. Along this line of justification, the validity of excluded-middle depends on a transcendental "hypothesis:" By this I do not mean a presupposition that can turn out to be false, but the expression, enjoying a priori validity, of an aspect of the intentional constitution of a domain of objects. According to this hypothesis scientific domains exist and are complete in themselves; the hypothesis is a priori true because scientific domains are so conceived. This hypothesis, Weyl recognizes, is "the guiding principle in building up

---

[34] *Op. cit.*, 51–2.

[35] *Op. cit.*, 54.

[36] See da Silva 2013a.

our formalism.[37] In fact, Weyl thinks that 'transcendental' mathematics of the sort of Hilbert's demands a "transcendental logic on which one is used to rely in geometry and analysis, including the free use of 'there exists' and 'all' with reference to the objects of the axiomatic system:"[38] For him[39]:

> Since 'there is' and 'all' are applied without hesitation to objects of all types, we must assume that the objects of any given type form an existential category, in the sense that the question whether there exists an element in that category of such and such a property *A* always has a meaning and that there either exists such an element or every element of the category has the complementary property.[40]

Given his interpretation of mathematical propositions, however, Weyl could not cope with this 'hypothesis,' at least as far as intuitive mathematics is concerned. In fact, he goes as far as blaming it for the 'paradoxes' of set-theory[41]:

> The deepest root of the problem lies elsewhere: a field of possibilities open into infinity has been mistaken for a closed realm of things existing in themselves. As Brouwer pointed out, this is a fallacy, the Fall and original Sin of set-theory, even if no paradoxes result from it.

But for Husserl being ontologically complete is an intentional trait of any *objective* domain of investigation. Objectiveness (which does not mean independent existing) *entails* ontological completeness. Objective scientific domains just happen to be so constituted and Husserl did not see himself as a reformer of science, but an interrogator of the sense of scientific formations. Mathematics, like any other objective science, has the right of positing its domains as ontologically complete domains.[42] This, of course, does not disqualify the idea of a constructive mathematics of the *effectively* intuitable, which, however, involves a shift of focus, from the noematic sphere to the noetic one, from the objective to the subjective. Husserl, contrary to Weyl, does not grant constructive, noetic mathematics the exclusive privilege of providing mathematical knowledge proper.

6. Indeed, Weyl clearly separated intuitive mathematics, which he believed can convey knowledge proper, from symbolic mathematics, which cannot. For him, the most developed form of symbolic mathematics are freely created purely formal axiomatic systems. Tertium-non-datur is allowed in formal contexts, but only as a presupposition (not a transcendental hypothesis). Since, for Weyl, formal-symbolic mathematics is epistemologically justified only as part of the symbolic reconstruction of reality, symbolic mathematics and its presuppositions have the same epistemological dignity of physical hypotheses. According to Weyl "a truly realist mathematics should be conceived, in line with physics, as a branch of the

[37]Weyl 1963, 220.

[38]*Op. cit.*, 230.

[39]*Op. cit.*, 233.

[40]Weyl's "existential category" is what I here call "ontologically complete domain."

[41]*Op. cit.*, 234.

[42]See Husserl 1970b (particularly §9) for "genetic" analyses of the constitution of scientific domains, including mathematical ones, and the transcendental presuppositions they involve.

theoretical construction of the one real world, and should adopt the same sober and cautious attitude toward hypothetic extensions of its foundations as exhibit by physics."[43]

Husserl too considered intuitive and symbolic mathematics as two different forms of mathematics. For him, however, although intuitive mathematics conveys knowledge proper, symbolic mathematics *per se* is not completely devoid of knowledge content. It conveys either an *anticipation* of knowledge, in case of non-intuitive or only partially intuitive contentual theories, or, in case of purely formal-symbolic theories, knowledge of *logical forms* that can *in principle* in-form *possible* domains of objects (or, what is the same, a priori formal knowledge of possible realms of being), thus providing *logical* knowledge.

As we can see, Weyl and Husserl have complementary views concerning the status and role of symbolic mathematics. Whereas the latter recognized its logical character, but was suspicious of its intromission in science, the former saw in its scientific uses the only way of justifying purely symbolic mathematics, apparently unimpressed by its logical dimension.

Weyl (let alone Brouwer) could not correctly evaluate, or accept, the *transcendental* character of the hypothesis associated with the validity of tertium non datur in mathematics, namely, that mathematical theories are concerned with domains of objects (objects proper or merely formal objects) that are *conceived* as ontologically complete. He did not see, or accept, that the *sense-formation* 'objective realm of objects' *requires* ontological completeness, and, consequently, missed the point of the logical-transcendental justification of classical logic and the epistemological vindication of classical mathematics *per se* that, I believe, is one of the accomplishments of Husserl's philosophy of mathematics.

In fact, Weyl *equated* transcendence with ontological incompleteness. Although agreeing with Husserl that real objects cannot be adequately intuited: "a real thing can never be given adequately, its 'inner horizon' is unfolded by an infinitely continued process of ever new and more exact experiences; it is, as emphasized by Husserl, a limiting idea in the Kantian sense,"[44] he derived from the transcendence of the real world idealist consequences that Husserl would not endorse: "[f]or this reason it is impossible to posit the real thing as existing, closed and complete in itself."[45] Whereas for Weyl a transcendent reality cannot be posited as existing in itself and be in itself completely determined, a view in harmony with his intuitionist tendencies, for Husserl the positing of an ontologically complete reality where everything is in itself determined (thus validating *tertium non datur* in the logic of science) is a move in the constitution of domains of *objective* science. Weyl's idealism amounts to a refusal of complying with all the intentional acts that, according to Husserl, are required for the constitution of the objective domains of science, empirical or mathematical. Whereas Husserl *analyses* scientific procedures

[43] *Op. cit.*, 235.
[44] *Op. cit.*, 41.
[45] *Op. cit.*, 41.

phenomenologically, Weyl *criticizes* them from the perspective of epistemological and ontological parti-pris.

7. Weyl's idealism is particularly evident with respect to sets (and, as we'll see later, the continuum). Since in set theory, he says, 'there exists' refers to sets, sets must exist; he however believes that sets exist only as objectified extensions of properties, the criterion of identity being extensional identity.[46] This view is, I believe, derived from a somewhat myopic interpretation of Husserl's notion of intentional constitution. For Weyl, sets exist only as objective correlates of acts whose intentional meaning is expressed by set-defining properties. He, moreover, believes that an intuitive theory of sets must necessarily be a theory of intuitable sets, ignoring the Husserlian distinctions between intuitive theories of objects, based on object-intuition, and intuitive theories of concepts, based on conceptual intuition. Cantorian set theory, to which Husserl never objected (Cantor was not only a friend but an important influence on him) is obviously not of the first type, but, one may argue, it is of the second, i.e. an intuitive conceptual theory, as Gödel believed it to be. But Weyl, certainly under the spell of Brouwer, believed differently; for him, set theory was a theory of objects that largely extrapolated the possibility of intuitive presentation.

Weyl objected in particular to 'existential absolutism' with respect to subsets of infinite sets, say **N**, and the set of all such subsets, which depends, he believed, on the 'elusive' notion of the "aggregate of all possible properties of natural numbers." For Weyl, the domain of properties of natural numbers is a non-extensional domain; they are stratified in levels, first, second, etc. (this is Russell-Poincaré's vicious circle principle). Accordingly, sets of numbers are also stratified in levels. "The obliteration by the existential absolutism of these differences in level ... constitutes an unquestionable vicious circle."[47] A way of escaping stratification, he thinks, is Russell's axiom of reducibility (which, however, he considers a sort of hara-kiri); another is the analysis of *Das Kontinuum*, which is content to consider only first-level real numbers (as definable sets of natural numbers).

However, Weyl says[48]:

> In spite of its paradoxical character, the idea of absolute existence in the domain of natural numbers and sets of natural numbers has so far not yet led to any contradiction [*but*] at the farthest frontiers of set theory actual contradictions did show up. But their root can only be seen in the boldness perpetrated from beginning in mathematics, namely, of treating a field of constructive possibilities as a closed aggregate of objects existing in themselves.

Surprisingly, Weyl passed in silence by the fact that Zermelo-Fraenkel's set theory, which allots sets in levels but commits the 'original sin' of taking the universe of sets as existing and completely determined in itself, has *also* not so far manifested any inconsistency.

---

[46] *Op. cit.*, 48–9.

[47] *Op. cit.*, 49.

[48] *Op. cit.*, 50.

In conclusion, despite the influence of Husserl's constitutive phenomenology, Weyl's philosophy of set theory seems closer to Poincaré's than to Husserl's, at least up to *Das Kontinuum*, where a distinctive trait can be detected that is traceable back directly to Poincaré, not to Husserl, namely, the intuitive character of the *domain* (not the *concept*) of natural numbers. From the publication of *DK* on, however, Weyl becomes even less Husserlian and more hardcore Brouwerian, abandoning the perspective of *DK* for a theory of the continuum based on the notion of choice sequences. He eventually brought himself to accept purely formal theories, such as Hilbert's axiomatic analysis, as we have seen, but only on pragmatic basis. Although arguably Weyl's intuition-based epistemology was inspired by Husserl (as we can infer from the preface of *DK*),[49] he seems to derive his conception of intuition from Poincaré and Brouwer rather than Husserl himself.

8. Let's dwell a bit longer on Weyl's views on the relation of mathematics to physics. As I've been insisting, despite his intuitionism, Weyl could not give up classical mathematics, which is, for the most part, purely symbolic and non-intuitive. Formal mathematics, he thought, in conformity with Hilbert, is a rule-based symbolic 'game' extending intuitive mathematics – not knowledge proper – which, however, can be logically vindicated by a proof of consistency (Hilbert's program). For Weyl, the Ackermann-Von Neumann proof of the consistency of arithmetic without the axiom of comprehension vindicates the approach of DK, where the set of natural numbers is taken as a closed totality (since this proof guarantees, he believed, the 'existence' of this set). The consistency of arithmetic with comprehension would, Weyl thought, vindicates the set of subsets of natural numbers (thus, classical analysis). But the limitations imposed by Gödel's theorem leaves Hilbert's formalism no alternative but remaining a somewhat risky 'game' with symbols.

But the raison d'être of mathematics, *even formal-symbolic mathematics*, Weyl reminded us, is to serve science. Scientific propositions, to the extent that they involve formal mathematics, however, do not have the intuitive character Brouwer requires of meaningful assertions; so, they can meet the tribunal of experience only as a system. Only scientific theories *as wholes*, not individual propositions, are empirically testable. For Weyl, only 'seeing' in the Husserlian sense of 'intuiting' can furnish knowledge, 'theoretical constructions,' on the other hand,[50]

> seem to be bound only to one strictly formulable principle, that of concordance ... which in mathematics, where the domain of sense data remains untouched, reduces to consistency ... the life of mind is light and darkness, contingency and necessity, bondage and freedom, and it cannot be expected that a symbolic construction of the world in some final form can ever be detached from it.

In short[51]:

---

[49] See da Silva 1997.

[50] *Op. cit.*, 61–2.

[51] Weyl 1932, 80.

> My opinion may be summed up as follows: if mathematics is taken by itself, one should restrict oneself with Brouwer to the intuitively cognizable truths.... But in the natural sciences ... we need no longer demand that when mathematics is taken into consideration into the process of theoretical construction in physics it should be possible to set apart the mathematical element as a special domain in which all judgments are intuitively certain; from this higher viewpoint which makes the whole of science appears as a unit, I consider Hilbert to be right.

The compromise solution found by Weyl, then, was to divest symbolic mathematics of any knowledge content, but to preserve it as a useful tool for the 'symbolic reconstruction of the world,' which is all that objective science can hope for.[52]

This solution, however, is not immune to criticism. If the rules-of-the-game sort of formalism is hard to swallow in pure mathematics (why would anyone be interested in this game, and why some and not other symbolic-mathematical games?) it is even more unpalatable in applied mathematics. *Why* should symbolic manipulations unconcerned with truth be of any relevance for science? *Why* can a symbolic game, which may not correspond to anything objectively real or experienceable, be effective in organizing our experience of the world to the point of allowing testable *correct* previsions? These questions were not even raised by Weyl, let alone answered. But Husserl did not treat this problem conveniently either. In fact, more conservative than Weyl, Husserl didn't even allow the free use of 'empty' symbolic mathematics in science for fear of compromising the precious link of science with the possibility of experience (for Husserl, the 'technization' of science, that is, the predominance of intuitive-free symbolization over intuition, is a form of alienation, a malaise of modern European science).[53]

Weyl, however, was able to appreciate something that has apparently escaped Husserl, contributing to the latter's inability to understand the role of purely symbolic manipulations in mathematical physics. Weyl realized that any *objective* science is necessarily *formal*, and that the *material content* of experience is necessarily confined to *subjectivity*. Weyl notes, correctly, that[54]:

> A science can never determine its subject-matter except up to an isomorphic representation. The idea of isomorphism indicated the self-understood, insurmountable barrier of knowledge. It follows that toward the 'nature' of its objects science maintains complete indifference. This ... one can only know in immediate alive intuition. But intuition is not blissful rest in itself from which it may never step forth, but it urges on toward the variance and venture of cognition. It is, however, fond dreaming to expect that by cognition a deeper nature than that which lies open to intuition should be revealed—to intuition.

As I read him, Weyl's argument for allowing the use of purely symbolic, materially empty mathematics in science (Husserl's 'technization'), despite his intuitionism, is as follows: the *objective* science of nature is (for it can only be) concerned

[52]"[S]cientific cognition ... does not state and describe states of affairs —'Things are so and so'—but ... constructs symbols by means of which it 'represents' the world of appearances" (Weyl 1934, 83).

[53]See da Silva 2013b.

[54]Weyl 1934, 96.

with *formal* aspects of reality, that which remains when the essentially private and subjective material content of (actual or possible) experience is removed; symbolic mathematics, then, as the science of possible *forms* can provide representations of the formal skeleton of reality. Weyl may also have believed that symbolic mathematics is useful, even when not concerned with representing formal aspects of reality, as an instrument for articulating theories internally. Unfortunately, he didn't dwell for long enough on this issue.

Husserl made it clear in *Crisis* that our experience of reality can be mathematized only because experience, or, rather, formal-abstract aspects of experience, can be idealized. He was seemingly less conscious of a fact that Weyl saw clearly, namely, that despite the fact that *only* the formal-abstract aspects of experience can be mathematized, this does not constitute a limitation, for these are the *only* aspects of experience that can be subjected to the scrutiny of *objective* science. To realize this, that empirical sciences are formal (i.e. concerned only with forms) to the exact extent that they are objective, is a prerequisite for understanding *how* and *why* mathematics can be so fruitful in empirical science.[55]

Abstraction, the intentional experience focusing on *abstract* (non-independent) aspects of reality, and idealization, focusing on *limits* of series of possible intuitions, do not necessarily sever the ties with intuitable reality completely. But formalization, by opening the possibility that forms abstracted from experience be further formally enriched, which means allowing empty symbols into the language of science, can rule out completely the possibility of fulfilling intuitions. This would constitute a problem for Husserl's epistemology. Weyl seems to have detected this limitation of the Husserlian analysis of the methods of the mathematical science of nature, and substituted Husserl's genetic approach by a freer dialectics between symbolic 'reconstruction' (cognition) and intuition characteristic of modern science.

The convenience and superiority of symbolization over intuition, Weyl claims, *even in mathematics*, is clearly displayed in analytic geometry. Only after substituting points by labels (ordered sequences of numbers)[56]:

> Can one think of representing the spectacle of the actually given world by construction in a field of symbols. *All knowledge, while it starts with intuitive descriptions, tends toward symbolic construction* [*my emphasis*].... The objectification by elimination of the ego and its immediate life of intuition [*however*] does not fully succeed, and the coordinate system remains as the necessary residue of the ego-extinction.... It took a long time for mathematics before it had acquired the constructive tools to cope with the complexity and variety of such intuitively given figures. *But once it had reached that stage the superiority of its symbolic methods became evident* [*my emphasis*].

Any given group Σ of point-transformations, Weyl explains, can be represented by a group G of coordinate-transformations in some system of reference *f* (both groups

[55]Since empirical science can only reach the formal surface of phenomena, as Weyl clearly saw, there is no reason why empirical reality cannot be conveniently investigated by mathematical theories, provided the abstract forms these theories are concerned with display relevant connections with the forms we managed to discern in experienceable reality.

[56]Weyl 1963, 75.

are isomorphic). Other *similar* systems of reference can be chosen, in the sense that they can be obtained from f by isomorphisms of Σ. Let $f^* = \gamma f$, p' = pσ, γ and σ in Σ, p a point; x and x' the coordinates of p and p' in f and y and y' their coordinates in f'. If Sx = x' and Cy = x (so, C and S represent respectively γ and σ and in f), the transformation $S^* = CSC^{-1}$ such that S^*y = y' represents σ in f'. In short, we are not 'stuck' with f, and can move freely from one to another similar reference frame. If *a* frame stands for *an* ego, the possibility of comparing representations in different frames account for the *objectivity* of what is being represented. "To fulfill the demand of objectivity we construct an image of the world in symbols."[57] *That is, objectivity requires symbolization, whereas intuition remains irredeemably private.*

The criterion of objectivity referred to above, i.e., the objective is what different subjects agree upon, that is, the intersubjective, is, of course, Husserlian through and through. Moreover, Weyl believed, the intersubjective must be linguistically expressible, which implies that only the formal aspects of intuitions can be effectively shared. In the end, for Weyl, objectivity required symbolization, and so *the demand of objectivity marks the limit of Brouwer's intuitionism.*

9. Although Weyl recognizes that Hilbert's 'transcendentalism,' in which intuition is left behind in a leap into the transcendent through the symbolic, answer to a natural desire of reaching for totality, he disqualifies the naïve realism which, he thinks, is unaware of the gap between the given and the transcendent.

About this he says[58]:

> It cannot be denied that a theoretical desire, incomprehensible from the merely phenomenal point of view, is alive in us which urges toward totality. Mathematics shows that with particular clarity; but it also teaches us that that desire can be fulfilled on one condition only, namely, that we are satisfied with the symbol and renounce the mystical error of expecting the transcendent ever to fall within the lighted circle of our intuition.

10. For Weyl, the rational reconstruction of experience advances in steps from the intuitively given to the symbolic. The rational analysis of the intuitive continuum, for example,[59]

> proceeds in three steps: (1) morphology, which operates with vaguely circumscribed types of forms; (2) topology, which guided by conspicuous singularities or even in free construction places into the manifold a vaguely localized but combinatorially exact determined skeleton [*a grid of subdivision of the continuum, capable of increasingly finer refinement according to a fixed schema JJS*]; and (3) geometry proper, whose ideal structures could only be carried with exactness into a real continuum after this has spun over with a subdivision net of a fineness increasing *ad infinitum*.

The final step into the purely formal-symbolic, where only empty forms remain, completes the ascension. Intuitive space and intuitive time, for example, he says,[60]

---

[57] *Op. cit.*, 77.

[58] *Op. cit.*, 66.

[59] *Op. cit.*, 90–1.

[60] *Op. cit.*, 113.

> *are* hardly the adequate medium in which physics is to construct the external world. No less than the sense qualities must the intuitions of space and time be relinquished as its building material; they must be replaced by a four-dimensional continuum in the abstract arithmetical sense.... What remains is ultimately a symbolic reconstruction of exactly the same kind as that which Hilbert carries through in mathematics.

Curiously, despite Husserl's discussion of the contrast between mereological and exact concepts in *Ideas I* Weyl preferred to direct his readers to O. Becker "for a more careful phenomenological analysis of the contrast between vagueness and exactness and of the limit concept."[61]

He continues[62]:

> *although* the distillation of this objective world, capable only of representation by symbols, from what is immediately given to my intuition, takes place in levels ... [a] systematic scientific explanation, however, will reverse the order; it will erect the world of symbols as a realm by itself and then, skipping all intermediate levels, attempt to describe the relations that holds between the symbols representing objective conditions on the one hand and the corresponding data of consciousness on the other.

In this quote, Weyl is seemingly contrasting Husserl's genetic approach with the method of symbolic reconstruction he advocates. Whereas the latter seeks to erect *free* symbolic 'reconstructions' of the world that must eventually be confronted as a whole with intuitive experience, the former seeks to establish how a scientific representation of reality can be erected on perception as a foundation. However, despite the differences, Weyl's analyses of the constitution of the scientific representation of reality has important points of contact with *Crisis*. For both him and Husserl what science takes for physical reality is *not* given by the senses, requiring further intentional elaborations of sensorial experience, which opens the possibility of mathematical intervention at different levels of the process. For both, science, or what Weyl calls the 'systematic scientific explanation' is only a *method* for organizing our experience. Neither believes that it provides a *faithful* description of a *transcendent* reality that necessarily falls off the scope of possible adequate intuition.

11. "Science," Weyl says, "concedes to idealism that its objective reality is not given but to be constructed (nicht gegeben, sondern aufgegeben)."[63] This idea, that physical reality is 'constructed' or, better, intentionally constituted also occupies, of course, a central place in Husserl's system of thought. But Weyl parts company with Husserl, or believes to be so doing, by refusing to give the experiences of pure consciousness objective validity. Weyl tends to consistently give Husserl's ego a solipsistic interpretation, as if it were indifferent to other egos and their experiences.

---

[61] *Op. cit.*, 91.

[62] *Op. cit.*, 113.

[63] *Op. cit.*, 117.

Weyl gives the following characterization of idealism: “Immanent consciousness is the starting point of the idealist. As the primarily given, he finds the images of my consciousness – just as I have them.”[64]

The ‘naïve’ idealist, he believes, will grant reality to these images only:

> By reflection I may become aware of my own perception, which then is turned into the objects of a secondary act of inner awareness, and only thus can the perception be analyzed, with such results as these, that it radiates from an ego and points towards an intentional object ... the radical idealist will doubt very much whether this ego and the intentional object are inherent parts of the given perception.... He takes pain to describe how by a sort of productive synthesis or ‘psychic chemistry’ the intuition of space constitutes itself in a number of steps.

But, Weyl asks, doesn’t this synthesis go against the idealist’s own creed?

“Does it not transcend that which is given? ... I believe that every attempt to vindicate on such a basis the belief in a real outer world and in the reality expressed by word ‘you’ has been a failure and must be a failure.” Is Husserl this radical solipsist idealist struggling to concoct an *objective* world out of *his own* data of consciousness? Of course, he isn’t, but Weyl seems to believe differently.

This is why he turns to the existential philosophy of Heidegger, which he believed overcomes solipsism by inserting the ego within the vicissitudes of actual existence[65]:

> Existential philosophy turns this whole construction upside down, without relapsing into a crude realism, and thus overcomes the old contrast of realism and idealism, which had grown stale in the course of centuries.... *Phenomenology had lunged out in all directions beyond the narrow boundaries of sensual data; but still its pure ego was like a spirit soaring above the waters, untainted by worldliness, and it was hard to conceive how it could ever surrender its immanence and become flesh, man among men* [*my emphasis*].

Husserl’s pure ego, Weyl believes, must make space for the Dasein, being-in-the-world. “[N]o other ground is left for science to build on than this dark but very solid rook which I once more call the concrete Dasein of man in his world.” Weyl seems to ignore that, for Husserl, in the constitution of an objective world the intentional ego is *not* the personal ego, but a community of individual egos, and that the constitution of an objective reality, and the science of it, are *communal* tasks whose accomplishments are sedimented *in history and culture*.

Now, in order to make room for the *other* and escape the claustrophobic domains of the Husserlian pure ego as he interprets it, Weyl turns to Fichte: “Thou art for thyself once more what I am for myself, conscious-existing carrier of the world of phenomena,” he says.[66] It is in the confrontation of the *I* and the *thou* that objective reality is constituted. But Weyl is not content with Fichte’s apriorism either, in the end it is in the dialects between theorizing and experience that goes on in theoretical physics that he recognizes the only way an objective world can be

---

[64] Weyl 1949, 189–92

[65] *Loc. cit.*

[66] Weyl 1963, 124.

constituted. According to Weyl, "Fichte states the basic position of epistemological idealism even more radically than Husserl."[67] Fichte stubborn apriorism, however, is 'preposterous'[68]:

> In the antithesis of constructivism and phenomenology my sympathies lie entirely on his [*Fichte's*] side; yet how a constructive procedure which finally leads to the symbolic representation of the world, not a priori, but rather with continual reference to experience, can really be carried out, is best shown by physics – above all in its two most advanced stages: the theory of relativity and quantum mechanics.

But, again, objectivity requires that the *material* content of intuitions be relinquished in favor of their *formal* aspects. So, the objective world can be represented and recreated only symbolically, which is the sole way the formal can be expressed. The most adequate way, then, to represent objective reality is by means of axiomatic systems, in which the formal only is given shelter: "axiomatics reveals itself once again as the method of a purified realism which posits a transcendental world but is content to recreate it in symbols."[69]

12. The relations of science to the life-world (the Lebenswelt) are central in Husserl's Crisis and it would be interesting to see how Weyl saw the matter. But before, some observations concerning this notion, as Husserl conceived it, are in order. To start, the life-world is not science-free; the practice of measuring, for example, belongs to it (supporting and giving sense to scientific practices). But products of science too, including some of its concepts, not only methods, migrate to the Lebenswelt and are absorbed into it. The theoretical disposition, however, and the scientific pathos do not; they belong to the scientific sphere and characterize it. Although for Husserl the philosopher's task is to drive science back to the Lebenswelt this does not mean to undo the work of science, its idealizations in particular, but to understand science for what science is, namely, or so he thinks, a method, whose sense of being springs from practices of the Lebenswelt and must necessarily refer back to it.

Weyl seems to agree, for he says that "no other ground is left for science to build on than this dark but very solid rook which I once more call the concrete Dasein of man in his world."[70] Although Heidegger instead of Husserl is brought in association with this idea of grounding science in the conscious man living in the real world and immersed in time with his needs and vicissitudes, the theme has an obvious Husserlian pedigree. This, for Weyl, stands in contrast to both idealism and empiricism, with their attempts, which he thought bounded to failure, to 'reconstruct' reality from either basic sensorial data only or a pure ego floating over and above the real world.[71] For Weyl, it is licit for science to 'reconstruct' the world symbolically, mingling maybe meaningless and meaningful symbols, thus

[67] Weyl 1955a, 214.

[68] *Op. cit.*, 215.

[69] Weyl 1963, 124.

[70] Weyl 1949, 192.

[71] Contra the pure ego, see Weyl 1949, 190.

producing a concept of phenomenon that "we can hardly approach by means of our ordinary language,"[72] provided that, in the end, the *entire* edifice of science, as a whole, meets the test of experience. At the tribunal of experience, measurements are the privileged witnesses, whose results, it is important to notice, "belong to the disclosed world of our daily experience; they are describable in terms of classical physics and we can understand each other, communicate and compare notes about them."[73] The experimental physicist, for Weyl, "is on the same level of existence as the carpenter or the mechanic in his shop, on the same level of understanding on which our rough and tumble daily life moves."[74]

For Husserl, the world of science, which short of philosophical clarification science tends to see as the *real* world, the only *truly objective* one, reveals itself, provided its character of a sense formation is brought to light, as a theoretical construction. As such, Husserl noted, the world of science is *in principle* impermeable to experience, since it is a pure *logical-mathematical* reconstruction of experience.[75] Hence, 'naïve' science, by enthroning the scientific world as the only truly real one, necessarily *degrades* the epistemological relevance of experience, despite its avowed empiricism (which, for Husserl, is only a cover up). If experience should, as it must for science to be consistent with empiricist tenets, remains the sole ground of justification of scientific theories, the Lebenswelt must regain its legitimate position as the foundation where science is edified and the court where it is validated, *for experimentation and the practices related to it belong to the life-world.* For Husserl, the Lebenswelt is, for the scientist, "the world in which he examines his instruments of measurement, listens to his chronometers, evaluates thickness, etc., the world, moreover, to which he knows he himself to belong, with all his activities and theoretical ideas."[76] Husserl's idea that the scientific image of empirical reality is a logical-mathematical *construction* rooted in the Lebenswelt resonates in Weyl's idea that science provides *symbolic reconstructions* of the world that must be validated according to practices (measurements in particular) that, as he explicitly says, are those of the man of the world.

13. Let's now turn our attention to questions of geometry and spatial representation, an issue on which Husserl and Weyl agree to a considerable degree.[77] For example, both believed that our representation of space originates from sensorial data but cannot be reduced to them or, more precisely, that space representation contains also intentional components accounting for the a priori truths of geometry. For Weyl, "from [qualities of sensation] the *mind* has to build up the presentation of spatial extension" (my emphasis).[78] Curiously, Weyl quotes Fichte, "I am originally

---

[72] *Op. cit.*, 188.

[73] *Loc. cit.*

[74] Weyl 1949, 188.

[75] See *Crisis* §34.

[76] *Loc. cit.*

[77] See da Silva 2012a.

[78] Weyl 1963, 126.

not only sentient but also intuiting,"[79] to make a point that Husserl makes more clearly, namely, that perception, particularly of bodies in space, is not a passive experience and extends beyond mere sensation. Also, instead of Husserl (whose unpublished notes on the problem of space constitution he couldn't have known) Weyl refers to O. Becker for "a more detailed phenomenological description of the constitutive levels of spatiality."[80] He says[81]:

> Since the mere Here is nothing by itself that might differ from any other Here, space is the *principium individuationis*. It makes the existence of numerically different things possible which are equal in every respect.... The four-dimensional continuum of space and time is the field of the a priori existing possibilities of coincidence.

Again, coherently with his holism, Weyl says that[82]:

> The geometrical statements, therefore, are merely ideal determinations, which taken in individual isolation lack any meaning verifiable by what is given. Only here and there does the entire network of ideal determinations touch upon experienced reality, and at these points of contact it must 'check.' That, expressed in the most general terms, may well be called the *geometrical method*.

This method, he says, is essential for science:"[*Its*] enemies ... are ... the empiricists, because any aprioristic construction is a thorn in their flesh; they fondly imagine it to be possible to grasp reality as a thing of one stratum, as it were, without aprioristic ingredients, by a purely descriptive approach."[83]

Here, Weyl opposes the 'empiricists' Bacon, Hume and Mach to the 'idealists' Galileo, Kant and Einstein, respectively. But he continues: "On the other hand, out of hatred for the freedom, the open field of geometrical construction, those metaphysicians oppose the method who build up a rigid dialectical world of contents as the true reality [*Hegel versus Newton*]."[84]

In few words, for Weyl, as for Husserl, against empiricists and 'metaphysicians,' physical space (or space-time) is an idealized field of points (idealized virtual positions) and sets of points, and relations among them, displaying both a priori and a posteriori features (the a priori coming in different guises, even the purely conventional). The theory of space contains a lot more than what is or can be given in immediate intuition (even *geometric* intuition, let alone *sensorial* perception).

There are further points of contact between Weyl's and Husserl's views on geometry and space representation. Like Husserl, Weyl distinguished between mathematical, physical and intuitive spaces. Mathematical spaces are those "whose laws are logical consequences of arbitrarily assumed axioms"; physical space is "the ordering scheme of the real thing, which enters as an integral component into the

[79] *Op. cit.*, 127.
[80] *Op. cit.*, 129.
[81] *Op. cit.*, 131.
[82] *Op. cit.*, 132.
[83] Weyl 1963, 132.
[84] *Loc. cit.*

theoretical construction of the world"; intuitive space is the space of perception. Perceptual space, they both believed, is Euclidean and physical space must be locally Euclidean so as not to conflict with intuitive space.[85] Husserl admitted also a still more fundamental sensorial space out of which perceptual space is constituted, and although Weyl does not spell out his views on the constitution of perception, it is safe, I think, to suppose that his largely implicit views on the matter are similar to Husserl's. Weyl also believed that the Kantian a priori can be subdivided into Husserl's a priori of *essential laws* and Poincaré's *conventional* a priori. Among the a priori aspects of physical space Weyl mentions the Euclidean-Pythagorean *nature* of the Riemann-Einstein metric, i.e. the *qualitative* aspects of the metric field; among the a posteriori, the orientation of the metric, i.e. the *quantitative* aspects of the metric field, which depend on the distribution of matter-energy in space.

The topological connectivity of physical space, Weyl thinks, is also a priori, "fixed once and for all, especially the dimension number 4."[86] Here Weyl's views seem to clash with Husserl's, who takes the more commonly held view that the dimensionality of space (or space-time) is a posteriori. But, for Weyl, the a priori character of the dimensionality of physical space-time follows from the fact that it enters by necessity into 'the theoretical construction of the world' and is not only a matter of natural law, which already presupposes that construction. In other words, as Weyl argued, the dimensionality of space-time has consequences for the causal structure of the world and the propagation of effects in it, consequences that are, he thought, *required* by the scientific *idea* of an empirical world. So, the dimensionality of space-time is, in this sense, a priori. In connection with this, we could bring up Husserl's notion of the material a priori, but Husserl himself didn't; for him, space (resp. space-time) is only contingently tridimensional (resp. four-dimensional). He didn't seem to believe, with Weyl, that the very *idea* of an empirical world to be mastered by science entails four-dimensionality (maybe because he did not consider the fact that this number of dimensions is so closely tied to aspects of the world that makes science possible, for example, a clear distinction between past and future and hence cause and effect).

As Weyl explained, four-dimensional Riemannian spaces admit 0, 1 or 2 *negative* dimensions (the dimensions of time). 0 implies no propagation of actions; 2, that past and future would melt; so, it can only be 1. The physical world, Weyl thinks, has only one negative dimension because a causal structure in the world in which past and future are clearly separated is a *scientific necessity*. So, one negative dimension is an a priori feature of space-time. Weyl reasons thus: there is no science without scientific laws; scientific laws connect causes to their effects; so, for science to be possible the word must have a causal structure, which requires that time has dimension 1; therefore, time *must* have dimension 1. That there are 3 spatial dimensions follows, for Weyl, from theoretical physical constructions. Maxwell equations, he says, are everywhere gauge invariant (scale invariant) only

---

[85] *Op. cit.*, 134–5.

[86] *Op. cit.*, 135.

if the world is 4-dimensional. So, for Weyl, the dimensionality of space (space-time) follows *by necessity* from the constitution of an external world to be mastered by science.[87] The a priori character of the nature of the metric, expressed in the Euclidean group of rotations (of all linear transformations $z_i' = \Sigma a_{ij} z_j$ such that $z_1'^2 + \ldots z_n'^2 = z_1^2 + \ldots z_n^2$), follows, for Weyl, from the fact that[88]:

> Only in the case of this particular group does the contingent quantitative distribution of the metric field, however that distribution is chosen within the framework of its a priori fixed nature, uniquely determine the infinitesimal translation, the non-rotational progression from a point into the world.... It may be that the postulate of the unique determination of 'straight progression' can be justified on the basis of the requirements posed by the phenomenological constitution of space.

According to Weyl, among the a posteriori aspects of space we must distinguish between what is necessitated by natural law and "what even under their rule remains free and thus appears contingent."[89] The a priori stands no matter which laws happen to govern the world, *provided it makes sense to talk of natural laws at all* (which, he believes, requires a world with a causal structure – hence, the dimension 1 for time). Nothing indicates that Husserl would accept causal structure as constitutive of the physical conception of an empirical world and if Weyl shared Husserl's views on this, he would had certainly classified dimensionality among the a posteriori aspects of physical space.

14. Weyl also discusses questions of geometry and space representation extensively, as one should expect, in his epoch-making book on Einstein's theory of relativity Space, Time, Matter, published in 1921 but written at the same time he was working on Das Kontinuum (in both books Weyl favored a 'genetic' approach characteristic of phenomenology). It is immediately clear to the reader, from its introduction, that in this book phenomenological ideas are at work. For example, that the physical world is an intentional construct ("the real world, and every one of its constituents with their accompanying characteristics, are, and can only be given as, intentional objects of acts of consciousness"[90]), that reflection on the intentionally given, as they are given, can provide intuitive axiomatic foundations for theories of the given, that the naturalist thesis of reality of objects of experience must be suspended in the philosophical attitude (this, of course, is Husserl's epoché), that "'pure consciousness' is the seat of that which is philosophically a priori,"[91] that since empirical reality is transcendent, sensorial perception is not apodictic, to stay only with those that come up at the very first pages. But it is,

---

[87] See *op. cit.*, 136–7.

[88] *Op. cit.*, 137.

[89] *Op. cit.*, 135.

[90] Weyl 1952, 4. "[T]his world does not exist in itself, but is merely encountered by us as an object in the correlative variance of subject and object. The world exists only as that met with by an ego, as an appearing to consciousness; the consciousness in this function does not belong to the world, but stands out against the being as the sphere of vision, of meaning, of image, or however else we may call it" (Weyl 1934, 83).

[91] Weyl 1952, 5.

I believe, Weyl's genetic approach (or his version of one, which allows for non intuitively based contributions to the genesis), in which a 'reconstruction' of reality is carried out in stages, from the intuitively given to ever higher levels of abstraction, idealization, and symbolization, from the intuitively founded representation of the affine structure of space, through the Euclidean representation of its metric structure, to the Riemannian representation of the inertial structure of space-time, that most evidently shows a general phenomenological frame of mind.

For Weyl, physical space, by which I suppose he meant *empty* physical space, is "the form of the external world."[92] This means that space is *necessarily* homogeneous and isotropic, for the *same* unaltered body cannot have a *different* form by simply occupying a different position or a different orientation in space (in matter filled space fields of forces may impose privileged positions and orientations). "Every material thing," he tells, "can, without changing content, equally well occupy a position in Space different from its present one."[93] This means that homogeneity and isotropy are *a priori* features of space. "Space is a form of phenomena, and, by being so, is necessarily homogeneous."[94] *Hence*, its curvature *must* be constant and its metrical groundform *must* have a *particular* form. Weyl says that[95]:

> Such a space is necessarily Euclidean, spherical or Lobatschefskyian. Under these circumstances not only have the line segments an existence which is independent of place and direction, but any arbitrary finitely extended figure may be transferred to any arbitrary place and put in any arbitrary direction without altering its metrical conditions, i.e. its displacements are congruent.

As Riemann suggested, and Einstein showed, the metrical groundform depends on the distribution of matter in space and so the curvature of matter filled space may not remain constant. However, Weyl claims, a rigid body can retain free mobility by carrying with it the metrical field it produces.[96] In other words, even if the dimensions of a 'rigid' body may change in displacement from the perspective of an 'outsider,' supposing it makes sense to talk of one, the groundform varies in such a way that the metric relations of the body remain unchanged if measured by an observer 'inside' space.

The concept of *rigid body*, that is, a body "which, however it be moved or treated, can always be made to appear the same to us as before, if we take up the appropriate position with respect to it"[97] is, for Weyl, a concept of intuition. Sensorial perception is the source of the concept of rigidity. This concept allows the notion of *congruence* to be defined thus: two positions in space are *congruent* if a rigid body that fills perfectly one can also fills perfectly the other if transferred there. A *congruent*

---

[92] *Op. cit.*, 6.

[93] *Loc. cit.*

[94] Weyl 1952, 96.

[95] *Loc. cit.*

[96] See *op. cit.*, 98.

[97] *Op. cit.*, 11.

transformation (or *congruence*) of space is one in which any region of space is congruent to its image. The congruent transformations form a group. A congruent transformation T is a *translation* if and only if, for every congruence g, there is a congruence g' such that $T^\circ g = g'^\circ T$. These basic notions (and those derived from them) are enough to express the *affine* structure of space, that is, those structural features of space that are invariant under linear transformations. So, for Weyl, affine physical geometry is grounded on intuition.

Metric geometry requires the *choice* of a metrical groundform. The notion of *scalar product* can be defined in the 3-dimensional (affine) space, that is, a real number associated with any pair of vectors (supposing that an association of numbers with points – the arithmetical individuation of points, which, intuitively, are undistinguishable – has already been established). Since we can express the *length* of a vector in terms of the scalar product, this suggests generalizing it by arbitrary metrical groundforms, i.e. symmetrical bilinear forms. If A and B are two points, $d(A, B) = |AB| = \sqrt{}Q(A, B)$, where $Q(x, y)$ is the symmetrical bilinear form and $Q(x) = Q(x, x)$ is the metrical ground form.

By definition of rigidity, a rigid body can move freely in space always occupying congruent spatial regions (free mobility). So, a particular body being chosen as the metric unit, we can carry out measurements at will. So, the metric structure of space can be investigated, but nothing *guarantees* that it will turn out to be Euclidian. So, the Euclidian character of space is not *a priori*; experience however may indicate that, as a matter of *fact*, space is indeed Euclidian, at least locally. The only thing we know a priori is that because of free mobility the curvature of space is constant:"The metric relations of space are not the outcome of space being a form of phenomena, but of the physical behavior of measuring roads and light rays as determined by the gravitational field."[98]

So, Weyl claims, "it is only the whole composed of geometry and physics that may be tested empirically."[99] That is, the true geometry of physical space, *i.e. physical space as considered in the mathematical science of nature*, cannot be decided independently of a decision concerning physical laws, *unless the whole of physics is also formulated, like geometry, for both Euclidian and the generalized Riemannian space*. In this case, Weyl claims, we can decide experimentally which variety of Riemannian space physical space is.[100]

As for the dimensions of space, Weyl says that "there is nothing in the mathematical structure of space to prevent it from exceeding the dimensional number 3."[101] Only the scientific notion of a world under the rule of scientific law, as we've seen, may require the three-dimensionality of physical space.

The centrality of the ego and its experiences in the description of reality, an idea dear to phenomenology, is embraced by Weyl as the *correct* perspective both

---

[98] *Op. cit.*, 102.

[99] *Op. cit.*, 93.

[100] See *op. cit.*, 93–4.

[101] *Op. cit.*, 19.

in geometry and physics (represented by Riemann's and Maxwell's approaches respectively). For Weyl: "We pass from Euclidian 'finite' geometry to Riemann's 'infinitesimal' geometry [distance is determinate for infinitesimally near points by a quadratic form of the coordinate differentials – JJS] in a manner exactly analogous to that by which we pass from 'finite' physics to 'infinitesimal' or 'contact' physics."[102]

With the development of science, Weyl reminds us, we have moved both in geometry and physics from a global to a local perspective centered on the ego. This, as Weyl may have believed, attests to the *correction* of Husserl's philosophical views concerning the centrality of the ego in the building of a scientific image of the world. In any case, Husserl was satisfied with Weyl's philosophical outlook, as testified by a letter to him dated June, 5, 1920, where he says, referring to *Space, Time, Matter*: "How this work does approach my ideal of a philosophical-spirited supported physics [!]." That the world must be described as it presents itself locally to a subject, and that, for the sake of objectivity, different descriptions must be equivalent demand that physical laws be expressed 'infinitesimally' and in tensor form (since only thus they will have the same form in different systems of representation – the identity of form imposes itself, for any difference in form necessarily involve terms that would not have *objective*, referential-free referents).

But Weyl believes that Riemann still retains one element of 'finite' geometry, namely, lengths are comparable even if far apart. For Weyl, the 'infinitesimal' approach requires that comparability be restricted to infinitesimally near lengths only. But Weyl goes even further; in order for gravitation and electromagnetism to have a unified theory, he required that formulas be invariant by any continuous change of coordinates and by the change of the fundamental metric tensor g_{ij} by λg_{ij}, where λ is an arbitrary function of position (gauge invariance). Weyl seems to reason thus: Maxwell's equations and the electromagnetic field tensor are gauge invariant; so, gauge invariance must belong to the *essence* of space (Weyl's gauge theory turned out to be unsustainable).

Obviously, considering that phenomenological epistemology requires that only what can be intuitively given to a localized ego shall count as primarily given, then a phenomenologically founded theory of space, based as it must be on intuition, cannot presuppose more than local comparability of lengths or count on more than *relative* comparability of lengths. Weyl's emphasis on infinitesimal geometry and gauge invariance follows from this. But, as the title of this paper emphasizes, influence is elusive and the centrality of the ego in Weyl's approach to physics and geometry may also be due to Einstein's influence. Einstein's fundamental epistemological principle is that we can only have *direct* experience of world-lines crossing ours, i.e. of events occurring where we are at our local time. So, Weyl's emphasis on locality may simply follow from this principle, whose truth Einstein's theories of relativity have emphatically proven. But also, a more likely hypothesis,

---

[102] *Op. cit.*, 91.

Weyl may have taken the theoretical adequacy of Einstein's physical principle as support for Husserl's more philosophical views concerning the centrality of the ego.

15. An important work where phenomenological ideas are detectable is Weyl's *The Continuum* (1918) in which mathematical analysis is developed on intuitive basis. However, despite the influence of Husserl's idea of an intuition-based theory, the notion of intuition itself is arguably closer to Poincaré's than to Husserl's notion.[103] Mathematical analysis is the study of real numbers, sets of real numbers and real valued functions; but Weyl's analysis deals only with definable sets of natural numbers. Nonetheless, he was able to obtain a context rich enough to represent a segment of classical real analysis. Weyl's procedure takes the parti pris that only properly intuited entities exist properly and can be object of an epistemologically well-founded mathematical theory. Every higher-order construct like sets, sets of sets, etc. of natural numbers, as the non-independent entities they are, do not exist, at least not as independent, self-subsisting entities and, consequently, quantification over them is not allowed (Weyl call this the restrict procedure). By so proceeding, Weyl may be following Husserl's "first principle of all epistemological principles," that knowledge proper is necessarily intuitively founded.

But Weyl admits the infinite realm of natural numbers as given from the start. Well, as a mere glance at his *Philosophy of Arithmetic* shows, Husserl does not allow this much. There are some – few – numbers that can be properly intuited, but most of them can only be indirectly given through symbols. But Weyl had also read the *Logical Investigations* and *Ideas I*, and probably learned, as we may conjecture, that there are other forms of intuition than the intuition of particulars (numbers in this case). Concepts and other higher-level categorial objectivities can also be intuited. And this may be a case in point. Let's consider this possibility.

Some (small) natural numbers can be properly intuited by formal abstraction as formal (quantitative) aspects of collections of objects of whatever nature effectively given either in sensorial perception or imagination. On the basis of these numbers, by imaginary variation, the concept of natural number can be properly intuited, since imaginary variation does not require the effective presentation of *every* natural number, only enough for the concept to come out in full clarity. But as soon as the subject sets out to imagine numbers freely by varying those he can actually intuit, he immediately realizes that numbers can be obtained from each other by adding or subtracting units and that this process can generate all numbers. In short, the subject is able to clearly intuit (by conceptual intuition) the concept of finite cardinal number and its essential features, expressible in Dedekind-Peano's axioms, the principle of finite or complete induction in particular. Incidentally, Poincaré too believed that the principle of complete induction is given in intellectual intuition, so my conjecture that Weyl is not insensitive to Poincaré's philosophy of arithmetic has some basis.

---

[103] See da Silva 1997.

Now, once the *concept* of natural number is intuited Weyl may have inferred that the *domain* of natural numbers is intuitively given *too*, and that, as a consequence, any number-theoretical assertion is intrinsically decided (and so either true or false). Weyl, in conflict with phenomenological teaching, supposing he was indeed under its influence, may have assumed that the intuition of a concept amounts to the *adequate* intuition of its extension. But, and this seems to me a more plausible hypothesis, despite being awaken from his 'positivist slumber' by Husserl's intuition-based epistemology, Husserl's notion of intuition in *The Continuum* did not come from Husserl, but from Poincaré.

Real numbers can be seen as sets of natural numbers (2^{ω} is an arithmetical model of the intuitive continuum). Now, can one intuit the concept of real number, or better, that of an *arbitrary* sequence of natural numbers in the same way one intuits the concept of natural number? Two problems stand in the way. First, not a single such sequence can be adequately intuited, since they are all infinite; so, there is no exemplar with which to start the process of imaginary variation. Second, and more important, there are no principles of generation and closure of the domain of numerical sequences; imaginary variation, in this case, does not disclose generation principles to consciousness. So, the concept of real number seems not amenable to direct intuition. Weyl, then (assuming always that the notion of conceptual intuition was operative in his approach), decided to remedy the situation by, firstly, admitting only definable sequences and, secondly, requiring definability to be restricted to an explicitly given language. Definable sequences, however, or so Weyl believed, although explicitly given, do not exist in themselves; they only exist for us, they are objectivities of the understanding. And so, the 'restrict procedure' (quantification is restricted to natural numbers) stands.

Of course, from the point of view of phenomenology, purely intentional existence is *also* a form of existence; so, there would be no phenomenological malaise in admitting *definable* real numbers as fully existing entities, introducing constants and variables for them in the language, allowing quantification over them and proceeding indefinitely with the definition of new real numbers. A phenomenological approach to mathematical analysis consistent with the notion of being as intentional being would require a hierarchy of definable real numbers. But Weyl refuses this procedure. He cuts an ontological abyss between intuitive existence and purely intentional existence, privileging the former not because phenomenology so requires, but because his intuitionism, derived not from phenomenology, but from preconceived metaphysical theses, so wanted. But the concept of *mathematical* existence, as Husserl has clearly told, is not restricted to intuitive existence. Intuition is one, but not the only intentional, object-positing experience. By privileging the intuitive over the merely intentional Weyl turns his back to phenomenology to embrace a more radical philosophical perspective. A consistent phenomenological perspective from where to ground a contentual (as opposed to formal) mathematics based on the notion of intentional existence—a theory of real numbers and sets and functions of real numbers in the case of mathematical analysis—in which, moreover, each object demanded as noetic correlate a proper definition, requires objects to come in levels, a hierarchy of them, coming into existence as language is progressively enriched by terms for previously defined objects. In the case of real analysis this takes the form of predicative analysis as developed after Weyl.

But mathematical analysis in the classical mold can also be phenomenologically justified, for it can be developed as a formal theory on the basis of intuitively empty, formal axioms. As Husserl told us, theories of this type belong to formal ontology, a chapter of formal logic, as theories of possible realms of being insofar as their form is concerned. So, Weyl's intuitionism is a personal choice that can be justifiable from the point of view of a strict adherence to the epistemological principle of the primacy of intuition (and be praised by Husserl for precisely this reason). But this is not the only way of doing mathematical analysis in accordance with Husserl's teaching, as Husserl himself made quite clear in *his* philosophy of mathematics. Intuitionism, in Weyl's or Brouwer's varieties is *not* a necessary outcome of phenomenological epistemology.

It seems to me that the letter Husserl wrote to Weyl (dated April 10, 1918) thanking him for the copy of *Das Kontinuum* Weyl had sent him, reveals not only Husserl satisfaction in being taken seriously by a mathematician, but also, although in a very tactful way, his opinion that object-intuition is *not* the only basis on which to erect mathematical theories. Husserl praises Weyl for understanding the necessity of taking a phenomenological perspective in the clarification of basic concepts (that of the continuum in Weyl's work) and going back to the 'native soil' of logical-mathematical intuition, the 'real source of foundation for mathematics.' In the letter Husserl mentions his own efforts towards a phenomenological clarification of basic logical and mathematical concepts (such as set, ordinal and cardinal numbers, order, etc.) but *also*, and this, I think, is important, his efforts in drawing a distinction between contentual and formal concepts, judgments and theories. Was he telling Weyl that phenomenological clarification is fine and highly desirable, but that mathematics also involves purely formal concepts and theories that cannot have intuitive foundations? A bit further into the letter Husserl mentions his concept of definite manifold, which as I have shown plays a relevant role in *his* justification of the role played by purely *formal* reasoning in *contentual* mathematics.[104] Husserl did not forget to mention also his attempts towards a formal metaphysics, a "general a priori theory of individuation," in which I recognize the idea of formal ontology, which, as we know, is tailor-made to accommodate *non-intuitive formal* mathematics as the theory of formal molds of possible objectual domains and their mutual relations. In short, in praising Weyl's work Husserl also makes clear that intuitive mathematics is not the whole of mathematics and that the former is not to be developed in detriment of only partially intuitive and purely formal mathematics.

15. To conclude, let's give Weyl a chance to tell himself the tale of his intellectual development, and the role Husserl played in it. He does that in a paper written shortly before his dead in 1955 entitled "Insight and Reflection." Weyl tells that after Kant had been defeated by the non-Euclidean geometries he flirted for a while with positivism, only to have this compromised by a real flirt with a deeply religious

[104] See da Silva 2000a, b.

Hegelian girl. "My peace of mind in positivism was shaken when I fell in love with a young singer whose life was grounded in religion and who belonged to a circle that was led philosophically by a well-known Hegelian."[105] The difference in intellectual outlook, Weyl tells, was partially responsible for the failure of this romantic attachment. Soon afterwards Weyl got married to an ex-student of Husserl, the philosopher who was to rescue him from positivism: "So, it came to be Husserl who led me out of positivism once more to a freer outlook upon the world."[106] But, and this is a major misinterpretation, Weyl identifies Husserl's philosophy with *metaphysical* idealism, not *transcendental* idealism as would be correct, and, to make things worse, a solipsist version of it too. To mend the situation, at least as far as solipsism is concerned, his sympathies turned to Fichte[107]:

> Metaphysical idealism, toward which Husserl's phenomenology was then shyly groping [*received in Fichte*] its most candid and stronger expression. It captured my imagination, even though I had to concede to my wife, who was more at home with Husserl's careful methodology than with Fichte's dash, that Fichte could not help being swept to ever more abstruse constructions by his stubbornness, which made him blind to facts and reality in the pursuit of an idea.

Although Fichte takes the other, the non-I, into consideration in the constitution of reality, he, according to Weyl, makes the entire affair a completely a priori one. The insurmountable conditions of contour imposed by reality were, Weyl believed, ignored. In the end, Weyl came to see the methods of mathematical physics as the only way a representation of reality can be obtained. Perception offers the initial conditions which, once devoid of material content for the sake of objectivity, are incorporated into a much richer symbolic reconstruction of the world that only as a system can face the test of experience: "Following Galileo, one may describe the method of science in general as a combination of passive observation refined by active experiment with the symbolic construction to which theories ultimately reduce."[108]

Leibniz is also mentioned as an influence, and it is not difficult to see where he came in. Weyl's symbolic reconstruction of empirical reality sits comfortably with Leibniz's idea of a mathesis universalis, understood precisely as a symbolic science (a *calculus ratiocinator* expressed in a *lingua characterica*) of the totality of being.

This is how Weyl proposed to reconstruct by a mathematical analogy the problem which, he thinks, faced Fichte and Husserl, i.e. to bridge the gap between pure and empirical egos.[109] Suppose, he proposed, that *points* are the objects and *systems of reference* are the subjects, coordinates in systems of reference are, then, how objects appear to subjects. Objects and subjects are on the same sphere of reality, they are geometric entities, but appearances are numbers. For naïve realism (Fichte's

---

[105] Weyl 1955a, 209.

[106] *Loc. cit.*

[107] *Loc. cit.*

[108] Weyl 1954, 195.

[109] See Weyl 1955a, 215–16.

dogmatism) objects (i.e. points) exist on their own, but it is possible to construct a theory of reality (algebraic geometry) based on appearances only (i.e. numbers). The absolute ego is the system of reference in which object and appearance coincide. For example, if an object is *defined* as a triple of numbers which adds to 1 (for it is how it *appears*) and the ego is a triple of points (a baricentric coordinate system), the coordinates of a point in this system is again a triple of numbers which add to 1. The triple that defines an object and the triple of coordinates of this object in a system S coincide if S is formed by the triples (1, 0, 0), (0, 1, 0) and (0, 0, 1). This is the absolute ego. For the absolute ego what is and what appears are the same. For the sake of objectivity, however, only what is preserved when passing from the absolute ego to *any* other system of coordinates counts (the equivalence of all egos). *Any* ego is the sense-giving ego.

For Weyl, this interpretation of the absolute ego falsifies a number of Husserl's theses, but unfortunately he didn't care to tell which. Weyl seems to ignore that, for Husserl too, objectivity is constituted in intersubjectivity and my personal ego is not a privileged ego among other egos.

Weyl also says that in order to acknowledge the absoluteness of the other we must relinquish content altogether (objects are no longer triples of numbers, systems no longer triples of objects) and be content with an axiomatic approach. That is, only what is formal is truly objective. This is, for him, the point of view of *transcendentalism*, which supersedes those of idealism and naïve realism, defined as follows: Realism = there is an independent reality beyond appearances; idealism = what is is what appears to the absolute ego; transcendentalism = there is an objective reality, but we must be content with modeling it in symbols (objective reality is then reduced to invariant abstract Forms).

The Husserlian idea, however, of 'grounding' science in the life-world reappears. The world of science (but also those of ethics or history), he says, is a 'prolongation' of the world of daily life[110]:

> Modern mathematics and physics may seem to move in thin air. But they rest on a quite manifest and familiar foundation, namely the concrete existence of man in this world. Science is not engaged in erecting a sublime, truly objective world in such pure material, as Democritus employed, above the Slough of Despond in which our daily life takes place. It simply endeavors to prolong a certain important line already laid out in the structure of our practical world. By no means does it pretend to exhaust our concrete existence.

Weyl draws a clear distinction between phenomena and measurements[111]:

> The former is woven of an airy stuff which we can hardly approach by means of ordinary language; only the symbolism is its adequate representation. But the results of measurement belong to the disclosed world of our daily experience; they are describable in terms of classical physics and we can understand each other, communicate, and compare notes about them.

[110] Weyl 1949, 188.

[111] *Loc. cit.*

Weyl is here pointing to an ontological difference between the world of the theoretical scientist and that of the experimental one, the latter "is on the same level of existence as the carpenter or the mechanic in his shop, on the same level of understanding on which our rough and rumble daily life moves."[112] For Weyl, and Husserl agrees, the experimental scientist lives in the life-world.

Regarding mathematics, Weyl's position, as stated in an earlier paper,[113] is also one of compromise. For him, the best option is "to proceed with caution and put more emphasis on explicit constructions. It is not a matter of black and white, but of grades." "I propose another principle," he says[114]:

> *Whenever you can settle a question by explicit construction, be not satisfied with purely existential arguments* ... general propositions ... and their alleged proofs [*are*] not statements of facts but rather instructions for procedure in broad outlines. I would not apply the theorems mechanically to a special case but would, following the instructions, go through all the steps of proof *in concreto*, and while checking them, make them as direct, economic, and constructive as possible.... The blending of constructive and axiomatic procedures seems to me one of the most characteristic and attractive features of present-day mathematics [*but*] before one can generalize, formalize, and axiomatize, there must be a mathematical substance.

So, although Weyl favors constructivist arguments (emphasis on the intuitive) he is not willing to give up non-constructive (symbolic) mathematics, which he 'saves' in two different ways: in mathematics itself, as a provider of 'instructions' and 'guidelines' (Husserl, I remind the reader, gives non-intuitive or partially intuitive *contentual* mathematical theories a similar role, that of *indicating* which situations in their respective domains can be expected to be intuitively presented eventually in an epiphany of truth) or abstract formal 'molds'; and in science, as part of the symbolic reconstruction of the real world. But there is no place in Weyl's system for the 'rules-of-the-game' sort of formalism: "Whatever analogies there are between the mental activities of a mathematician and a chess player, the problems of the former are serious in the sense that they are bound with truth, truth about the world that is and truth about our existence in the world."[115]

References

S. Centrone, *Logic and Philosophy of Mathematics in the Early Husserl* (Springer, Berlin, 2010)

J. Da Silva, Husserl's phenomenology and Weyl's predicativism. Synthese. **110**, 277–296 (1997). Reprinted in Hill, Da Silva (2013), pp. 265–284

J. Da Silva, Husserl's conception of logic. Manuscrito **22**(2), 367–397 (1999)

---

[112] *Loc. cit.*

[113] Weyl 1946.

[114] *Op. cit.*, 170–1.

[115] *Op. cit.*, 172.

J. Da Silva, Husserl's two notions of completeness: Husserl and Hilbert on completeness and imaginary elements in mathematics. Synthese. **125**, 417–438 (2000a). Reprinted as 'Husserl and Hilbert on Completeness and Imaginary Elements in Mathematics' in Hill, Da Silva (2013), 115–136

J. Da Silva, The many senses of completeness. Manuscrito. **23**(2), 41–60 (2000b). Reprinted in Hill, Da Silva (2013), 137–150

J. Da Silva, Beyond Leibniz, Husserl's vindication of symbolic mathematics, in *Phenomenology and Mathematics,* ed. by M. Hartimo (Springer, Dordrecht, 2010), pp. 123–145. Reprinted in Hill, Da Silva (2013), pp. 61–80

J. Da Silva, Husserl on geometry and spatial representation, Axiomathes. **22**, 5–30 (2012a). Reprinted in Hill, Da Silva (2013), pp. 31–60

J. Da Silva, Away from the facts, symbolic knowledge in Husserl's philosophy of mathematics, in *Symbolic Knowledge from Leibniz to Husserl*, ed. by A. L. Casanave (Ed), (College Publications, London, 2012b), pp. 115–135

J. Da Silva, Husserl and the Principle of Bivalence, in Hill, Da Silva (2013a), pp. 285–298

J. Da Silva, Mathematics and the Crisis of Science, in Hill, Da Silva (2013b), pp. 345–367

M.H. Hartimo, Towards completeness: Husserl on theories of manifolds 1890–1901. Synthese **156**, 281–310 (2007)

C.O. Hill, Husserl and Hilbert on completeness, in *From Dedekind to Gödel*, ed. by Hintikka. Synthese Library v. 251 (Kluwer, Dordrecht, 1995). Reprinted in C.O. Hill, G. Rosado Haddock, *Husserl and Frege. Meaning, Objectivity, and Mathematics* (Open Court, Chicago/La Salle, 2000), pp. 179–198

C.O. Hill, J. da Silva, *The Road Not Taken. On Husserl's Philosophy of Logic and Mathematics*, Texts in Philosophy (21) (College Publications, London, 2013)

E. Husserl, *Die Krisis der europäischen Wissenschaften und die transzendentale Phänomenologie*, Husserliana Vol. VI, Martinus Nijhoff, The Hague, 1954 (1936). English translation: *The Crisis of European Sciences and Transcendental Phenomenology* (Northwestern University Press, Evanston, 1970a)

E. Husserl, *Logische Untersuchungen.* Bd I: *Prolegomena zur reinen Logik* (Max Niemeyer, Halle, 1900), ([2]1913). Bd II: *Untersuchungen zur Phänomenologie und Theorie der Erkenntnis* (Max Niemeyer, Halle a.d.S., 1901) ([2]1913–1921). English translation: *Logical Investigations*, J. N. Findlay (ed.) (Routledge, New York, 1970b)

E. Husserl, *Erfahrung und Urteil* (Meiner, Hamburg, [5]1975) [original ed. 1939, first Meiner ed. with preface by Lothar Eley 1972]. English translation: *Experience and Judgment* (Routledge and Kegan Paul, London, 1973)

E. Husserl, *Ideen zu einer reinen Phänomenologie und phänomenologischen Philosophie*, Jahrbuch für Philosophie und phänomenologische Forschung, I, 1913, 1–323 [as a separate volume: Niemeyer, Halle a.d.S. 1913; [2]1922; [3]1928]. English translation: *Ideas Pertaining to a Pure Phenomenology and to a Phenomenological Philosophy. First Book: General Introduction to a Pure Phenomenology*, in *Edmund Husserl Collected Works* vol. II (Springer, Berlin, 1982)

E. Husserl, in *Studien zur Arithmetik und Geometrie (1886–1901)*, ed. by I. Strohmeyer. Husserliana Vol. XXI (Martinus Nijhoff, The Hague, 1983)

E. Husserl, *Early Writings in the Philosophy of Logic and Mathematics* (Kluwer, Dordrecht, 1994)

E. Husserl, *Ding und Raum. Vorlesungen 1907*, ed. by U. Glaesges. Husserliana Vol. XVI (Martinus Nijhoff, The Hague, 1973), [Meiner ed. 1991]. English translation: *Thing and Space: Lectures of 1907* (Kluwer, Dordrecht, 1997)

E. Husserl, *Philosophie der Arithmetik. Logische und psychologische Untersuchungen*, Bd. I, Pfeffer, Halle a.d.S. 1891. English translation: *Philosophy of Arithmetic, Psychological and Logical Investigations with Supplementary texts from 1887 – 1901*, D. Willard, (ed.) (Kluwer, Dordrecht, 2003)

U. Majer, Husserl and Hilbert on completeness: A neglected chapter in early twentieth century foundations of mathematics. Synthese **110**, 37–56 (1997)

M. Odada, Husserl and Hilbert on completeness and Husserl's term rewrite-based theory of multiplicity, invited talk, in *24th International Conference on Rewriting Techniques and applications (RTA'13)* (Eindhoven, 2013)
D. van Dalen, Four letters from Edmund Husserl to Hermann Weyl. Husserl Stud. **1**, 1–12 (1984)
H. Weyl, Levels of infinity (1930, in Weyl (2012), pp. 17–31
H. Weyl, The open world: Three lectures on the metaphysical implications of science (1932), in Weyl (2009), pp. 34–82
H. Weyl, Mind and nature (1934), in Weyl (2009), pp. 83–150
H. Weyl, The mathematical way of thinking (1940), in Weyl (2012), pp. 67–84
H. Weyl, Address at the Princeton bicentennial conference (1946), in Weyl (2009), pp. 162–174
H. Weyl, Man and the foundations of science (ca. 1949), Weyl (2009), pp. 175–193
H. Weyl, *Space, Time, Matter* (Dover, New York, 1952), translated by H. L. Brose from the 4th ed. of *Raum, Zeit, Materie* (Springer, Berlin, 1921)
H. Weyl, Axiomatic versus constructive procedures in mathematics (1953). Weyl (2012), pp. 191–202
H. Weyl, The unity of knowledge (1954), in Weyl (2009), pp. 194–203
H. Weyl, Insight and reflection (1955a), in Weyl (2009), pp. 204–221
H. Weyl, Why is the world for-dimensional? (1955b), in Weyl (2012), pp. 203–216
H. Weyl, *Philosophy of Mathematics and Natural Science* (Atheneum, New York, 1963) (revised and augmented English edition based on a translation by Olaf Helmer). Parts of the book originally published in *Handbuch der Philosophie* as 'Philosophie der Mathematik und Naturwissenschaft' (Oldenbourg Verlag, München, 1927)
H. Weyl, *The Continuum, A Critical Examination of the Foundation of Analysis* (Dover, New York, 1994), translated by S. Pollard, Th. Bole, originally published as *Das Kontinuum. Kritische Untersuchungen über die Grundlagen der Analysis* (Veit, Leipzig, 1918)
H. Weyl, *Mind and Nature. Selected Writings on Philosophy, Mathematics, and Physics*, ed. by P. Pesic (Princeton University Press, Princeton, 2009)
H. Weyl, Levels *of Infinity. Selected Writings on Mathematics ad Philosophy*, ed. by P. Pesic. (Dover, New York, 2012)

Chapter 14
Paradox, Harmony, and Crisis in Phenomenology

Judson Webb

Abstract Husserl's first work formulated what proved to be an algorithmically complete arithmetic, lending mathematical clarity to Kronecker's reduction of analysis to finite calculations with integers. Husserl's critique of his nominalism led him to seek a philosophical justification of successful applications of symbolic arithmetic to nature, providing insight into the "wonderful affinity" between our mathematical thoughts and things without invoking a pre-established harmony. For this, Husserl develops a purely descriptive phenomenology for which he found inspiration in Mach's proposal of a "universal physical phenomenology." To account for applications to any domain, Husserl envisages a theory of all possible deductive systems, which he develops extensively in his Göttingen lectures wherein he engages with Hilbert's work on deductive systems for geometry, real arithmetic, and physics. This leads Husserl to formulate claims of decidability and proofs of completeness for various arithmetics that result from his analysis of Kronecker's general arithmetic. Careful attention to these proofs seem to show that Husserl was not oblivious to problems that underlie our incompleteness theorems, namely that of showing that some inversions of his algorithmic arithmetic are undefined. His growing preoccupation with the issue of a pre-established harmony between mathematical thought and reality motivate his pursuit of a "supramathematics" of all possible complete theory forms to demystify such harmony, by having such a form on hand for describing any empirical domain. He soon decides that a transcendental idealism of nature will reveal the wonderful affinity of thoughts and things comprising such harmony, to be a wonderful "parallelism of objective unities and constituted manifolds of consciousness." But the paradoxes of logic and set theory cloud the clarity of mathematics, which Weyl would restore with Brouwer's intuitionism and Hilbert with his metamathematics. Husserl informed Weyl that his student Becker had formulated a phenomenological foundation not only for Weyl's generalization of relativity theory but also for the Brouwer-Weyl continuum. But Weyl eventually rejected much of Becker's work, especially when it became clear that his phenomenological intuitionism could not account for the success of Hilbert's transfinite mathematics in quantum physics. Becker responded to this

---

J. Webb (✉)
Department of Philosophy, Boston University, Boston, MA, USA
e-mail: ahooper4@verizon.net

S. Centrone (ed.), *Essays on Husserl's Logic and Philosophy of Mathematics*,
Synthese Library 384, DOI 10.1007/978-94-024-1132-4_14

"crisis of phenomenological method" with his mantic phenomenology celebrating the magic of mathematical mysticism, which Husserl finally rejects in favor of a pluralistic phenomenology of mathematics and nature.

Keywords Edmund Husserl • Computability • Completeness • Pre-established harmony • Paradox • Idealism • Crisis • Oskar Becker

14.1 Introduction

Husserl's last work limns a crisis gripping modern science and mathematics. He admits that neither the puzzling price paid to overcome the "paralysis" of classical physics nor the "continuing conflict over the genuine form of construction for pure mathematics" means that physics and mathematics were unscientific: both obtained "convincing insights" despite "certain unclarities and blindspots."[1] Conceding that one may doubt such a modern crisis, Husserl invokes "the general lament about the crisis of our culture and the role here ascribed to science." He argues that the perennial difficulties of psychology are crucial "both for the appearance of puzzling, insolvable obscurities in modern, even mathematical sciences and … for the emergence of a set of world enigmas … unknown to earlier times."[2] Their common source is "the paradox of human subjectivity: being a subject for the world and at the same time an object in the world."[3] Only a radical epoché can give us "the attitude *above* the subject—object correlation which belongs to the world," allowing us to see by self-reflection that "our world in its being … takes its ontic meaning entirely from our intellectual life through *a priori* accomplishments that can be exhibited rather than argumentatively constructed or conceived through mythical thinking."[4] Husserl admits that

> the phenomenologist lives in the paradox of having to look upon the obvious as questionable, as enigmatic, and of henceforth being unable to have any other scientific theme than that of transforming the universal obviousness of the being of the world—for him the greatest of enigmas—into something intelligible.[5]

If his paradox is unsolvable, radical epoché would be impossible "for the purpose of a science bound to it."[6] This would be a crisis for phenomenology itself, leaving it unable to clarify those "puzzling insolvable obscurities in modern, even mathematical sciences," among which Husserl includes the paradoxes of set theory. In his *Britannica* article he claimed that "for the a priori disciplines (e.g.

[1] *CoESP*, 4.
[2] loc. cit., 5.
[3] loc. cit., 178.
[4] loc. cit., 181.
[5] loc. cit., 180.
[6] loc. cit.

mathematical sciences) which are founded within phenomenology there can be no 'paradoxes' and no 'crisis of foundations.'"[7] In his third draft he wrote that the foundational crisis of the sciences was the "most severely felt in the pure mathematical sciences underlying the exact natural sciences. The strife over the 'paradoxes' concerning the legitimacy or illusory obviousness of the basic concepts of set theory, arithmetic, geometry, chronometry, etc." have revealed that "none of these disciplines are genuine sciences in the sense of being transparently justified."[8] One aim of this paper is to gauge how the emergence of the paradoxes of logic and infinity factored into the evolution of Husserl's sense of crisis.

In Sect. 14.2, I analyze his remarkable theory of computation in the *Philosophy of Arithmetic* resulting in his canvas of the "totality of conceivable arithmetical operations." Stefania Centrone has argued persuasively that the totality of functions computed by them is extensionally equivalent to the partial recursive functions, while admitting that Husserl advanced no thesis comparable to Turing's about their significance.[9] I argue that he was motivated to explore Kronecker's program of a 'general arithmetic' that unified number theory with algebra to reduce analysis to algorithms for calculating with integers. But Husserl's philosophical critique of nominalism led him to ignore Cantor's mathematical critique of it, whose cardinality argument against it would help orient Turing's analysis. The critique of nominalism also prompted his search for a principled account of the applicability of mathematics, leading in the *Logical Investigations* to a vision of all possible deductive systems and the possible fields of knowledge to which they apply. Philosophy must achieve insight into the "wonderful affinity" between thought and the things, events, and causes studied in the sciences, an affinity some took for a pre-established harmony, and for the analysis of which Husserl develops a descriptive phenomenology which appears to be inspired by Mach's vision of "universal physical phenomenology."

In Sect. 14.3, I examine Husserl's theory of complete deductive systems and his use of it to solve what he takes to be a most important paradox: the fruitlessness of imaginary arithmetical concepts. This leads to a logical analysis of Kronecker's theme of the conservativity of analysis over number theory. It also leads Husserl to make claims of the deductive completeness of various arithmetics, which seems at odds with our post Gödelian landscape, but I argue that he had actually caught a glimpse of the difficulty that underlies incomplete formal arithmetics in his theory of conceivable arithmetical operations, namely that of demonstrating that the inverses of such operations are defined for particular numbers or even pose consistent problems. On the other hand, Husserl was even more aware of Kronecker's success

[7] Husserl 1981, 32.

[8] Husserl 1968, 520. Husserl's late acknowledgement of the failure of phenomenology to clarify the mathematical sciences contrasts sharply with the confidence he had in his pre-phenomenological theory of complete deductive systems to clarify not only his own paradox of the fruitlessness of imaginary arithmetical concepts but also the perennial puzzle of their harmony with nature.

[9] Centrone 2010.

in developing algorithms for calculating the roots of algebraic equations, which clearly fed his belief in the decidability and completeness of various "arithmetics." Husserl also engaged himself with Hilbert's work on geometry and the real ordered field, which offered vivid glimpses of both a class of logical problems seemingly impervious to algorithmic treatment, as well as another class that clearly was. But only the latter seems to have caught his attention.

In Sect. 14.4, I trace Husserl's early preoccupation with the issue of a pre-established harmony between logical thought and reality, and his vision of a supramathematics of all possible theory forms intended to demystify such harmony by having such a form on hand for application to any empirical domain. I examine his phenomenology of set theory, its paradoxes, Zermelo's axioms for it, and his perusal of Weyl's critique of it, which seems to persuade Husserl that only constructivism can eliminate paradoxes. We then analyze his initial foray into a transcendental idealism of nature.

In Sect. 14.5, we begin with what appears to Husserl as Becker's phenomenological basis for Weyl's version of general relativity as well as for the intuitionist continuum. But Hilbert initiates a *Grundlagenstreit* with his new metamathematics, which also appears to be compatible with phenomenology. Weyl's critique of Becker's attempt to find a phenomenological basis in Husserl's levels of consciousness for the recursive ordinals formulated by Hilbert in his attempt to prove the continuum hypothesis, together with the success of Hilbert's transfinite mathematics in quantum mechanics, leads Weyl to envisage a decisive defeat for phenomenology. For Becker this challenge comprises a crisis of phenomenological method, which he meets with his mantic phenomenology celebrating mathematical mysticism. Husserl meets this challenge by embracing a pluralistic philosophy of mathematics and natural science that rejects mysticism and the magical origins of science.

14.2 Computability, Nominalism, Application

Husserl saw arithmetic as deriving numbers from numbers, either by "an *essentially conceptual operation*," that, using signs sparingly, is abstract, limited, and laborious; or by "an *essentially sense perceptible operation*" that derives one number sign from another by fixed rules, where only the final result designates a concept. This method is concrete, convenient, and "there is no conceivable problem which it would not be capable of solving."[10] It is thus *the* logical method of arithmetic, whereupon calculation is "any symbolic derivation of numbers from numbers substantially based on rule governed operations with sense perceptible signs."[11] Symbolic formulations of numbers with function symbols are reduced to "systematic numbers" without them, and Husserl seeks a taxonomy of "all

[10] *PoA*, 272.

[11] loc. cit., 272–273.

conceivable modes of formations of numbers" and efficient rules for such reduction. Those for addition and multiplication yield sequences of steps for sums and products he later calls "fulfillment chains," but problems of the partial functions of subtraction and division "do not always have a sense and solution.".[12]

The issue is critical for "higher operations" such as powers and their inverses. Husserl considers the sequence "a + b, a•b, a^b, etc.," explaining that "through counting how often a number has been raised iterately to a power there arises a new type of symbolic number characterization, that of *elevation*; through the counting up of the iterations, a new one again, and so on *ad infinitum*."[13] But as product led to quotient,

> so each of these new forms lead to corresponding inverses ... the symbolic formulation a^b points to a certain number *c*, where $a^b = c$. But now *b* is in a certain way characterized by *a* and *c* and likewise *a* by *b* and *c*. ... We therefore have acquired two new ways of indirectly symbolizing number formations (in symbols $\log_b a$ and $\sqrt[b]{c}$) through the inversion of the concept of power.[14]

And similarly for elevation. Unlike inversions of addition and multiplication, the question of whether *these* always have meaning for every pair of numbers "could not be directly decided without deeper analysis ... prior to beginning the calculation, one must know whether it could yield any result at all, whether the number problem posed ... does not involve an impossibility a priori."[15] One can always follow the progression of steps: "But logical interests once again require that ... one must take thought to guarantee beforehand the possibility of the execution (thus, that the problem is free of contradiction).".[16]

Husserl also considers indirect characterizations of numbers as *solutions of equations* and the difficulty of "unraveling complicated number interrelationships into which the unknown number itself is interwoven."[17] He notes that neither negative, imaginary, fractional, or irrational numbers are thus introduced until we consider equations such as $ax \pm b = c$ and $ax^2 + b = c$, and that whether they lead to number forms "that are ultimate and not further reducible—and whether what they yield is free of contradiction under all circumstances or only certain ones ... require separate investigations."[18] Such problems are treated by algebra and

[12]loc. cit., 285.

[13]loc. cit., 292.

[14]loc. cit., 293. Later Husserl uses these formulations to illustrate the power of formal laws in the science of meaning, explaining that "There are definite forms of synthesis, through which, quite in general or, in certain definite conditions two numbers give rise to new numbers. The 'direct operations' a+b, a•b, a^b yield resultant numbers unrestrictedly, the 'inverse operations' a-b, a/b, $\sqrt[b]{a}$, ${}^b\log a$, only in certain conditions. That this is the case must be laid down by an *assertion* or rather a *law of existence*, and perhaps proved from certain primitive axioms." (Husserl 1970, 518). In effect, Centrone has proved Husserl's assertion.

[15]loc. cit., 293–294.

[16]loc. cit., 294.

[17]loc. cit., 297.

[18]loc. cit., 298.

evince "the *need of a general arithmetic* in the *sense* ... of a general theory of *operations*." This is just what Kronecker needed to reduce the infinite decimals of irrationals to an "arithmetical law of their terms" and thus to "finite expressions" of methods of calculation.[19] Husserl admits that while his own arithmetic agrees with those of Weierstrass, Cantor, and Dedekind, he "feels a certain harmony" with that of Kronecker. For any sound development of general arithmetic, says Husserl, must "run completely parallel" to his own and can clear up all the difficulties he is facing, "provided only that it considers the domain of 'positive whole numbers' to be the first and most underivate domain, the sole foundation of all remaining domains."[20] This is precisely the view of Kronecker and in fact, the logical difficulties that motivated the theories of irrationals of those three distinguished mathematicians are still unresolved, says Husserl: "For how else would it be possible for an equally eminent mathematician to seriously try to banish the irrationals that give rise to these difficulties from analysis, and reduce them entirely to calculations with whole numbers? Kronecker."[21] By introducing the "indeterminates" of Gauss' modular arithmetic into algebra, says Kronecker, "the special theory of whole numbers is extended to a general arithmetic of whole number functions of indeterminates, allowing all concepts foreign to true arithmetic, negative numbers, the real and imaginary algebraic numbers, to be eliminated."[22] Having refined Sturm's theorem that calculates the number of roots of a polynomial in an interval, Kronecker argues that "the so-called existence of the real irrational roots of an algebraic equation is purely and simply based entirely on the possibility of isolating them, hence on the possibility of a number."[23] Kronecker concludes that "all the deepest results of mathematical research must ultimately be expressed in the simple form of properties of whole numbers." Husserl admits that if Kronecker's reduction succeeds he will have "achieved a work of epoch making significance.".[24]

This brings us to Husserl's critique of the nominalism of Hemholtz and Kronecker. Hemholtz derives formulas for calculating with numbers from equations such as (a + b) +1 = a + (b + 1), says Husserl, "keeping in mind that they all appear as mere equivalences between certain complexes of symbols.... But what is this empty play on symbols, one will ask in astonishment?"[25] Hemholtz admits that arithmetic "would be pure game of ingenuity with dreamt up objects ... if it

[19]Becker 1975, 327. Husserl also considers Kronecker's method of determining numbers by asking "which numbers x satisfy the congruence a • x ≡ b mod p", or for "the necessary and sufficient conditions that an equation of the second degree is solvable" (Husserl 1891, 519). He says that "all such problems require a general arithmetic", for they all presuppose an account of various operations.

[20]Husserl 1994, 1–2.

[21]Husserl 1983, 240.

[22]Kronecker 1887, 345.

[23]loc. cit., 353.

[24]Husserl 1983, 40. A strong case for Kronecker's influence on Husserl is made by Gauthier 2004.

[25]*PoA*, 181.

did not contain such extraordinarily useful applications."[26] This requires continuous magnitudes for measuring the effects of real bodies with any degree of precision. Of Hemholtz' number signs Husserl writes: "By means of the store of designations, assumed to be unlimited, we then can symbolize the elements of any conceivable group," but says that this must be a "senseless act" for the nominalist.[27] Husserl ignores Cantor's refutation of their nominalism, which denies just what Husserl apparently grants, namely, that an unlimited 'store of designations' can symbolize the elements of any conceivable group. Cantor doubts that the numbers resulting from Kronecker's unification project apply to geometry and mechanics, for he writes

> I believe that I can say without any doubt in advance that it will not be possible for him, with the 'ideal supply' of his symbols to describe the *actual infinite supply* of points of the continuum completely ... my conviction depends on my having proved the theorem that the power of the continuum is greater than that of the set of all finite whole numbers.[28]

And so greater than the set of finite expressions for calculating numbers.

Husserl knew Cantor's 1887 paper from which we have quoted this argument, but his theorem had as yet only appeared under the radar in his 1874 paper on "A Property of the Set of Algebraic Numbers."[29] Cantor first proved the countability of the algebraic numbers by careful analysis of the possible algebraic equations and their roots, and then the uncountability of the real numbers by an ingenious compactness argument. Cantor concluded that transcendental numbers must exist, confirming Liouville's lengthy proof of their existence, which Kronecker rejected because no method for calculating their expansions was given. Cantor addressed this in 1882, explaining that a set is "well defined" if its definition together with the law of excluded middle "internally determined" whether an object belonged to it or not. The definition of an algebraic number, he says, "internally determines whether a given definite number n belongs to the algebraic numbers or not, though the problem of actually effecting this decision is often among the *most difficult*; thus it has long been an open question of great importance whether the number π is algebraic or, as is very probable, transcendental."[30] Such was Cantor's reply to Kronecker's demand for effective decidability whereas H.A. Schwarz challenged Kronecker to back up his objection to Weierstrass' theory of irrationals by "giving a

---

[26]Hemholtz 1977, 74.

[27]loc. cit.,182. And similarly for Kronecker's symbols.

[28]Cantor 1932, 384.

[29]Cantor's title deliberately camouflages his more significant result of the uncountability of the real numbers, fearing opposition of Kronecker (then editor) to much analysis. See Dauben 1979, 66–70.

[30]loc. cit., 150–151. That the difficulty of proving the transcendence of π was related to the uncountability of non-algebraic numbers was vaguely felt by Leibniz when he claimed that his infinite series for π/4 "cannot be expressed in finitely many rational numbers", but asks if "a precise abbreviated form of this series can be found. But finite expressions—and especially the irrationals, if one goes out to the supersurds—can vary in so many ways that one cannot enumerate them, and easily determine all the possibilities" (Leibniz 1981, 376) Leibniz also calls supersurd numbers "transcendental."

provably undecidable example" of an irrational not meeting his demand. While no one was in a position to give such an example, Husserl was clearly very interested in Kronecker's project, which brings us to Centrone's remarkable demonstration that the class H of functions computable by Husserl's totality of arithmetic operations is extensionally equivalent to the partial recursive functions, as well as her question of a possible 'thesis' motivating his canvas of this totality.

Husserl never ruled out the possibility that Kronecker's project of reducing all number forms to algorithms for calculating with integers might succeed, something he said would be "epoch making." A desire to clarify such a possibility was surely a good reason for canvassing his totality. Cantor's cardinality argument against it implies that there must exist real numbers for which Kronecker could provide no algorithm for calculating their decimal expansions, and assumes that his "finite expressions" of algorithms he could muster from his supply of symbols must be countable. This argument could have alerted Husserl that there might be functions falling outside his class H, which by Centrone's result is countable, something Husserl was not prepared to establish. It has been objected to Centrone by way of questioning the historical significance she ascribes to Husserl's canvass of H, that he advanced no 'thesis' comparable to Turing's that the computable functions were just those computed by his machines.[31] But Husserl did understand the burden of Kronecker's program to clarify the last two words of the first sentence of Turing's classic paper: "The 'computable' numbers may be described briefly as real numbers whose expression as a decimal are calculable by finite means."[32] Centrone grants that Husserl was not motivated by anything comparable to Turing's specific desire of showing Hilbert's *Entscheidungs problem* to be effectively unsolvable. To prove his halting theorem he had only to satisfy himself that Cantor's diagonal argument for the uncountability of the reals could not be used to prove the uncountability of the *computable* real numbers. Turing was guided by his clear understanding of Cantor's distinction between the countable and uncountable sets and his arguments for it, something not to be found in Husserl.[33] This left him, despite his insight that a "deep analysis" may be required to determine whether a function in H resulting from inversion is defined for certain numbers, with neither motivation nor means to investigate whether H contains the totalization of every partial function it contains, or any universal functions. That is, to study the extraordinary closure properties of H that would encourage one to believe that it contained all functions computable by 'finite means'.

An uncertainty that clouds this analysis is Husserl's view of infinite sets in general and the distinction of their powers, as well as his awareness of paradoxes associated with them. He was influenced by Bolzano's *Paradoxes of the Infinite* from which he took the notion of "propositions in themselves" whose truth value is independent of our thought. Such propositions are crucial to Bolzano's proof of the

[31] van Atten 2012.

[32] Davis 1965, 116.

[33] I have been unable to find any indication of Husserl's awareness of Cantor's theorem.

existence of infinite sets, which he defines as a "multitude so constituted that every single finite multitude represents only a part of it."[34] In fact, "the set of all absolute propositions and truths is easily seen to be infinite. For if we fix our attention on any truth taken at random and label it A, we find the proposition 'A is true' is distinct from the proposition A itself, since it has the complete proposition A for its subject."[35] And then, "'A is true' is true" is a further distinct proposition, and so on without end. Bolzano found "the very remarkable peculiarity" of infinite sets that they can be in 1–1 correspondence with a proper subset, a fact overlooked he says, "to the disadvantage of our insight into many truths of metaphysics and physics and mathematics."[36] He admits his account "will sound so paradoxical that we shall do well to linger somewhat over its investigation" and concludes presciently that "an air of paradox clings to these assertions" only because being in 1–1 correspondence is sufficient for finite sets to be "perfectly equinumerous."[37] But he does not ask whether infinite sets all stand in such correspondence.

In the *Philosophy of Arithmetic* Husserl denied that we can actually represent either the infinite number series, the points on a line, or the continuum. He analyzed small numbers by appeal to "psychic acts" of collecting multiplicities into a unity and large numbers that cannot be authentically given in terms of symbolic operation on signs. While criticizing Cantor's definition of numbers as the abstract power of a set, Husserl uses his language of sets in order to improve his own psychological analysis of number. His 1891 study "On the Concept of Totality" formulates six axioms for finite sets. He begins:

> Let A, B, C ... designate any objects whatever whether intuited or thought, existing or imaginary, so long as they are compatible with each other. Then the expression 'A and B and C ...' taken in its general sense yields a definition of the term 'totality' of objects A, B, C.... The objects totalized we call the members of the totality.[38]

Husserl's first axiom states that "[f]or every totality, except the one which includes everything representable in the widest sense, there is a possible further object which is not contained in it."[39] This universal set comprising the sole unextendable set was evident to Husserl, who criticized Schroder for rejecting "the *Boolean concept* of the class of all possible objects." But this class exemplified what Cantor will call an 'inconsistent set', since the application of his theorem to it yields his paradox: it contains more subsets than the set of all possible objects of thought. As yet unaware of this manifestation of the paradoxical nature of infinite sets foretold by Bolzano, Husserl proceeds to base theorems about finite sets and numbers on his axioms.[40]

---

[34] Ewald 1996, 254.

[35] This assumption that propositions are individuated by their subjects is also crucial to Dedekind's proof of infinite sets.

[36] loc. cit., 266.

[37] loc. cit., 268.

[38] *PoA*, 360.

[39] loc. cit., 361.

[40] Centrone (loc. cit., 89) has found a quote that suggests an inkling of inconsistent sets.

Theorem 3 states that "If a is a finite number, then a+1 is a finite number," which he proves by contradiction, using Cantor's definition of infinite sets. Husserl defines the natural number series by an infinite "chain of definitions 2= 1+1, 3= 2+1, 4= 3+1 ..." and proves that "the natural number series has a beginning but no end," appealing to his first axiom, from which he infers that every set except his universal set can be expanded by a unit. Finally he proves the existence of the infinite set of natural numbers.

Husserl explains the concept of 'finite number' by appeal to a 'finite number of iterations' of adding a unit, something Frege avoided by appeal to his logic of ancestral relations and his Basic Law V. He gave up his logical reduction of the natural numbers on being shown that this law leads to paradox: it implies a 1–1 correspondence between concepts and objects, in contradiction with Cantor's theorem which states, in Frege's terms, that there are more concepts definable over a set than there are objects in the set.[41] His system foundered on the same difficulty Cantor said Kronecker could never overcome: while his theorem implied there are uncountably more reals than algorithms for computing them, it now implies there are uncountably many more concepts than objects. Remarkably, Frege appeals to this theorem in his theory of real numbers, which he intended to be *a priori* applicable to any "quantitative domain." For him a mathematical theory is raised above the level of a game by its applications, which however, *must somehow be built into it.* Frege's account of natural numbers did guarantee mundane uses of them, but his theory of real number assumed that all quantitative domains must be continuous.[42] This was also presupposed by Helmholtz to account for the applicability of his formal arithmetic to physics.

Husserl claimed that Helmholtz' nominalism left him unable to explain "the objective sense of the fact that we express relationships between real objects by means of concrete numbers as magnitudes."[43] He complained to Stumpf that *no* one has explained how arithmetic can deal with mere signs and still have applications. When its signs are applied to discrete magnitudes, all but the whole numbers lose their meaning, whereas:

> In relation to continuous magnitudes (e.g. weight) fractions and irrational numbers are meaningful signs ... nevertheless, in the 'theory of numbers' (dealing exclusively with whole numbers), the integral and differential calculus, and possibly its irrationals, are applied—with success! Each theorem obtained is an actual identity, if one calculates it out. How is this possible?[44]

---

[41]For a good discussion of Frege's collision with Cantor's theorem, see Burgess 1995,101–104.

[42]An assumption for which he has been criticized by Dummett 1991, 278 ff.

[43]*PoA*, 182. His sole argument was that when I count four apples I do not "have in mind" the ordinal but rather the cardinal concept of number.

[44]Husserl 1994, 15. Husserl alludes here to the success of analytic number theory in providing theorems about the integers, as in Dirichlet's remarkable generalization of Euclid's theorem on the infinity of primes, where he uses analysis to prove that every arithmetical sequence a+b, a+2b, a+3b, ... a+nb where a and b are relatively prime contains infinitely many primes. Kronecker

Husserl wanted an account of the application of imaginary numbers to properties of real bodies in physics as well as to those of whole number in analytic number theory, without having to regard them as "impossible concepts." The case of arithmetic will lead him to explore Kronecker's theme of how conservative higher mathematics is relative to its elementary branches, while physics will eventually confront him with the issue of a pre-established harmony between mathematical thought and the world. Husserl was nagged by the fact that: "Arithmetic operates with remote symbolic representations whose real correspondents are inaccessible to us. How is it, that this circumstance does not hinder the application of arithmetic?"[45] Making such representations of modern mathematics and physics accessible will pose a test for his later phenomenology.

In the *Prolegomena* Husserl ponders how thought adapts to nature: "Even Hume spoke in this connection of 'a sort of pre-established harmony' between the course of nature and the sequence of our ideas ... and the modern theory of development has prompted ... a detailed investigation of the relevant teleologies of our mental constitution."[46] In this regard Husserl quotes the 'excellent remark' of Mach that "To do mathematics without achieving clarity in this regard is often to have the disagreeable impression that one's paper and pencil are cleverer than oneself."[47] This betokens to Husserl our mental limitations, leaving us only "a narrow sphere within which complex abstract notions can be fully understood."[48] Our sphere of "full comprehension, operating with thoughts themselves" is so small that science would be impossible without symbolic processes unaccompanied by intuition or insight. But the purely mechanical use of symbolic thinking produces "an economy which leads imperceptibly to formal generalizations of our original thought trains", leading almost effortlessly to infinite deductive horizons: "Fully conscious reflection now takes over and the pure theory of manifolds emerges as a further extension. In its form this covers all possible deductive systems: the form-system of formal arithmetic is merely one of these special instances."[49] Husserl insists we cannot fully clarify the "economic achievement" of this theory of deductive systems or it's debts to systematic thinking by "vague talk of the surrogative function of signs, of mechanisms which save energy, etc.: deep-going analyses are everywhere needed.".[50]

succeeded in proving this result in his general arithmetic, confirming to his satisfaction that analysis was conservative over his arithmetic.

[45] Husserl 1983, 257.

[46] Husserl 1970, 199. If we see our mental performances "as promoting survival we can treat them from an economic standpoint and test man's actual performances from a teleological angle."

[47] loc. cit., 201. Husserl says the continuation of Mach's observation is also worth quoting: "Mathematics taught in this manner has barely more educational value than absorption in the Kabbala or in mystic squares. It necessarily breeds a mystical tendency, which on due occasion bears its fruits" (201). We shall return to Mach's observatons.

[48] loc. cit., 201.

[49] loc. cit., 203.

[50] loc. cit., 203.

Husserl envisioned not just all possible deductive systems but the possible fields of their application: "the *objective correlate* of the concept of a possible theory, definite only in respect to form, is the *concept of a possible field of knowledge over which such a field will preside*."[51] Only in the "division of labor" proposed by Husserl for realizing his vision do we see how high he aimed. Logic is left to mathematicians, for "the mathematical treatment of inference is the only one that offers us systematic closure and completeness, and a survey of all possible questions together with the possible forms of their answers."[52] But the mathematician is "not really the true theoretician, but only the ingenious technician ... who looking only to formal interconnections, builds his theory ... without ultimate insight" into its concepts and laws.[53] This is only provided by the philosopher who "does not seek to meddle in the work of the specialist, but to achieve insight in this regard to the sense and essence of its achievements as regards method and manner."[54] Not content with the *success* of scientific predictions,

> he wants to clarify the essence of a thing, an event, a cause, an effect, of space, of time, etc., as well as the wonderful affinity which this essence has with the essence of thought which enables it to be thought, and with the essence of knowledge which makes it knowable, with meanings, which make it capable of being meant etc.[55]

Husserl wanted not only the construction of a deductive theory of the appropriate form for application to any possible field of knowledge, but to understand the "wonderful affinity" between thought and things that makes a mathematical science possible. He seeks an explanation of this affinity which others have been content merely to label pre-established harmony between our thought and reality.

To compare essences of thought and things Husserl develops a descriptive phenomenology he encountered in Mach's essay "On the Principle of Comparison in Physics" which invoked "a universal physical phenomenology" to explain the rising tide of physical facts.[56] In reviewing it Husserl says that "this brilliant essay is much more general than the title might lead us to think" and summarizes its main lesson:

> We are in no position forthwith to describe every fact. The abundance of facts is too great. We advance by degrees only again and again comparing the new with the already known. The domain of facts compared becomes evermore inclusive and the concepts evermore abstract. In this manner there will at some point develop 'a general phenomenology of the physical, encompassing all physical domains.'"[57]

---

[51] loc. cit., 241.

[52] loc. cit., 244.

[53] loc. cit., 244–245.

[54] loc. cit., 245.

[55] loc. cit., 245.

[56] Mach 1895, 250.

[57] Husserl 1994, 195.

Husserl later recalls how his own phenomenology developed by a "certain radicalizing of an already existing phenomenological method which natural scientists and psychologists had previously demanded and practiced."[58] Mach had reacted, said Husserl, against "the threatening groundlessness of theorizing in the exact natural sciences ... against a mode of theorizing in mathematical speculations and concept-forming which is distant from intuition."[59] In the review Husserl had written that if Mach's ideal is attained for a domain, "then description accomplishes all that the investigator can demand. Description is a construction of facts in thought ... Our conceptual formulations serve us as almost a complete substitute for facts."[60] In the *Logical Investigations* Husserl introduces "pure phenomenology," dealing "exclusively with experiences intuitively seizable and analyzable in their pure generality of the essence," which it "must *describe* in terms of their essential concepts and their governing formulae of essence, the essences which make themselves known in intuition, and the connections which have their roots purely in such essences."[61] Nothing less would do for revealing the wonderful affinity between thoughts and things.

In support of his phenomenology Husserl develops an elaborate theory of meanings and their fulfillments, but the problem of meaninglessness confronts him with "a most unwelcome ambiguity." The concept of a round square is absurd, but if we call it senseless because it applies to no object,

> we should, in consistency, have to apply the word 'senseless' not merely to expressions immediately absurd, but to those whose absurdity is mediate, i.e. to the countless expressions shown by mathematicians, in lengthy indirect demonstrations to be objectless *a priori*. We should likewise have to deny concepts like *regular decahedron*, etc. are concepts at all.[62]

Husserl thus argues, that one must distinguish the true meaninglessness of words like 'abracadabra' from "*the* a priori *impossibility* of a *fulfilling sense*. An expression has a meaning in this sense if a possible fulfillment, i.e. the possibility of a unified intuitive illustration corresponds to its intention."[63] In this sense an expression is meaningless if "we apprehend the real impossibility of meaning fulfillment through an experience of the incompatibility of the partial meaning in the intended unity of fulfillment."[64] This mediate sense of meaninglessness is appropriate for Husserl's theory of computation, where he warns that whether an arithmetical operation has meaning for given numbers, i.e. whether it involves "an impossibility *a priori*" may need a deep analysis to determine whether "the problem posed is free of contradiction."

---

[58] Husserl 1968, 302.

[59] loc. cit., 302.

[60] Husserl 1994, 195.

[61] Husserl 1970, 249.

[62] loc. cit., 293–294.

[63] loc. cit., 294.

[64] loc. cit., 294.

14.3 Deduction, Decidability, Completeness

Husserl came to Göttingen brimming with ideas about deductive systems sketched in the *Prolegomena* on which he lectured the Mathematical Society in 1901, calling mathematics at its highest level "science of theoretical systems in general, in abstraction from that which is theorized in various sciences."[65] As the old mathematics of quantity was "the greatest instrument of natural scientific investigations, namely, the instrument of deductive theorizing for various domains of physical knowledge ... so the new formal mathematics intends to accomplish not only the same, but rather very much more."[66] It would "render superfluous" all previous work on methodology of "substantive" mathematics: "But the difficulties lie precisely in the relationship between formal mathematics and its employment in substantive mathematics or in the particular domains of knowledge."[67] Problems about this relation are of great interest to philosophers, says Husserl, because their solutions require "the understanding of the general essence of the deductive sciences and of theories in general." He explains that such theories are "defined by a totality of formal axioms, i.e. by a limited number of purely formal basic propositions, mutually consistent and independent of one another."[68] Theorems result from systematic deduction

> in a purely logical manner, i.e. purely according to the principle of contradiction ... but the object domain is defined through the axioms in the sense that it is determined as a sphere of objects in general irrespective of whether real of ideal, for which basic propositions of such and such form hold true.

An object domain thus defined we call a "determinate but formally defined manifold."[69] Clearly wanting to account for the application of formal mathematics to "various domains of knowledge," Husserl first poses the problem of imaginaries.

One knew that " the tendency towards formalization built into algebraic calculation led to forms that were arithmetically meaningless," but if one followed the rules of operation, almost "every result of calculation free of the imaginaries could be claimed as correct, as one could empirically establish by means of direct verification."[70] Hence the problem: "With what justification can the absurd be assimilated into calculation—with what justification, therefore, can the absurd be utilized in deductive thinking—as if it were meaningful?" After rejecting four solutions to this problem Husserl considers a fifth:

> We raise ourselves, according to the principle of permanence above the particular domain, pass over into the sphere of the formal, and there can freely operate with $\sqrt{-1}$. Now the

[65] *PoA*, 410.

[66] loc. cit., 411.

[67] loc. cit., 411.

[68] loc. cit., 410.

[69] loc. cit., 410.

[70] loc. cit., 412.

> algorithm of the formal operation is indeed broader than the algorithm of the narrower operations, which alone are really presupposed in a given conceptual domain. But if the formal arithmetic is internally consistent, then the broader operations can exhibit no contradiction with the narrower. Therefore what I have formally deduced in such a way that it contains only signs of the narrower domain must also be true for the narrower domain.[71]

Writing A_n for the narrow algorithm and A_B for its formal extension, Husserl takes the Principle of Permanence to claim:

(PP) If the extension A_B of A_n is consistent, then A_B is a conservative extension of A_n.

Against this Husserl raises the question of "under what conditions one, untroubled about the possibility or the absurdity of the concepts, can compose them and freely apply the axiomatic propositions to the composites without ever arriving at a contradiction. And then, further, under what condition the propositions that are free of absurdity are also actually valid."[72] Husserl explains that a proposition is "valid for the narrow domain" if it "holds true on the basis of the axioms of that domain ... a purely logical consequence of those logical axioms."[73] But we can only know that it *is* such a consequence, he will argue, if the system of those axioms is a "definite" one defining a "definite manifold." Husserl's condition is that (PP) holds only if A_n is definite. What he meant by a definite formal system has been a matter of much debate, centering largely on the question whether he intended it as a syntactical or a semantical completeness property. Any attempt to clarify this brings us to the question of the relation of Husserl's work to that of Hilbert, who in the sixth problem of his Paris lecture called for an axiomatization of physics, for which "as he has in geometry, the mathematician will not merely have to take account of those theories coming near to reality, but also of all logically possible theories. He must be always alert to obtain a complete survey of all conclusions derivable from the system of axioms assumed."[74] Husserl had found a new colleague equally bent on canvassing all possible deductive systems, even more so those "coming near to reality."

The term "completeness" first enters Husserl's lecture "with reference to Hilbert's requirement that consistency and completeness be proven for Euclidean geometry."[75] But proven it must be, for "on my view, completeness is never an axiom—but rather a theorem, for definite axiom systems and manifolds."[76] Husserl alludes here to Hilbert's "completeness axiom" for real ordered fields and proceeds to define a manifold as "absolutely definite if there is no other manifold which has the same axioms (all together) as it has." For such a manifold, says

---

[71] loc. cit., 418.

[72] loc. cit., 418.

[73] loc. cit., 419.

[74] Gray 2000, 258.

[75] loc. cit., 425.

[76] loc. cit., 426.

Husserl, “every proposition meaningful according to it is decided in general”, and is therefore “complete, in Hilbert’s sense.”[77] But this sense is “spurious,” says Husserl, since “we can make any axiom system, any system of first principles for possible deductions, quasi-complete by means of such a negative closure axiom, and therefore this “completeness” is of no use at all to us here.”[78] Nevertheless, Husserl admits that an axiom system closed in this “spurious manner, already has the property which we have in mind, namely it can be read off of each proposition whether it is or not a consequence of the axiom system.”[79] So apparently Husserl believed then that it was effectively decidable whether any proposition logically followed from Hilbert’s axioms after all, and raises the question whether this is also the case for systems without any closure axioms, that is whether they could “permit it to be read off of each proposition whether that proposition belongs within the sphere of its deductions as to its truth and falsity.”[80]And he answers this affirmatively for *every* arithmetic, be it of whole numbers, rationals, or reals, claiming that every arithmetic is defined by an axiom system in which every proposition is either a consequence of axioms or in contradiction with them: “The proof of this assertion lies of course in the fact that every defined operational formation is a natural number, and that each natural number stands to every natural number in a relation of magnitude determinable on the basis of the axioms.”[81] For a numerical equation is true, says Husserl, if “the execution of the operations” transforms it into an identity a = a, otherwise it is false: “Every algebraic formula is, then, also decided, for it is decided for each numerical case.”[82] This seems to collide with Gödel’s theorem but Husserl was not thinking of systems like Peano arithmetic but rather of Kronecker’s general arithmetic built around his effective sharpening of the fundamental theorem of algebra by means of his generalization of Sturm’s theorem.[83] This is not to say however, that Husserl had no reason to doubt the completeness of his arithmetics.

The above proof assumes “every defined operational formation is a natural number.” A formal arithmetic must deduce numbers from its expressions of arithmetical operations, and an adequate set of axioms should enable it to express any such operation of Husserl’s totality of conceivable operations. This would allow the deduction of the correct numerical value from the application of such

---

[77]loc. cit., 427 Husserl distinguishes his absolute sense of definite systems from those which are “relatively definite” systems where “every proposition meaningful according to it is decided under restriction to its domain.”

[78]loc. cit., 429.

[79]loc. cit., 429–430.

[80]loc. cit., 430.

[81]loc. cit., 430.

[82]loc. cit., 430.

[83]In an 1894 manuscript Husserl writes that “All the axioms of arithmetic have the form: ‘If something corresponds to the concepts $b_1, b_2, \ldots$ then it is valid that $f(b_1, b_2, \ldots) = 0$’” (Ierna 2011, 221). This and similar indications in the Double Lecture show he was not thinking of axioms like those by Dedekind and Peano for the recursive structure of the natural numbers, but rather of an algebraic theory like Kronecker’s ‘general arithmetic’.

an operation to a pair of numbers. But for the inverses of direct operations, which Husserl said repeatedly, may require a "deep analysis" to determine whether they are defined for a given number, or even whether the operational problem they pose is free of contradiction, for *those* he could hardly be confident that such axioms could deductively settle the results of all such problems. He was well aware that

> [n]ew problems arise through inversions. We have no *a priori* guarantee that these problems are solvable. Induction from the simplest direct combinations leads to limitations on solvability. So must it be generally. Of every form we invert, we could say that they must have solutions: in general, whether for every system of specific values is very questionable.[84]

That such consideration still figured in his thinking is suggested by his reformulation of the completeness proof above, presumably in response to what he calls "Hilbert's objection" to it on grounds that it provided no principled choice of a *logic* to deduce theorems from the axioms. We shall come back to this point.

Husserl prefaces his new proof by remarking that a new axiom cannot be added to a "definite" system "without evoking contradiction," that is every proposition using besides concepts of the domain "only absolute concepts, i.e. those which are purely logical ... is true or false in the domain."[85] He knows this because:

> Every *direct* operational combination, however often it may contain each operation, is equal to a number. That will be proven. Therefore every proposition which asserts two algebraically general, closed expression to be equal—and likewise every mixed equation built up from algebraic and number signs—will of course have to be necessarily true or false on the basis of the axioms.[86]

For any substitution of numbers for variables of an algebraic equation results in a "determinate number" on each side of the equation allowing one to decide whether the equation holds or not: "It suffices that I can demonstrate from the axioms that every expression is a number, and consequently it is evident that two expressions either always present the same number or different numbers."[87] Husserl can be confident of this under his restriction to direct operations, which he presumably thought overcame the doubts he had about solving computational problems posed by inversions. In any case, if we are tempted to read Husserl's argument as a groping anticipation of Tarski's result on the completeness and decidability of the elementary algebra of real closed fields, we should recall that as Tarski explained, the key theorems on which his elimination of quantifiers rests are "from a mathematical point of view ... closely related to the well known theorem of Strum" and to Kronecker's extension of it.[88] Let us try now to describe broadly how Husserl's achievements, beliefs, and goals with respect to arithmetic stack up against current metamathematical wisdom.

---

[84]Husserl 1891, 540.

[85]*PoA*, 466.

[86]loc. cit., 466. Emphasis added.

[87]loc. cit., 467.

[88]Tarski 1951, 50, 52.

In the *Philosophy of Arithmetic* he developed the traditional *algorithmic* conception of arithmetic that pursues evermore complicated mixtures of the four arithmetic species of calculation, finally canvassing a class of algorithms comprising what he called the "totality of conceivable arithmetical operations." Subsequently in his Göttingen lectures he cultivated the more theoretical deductive conception of arithmetic that aims at the logical deduction of the arithmetical theorems from axioms. These two conceptions have their respective goals of algorithmic completeness and deductive completeness. Centrone has shown that, unbeknownst to Husserl, his totality of arithmetical operations is algorithmically complete by our current standards. His motivation, I have suggested, could well have been to gain some mathematical perspective on Kronecker's 'calculable real numbers' which would replace the theories of Weierstrass, Cantor, and Dedekind, a possibility Husserl said would be epoch making. Kronecker's unification of number theory and algebra resulted in a 'general arithmetic' allowing for the effective calculation of the real roots of algebraic equations, something Husserl clearly saw and resourcefully incorporated into a deductive logical framework. His claim that all arithmetics have complete axiom systems may seem to leave him oblivious of the undecidability and incompleteness that pervade our landscape of formal systems, but we have seen that he did harbor serious doubts about such decidability, doubts which are quite appropriate for his landscape. The central fact of our metamathematical wisdom is that the algorithmic completeness of a formal arithmetic is incompatible with its deductive completeness. Such an algorithmically complete system must, by Centrone's result, contain true unprovable sentences expressing the proposition that the function computed by one of Husserl's conceivable operation is not defined for certain numbers. He did not know this nor was he in a position to prove it, but as his words we quoted show, he certainly did catch a glimpse, however fleeting, of the underlying difficulty. If we try to relive this glimpse, we will wonder whether it was reinforced by a glimpse he surely also caught if not record of staggering growth rates when he broached Dedekind's sequence of primitive recursive functions and stopped at elevation, the iteration of exponentiation. Such a reason for curtailing his exploration of this avenue of calculability would have been compatible with Kronecker's practical approach, where he did not generally insist on explicit calculations but only on some assurance of their possibility. This brings us to the glimpse Hilbert caught of this aspect of calculability when he tried to comply with Kronecker's demand.

Hilbert rationalized his non-constructive proof of his Basis theorem for algebraic invariants by distinguishing three questions about any mathematical assertion: (i) can it be proved? (ii) the question "whether it can be determined how many operations are needed to carry out the assertion of the theorem. Kronecker has particularly emphasized the question of whether one can carry it out in a finite number of steps."[89] (iii) whether one can *actually* carry it out. As an example Hilbert poses the problem:

[89]Hilbert 1993, 133.

> Is there a place somewhere in the decimal expansion of $\pi = 3.14159 \ldots$ at which there appears ten consecutive ones 1111111111? It is not improbable that this may be the case. If we assume one can prove this in some way, then we can ask the second question: can one find a number N of which one knows that there are at least ten consecutive ones before the Nth decimal of $\pi = 3.14 \ldots$? The number N can be much too large as long as one can only prove the assertion of it. Third, one would then actually have to calculate the number N, so that the Nth up to the $(N+9)^{th}$ decimal are all ones, and so that there are no ten consecutive ones appearing earlier."[90]

Such questions arise in the theory of invariants, says Hilbert:

> We have proven that a binary form of order n has a finite full invariant system. Suppose that the degree of the invariant of the highest degree in the simplest full invariant system is N; then the question arises: If we are given the number n, can we calculate an upper bound for N? Our proof has settled the finiteness question only in principle; there is not the slightest indication that we can actually calculate such a number N. If one considers n=307, for instance, then the proof does not indicate any number of which one would know that it is bigger than N.[91]

The third question of "actually calculating the invariants which form the full invariant system", says Hilbert, "is pointless for base forms of higher order." In fact, we now know that Hilbert's Basis theorem is equivalent to the well-orderedness of ω^{ω}, and hence to proving the totality of the Ackermann function.[92] Hilbert no more knew of this theorem than Husserl of undecidability theorems, but he clearly caught more than glimpses of how pointless it could be to try to actually satisfy Kronecker's demand.

Kronecker's program was important for both Husserl and Hilbert, and though each have their own differences with him, both reject his view of geometry as an impure science with which the pure arithmetical algorithm should not be mixed.[93] To avoid a perceived weakness in Kronecker's account of irrational numbers Hilbert proposed a novel continuity axiom, his so-called completeness axiom, which Husserl said was "spurious" for postulating what should be proved. But he must at least have appreciated the rationale Hilbert offered for this axiom, namely, that it relieved us of having "to imagine, say, the totality of all possible laws according to which the elements of a fundamental sequence can proceed."[94] Husserl's own attempt to formulate this 'completeness' property had led him to the ambiguous claim that:

---

[90]loc. cit., 133–134.

[91]loc. cit., 134.

[92]Simpson 1988.

[93]Husserl 1983, 41.

[94]Ewald 1996, 1095. As did Cantor's continuity principle in terms of his fundamental sequences. Dedekind's continuity axiom postulating "cuts" among the rationals did not invoke such laws, but Hilbert knew that Stoltz had shown that this axiom logically implies the Archimedean axiom, which would undercut his goal of an independent axiom system. However, Hilbert formulates these continuity axioms for geometry only in order to show that they can be *eliminated*, i.e. that a 'complete' geometry providing its own analytic representation is possible without them.

> Every incommensurable magnitude is characterized by a law which determines which rational units it contains and how many. . . . If we assume that every conceivable law also determines a magnitude, then we call the system a complete system of magnitudes. *A priori* no law is distinguished from another, thus we can *a priori* allow that to every law a magnitude corresponds. After all, were this not so, we could generally still calculate with the laws themselves.[95]

Hilbert's problem was that Kronecker's demand for finitely expressed laws implied their *countability*. Husserl might have felt that he *had* already imagined all possible laws when he canvased the totality of conceivable arithmetical operations. We are back to the question of his awareness of Cantor's proof of the uncountability of the real numbers, at once a springboard for new mathematical results and a breeding ground for new logical paradoxes. Hilbert found a paradox so starkly mirroring Cantor's diagonalization proof of this result that he feared for a while that it "called into question the most beautiful proof in set theory."[96] These paradoxes do not disturb Husserl's vision of deductive systems and attract only muted allusions from him for several years, though we saw he later acknowledges them as significant ingredients of the crisis of science he envisioned.

Hilbert claimed that his completeness axiom rendered his system "*finite and closed* . . . and about which new statements are valid only if one can derive them from the axioms by means of a finite number of logical inferences."[97] This seems to assert semantic completeness, which would also presuppose a complete system of logical inference, but he may have only meant to *call* a statement 'valid' if it can be so derived. Husserl called Hilbert's system "absolutely definite" despite its spurious completeness axiom, even suggesting that "it can be read off each proposition" of this system whether it is a consequence of the axioms. This seems like a claim of decidability resulting from syntactical completeness of the axioms as well as the completeness of the logic used. But we cannot identify anything in the work of these writers clearly indicative of 'our' distinction between syntax and semantics, much less a presumption of the completeness of logic, without knowing just what they took the nature and scope of logic to be.

In his notes on a lecture by Hilbert on axioms that might yield a "closed" system, Husserl suddenly mentions:

> Hilbert's objection: Am I justified in saying that every proposition containing only the whole positive numbers is true or false on the basis of the axioms for whole positive numbers? Here the following would have to be added: if we assert that a proposition is decided on the basis of the axioms of a domain, what may we thereby use besides those axioms? All that is logical. What is that? All propositions that are free of all particularity

[95] Husserl 1983, 82. What Hussel seems to miss in Cantor's set theory is the insight that the "completeness" of the real numbers is comprised of their uncountability, and hence in the incompleteness of any countable list of them.

[96] See Peckhaus and Kahle 2002, 157–175.

[97] Ewald 1996, 1095.

> of a knowledge domain, of that which is valid independently of all 'special axioms', of all matter of knowledge.[98]

This passage hints at doubt that logical rules can be found that would be free of material knowledge and still allow the deduction of all propositions true of the whole numbers. One might think this was easy:

> But here one falls into a fine dilemma: In the domain of an algorithmic logic, in the domain of cardinal number, in the domain of combinatorics, in the domain of the general theory of sequences and ordinal numbers. And finally, is not the most general theory of manifolds itself purely logical?[99]

The objection then appears to be that the completeness in question depends on which of these domains was adequate while qualifying as purely logical, but that no principled choice of them was evident for application to the "axioms for whole numbers." But the completeness proof discussed above that Husserl proceeded to reformulate does not identify any such *axioms*, certainly none like Peano had already formulated for such numbers. It proceeded as if applying an algorithmic logic to the solution of algebraic equations.[100] Did Husserl regard any of the other candidate domains of logic he mentions as having an equal claim to being purely logical?

Husserl's 1896 lectures on logic present an impressive system of logical axioms and rules, especially for propositional inferences. But infinitely many logical laws are possible, and to gauge the prospects for a complete system Husserl considers how the arithmetician has fared in his quest for algorithms and axioms: "Has he any a priori conceptions assuring him of the completeness of derived arithmetical laws? None at all."[101] He bases certain operations on the number concept, such as addition, multiplication, and exponentiation, and their inverses, from which he distills laws that he prunes to a few simple ones: "But how does the arithmetician know that the operations he has enumerated are all those that are conceivable on numbers?" This is not evident, says Husserl, for as many mathematicians look for such operations, we cannot be certain new ways of combining them will not be found. He was conceding that he could not be certain that his own totality of conceivable arithmetical operations is complete. As for axioms: "Arithmeticians have taken infinite pains to establish the minimum number of arithmetical axioms. Yet it is still in question whether one of the axioms is merely a consequence of the others, without their having noticed it. At least, we still lack any systematic

98 *PoA*, 465.

99 loc. cit., 465.

100 Husserl explains that, since the "sphere" of the logical calculus is comprised of truth-values and their combinations, "The logical calculus is also definite. For every letter symbol is either = 0, or = 1, and consequently it is *apriori* determined, for every relation presenting itself as a formula, whether it is satisfied. It is, in general, satisfied, if it, in general, yields either 0=0 or 1=1, and otherwise it is false. Likewise for every "equation": either there is a truth-value which satisfies it, etc." (*PoA*,491) Husserl's algorithmic logic is a fragment of propositional logic.

101 Husserl 2001a, 243.

proof."[102] This is the sort of difficult problem advertised by Hilbert in his paper on the number concept wherein he reported that the commutative law of multiplication was not a consequence of the axioms of combination, calculation, and ordering, unless the axiom of Archimedes was added, which he said, "has special significance for the foundations of geometry.".[103]

Husserl's perusal of Hilbert's work on geometry and Frege's critique of it allowed him a view of a class of logical problems that *prima facie* defied algorithmic treatment. In a study of sub-structures of domains Husserl asks whether every sub-structure of a closed domain must itself be closed:

> Consider a plane in space. In it there are various kinds of sub-structures and we can imagine a geometry set up for this plane. Can some proposition or other now be demonstrated from the geometry of space which concerns the geometry of this plane and yet is not deducible within that geometry itself?[104]

His inconclusive discussion shows he had not yet read Hilbert's *Festschrift* wherein this question is answered by showing that whereas Desargues' theorem can be proved from the space incidence axioms, it cannot be proved in the plane from the axioms of incidence, order, and parallels without the congruence axiom. The model Hilbert originally constructed to prove this was complicated, and some are still struck by the "stupefying magic" of his discovery of such models.[105] This proof exemplifies what Hilbert called "the modern principle: the proof of unprovability."[106] Such impossibility proofs were for him the hallmark of modern mathematics, and their discovery, after prolonged futile searches for positive results, persuaded him that all well posed mathematical problems were eventually solvable in principle. But their very difficulty will also motivate his *Entscheidungsproblem*.

To prove the result he alludes to above, he constructed a non-Archimedean model to show that Pascal's theorem cannot be proved from the axioms of incidence, order, and parallels even if Desargues' theorem is assumed, contravening his own belief that Desargues' theorem was stronger than Pascal's.[107] Years later, after distinguishing his solvability thesis from the problem of deciding a mathematical question in a finite number of steps, Hilbert admitted that "my proof theory cannot

[102]loc. cit., 243– 244. Husserl was persuaded that "just the same" will hold for logic: "We cannot guarantee completeness" (244).

[103]Ewald 1996, 1095.

[104]*PoA*, 462.

[105]See Pambuccian 2013.

[106]Hilbert 2004, 284.

[107]A conviction shared by Peirce, Schur, and others. Schur wrote to Hilbert in 1900 that this result was the most important of the *Festschrift*, but that he found Hilbert's proof of it "exceedingly difficult." See M. Toepell 1985, who marshals evidence for Hilbert's prior belief that Desargues' theorem was stronger than Pascal's. I shall hereafter abbreviate Desargues' theorem (D) and Pascal's theorem (P).

specify a general method for solving every mathematical problem; that does not exist."[108] He and Ackermann motivated the *Entscheidungsproblem* by recalling that

> [i]n the investigation of the relations of logical dependence between various groups of axioms of geometry there is one especially important and interesting result, that Pascal's theorem, which plays a central role in grounding the theory of proportion without the use of continuity axioms, cannot be proved from the axioms of incidence, order, and parallels alone.[109]

They stress that a positive solution would "give a method by which this unprovability must be established at least in principle; though perhaps the intricacies of the method would make its execution illusory." Such a method would have enabled Hilbert to find the non-Archimedean model for proving this result mechanically by calculation, something he would be loathe to believe given the difficulty he experienced in just sorting out the logical relations between (D) and (P), let alone concocting the model to demonstrate the impossibility of deriving the full congruence axiom for triangles from that restricted to triangles of the same orientation without continuity. No mathematician acquainted with these and other such models would dream of being able to crank them out by a fixed algorithm. As Weyl remarked, Hilbert's construction of models was "a wonderful trick to avoid analyzing the deducible mechanism itself", but still required "all his amazing wealth of invention."[110] Hilbert's use of models was semantical insofar as it presupposed the *soundness* of the logic he was using; but he did not presuppose its completeness, as he never assumed *that models always existed* to demonstrate that a geometrical proposition was not derivable from given axioms. The challenge to prove the impossibility of deriving a contradiction from the axioms of geometry syntactically by an analysis of 'the deductive mechanism itself' was posed by Frege.

When Husserl arrived in Göttingen, Hilbert was engaged with Frege in correspondence over the legitimacy of his *Festschrift* which he shared with Husserl. In his extensive notes on these letters Husserl *quotes* Frege's objection to Hilbert's consistency proof that "from the truth of the axioms it follows that they do not contradict one another. That therefore requires no further proof." Husserl observes that "Frege does not understand the sense of Hilbert's 'axiomatic' grounding of geometry. Namely, that it is a matter of a purely formal system of conventions, which coincides, as to the form of the theory, with the Euclidean."[111] Husserl explains that we can be certain that deductions proceed by purely logical means:

> Only when we express the matter symbolically and raise ourselves to the level of a formal system, to a theory form, which is defined through sentence forms of the basic

[108] van Heijenoort 1967, 384.

[109] D. Hilbert & W. Ackermann 1928, 74.

[110] See Weyl 1944, p. 632.

[111] *PoA*, 469–470. Neither Frege nor Husserl notice that for Hilbert the important point was that his model for euclidean geometry was *countable*.

> material principles of the domain. Only so can we resolve the questions of dependence and independence, etc. In a formal deductive system (an algorithm) there are no "explanations" in Frege's sense.[112]

Such questions were essential to Hilbert's stated goal of the *Festschrift* of "a logical analysis of our spatial intuition", that is, to decide whether Euclid's plane geometry logically depends on either explicit or implicit appeals to space: explicitly in the use of our intuition of space, or implicitly in the use of rotation to prove the congruence of oppositely oriented triangles.[113] The unprovability of (D) from the plane incidence, order, parallel axioms shows that without the concept of rigid motion expressed in the congruence axioms, it depends on our *spatial intuition of the properties of planes*, while the unprovability of (P) from (D) and these other axioms shows that such space intuition alone is not sufficient for analytic geometry, i.e. that the commutativity of multiplication in the segment calculus based on (D) cannot be proved without it. Husserl further quotes Frege: "It seems to me that you wish to completely separate geometry from our spatial intuition and make it into a purely logical science like arithmetic."[114] If he followed Hilbert's logical analysis of such intuition, Husserl could only conclude that such a "separation" need only be a temporary epoché while reinterpreting various sentences to construct models showing that they do not logically follow from others. This would have given him a more vivid glance of problems possibly impervious to algorithmic solutions than those posed by the inversions of his algorithmic operations.

In view of Husserl's interest in sub-structures of geometrical systems and his descriptions of Hilbert's system as "an algorithm," it is more probable that it was the "mechanization theorem" that caught his attention. Hilbert found that to prove the commutativity of multiplication from his other axioms of the real ordered field one must use the Archimedean axiom together with mathematical induction. Since (P) can be used to construct a field, it comprises, in the context of elementary geometry, a "finite" substitute for number, induction, and continuity. To pursue this Hilbert formulates the idea of a "pure point of intersection theorem", or Schnittpunktsatz, dealing with a finite number of points and lines and their incidence relations.[115] (D) is such a Schnittpunktsatz but cannot be proved in the plane without the

[112]loc. cit., 470.

[113]Hilbert begins by quoting Kant: "All knowledge begins with intuitions, then proceeds to concepts, and ends with ideas." Recall that the only example he gives of an intuitive axiom in the Doctrine of Method is that a plane passes through any three points in space, which is essential for the proof of (D) in space.

[114]loc. cit., 472. A curious remark, given that Hilbert organized his system around the analysis of (D), the most direct expression of spatial intuition. As Rota put it: "The relevance of a geometric theorem is determined by what it tells us about space and not by the eventual difficulty of the proof. The proof of Desargues theorem of projective geometry comes as close as a proof can to the Zen ideal. It can be summarized in two words: 'I see!'" (Rota 1997, 189). Hilbert observed more pointedly that to see the proof of (D) in space one need only view its plane figure "räumliche", that is, in perspective (Hilbert 2004, 171).

[115]These theorems take the following form: "Choose an arbitrary set of finite number of points and lines. Then draw in a prescribed manner any parallels to some of these lines. Then, if connecting lines, points of intersection and parallels are constructed through the points already existing in

"foreign idea" of congruence, while (P) is another, but cannot be proved from (D) without continuity. Could one find Schnittpunktsätze from which all others could be logically proved without using either congruence or continuity? Hilbert finds that (P) alone suffices: he calls a geometry Pascalian if the axioms of incidence, order, and parallels as well as (P) holds in it and proves that any Schnittpunktsatz that holds in a Pascalian geometry can be proved "by the construction of suitable auxiliary points and lines from a finite number of Pascalian configurations."[116] This leads to a decision procedure for *Schnittpunktsätze* and is known in the literature of computer theorem proving as "Hilbert's mechanization theorem." Hilbert's original proof of it reduced the conditions for a *Schnittpunktsatz* to hold to the vanishing of a certain algebraic expression for all values of its parameters, which could be decided "by means of the *formal rules of calculation* based on Desargues and Pascal."[117] This algebraic proof broadly resembles Husserl's proofs for the completeness of his arithmetics and shows that the sub-system of Schnittpunktsätze comprises a definite system in Husserl's sense, indeed a decidable one.

14.4 Pre-established Harmony, Paradox, Idealism

In the 1901 lecture Husserl posed the goal of the new formal mathematics of deductive systems to illuminate the applicability of mathematics to "domains of knowledge." In the lectures of 1902/03 he addressed the issue of the relation of logic to epistemology by asking: "How is a cognizing representing, a knowing of objects possible, since the objects are for themselves and cognizing consciousness for itself? The cognizing is an imminent condition of cognizing consciousness. It does not go out of itself to grasp the objects outside to draw them into consciousness."[118] This was Husserl's take on Kant's question to Herz of how our cognitive representations can refer to an object without being affected by it, which he would take a step further:

> How is it to be understood that a subjective train of acts of thinking as it plays out in our mind in a demonstration, ending in a proven theorem, acquires objective significance? How are we to understand that the result of a logical action of thought does coincide with the nature of things? Is the course of nature, that we calculate logically in advance, directed at our thought and calculation, in accord with this course of subjective experiences, or does a harmony pre-established by God between the course of thought and the flow of reality exist?[119]

---

the prescribed manner, a definite set of finitely many lines is eventually reached, about which the theorem asserts that they either pass through the same point or are parallel." (Hilbert 1971, 97)

[116]loc. cit., 97–98.

[117]Hilbert 2004, 283. This was before Hessenberg discovered in 1905 how to derive (D) from (P), allowing Hilbert to formulate the theorem above.

[118]Husserl 2001b, 16.

[119]loc. cit., 17. For an absorbing account of the hold on some of the best scientific minds of Husserl's Germany exercised by the doctrine of pre-established harmony, see Pyenson 1982.

Since God's existence can only be presumed, how could we know of such a harmony?

Examples of mathematical theories that find unanticipated applications to reality raise the question of what can be said *a priori* about such use of mathematics. Husserl notes that, since all the pure mathematical concepts, such as unity, plurality, number, order and ordinal number, manifold, combination, permutation, etc., are purely logical and hence apply to objects generally, "there is no science for which the possibility of using mathematics would be in principle excluded, though because of the special nature of its domain mathematics may play a greater role in one, and a smaller in another."[120] But then again:

> How is it that subjective actions of the human mind harmonize with the course of reality? Is perhaps the course of actual things directed to the process of my subjective thought ... how can I by thinking, be it with evidence, assert that a theory I have found *a priori* must agree with nature?[121]

One can either hold that logical thought has absolute significance, or that it has only subjective or anthropological significance. The former position invokes a parallelism established by God between the logical and the factual and tries to order the facts so that they explain logically the nature of being. Or one may conceive thinking as a biological phenomenon of adjustments to nature fitting the course of human development to the harmony of things. Husserl seeks to develop and deploy his theory of manifolds as theory forms to uphold such parallels without appeal to God.

In his 1906/07 lectures on logic and epistemology Husserl argues that mathematics and logic provide an infinite abundance of "pure truths of the understanding" containing nothing of the specific subject matter of any special fields of knowledge. "Rather they are thoroughly, purely grounded in mere 'thinking forms,' in the essence of the formal meaning and object thoughts that resemble molds that must first be filled with some substance so that material thoughts referring to material objects can result."[122] Having established a "correlation of the concepts of proposition and state of affairs, apophantic logic also contains a formal ontology," allowing Husserl to claim that "If in virtue of the correlation between meaning and objectivity, we then undertake a *change of standpoint*, if we now therefore adopt the standpoint of the side of objectivity, *then corresponding to the theory of forms of propositions is a theory of forms of the corresponding states of affairs*."[123] He is gradually laying the groundwork for a correlation between logical thought, understood in the broadest sense to include mathematics, and states of things.

In developing his vision of a theory of manifolds as "a science of theory forms" Husserl stresses the need for greatly expanding the "determinate domains" for which formal theories have been found, such as the lines on a plane, the domain

---

[120]loc. cit., 32.

[121]loc. cit., 42.

[122]Husserl 2008, 59.

[123]loc. cit., 71.

of numbers, or the domain for forces: "Here the fixed pre-established nature of the matter determines which connections and relations it allows and which basic principles are valid for it. And, with the concepts and propositions, the forms of the broader theory are pre-established."[124] Nevertheless, "one can *construct* manifolds ... conceived in indeterminate universality that are the bases of a possible mathematics, *by arbitrary definition*," thereby opening up "a domain of creatively free mathematical investigation ... that is precisely the universal theory of manifolds (or science) of theory forms."[125] If we can canvass "all possible forms of deductive disciplines," says Husserl,

> we are then in possession of all theories in advance, before we even know the fields in which these theories will formulate and solve the problems of the field. So, just as when physicists come to a linear differential equation, they immediately know that it is a mathematical form that has been completely dealt with by mathematicians, ... so in the future, were the ideal attained of constituting a system of propositions in a field as basic principles, then all theoretical work would already be at an end.[126]

If one expresses its propositions "algebraically" one is assured that "mathematicians have already derived all the possible formal consequences in formal universality." Indeed, if we construct "all possible spheres of theories," we are assured of having on hand in our "supramathematics," as Husserl calls it, an appropriate mathematical theory to fit "any pre-established purely logical or extralogical domain that in general admits of a mathematics, but is to be obtained by simple subsumption under the corresponding theory forms."[127] But he knows that in view of the lack of clarity and agreement among mathematicians in 1906, such supramathematics remains a distant utopian goal.

Husserl notes that when mathematicians are asked about "the actual objects of their theorizing, all unanimity and clarity cease. They all have their private opinions there, and these are most often in glaring opposition."[128] Their lack of clarity is also shown by controversies over irrational and imaginary numbers, and "as long as we do not have ultimate clarity about the meaning and significance of mathematics, we do not know what mathematics accomplishes and in what sense claims for it can ultimately be made."[129] Husserl is clear that "in set theory, we make judgments universally about sets that in a certain way are higher-order objects. We do not make judgments directly about elements, but about whole totalities of elements and arbitrary elements, and the whole totalities, the sets to be precise, are the objects-

---

[124]loc. cit., 83.

[125]Husserl says that this is how non-euclidean and higher dimensional manifolds should be regarded. Hilbert's just published spectral theory of operators in an infinite dimensional space comprised a striking example of such a construction of a new kind of domain, but it was not "constructive" in the same sense as Kronecker or Brouwer.

[126]loc. cit., 86.

[127]loc. cit., 90.

[128]loc. cit., 158.

[129]loc. cit., 159.

about-which."[130] Apophantic logic decides which objects are conceivable as states of affairs and looks for laws for states of affairs that are valid by virtue of their form. Set theory rather looks for laws that are valid for its higher-order objects, such as the formation of the union of two sets, or their intersection if neither is included in the other:

> *Such truths and those derivable from them do not belong in apophantic logic, in the logic of prediction as such.* They form a field of their own: set theory. And likewise, grasped objectively, cardinal number as the formal determination form peculiar to sets gives a field that does not belong within the framework of a simple *theory of objects in general and the forms of their states of affairs.*[131]

It seems then that Husserl saw no problems with set theory as such in 1906, certainly he mentions no contradictions or paradoxes. But he did object to the unclarity he saw in attempts to axiomatize the "theory of manifolds," yet a third layer of formal ontology he would place above apophantic logic and set theory. For when "modern mathematicians" refer to axiomatic theories,

> the usual concept of axiom that has been commonly and universally used in philosophy from time immemorial is in this case completely abandoned, and the term theory of manifolds is also questionable, since set theory is also always included under it, which according to our careful discussion belongs in the other layer.[132]

For Husserl the theory of manifolds deals with all possible deductive theory forms, whereas an axiomatic set theory would be just one such theory. But was there such a theory whose axioms were clearly valid for sets in Husserl's eyes?

Zermelo had been lecturing on mathematical logic and the axiomatization of set theory since 1901 and had by 1906 formulated four axioms for it, the first of which reads: A well defined set never contains itself as a member. $M \notin M$. His second axiom posited unit sets and restricted unions of the kind that were accepted above by Husserl as clearly valid for sets. His third axiom of separation states that: If M is a well defined set, and P is any property of elements of M for which it is determined "without arbitrariness" whether m has P or not, then the set of elements having P is a well defined subset of M. Finally, his fourth axiom states that the power set of all subsets of a given set is also a well defined set.[133]

The paradoxes of set theory had been discovered in Göttingen before 1900, and Husserl had to know that the problem of solving them by the axiomatic method was being pursued by Zermelo, who in 1902 showed him a proof that a set M which contains all its subsets as members is inconsistent.[134] Some of the controversies Husserl complains of in his own lectures were likely disagreements about how to

---

[130] loc. cit., 73.

[131] loc. cit., 74. Later Husserl mentions "the theory of cardinal numbers, or as mathematicians say lately, the theory of powers" (445).

[132] loc. cit., 76.

[133] See Peckhaus 1990, 95–97.

[134] Husserl 1994, 442. Zermelo's proof considers the set M_0 of subsets of M which are not members of themselves.

solve these paradoxes. But he does not explicitly deal with them until 1912, in private notes that are still unpublished, wherein he analyzes what he calls 'Zermelo's Paradox.'[135] He presents the proof Zermelo had given him and concludes that: "If it essentially belongs to the concept of set that (without contradiction) no set can contain itself as an element, then M_0 and M are identically the same set, and we show that the whole reasoning was untenable."[136] Hill observes that "Given Husserl's conviction that logical, mathematical laws are laws of essence, it is not surprising to find him arguing over and over that the set theoretical paradoxes must involve some violation of the essence of set."[137] Membership being an essence relation for Husserl, he repeatedly affirms that "It belongs essentially to the concept of set that (without contradiction) no set can contain itself as an element."[138] Since this is just what Zermelo's first axiom of 1906 claims, it must be valid for Husserl, as was also his second axiom of elementary sets. But what about his separation and powerset axioms?

Hill informs us that although Husserl's notes are rife with reflections on the essence of sets, "he also expresses his conviction that we do not yet by any means have the real and genuine concept of set that the mathesis and logic need. The paradoxes, he declared, only demonstrate that a general logic of sets in general, of totalities is still lacking."[139] Most importantly, she explains, he contemplates "reforming the mathematical theory of manifolds by transforming it into a transcendental theory of manifolds that consciously captures the formal essence of a genuine constructible totality," for "each field of an exact science, he maintained, is a constructible totality."[140] In the *Ideas* sets emerge from a transcendental process of nominalization that yields the plural thesis: "Through the nominalizing turn, it becomes the object designated as a 'set' and thus emerges the basic concept of *set theory*. In the latter, judgments are made *about* sets as objects that have their own distinctive kinds of properties, relations, and so forth."[141] The phenomenologist is not concerned with the developments of this mathematical discipline: "Only

---

[135]See Rosado-Haddock 2006, who illuminates some quotations from Husserl's notes with an incisive commentary. More extensive quotations from Husserl's notes are given by Hill, 2002. Still further quotations from them with commentary are found in Schmit 1981.

[136]Quoted in Hill, Op. cit., 213.

[137]loc. cit., 214.

[138]Husserl also claimed that: "An essential part of our intention in speaking of a set, plurality ... is that from objects of thought something *new* should be formed that is determined by these objects, but arises only with this formation. It contradicts this conception, that among these objects the result of this formation could be found." (Quoted in Schmit, loc. cit., 115.)

[139]loc. cit., 215.

[140]loc. cit., 215–216.

[141]Husserl 2014, §148. Husserl refers to the transcendental formation of the plural thesis in synthetic consciousness wherein "the pure ego 'in' it is directed at what is objective *via multiple beams* [of consciousness]; the single thetic consciousness does so in a single beam. Thus the synthetic collecting is a 'plural' consciousness: one and one and one are taken together. In other words, it is essentially possible for the plural consciousness to be converted into a singular consciousness that takes from it the plurality as one object, as something individual" (§119).

the axioms and their conceptual composition are of interest to phenomenology, as headings for phenomenological analyses."[142] For such an analysis, says Husserl, we do not turn to empirical philosophers

> but to the consciousness in which, thinking mathematically, we comprehend the axiomatic states of affairs in a fully discerning way. If we hold ourselves purely to this Intuition, then it is not subject to any doubt that pure, essential connections come to expression in the axioms, without the slightest addition of facts of experience.[143]

What then of the phenomenological content of Zermelo's axioms?

By 1908 Zermelo had published his full system of definitions and axioms, which bear the possible influences of Husserl. His first definition explains that "set theory is concerned with a *domain* B of individuals, which we shall call simply *objects* and among which are the *sets*."[144] His fourth definition replaces an earlier supplementary "axiom of Definitheit" and explains that "a question or assertion P is said to be *definite* if the fundamental relations of the domain, by means of the axioms and the universally valid laws of logic, determine without arbitrariness whether it holds or not."[145] The separation axiom now states that "whenever the propositional function P(x) is definite for all elements of a set M, M possesses a subset M_P containing as elements precisely those elements x of M for which P(x) is true."[146] It has been claimed that Zermelo borrowed the word "definite" from Husserl.[147] While this would explain Husserl's complaint in §72 of *Ideas*, it highlights a puzzling lacuna in his notes on paradoxes: the absence of any analysis of the Richard paradox that Zermelo saw as the real threat to the consistency of his axioms and could only avoid by arguing that the condition 'definable by a finite number of words' is not definite and hence cannot be used to formulate it.[148] Russell's paradox he solves simply by using it to prove the theorem that every set M has a subset that is not an element of M, and hence that the domain B is not a set, which he says "disposes of the Russell antinomy as far as we are concerned."[149] Both theorem and consequence

[142] loc. cit., §148.

[143] loc. cit., §25.

[144] van Heijenoort, 201.

[145] loc. cit., 201.

[146] loc. cit., 202.

[147] By Ulrich Felgner in his introductory note on Zermelo's 1908 paper in *Ernst Zermelo*, 2010. As evidence he cites Husserl's complaint in §72 of *Ideas* that some of his ideas about definite deductive systems had "found their way into the literature without reference to their original source."

[148] The Richard paradox is the contradiction arising when one argues that the set of all real numbers definable in a finite number of words can be enumerated in a sequence, from which one then obtains a new real number defined by diagonalization in a finite number of words. Zermelo did argue that the notion of 'finitely definable' was relative to the chosen language, and that the countability of all finitely definable objects "holds only if one and the same system of signs is to be used for all them, and the question whether a single individual can or cannot have a finite designation is in and of itself meaningless since to any object we could, if necessary, arbitrarily assign any designation whatever" (op. cit., 192).

[149] loc. cit., 203.

seem prima facie congenial to Husserl's view of sets which, as Rasado-Haddock informs us, "is clearly related to the view of his friends Cantor and Zermelo on the iterative notion of set. . . ."[150] And yet Husserl's notes focus exclusively on alleged failures to solve Russell's paradox, while ignoring the notion of definite, allegedly appropriated from Husserl.[151] Before trying to understand this we look at the doubt cast on set theory by the Frenchman's paradox.

In 1900 Hilbert attributed the "numerous and surprising analogies and the apparently pre-established harmony" between the various branches of mathematics to the "recurring interplay between thought and experience" and proposed 23 problems whose solutions he thought would enhance the "organic unity" that had made mathematics the "foundation of all exact knowledge of natural phenomena."[152] Such problems would have to be difficult but still solvable, and Hilbert urged that all well posed problems were solvable, which he motivated by noting the increasing number of old hard problems that had finally found rigorous solutions by proofs of the impossibility of the positive solutions sought for them. His first problem was to prove Cantor's hypothesis that every infinite set of real numbers is either equivalent to the integers or to the set of all reals and hence to the continuum. His motivation was that: "From this theorem it would follow at once that the continuum has the next cardinal number beyond that of the countable set; the proof of this theorem would, therefore, form a new bridge between the countable set and the continuum.".[153]

The first responder to Hilbert's solvability thesis was Jules Richard who argued that it presupposed that all true mathematical propositions were provable, but he doubted that 'true' and 'provable' have the same meaning.[154] Hilbert asked for rigorous proofs that "establish the correctness of the solution by means of a finite number of steps based on a finite number of hypotheses." Richard suggested that the argument for the countability of all finitely definable real numbers could also

---

[150] loc. cit, 219. Rasado-Haddock bases this somewhat contentious claim on having elsewhere shown that "neither Russell's nor Cantor's sets can be obtained in the iterative hierarchy of mathematical objects propounded in the Sixth Logical Investigation" (220).

[151] Rasado-Haddock, 2010 quotes Husserl's claim in his 1920 notes that the paradoxes are only avoided by a constructive axiomatization of set theory, providing an existence proof for every mathematical totality, where "a 'manifold' here must mean a formal as constructively (definite) characterized region of objects . . . by determinately formed operations that can be iterated into infinity. The axioms must be so chosen as to found *a priori* the constructability." (29) As Rosado-Haddock says, "Husserl uses Zermelo's expression 'definite' as if it were synonymous with the expression 'constructible'" (30). Would this not make Zermelo's separation axiom constructive? Or did Husserl think it spuriously *postulated* the existence of subsets of a given set? Until his notes are published it will be difficult to know just what he thought.

[152] Gray 2000, 282.

[153] Ewald 1996, 1103.

[154] Richard 1903, 106–113. To Hilbert's claim that proving consistency of the axioms for the real numbers would establish their existence, Richard responded that "concerning any system of propositions containing no contradictions, it does not follow that there are objects satisfying this system and not satisfying any property other than those one can derive from them" (110–111).

establish the countability of all "complete and rigorous *proofs*."[155] He accepted Jules Tannery's conclusion that there must exist real numbers that cannot be defined in a finite number of words. If we now assume with Poincaré that any possible object of thought must be so definable, Cantor's theorem would imply the existence of real numbers that could not be objects of thought. This would pose a paradox for any idealism holding that all mathematical objects were constituted in and accessible to consciousness.

Poincaré dealt with these issues in a lecture in 1909 in Göttingen, in which he criticized Zermelo's axioms for set theory and gave a complicated critical version of what in his view Cantor has "really proved" in his theorem.[156] He especially rejected Zermelo's separation axiom, arguing that his appeal to "definite" conditions for separating subsets of a given set was too vague and indefinite to avoid Richard's paradox. Poincaré concluded from it that "there could not be any evident axiom concerning infinite numbers."[157] If Husserl was present, he left no written trace of his reaction to it all. The lecture made a great impression on Weyl who took up the issue in his *Habilitation*, wherein he proposed essentially that "definite" conditions be those expressed by first-order logical formulas with equality and membership as primitive relations. This leads to what he calls the "true kernel" of Richard's antinomy, namely, that "in set theory or logicist mathematics one has only to do with countably many relations, by no means however with only countably many things or sets."[158] In other words, in axiomatic set theory there are now only countably many conditions available to prove the existence of a real number by separation, but one can nevertheless prove the theorem that the set of real numbers is uncountable.[159]

That there is no law for constructing the reals may explain why Husserl never mentions them as a totality; in his notes of 1912 the totality of natural numbers is "given originally" by means of the construction law of adding one to any number and the evidence that its iteration is always possible and reaches every number: "A totality in the true sense is thus constructible by an evident law of iteration, constructible with respect to each of its terms, so that each is brought to actual

[155]Later he applies Cantor's diagonal argument to the finitely defined real numbers to get his paradox, setting the stage for the attempts of later logicians to use it to construct true but unprovable propositions in formal systems, something finally accomplished by Gödel, who explicitly notes the analogy of his proof with Richard's paradox. But Gödel still endorsed Hilbert's solvability thesis. For a lucid exposition of the incompleteness theorem as the result of formalizing Richard's paradox, see Church 1934. Here the undecidable sentence is seen as just the inability of the formalism to prove the totality of a function defined for all numbers by diagonalization. This is the kind of problem that I have suggested above could have crossed Husserl's mind in pondering whether one of this arithmetics could prove that an inverse operation was undefined for some number.

[156]Poincaré 1963, 55*ff*.

[157]loc. cit., 62.

[158]Weyl 1910, 112. Weyl's article was reviewed by Skolem who would soon try to develop this conclusion into an incompleteness result for axiomatic set theory.

[159]When one exploits the countability of these conditions to prove the existence of a countable model for Zermelo's axioms, assuming their consistency, one has Skolem's paradox.

givenness."[160] This would seem to exclude any talk of the totality of real numbers, or the totality of the subsets of the natural numbers. Later in a note of 1920, after he became acquainted with Weyl's constructive analysis in *Das Kontinuum*, Husserl says of the natural number series that

> it has no end, it is not fixed as a finished product, every step is a step in progress, every productive act is an act in a connected production comprising a mental becoming; and as this kind of becoming bears a sense of the infinite itself ... it is a becoming 'iterative' production and in this lies an intentional horizon of 'and so on', the endless free repeatability, the endless series 'I can always again.'[161]

These texts have reinforced the inclination of some of his commentators to regard Husserl as a champion of a constructionist philosophy of mathematics, but his formulations of constructability are vague, and he never tries to impugn the validly of any theorem of classical mathematics, certainly not Cantor's theorem which, as far as I am aware, he never mentions, much less criticizes. Husserl is more concerned that his supramathematics be able to theorize any material domain of nature, which he says calls for an inventory of *all possible* deductive theories.

A supramathematical storehouse of theories still left him with nagging questions: "What does reality, which exists in itself, care about the activities of our thinking? How is one to understand the fact that if the subjective activities of our thinking proceed in accordance with logical laws, they finally concur, and even must concur with the nature of things."[162] Husserl has no doubt that logical thought *does* concur with the nature of things, but still wants an account of the "wonderful affinity" between thought and things that would explain this concurrence or harmony. Despite the lack of ultimate clarity in its foundations:

> Mathematics is the great tool of knowledge of nature. By what right, though, is it applied to knowing reality if numbers are simply laws of thinking? How is it that what we have arithmetized in our thinking tallies with actual reality, nature existing in itself? What does nature care about our colligating and counting? If arithmetic belongs *a priori* to the thinking mind, how is the right to apply it to what is *a posteriori* in nature to be understood?[163]

Unless we can somehow suspend our natural belief in the existence of a *natural world independent of our thought*, we can only *assume*, as Hilbert wrote in his Mathematical Notebooks, that "Between thought [Denken] and event [Geschehen] there is no fundamental and no quantitative difference. This explains the pre-established harmony [between thought and reality] and the fact that simple experimental laws

---

[160]Quoted in Schmit, Op. cit., 116.

[161]On receiving his book Husserl wrote to Weyl in1918: "Finally a mathematician who understands the necessity of phenomenological considerations in all questions regarding the clarification of fundamental concepts and who finds himself on the original ground of mathematical logical intuition, the only ground on which a really genuine foundation of mathematics and insight into the sense of its achievement is possible." (van Dalen, 1984, 3).

[162]Husserl 2008, 150.

[163]loc. cit., 159.

generate ever simpler theories."[164] Husserl eventually decides that only a form of *transcendental idealism* in which all objects are constituted in consciousness and which bestows absolute clarity on mathematics, can answer his questions without this assumption.

Husserl's initial foray into such idealism poses the problem of "the 'resolution' of empirical 'being' in connections of absolute consciousness."[165] Our natural cognition poses no difficulties as long as it does not reflect on the relation between cognition and being, but such reflection raises the question of "what wonderful affinity all other things, all objects in themselves have with thought?" By means of our logical thought, intuition, and insight we determine a "something" which is not subjective, but "in itself." Natural science discovers objective things, connections of things, laws of things, mathematical laws, manifolds:

> But if they are 'in themselves' how can I grasp them? How can I approach them? If they are in and of themselves, then they are cut off from thought. If they then become something for me, if they are thought, do they 'enter into thought'? How is this relation to be understood, this entering into cognition, this being grasped, determined, and cognized in perception, experience, and thinking? What kind of wonderful affinity do all other things, all objects in themselves and the thinking have?[166]

Husserl cannot answer this by reducing things *in themselves* to consciousness: rather he holds that

> being in the sense of the objective sciences is 'no ultimate being', but rather resolves itself into consciousness. The thing *itself* does not resolve *itself* into consciousness. It resolves itself into atoms and molecules. However, 'a thing exists in reality' and 'there is reality' and similar cognitions refer back to cognitive connections, to formations of consciousness, wherein the being of the thing and the being of all thinglike states of affairs acquire their sense.[167]

Indeed, "the world constitutes itself in consciousness. It is what it is only in relation to consciousness."[168] Husserl's idealism seeks to exorcise any unknowable "things in themselves" from the epistemology of the natural sciences. Such unknowable things he would resolve into atoms and molecules.

A corollary of the constitution of the world in consciousness is that "[t]he being of nature is thus completely and absolutely contained in consciousness and in entirely the same sense as tones in tone consciousness."[169] Consequently, "objects of nature are obviously true objects, their being is true being, nature is actuality in a genuine and full sense."[170] And hence "to say natural science has nothing to do with nature, that the true objects it concerns itself with are sensations and

[164] Thiele 2003, 18.
[165] Husserl, 2003, 21.
[166] loc. cit., 24.
[167] loc. cit., 28.
[168] loc. cit., 29.
[169] loc. cit., 68.
[170] loc. cit., 70.

what we call things, atoms, etc. are mere thought economizing abbreviations for connections of sensations is the height of folly."[171] Husserl was thinking of Mach whose "phenomenological physics" he had earlier admired but which led him to reject atomism. Now that atoms had been discovered Husserl blamed this on Mach's positivism, unaware that his rejection had reasonably resulted from the failure of his attempts to establish a molecular atomic theory of spectral analysis.

The physicists in Göttingen were split between atomists and phenomenologists. Among the latter was W. Voigt, a respected physicist who still doubted the usefulness of spectral analysis and the quantum hypothesis. As he put it in his rectorial address to the university in 1912:

> A solved question only gives birth to ten new ones and the mysteries become more and more enigmatic. To mention a single one. All those new results that are accumulated in huge quantities make the constitution of the atom nothing but more incomprehensible as they show the most unexpected phenomena. What kind of an object of inconceivable smallness is it that has the ability to execute and emit thousands of different but at the same time fully determined oscillations that are characteristic of the substance?[172]

That each species of atoms has its characteristic spectrum allows us to see that stars are composed of the same substances found on earth. Husserl saw this use of spectral analysis as analogous to how phenomenology can solve the problem of the correlation between being and consciousness:

> Think now of problems about the natures of the fixed stars and the spectral analysis which alone is called upon ... to comprehend these problems scientifically. Will one say that spectral analysis belongs already to antiquity, since the ancients concerned themselves considerably with the problem of the nature of the firmament of the fixed stars?[173]

Husserl compares such problems to the old philosophical ones:

> No one will put these ancient speculations about the fixed stars on a level with epistemological speculations about them, or even with those of modern times before the establishment of phenomenology. Just as spectral analysis must first be available and scientifically developed extensively ... so is it with phenomenology in relation to those old puzzling questions of the philosophical firmament.[174]

Husserl never loses sight, however, of how the considerable knowledge of the firmament achieved from the ancients to Galileo depended wholly on geometry.

In §72 of *Ideas* Husserl focuses on geometry to illustrate the completeness any "material mathematics" of his supramathematics must have for "mastering all possibilities" of its domain, so that it comprises a "definite manifold." A finite number of concepts and axioms must be gathered from the essence of the domain that "*completely and univocally determine the totality of all possible configurations*

---

[171] loc. cit., 71.

[172] Quoted in Schirrmacher 2003, 8. Husserl served with Voigt and Hilbert on Grelling's defense of his dissertation on axiomatic number theory, and one wonders whether he ever chanced to discuss such problems of physics with either of them.

[173] Husserl 2007, 6.

[174] loc. cit., 6.

of the domain in the manner of a purely analytic necessity—so that consequently, *as a matter of principle, nothing more* remains open to it."[175] Husserl asks how *phenomenology*, which deals with concrete essences of experience, fares in this respect: "Do the supreme genera pertaining to these abstract aspects form domains here as well for definite disciplines, for 'mathematical' disciplines after the manner of geometry? Do we, accordingly, have to search here as well for a definite system of axioms and to build deductive theories on it?"[176] Husserl explains that this is actually equivalent to the question: "Is the stream of consciousness a genuine mathematical manifold? Is it, taken in its facticity, like physical nature that, indeed, would have to be designated as a concrete definite manifold, if the ideal ultimately guiding the physicist is taken in a valid and conceptually rigorous way?" Would the physicist ideal guarantee the applicability to nature of the complete deductive systems of Husserl's supramathematics without appeal to a pre-established harmony arranged by God?[177]

Elsenhans questioned whether Husserl's analogy between mathematics and phenomenology could match Kant's account of how the pure non-empirical intuitions of geometry necessarily applied to reality. Nor could he see how any "deductive theorizing" could play a role in phenomenology, much less how a "mathematically definite manifold" comes about in phenomenology as a descriptive science.[178] Husserl responded that consciousness comprises a domain of systems and laws ruled by intentionality: "Just as nature is an ordered system in the mathematical manner of a definite manifold, so consciousness is also an ordered system. But not a mathematically ordered system for clearly determining all being according to mathematical laws, but a system standing under laws of reason and laws that concern the conditions of possibility of laws of reason."[179] Hence the "causality of consciousness" is not one governed by mathematical laws, but a wholly different "motivational causality."

Apparently Elsenhans stimulated Husserl to revisit the question of whether Kant's idealism, without invoking an archetypical intellect, insured the harmony of his "axioms of pure reason" with things without borrowing it from experience:

---

[175]Husserl 2014, 130.

[176]loc. cit., 132.

[177]Husserl says that, upon the suspension of God's transcendence: "The reduction of the natural world to the absoluteness of consciousness yields *factual* connections of specific kinds of experiences in consciousness ordered in prominent, rule-governed ways, within which a morphologically ordered world in the sphere of empirical intuitions constitutes itself as an intentional correlate ... a world for which there can be classificatory and descriptive sciences." (loc. cit., 106). Having purged his idealism of things in themselves that cannot be known, Husserl can further argue that "this very world can be determined as the 'appearance' of a *physical nature*, standing under exact natural laws, in the theoretical thinking of the mathematical natural sciences. In all this there lies a wonderful *teleology* since the *rationality* realized by the *factum* is not the sort demanded by its essence." That is, the rationality realized by factual connections in empirical intuition yielded by the reduction is not necessary but contingent.

[178]Elsenhans 1915, 230–237.

[179]Husserl, 1986, 326.

"Should we appeal to God and a harmony pre-established by him of our intellectual creations and their empirical consequences with the things themselves?"[180] The outlook that raises such difficult questions in earnest also leads to questions about the real significance of logic: "There are enough questions taken seriously scientifically and nevertheless in the most difficult sense paradoxical, as every expert in history, even the history of the most exact science, mathematics, knows."[181] He also knows that by means of an "imminent teleology," they are eventually solved with profit to science. Husserl tries to solve the problem of such harmony by reflection on the "parallelism of objective unities and constructed manifolds of consciousness." The wonderful affinity between thought and things he previously tried to grasp is now what he calls this "wonderful parallelism." Objects of experience constituted theoretically through experience already comprise a system:

> If we ascend to the level of the 'idea,' we have this great and manifestly multifaceted parallelism of ideas: on the one side nature itself, as the universe of real being itself and the ideally complete science of nature: the ideally closed system of true propositions and theories, which coming together with the unity of the true and complete theory of nature, constitutes its theoretical parallels. On the other side, we have again as an idea the complete system of possible harmonious, thoroughly confirmed experiences of nature.[182]

Finally, in "consciousness we have boundless infinities, since to every unity of nature ... itself corresponds various infinities of consciousness as 'constituted' according to noesis and noema, and they represent respectively, as it were, a focal point in which an infinite bundle of rays converge."[183] So by 1917 Husserl could believe that, in principle, his idealism can ground a complete theory of nature as a definite manifold in consciousness, even though consciousness itself is not such a manifold. Could he expect all concepts of such a theory to be phenomenologically accessible?

14.5 Crisis, Mysticism, Pluralism

By 1920 the one mathematician whose grasp of phenomenology had satisfied Husserl had produced a remarkable version of general relativity that went beyond Einstein's to include electromagnetic phenomena, while invoking phenomenological themes for its foundations. Husserl was impressed by Weyl's introduction of pure infinitesimal geometry in relativity theory, for which Becker pursued a phenomenological basis in his *Habilitation* on geometry. In 1922 he wrote to Weyl that:

---

[180]loc. cit., 143.

[181]loc. cit., 144.

[182]loc. cit., 146–147.

[183]loc. cit., 147.

> It is nothing less than a synthesis between Einstein's and your discoveries with my phenomenological investigations on nature. It aims by deep and original means to prove that Einstein's theory, but only when completed and recast through your researches into infinitesimal geometry, represent those forms of the 'structural lawfulness' of nature, which must be presumed necessary on the deepest *transcendental-constitutive* grounds.... What will Einstein say, when it is proved that nature seeks a relativistic structure on the *a priori* grounds of phenomenology and not on positivistic principles, and that only in this way is a fully comprehensible and exact science possible.[184]

In view of his belief that Becker *had* provided such grounds from Einstein's theory, it seems that in 1922 Husserl saw nothing of the crisis in physical science he will deplore a decade later. Nor in mathematical sciences: he further related to Weyl that Becker was also able to

> portray a constitutive theory of the continuum (a rational recomposition of the vague continuum by means of limits and approximations). In this way he looks for a demonstration that only Brower's and Weyl's theories are coherent with the definite and fundamental demands of a program of research constituted in its sources phenomenologically.[185]

What room was there then in Husserl's mind in 1922 for a serious crisis of science?

To repeat Husserl's claim from the *Britannica* article, "for the a priori disciplines (e.g. mathematical sciences) founded in phenomenology there can be no 'paradoxes' and no 'crisis of foundation'." Weyl's paper "On the New Foundational Crisis of Mathematics" spelled out the intuitionistic revolution of Brouwer he had joined and which, Husserl assured him, Becker had founded in phenomenology. The crisis in question resulted from trying to base mathematics on set theory, whose paradoxes, said Weyl, "reveal what is hidden by the outwardly shining and frictionless operations in the center: namely, the inner groundlessness of the foundations upon which rest the superstructure of the realm."[186] But Husserl knew that Hilbert was convinced that set theory was in good hands with Zermelo, and he was happy to inform Weyl that, "as Courant writes to me, Hilbert has formulated a new way of founding mathematics 'completely in the phenomenological spirit.'"[187] And Husserl certainly wanted to believe this, for he had just written to Hilbert to convey his deepest appreciation,

> not only for the great significance that your mathematical creations may claim for philosophy, but also for the memory of the very humane and intellectually rich communion I had with you all those years in Göttingen, always grateful for the interest and confidence

[184]Tonietti 1988, 348. As for Einstein, he said that Weyl's theory was like a beautiful symphony but that it could not explain the stability and distinctness of the spectral lines. He argued Weyl's theory made the radiation frequencies of, say, two hydrogen atoms dependent on their histories, and thus that that of atomic clocks would depend on their history, excluding stable spectra. For this and Weyl's response, see Scholz 1994, 203–230.

[185]loc. cit., 369–370.

[186]Mancosu 1998, 86.

[187]Tonietti, op. cit., 373.

> you showed in my philosophical endeavors at a time when they had found little sympathetic understanding.[188]

So when Husserl sent his 1922 letter to Weyl, he had reason to believe that his phenomenology was not only providing *a priori* grounds for relativity theory, as well as for theories of the continuum of Brouwer and Weyl, but that Hilbert may now have recognized its need to justify his belief in pre-established harmony. At least until he looked at the paper Courant recommended to him.

Hilbert's 1922 paper both inaugurated modern proof theory and initiated a *Grundlagenstreit*, proclaiming that Weyl and Brouwer were pursuing a "false path" to the solution to the paradoxes: "by following the erstwhile path of Kronecker: they seek to ground mathematics by throwing overboard all phenomena that make them uneasy and by establishing a dictatorship of prohibitions *à la* Kronecker."[189] In Hilbert's conception of mathematics as its own subject matter, proofs themselves are studied as objects in a formal system representing a branch, to which he adds a new *metamathematics* to protect it from unnecessary prohibitions and paradoxes:

> In this metamathematics—in contrast to the purely formal modes of inference in mathematics proper—we apply contentual inference; in particular to the proof of the consistency of the axioms. The development of mathematics accordingly takes place in two ways that constantly alternate: (i) the derivation of new 'provable' formulae from the axioms by means of formal inference; and (ii) the adjunction of new axioms together with a proof of their consistency by means of contentual inference.[190]

The resulting relativity of the concept of 'provable' to the underlying formal system, says Hilbert, "is natural and necessary, and causes no harm, since the axiom system is constantly being extended, and the formal structure, in keeping with *our* constructive tendency, is always becoming more complete."[191] This relativity of proof may seem at odds with the completeness Husserl expects for the systems of his supramathematics, but it also finds a role for the "constructive tendency" precisely in making formal systems ever more complete, without having to reject useful theorems whose proof was lacking this tendency. For someone like Husserl, who was never inclined to give up any part of the classical mathematics he was trained in but was now convinced, as his notes show, that the constructive tendency is essential for avoiding paradoxes, this metamathematical option could not be ignored. But was its philosophical motivation respectable, if not phenomenological?

---

188 *Briefwechsel*, vol. VII, 119.

189 Ewald 1996, 1119. We know Husserl owned and bookmarked this paper, thanks to Hartimo 2016.

190 loc. cit., 1132.

191 loc. cit., 1127. This is my emphasis, to remind us that, as Ewald puts it: "Despite Hilbert's fiery polemic against Kronecker, Weyl, and Brouwer, it should be observed that the entire controversy is an internal feud among constructivists" (1116). Of course, the axioms systems that Hilbert is constructively extending can be formalizations of theories as non-constructive as you like. Weyl calls them "transfinite formula games."

Husserl left this question to his trusted student Dietrich Mahnke, who had studied mathematics with Hilbert. Mahnke composed a fascinating study of Hilbert's paper in which he basically confirmed Courant's judgment, agreeing that "it is precisely this latest and formal-mathematically most complete 'foundation' of mathematics which comes closest to Husserl's phenomenological foundation; a fact, by the way, which will also be a good indication of the correctness of the corresponding results of their respective theoretical researches, having emerged from completely different motives."[192] Mahnke deftly limns the critical source of Hilbert's metamathematical turn, "equivalent to the Kantian one from the object to the subject of mathematical knowledge or, in Husserl's more precise terminology, by way of going back from noëma to noësis, from the intended objects to the intending acts of reason."[193] And despite some infelicitous formulations of the givenness of the numerals as objects of arithmetic, "Hilbert is phenomenologically correct when he demands the givenness of 'extra-logical discrete objects' as the foundation of the arithmetically evident."[194] In his insistence that existence proofs deal not with 'things' such as numbers but with 'operations' comprising arithmetical 'proofs', Hilbert does distinguish symbolic operations from acts of reason: "But these fundamental distinctions achieve full clarity first only in Husserl's theory of the correlation between objects of reason and the objectifying acts of reason, between the *noëma* and the *noësis*, and in his method of basing noëmatics upon noëtics."[195] Finally Mahnke asks whether Hilbert is justified in deriving "the objective existence of the set of natural numbers from the investigation of subjectively human ways of drawing inferences. To start with, do things conform to our thoughts?"[196] His answer to this perennial Husserlian question is that "the worldly course of events can never contradict the logical-mathematical laws of form, which do not state anything concerning sense contents at all." Thus phenomenological scrutiny not only gives Hilbert's metamathematics high marks, but is consistent with his belief in pre-established harmony.

Mahnke also gives his phenomenological blessing to the requirement of proving consistency when he says: "Hilbert succeeds in showing that if a=b there is no way to obtain the formula which 'appears' as a≠b through operations with formula signs. Thus the non-existence of a contradiction becomes available to 'inspection' in the true sense of the word."[197] But Mahnke passes over Hilbert's claim that "We can never be certain in advance of the consistency of our axioms if we do not have a special proof of it. Axiomatics therefore compels us to take a stand on this difficult epistemological problem."[198] But Weyl and Becker judged Hilbert's consistency problem to be rather banal, comparing it to a proof that one could not

[192]Mahnke 1977, 77.

[193]loc. cit., 79.

[194]loc. cit., 81.

[195]loc. cit., 82.

[196]loc. cit., 82.

[197]loc. cit., 78.

[198]Ewald 1996, 1120.

legally reach certain positions in chess, thereby motivating the myth that Hilbert was reducing mathematics to "formula games." Weyl played his 'ten queens' gambit, comparing the consistency proof, which Hilbert said would be difficult, to proving the impossibility of ten queens of the same color in a legal chess position, which is obvious and easy to prove: "At bottom a triviality rather than a paradox", as Becker put it.[199]

Becker had planned an extended study of the infinite and the continuum, as well as an immanent critique of Hilbert's metamathematics of 1922, but after he had written most of it, Hilbert's lecture "On the Infinite" appeared with its fuller and more developed formulations. To show that his proof theory could solve important problems he proposed to prove Cantor's continuum hypothesis by associating the ordinals of Cantor's second number class with number theoretic functions defined by higher and higher type recursions obtained by diagonalization. Becker was fascinated by Hilbert's attempted proof in which he saw a golden opportunity to show how a mathematical theory can be grounded in Husserl's theory of iterated reflection on the noematic levels of consciousness, which Becker extended to transfinite levels. Specifically, he takes the iteration of a reflection as the phenomenological analogue of Cantor's successor ordinals and Husserl's 'horizon phenomena' as the analogue of the production of limit ordinals. Thus does Becker try to ground a large part of the second number class in "transfinite structural complications of consciousness," while claiming that in building up his ordinal variable-types Hilbert was implicitly appealing to the phenomena of "the nesting of intentional structures of consciousness." Indeed Hilbert's work on the continuum problem, he writes to Weyl, "is an astonishing confirmation of my conception.".[200]

Weyl raised a number of objections to Becker's phenomenological account of the ordinals concerning their finite definability and his treatment of critical numbers by reflection on all previous ordinals. Weyl claimed that it was a mistake to even attempt an intuitive interpretation of all ordinals. Becker replied that:

> You see neither the necessity nor the possibility of exhibiting a special (transfinite) horizon of consciousness. On this point I must decisively object. Unfortunately phenomenological descriptions have different persuasive forces for different people, and there is no way to make them compelling. But I claim to have exhibited this consciousness horizon very clearly and I can also appeal to Husserl's agreement with me on this matter.[201]

---

[199] Becker 1927, 53. In fairness to Weyl, he did soon see the depth and importance of Hilbert's problem. On studying von Neumann's proof for a restricted system, Weyl conceded that: "We are here dealing with a concrete mathematical problem which is not trivial, but at the same time is solvable, and I cannot imagine that any mathematician can find the courage to elude its honest solution by means of a metaphysical dogma." (Weyl, 1929,265).

[200] Mancosu &. Ryckman 2002, 177. As for Hilbert, he was trying to build that "new bridge from the countable set to the continuum" he said a proof of Cantor's hypothesis should provide, but the hierarchy of recursive functions he was building closes off at a small countable ordinal.

[201] loc. cit., 142. Husserl left no written evidence of his agreement with Becker's project of constructively grounding Cantor's transfinite ordinals in his own sketchy account of iterated reflections on the noematic levels of consciousness. We may recall, however, that in his response to Elsenhans' criticism Husserl denied that consciousness was a mathematical manifold.

Becker could not make up his mind about Weyl's claim that "in the course of its recent history physics is moving more and more away from the intuitively graspable, and that the time has come when the last rope is cut that keeps it at the phenomenal shore." Finally, Weyl argued that:

> If Hilbert's view prevails over intuitionism, as appears to be the case, then I seen in this a decisive defeat for the philosophical attitude of pure phenomenology, which thus proves to be insufficient for the understanding of creative science in the area of cognition that is most primal and most readily open to evidence—mathematics.[202]

This prompted Becker to write to Mahnke that "we are facing here a certain crisis of phenomenological method itself. If Weyl is right that Brouwer's intuitionism cannot support theoretical physics, then phenomenology in Husserl's 'classical' sense must be seen as incapable of securing the modern form of knowledge of nature—and making it understandable."[203] Before trying to see whether Husserl himself recognized such crisis in his method, perhaps as part of the larger crisis of science he would eventually deplore, let us look at the specific development of "creative science" that Weyl had in mind.

Weyl granted that, in addition to Brouwer's way, which he compared with Mach's phenomenalism, there is a theoretical need for Hilbert's with its "creative urge directed upon the symbolic representations of the transcendent, which demands to be satisfied."[204] Only by its fusion with physics in the theoretical construction could his transfinite formula game acquire meaning, but one radically different from intuitive insight. Weyl mentioned in this context the defeat of Mach's phenomenalism by the discovery of atoms, leaving however the extraordinarily difficult problem of explaining their phenomenal spectra by divining their inner mechanics. Hilbert formulated in 1906 a non-constructive spectral theory of eigenvalues of linear operators in an infinite dimensional function space solely from the mathematical motive of unifying algebra and analysis, but which was found in 1926 to be precisely the mathematics physicists needed for the quantum mechanics of the atom.[205] That such recently created mathematics became available to them just at the time of their

[202]Van Heijenoort, 1967, 484. One should bear in mind here that it is Becker's version of phenomenology, influenced by Heidegger and committed to intuitionism, that Weyl would decisively reject. But given Becker's claim that Husserl was in agreement with him, something Weyl could easily believe in view of Husserl's letters to him, he could understandably envision a defeat of the attitude of "pure phenomenology," period. I am indebted to e-mail correspondence with Mirja Hartimo for bringing home to me that Weyl's dissatisfaction with phenomenology was less with Husserl's own formulations than with Becker's.

[203]Mancosu and Ryckman 2002, 147.

[204]Mancosu 1998, 141.

[205]Hilbert saw this as just the latest example of pre-established harmony: "Recently cases have been piling up in which precisely the most important mathematical theorems, the ones that stand at the center of mathematical attention, are at the same time the ones that are needed in physics. I had developed the theory of infinitely many variables from pure mathematical interest, and had even used the term spectral analysis, without any inkling that it would one day be realized in the actual spectrum of physics." (Ewald 1996, 1160) Below we look at Husserl's transcendental gloss on such harmony.

own theoretical crisis seemed miraculous to some of them. As he was discovering the equivalence of his own wave mechanics to the algebraic matrix mechanics of Born, Jordan, and Heisenberg, Schrödinger wrote in 1926 to Wien:

> Time is flying. Each second or third day brings with it a small novelty—*it* works, not I, and that 'it' is the magnificent edifice of classical mathematics and Hilbert's mathematics, the wonderful edifice of the theory of eigenvalues. These unfold everything before us so clearly, that all we have to do is take it, without any labour and bothering: because the correct method is provided in time, as soon as one needs it, completely automatically. I am so happy to have escaped the terrible mechanics including its action and angle variables and the perturbation theory which I have never really understood. Now everything has become *linear*, everything can be superposed; one computes as easily and comfortably as in good old acoustics.[206]

Weyl was there with Schrödinger advising him on mathematics and will later call this windfall application of Hilbert's theory "a sort of miracle." This striking success of non-constructive mathematics was surely on his mind when he envisioned the defeat of pure phenomenology. How did Husserl see things?

The first thing to note is that Schrödinger's words are a striking echo of the formulation Husserl quoted from Mach on how the use of symbolic mathematical cognition can easily give one the impression that one's pencil is smarter than oneself. Husserl knew the success of symbolic mathematical cognition can precede insight into its *a priori* grounds. He would ascribe the success extolled by Schrödinger to the "wonderful teleology" extolled in the *Ideas*. But whether he realized that Weyl's disappointment with phenomenology was with Becker's version of it, and how he viewed the latter's "crisis of phenomenological method itself," all depends on when and how extensively he read Becker's opus on mathematical existence.[207]

Becker acknowledged that modern physics had left intuitive geometric space for "the many dimensional spaces of quantum theory (Schrödinger) and Hilbert's pure 'ideal' transfinite statements which despite their formal-ontological incomprehensibility still refer to 'reality,' but only as components of a system also containing

---

206 Mehra & Rechenberg 1987, 583. The linear superposition property celebrated by Schrödinger proved to be not only the strength of quantum mechanics but also the source of some of its paradoxes. Jordan 1973 recalled that: "Hilbert especially admired the fact that Schrödinger, with his wave equation of the hydrogen atom, had found a very simple example of an eigenvalue problem showing on the one hand discrete eigenvalues, and on the other hand a continuous interval of eigenvalues. Hilbert himself had known from his fundamental researches on integral equations and the quadratic forms of infinitely many variables that such occurrences must exist. Only nature itself, as studied by the quantum physicists, held in store simple examples of this kind, at that time not yet detected mathematical fantasy." (298). Husserl might well have viewed this circumstance as exemplifying his "wonderful parallelism" between "objective unities and constituted manifolds of consciousness."

207 Schumann 1977 says that only in March 1937 did Husserl read larger portions of *Mathematische Existenz*, but I have to agree with Gethmann 2003 when he says "It is safe to assume, however, that as coeditor of the *Jahrbuch* Husserl looked at the manuscript already well before 1927 and discussed with Becker in this period questions of the foundations of logic and mathematics" (147).

'real' statements."[208] He is convinced that an examination of modern mathematics is possible only in its connection with modern physics. He cites passages in the *Ideas* as evidence that Husserl's idealism is akin to Weyl's claim that the question of the reality of the world "is inseparable from its lawful mathematical harmony."[209] A world must be harmonious for our finite minds to grasp it, but that a world really exists is "accidental", says Becker, and "the fact of the existence of modern physics shows that the 'symbolic,' unintelligible transcendent-transfinite mathematics of Hilbert unveils the harmony of the world. Thus has that empty formula game a mysterious reference to the metaphysical structure of the cosmos."[210] Only in his own "mantic phenomenology," analogous to the "hermeneutical" conception of Heidegger, could Becker find a place for "the modern 'mathematical mysticism' of Weyl and Hilbert."[211] Husserl read this, for in two notes to Heidegger in May 1927 he asks "Are you getting the galleys of Becker's text, which is especially interesting for you towards the end?" and 2 weeks later, "Have you also read Becker's work? Direct application of Heideggarian ontology."[212] So Husserl could see that the phenomenology rejected by Weyl was not his own, but also that phenomenology had to face the success of the empty formula game of Hilbert in physics.

Becker formed a phenomenological substratum for Cantor's transfinite ordinals in Husserl's theory of iterated reflection on levels of consciousness, for which he found a "remarkable confirmation" in Hilbert's attempt to prove the continuum hypothesis. His analysis of Hilbert's effort resulted in his realization that Hilbert was actually trying to prove the *consistency* of Cantor's hypothesis, and that *this* is precisely the sense in which his mathematics deals with "formula games." Becker analyzed the 'Brouwer-Weyl idea' that the continuum hypothesis was meaningless, which he rejected on the grounds they had not formulated "an admissible concept of the most general law for a sequence."[213] He concluded that Brouwer's idea of choice sequences was useful for analyzing the *intuitive* continuum, but contributed

[208] Becker 1927, 265.

[209] loc. cit., 326. He cites the passage in §58 quoted above in footnote 177.

[210] loc. cit., 327.

[211] loc. cit., 328. The classic study of Becker's mantic phenomenology is by Poggeler 1970, who clearly exposes the Pythagorean pedigree that distinguishes it from 'Heidegger's hermenutical phenomenology. Becker was influenced by Weyl's account of the recent success of quantum mechanics, by which "the key was manufactured that unlocked the secret of the amazing regularities governing the series of the spectral lines which are emitted by radiating atoms and molecules. The success was most striking in the simplest, that of the hydrogen atom," concerning which Weyl approvingly quotes Sommerfeld's claim that "our spectral series, dominated as they are by integral quantum numbers, corresponds, in a sense, to the ancient triad of the lyre, from which the Pythagoreans 2500 years ago inferred the harmony of natural phenomena; and our quanta remind us of the role which the Pythagorean doctrine seems to have ascribed to the integers, not merely as attributes, but as real essence of physical phenomena" (Weyl 1949, 185).

[212] Quoted in Kisiel 1993, 487.

[213] loc. cit., 165.

nothing to the solution of the *mathematical* problem of the continuum.[214] Since Hilbert was only after the consistency of the hypothesis, says Becker, "one sees here the deep opposition between Brouwer's basic intuitive idea, truth, and the formalistic view of Hilbert that requires mere consequence (consistency), and for which truth in formal mathematics is a meaningless concept. Here is where Hilbert's characteristic attitude becomes important."[215] Becker's accommodation for intuitionism and formalism probably influenced the tolerance for both of them shown by Husserl in his opus.[216]

Husserl will reject, however, the mystical account of physical science embraced by Becker's mantic phenomenology, though it does emphasize one feature of classical science he was loathe to give up, the central role of geometry. In a paper elaborating on the "mathematical mysticisms" of Weyl and Hilbert, Becker explained that quantum mechanics comprised an essentially greater break than relativity from intuitively accessible physical science. Classical physics described processes occurring in intuitive geometrical space, and even relativity theory still had a strong "phenomenological bent" due to Weyl's formulations; but now we have a "complete lack of intuition in quantum theory, where even the relatively 'visualizable' theory of Schrödinger already works with wave equations in many dimensional spaces.".[217]

For Becker this results from the collaboration of a symbolic mathematics that transcends intuitionism with symbolic physics, which he exemplifies by quoting from Hilbert's introduction to a joint paper on the axiomatization of quantum mechanics. Hilbert insisted that one cannot understand a physical theory if the formalism and its physical interpretation are not clearly separated, even though we do not yet have a complete axiomatization: "But what is definite by now is the analytical apparatus, which will not admit of any alterations in its purely mathematical aspects. What can, and probably will, be modified is its physical interpretation, for it allows a certain freedom of choice."[218] Becker's mantic reading of this formulation is that:

> One leaps as it were with the complete, ontologically incomprehensible 'mathematical apparatus' into the 'interpretation' of nature: the apparatus is like a magic key that opens

---

214loc. cit., 166.

215loc. cit., 177. Becker's assimilation of the logic of consequence and non-contradiction to Wittgenstein's logic of tautologies and contradictions will be accepted by Husserl as a clarification of his own account of such logic and incorporated into Husserl 1969, 338–340.

216Hartimo 2012 makes a very persuasive case for Husserl's embrace of a pluralistic philosophy of mathematics. She shows that he finds kernels of truth in both the intuitionistic and formalistic philosophies that cannot reasonably be ignored.

217Becker 1928, 346. Husserl owned this paper and it is hard to believe he didn't read it. Even Schrödinger will give up his quest for an intuitive theory. See Schrödinger, 1934.

218loc. cit., 347. This quote is from Hilbert, Nordheim, and von Neumann 1928. Here Hilbert generalizes his reply to Frege that Husserl quoted concerning the tact needed to apply Maxwell's electromagnetic equations to reality.

> up the domain of physical problems; but it does so only in the sense of a symbolic representation, not in the sense of an interpretation that really 'discovers' the phenomena.[219]

Becker explains that he is not speaking of symbolic cognition in physics but rather trying to indicate "how by a quasi pre-established harmony the development of mathematics and physics towards symbolic methods leads to their reunion precisely here where the *'regulative use' of the idea of Hilbert's transfinite* becomes concretely graspable. The basic aim of this symbolic approach comes from time immemorial, archaic, and even pre-historic: the most modern exact science returns again to the magic from which it originally descended."[220] Mantic phenomenology sees the formalist approach to quantum mechanics as invoking an incomprehensible formalism to symbolically represent a transcendent reality that it magically divines rather than discovers or explains, precisely as does ancient number mysticism. But its success manifests only a "quasi" pre-established harmony, that is, a harmony resulting from the confluence of the *historical* development of two disciplines, as *opposed to the transcendental basis for such harmony* Husserl is exploring.[221]

A theme of Husserl's logic is that "the relation to a possible application is part of the logical sense of formal mathesis." When cognition has an interest,

> we stand within the horizon of science; and in the case of a formally universal consideration, we stand within *logic*, whose two-sided theme is all possible sciences as such: on the subjective side, the possible forms of the actions, production of ... scientific cognitional formations; on the objective side, these formations themselves.[222]

A *developing* mathematics need not worry about its "ideally possible applications" and may be indifferent to the actual use of its formations in applied mathematics, as when "they function in theoretical physics as components of physical-scientific determinations."[223] Theoretical physics purports to discover harmony in nature as we currently experience it:

---

[219]loc. cit., 347. With precisely Becker in mind, Hilbert writes that "If we just think of all the applications and make it clear to ourselves what a multitude of transfinite inferences of the most difficult and arduous sort are contained in, for example, the theory of relativity and quantum theory, and how nature nevertheless precisely conforms to these results—the beam of a fixed star, the planet Mercury, and the most complicated spectra here on earth and at a distance of hundreds of thousands of light years—how, in this situation, could we even for a moment doubt the legitimacy of applying the *tertium non detur*, just because of Kronecker's pretty eyes and just because a few philosophers disguised as mathematicians have put forward reasons that are utterly arbitrary and not even precisely formulable?" (Ewald 1996, 1151). I am convinced that Husserl eventually endorses this case against Becker's intuitionism, which he extends to a rejection of his mantic phenomenology.

[220]loc. cit., 347–348.

[221]For an excellent critical account of current views of pre-established harmony, more or less in the spirit of Husserl, see da Silva 2010. Another paper by this author I have found illuminating is da Silva 2013.

[222]*Formal and transcendental Logic*, 108.

[223]loc. cit., 109.

> Evidently 'nature itself' thus acquires the value of a concept that is being constituted synthetically throughout the course of such partial installments ... abandoned determinations being excluded and corrective ones inserted during the process; and as so constituted this concept is the categorial correlate of the idea of a process of judgment that can be prolonged harmoniously *ad infinitium*.[224]

A thesis of transcendental logic is that everything that exists is constituted in consciousness. Scientific experience is the consciousness of seizing things directly: "But experience is not an opening through which a world, existing prior to all experience, shines into a room of consciousness; it is not a mere taking something alien to consciousness into consciousness."[225] Husserl has to argue that if the object experienced has the sense of something transcendental then it is only the experience that constitutes this sense. If the experience of the object is imperfect, appearing only one-sided or in a distant perspective, it is consciousness that tells me that the object "is more than what is actually itself grasped; there is more of the same object to be experienced."[226] This "more that could be experienced" depends on nothing else than "the actual and potential performances of my own consciousness," whereby I bring into play "synthetically connected sequences of consciousness, with the unitary effect that I should continue to be conscious of the same object."[227] Husserl infers that "this same object would show itself and become determined in continuous harmony: thereby continuously confirming its actual being."[228] This implies that any constituted object of nature "points back ... to a correlative *essential form* of manifold ... which is constitutive for that objectivity."[229] So Husserl's transcendental consciousness does constitute an eventually harmonious world.

Husserl's harmonious world, however, is not one in which there is room for the magical pre-established harmony recognized by Becker between mathematical thought and atomic physics, for Husserl does not regard Hilbert's non-constructive mathematical apparatus as unintelligible. More importantly, he is not sold on atomism after all. In its miracle year of 1926, Husserl reflected that:

> We try to go back to 'atoms', to ultimate elementary realities. The world is based on the universal structure that we call physical nature. In the idealization, which should make rigorous objectification possible, we come in physical nature upon no ultimate elements that are indicated *a priori*. Everything is divisible *ad infinitum*. On empirical grounds, science indeed leads, to be sure, to 'atoms', as to molecules, but also to electrons, etc. But they are

224 loc. cit., 117. For informative analyses of Husserl's constitution of harmony in experience by rules of synthesis, see Byers, 2002.

225 loc. cit., 232.

226 loc. cit., 233.

227 loc. cit., 234.

228 loc. cit., 234. This inference would become especially strained in the case of quantum mechanical atoms.

229 loc. cit., 286.

> not actually ultimate units and turn out to be morphological, which then naturally refer back to absolute sub-units."[230]

Husserl grants that a resolution of the world's structure into absolutely indivisible units is "conceivable," but believes that our experience of repeatedly finding smaller units speaks against it. In his later work on the "origin of the spatiality of nature," Husserl is still pondering the great problem of a physical science of nature operating in astronomical infinity and "the problem of an inner infinity, the infinity of the continuum and the way to atomize or quantify—atomic physics—in the open endlessness of infinity."[231] This brings us to the question of whether Husserl agreed with Becker's claim that the idealizations of mathematics cannot be justified apart from their applications to physics.

We have seen that Husserl once envisaged a blanket solution to the puzzle of how mathematical systems apply so precisely to real domains: an idealized supramathematics of all possible such systems could be expected to provide a formal system to fit any domain we can describe precisely. In a further idealization these formal axiom systems are assumed to be complete in order to comprise a "*nomological science* ... of an *infinite province* ... just as, in the *ideal Euclid*, there is no truth about space that is not deducibly included in the 'complete' system of space axioms."[232] We saw that Husserl did glimpse an elementary core of Hilbert's axioms for the real ordered field for which, he said, "it can be read off of each proposition whether it is or not a consequence of the axiom system." But when he now says that in adding a separate completeness axiom Hilbert intended a definite manifold, "even if the inmost motives that guided him mathematically were inexplicit," Husserl again ignores the one motive Hilbert explicitly stated for his axiom system, namely, that it relieved him of having "to imagine, say, the totality of all possible laws according to which the elements of a fundamental sequence can proceed." This was his way of avoiding the gulf between the countability of finitely expressed laws and the uncountability of the domain of reals, a quintessential idealization of classical mathematics.[233] To appreciate the fate of Husserl's ideal of

---

230 Husserl 2012, 284–285.

231 Husserl 1981, 229.

232 Husserl 1969, 96.

233 Lohmar 2000, 144, suggests that Cantor's diagonal argument for the uncountability of the reals may transcend the justifiable limits of idealization, but Husserl's failure to even mention, let alone analyze, this proof leaves us unable to evaluate this suggestion. Von Atten, 2011, argues pointedly that for an accommodation of classical mathematics in Husserl's logic, "general considerations about intentionality, meanings, essences, idealizations, and perhaps, non-revisionists will not do: what is still wanting is a concrete and detailed phenomenological foundation of even just one characteristically classical alleged truth, such as the existence of the power set of N."(860) Again, Husserl never mentions Zermelo's axiom. But *were* he to say that, on seeing there are infinitely many subsets of N on the horizons of consciousness, he forms the power set of N by appeal to the transcendental formation of the plural thesis in synthetic consciousness, I do not see how we press the constructionist objection against him without begging the question, especially in view of his denial that consciousness is a mathematical manifold.

a complete system, we can best look at his penetrating analysis of other logical idealizations that underlie classical mathematics, in particular, "the presupposition that every judgement can be decided."

Husserl denies that "a claim to truth" is part of the essence of a judgement, since "it is not necessary for the judger to co-objectivate truth, whether intuitively or emptily."[234] People miss this because they do not distinguish two senses of the word assertion. The usual emphatic sense is where one says: "I vouch for that, it is true; one can legitimate it any time by an adequation." But in the usual narrower sense judgement "is non-modalized, categorial certainty—and not *per se* a 'having-convinced-oneself' by any witnesses or testimonies, in particular, the ultimately decisive ones: the affairs themselves."[235] It is precisely the narrower non-modalized sense in which Hilbert asserted in 1900 that all well posed mathematical problems are solvable. He could not vouch for this or offer an adequation; he motivated his belief by reminding his audience of one important reason many mathematicians were coming to share his optimism, namely, the number of difficult old problems that were finally being solved by proofs of the impossibility of the positive solutions that had long been sought for them. Such proofs had only succeeded by making precise the "incorrect sense" in which the problem had initially been taken. Hilbert's optimism, however, pales before Husserl's belief that: "Thanks to evidence, the life of consciousness has an *all-pervasive teleological structure*, a pointedness towards 'reason' ... towards the discovery of correctness ... and towards the canceling of incorrectness."[236] Hilbert also exemplifies Husserl's diagnosis that

> because the scientist, living in the will to cognize, strives to reach a decision regarding the correctness of every judgement not yet decided by evidence ... a *fundamental conviction already awaits* the logician and logic in the state of positivity, the unspoken conviction that guides every scientist in his province: his settled belief in *truth-in-itself* and *falsity-in-itself*.[237]

It should follow that "in itself every judgement is decided" and hence that either truth or falsity "belong" to its essence, despite not being a "constituent mark" of it, which Husserl finds "very remarkable."

To expose the "astonishing *a priori*" behind this intrinsic decidability Husserl argues that

> as logicians we stand on the certainty of available identical judgements. But now they are supposed to be 'decided in themselves.' This surely signifies: by a '*method*,' by a course of *cognitive thinking*, a course existing in itself and intrinsically pursuable, which leads immediately or mediately to an adequation, a making evident of either the truth or the falsity of any judgement.[238]

---

[234]loc. cit., 196.

[235]loc. cit., 196.

[236]loc. cit., 160.

[237]loc. cit., 197.

[238]loc. cit., 197.

But Husserl's notion of 'method' is too vague and indeterminate to arouse astonishment. Hilbert was surely *not* claiming any 'method' for solving all his problems nor even a separate method for solving each of them individually. Indeed, his tenth problem on the solvability of diophantine equations is precisely "to devise a method according to which it can be decided *in a finite number of operations* whether the equation is solvable in rational integers."[239] In 1917 Hilbert distinguished the "problem of the *solvability in principle of every mathematical question*" from that of "the *decidability* of a mathematical question in a finite number of operations," which he says "goes to the essence of mathematical thought."[240] In 1926, as we have seen, he flat out denied that a "general method" for solving every mathematical problem existed, by which he may well have intended that his *Entscheidungsproblem* had no positive solution, which truly *would* have been astonishing.[241] Meanwhile, as solutions to his Paris problems mounted, Hilbert proposed in 1928 another list of problems stemming from his maturing proof theory, which he again asserted were all solvable and which provide a good context for focusing Husserl's idealization of a complete deductive system.

To clarify the much mentioned problem of the bearing of Gödel's theorem on Husserl's theory of complete systems of arithmetic, let us look at the Hilbert problem that Gödel solved. Hilbert poses the problem by observing that:

> It is generally maintained that the axiom system for number theory ... is complete. However, the usual argument with which one shows that any two realizations of the axiom for number theory ... must be isomorphic, do not satisfy the demands of finitistic rigor. What is at issue here is to transform the usual proof of isomorphism finitistically ... so that the following is thereby demonstrated: If for a proposition S the consistency with the axioms of number theory can be proven, then the consistency with those axioms cannot also be proven for ∼S.[242]

Gödel constructed self-referring sentences G for which both G and ∼G were consistent with the arithmetic axioms, a result which can be plausibly stated in Husserl's terms: If one takes his "absolutely definite" systems to be those whose models are isomorphic, and his "relatively definite" systems to be our syntactically complete ones, then Gödel has shown that Husserl's absolutely definite systems cannot be transformed in this way into relatively definite systems. This is hardly

---

[239] It took 75 years to prove the impossibility of such a method, half of which was needed just to make the idea of such a 'finite method' precise enough for such a proof.

[240] Hilbert 1918, 1113.

[241] Von Neumann 1927 asserted without proof the undecidability of mathematics generally and first-order logic, remarking that "On the day that this undecidability ceases, mathematics as we know it today would cease to exist; an absolute mechanical rule would take its place, with whose help anyone could decide of any given statement, whether or not it can be proved. Thus we must hold the view that it is in general undecidable whether a given normal formula is provable or not." (10) One should keep in mind that the undecidability of first-order logic implies the incompleteness of both first-order arithmetic and second-order logic.

[242] Hilbert 1928, 211

a result that would undermine Husserl's case for the applicability of his formal systems to nature.

Paradoxes of logic, set theory, and physics persuaded him, however, that the basic concepts of these domains lack the originary clarity needed for phenomenology of nature. For, "the truth is that sciences that have paradoxes, that operate with fundamental concepts not produced by the work of originary clarification and criticism, are not sciences at all, but with all their ingenious performances, mere theoretical techniques."[243] For many the success of Zermelo's axioms in dissolving the paradoxes of Russell and Cantor was now clear. But was it clear to Husserl? In 1923 he published the paper on the paradoxes he recommended to Weyl by his student Hans Lipps, which begins by admitting that: "Of course an axiomatization has been assembled by Zermelo that excludes the formation of paradoxical sets, and which suffices for the derivation of the mathematically important propositions of set theory. Thereby the paradoxes in question are only evaded. The root of their dialectical illusion remains unexposed."[244] Husserl may well have still agreed with this assessment. In any case, he soon makes the stronger claim that, due to a failure to uncover intentional horizons, "we have, and precisely at the high level attained by modern positive sciences, problems of fundamentals, paradoxes, unintelligibilities."[245] But an important source of these difficulties is the very lack of agreement between logicians, mathematicians, and physicists about paradoxes in the foundations of their disciplines, which tempts Husserl to dream of a "communalization of constitutive intentionality." This would establish a transcendental intersubjectivity whereupon: "*the constitution of the world essentially involves a 'harmony' of the monads*: precisely this harmony among particular constitutions in the particular monads; and accordingly it involves also a harmonious generation that goes on in each particular monad."[246] But the price for such harmony would be truly paradoxical and Husserl eventually returns to earth, seeking to ground the sciences in our life world, which brings us back to where we started.

However grave a crisis of science Husserl envisioned in his last work, he does satisfy himself that one in-house crisis of phenomenology resulted from an irrational deviation from his own doctrine. He begins on his first page by warning that philosophy itself "in our time threatens to succumb to skepticism, irrationalism, and mysticism."[247] And considering his essay on the origin of geometry as part of his text, he concludes on his last page that

> we can now recognize from all this that historicism, which wishes to clarify the historical or epistemological essence of mathematics from the standpoint of the magical circumstances ... of a time-bound civilization, is mistaken in principle. For romantic spirits the mythical-magical elements of the historical and pre-historical aspects of mathematics may be

---

[243] Husserl 1969, 181

[244] Lipps 1976, 158

[245] Husserl 1960, 153.

[246] loc. cit., 107.

[247] *CoESP*, 1.

> particularly attractive; but to cling to this merely historical factual aspect of mathematics is precisely to lose oneself to a sort of romanticism and to overlook the genuine problem the internal-historical problem, the epistemological one.[248]

This could stand for an open protest against Becker's mantic phenomenology.

At the heart of his text Husserl considers "the possibility of a hidden truth in Kant's transcendental philosophy" which he represents as a hidden dimension that cannot be seen by those bound to the "life of the plane." To focus what this analogy leaves unanswered, Husserl raises three rather remarkable questions:

> How could the development of the positive sciences purely upon the 'plane' appear for so long in the form of a superabundant success? Why was it so late before, in the need for complete transparency in its methodological accomplishments, the difficulties, indeed incomprehensibilities, announced themselves, such that not even the most painstaking construction of logical technique could improve the situation? Why did the later attempts at an 'intuitionistic' deepening, which in fact touched upon the higher dimension, and all efforts to clarify the situation in this way, not lead to unanimously accepted, truly compelling scientific results?[249]

Husserl was thinking of his own attempts, emboldened by Weyl, Brouwer, and Becker, to formulate what he called "phenomenological intuitionism," but which, when confronted with the success of Hilbert's transfinite mathematics in quantum mechanics, Becker perverted into mysticism.

Husserl had already in 1912 cautioned, in connection with foundational problems being passed over in "the century of technology," that philosophical intuitionism "had reacted with complete justification against the one-sided surrender of mankind to the expansion of the sciences as techniques of thought.... But, of course, Intuitionism must not degenerate into mysticism instead of approaching sober tasks that are forthcoming from the situation described."[250] By 1922 we saw that Husserl had reason to hope that, on the basis of Becker's work, his phenomenology provided the philosophical justification of the intuitionism of Brouwer and Weyl that enabled it to ground *a priori* modern mathematics and relativity theory. But then Weyl melodramatically announced a crisis posed by set-theoretic paradoxes and a revolution needed to overcome it, which Hilbert immediately branded as an artificially motivated attempted coup. The influence on Husserl of this apparently fundamental disagreement, which he examined with some care, can be seen in his lecture of 1923, in which he admits that foundational research had as yet attained its "ideal of exactitude" only in the direction of deductive theorizing:

> The ardent pains taken to provide here authentic foundations, leads to known attempts at revolutionary reform, which are supposed to have profound changes in the content of science itself as a consequence. We think here of the conflict over the reshaping of geometrical and basic physical principles and Einstein's theory of relativity, and also of the strife over the reconstruction of the basic concepts of pure analysis (set, number, continuum) and of the revolutionary theories of Brouwer and Weyl. This state of affairs is the most

[248]loc. cit., 378.

[249]loc. cit., 119–120.

[250]Husserl 1980, 83.

> paradoxical insofar as, for all that, it is still a question of axiomatics, which pretends to be consulting apodictic evidence. And, in fact, who could distance himself from, e.g. the evidence of geometrical axioms? And yet subsequent scientific reflective critique requires new foundations.[251]

It seems then that, when he wrote to Weyl, Husserl may have regarded their alleged revolution as more of a paradox than a response to a crisis. In any case, he had by 1936 dissolved to his satisfaction the crisis of phenomenology invoked by Becker.

Granted, Becker did clearly recognize the challenge that quantum mechanics undoubtedly posed for any kind of phenomenological insight, but I will give Husserl the last intriguing word on this:

> All natural being in the world is in itself decided, and as nature has universally the sense of being a definite manifold, so is every individual definite. If in subatomic processes unambiguous calculability does not exist in the old sense, nevertheless everything there is still regulated, namely, according to mathematical probabilities. So great a revolution as this signifies for physics, philosophically it makes no essential difference, nor figures generally into philosophical formulations of problems.[252]

Husserl was unaware, however, that to solve the measurement problem for quantum mechanics, physicists had felt the need to consider the role of consciousness in observing the results of measurement, some of whom even turned to Husserl's writings for enlightenment about consciousness. Now philosophers are exploring this possibility.[253,254]

References

O. Becker, *Mathematische Existenz*, 2nd unrevised edn. (Max Niemeyer, Tübingen, 1927)

O. Becker, Das Symbolische in der Mathematik. Blätter für deutsche Philosophie **1**, 329–348 (1928)

O. Becker, *Grundlagen der Mathematik in geschichtlicher Entwicklung* (Suhrkamp, Frankfurt, 1975)

B. Bolzano, *Paradoxes of the Infinite* (1851), in Ewald 1996, vol 1, pp. 249–292

J. Burgess, Frege and Arbitrary Functions, in *Frege's Philosophy of Mathematics*, ed. by W. Demopoulous (Harvard University Press, Cambridge, 1995)

D. Byers, *Intentionality and Transcendence* (University of Wisconsin Press, Wisconsin, 2002)

G. Cantor, *Gesammelte Abhandlungen*, ed. by E. Zermelo (Springer, Berlin, 1932)

---

251 Husserl 2002, 297–298.

252 Husserl 1992, 175. In *CoESP* itself Husserl says rather that, "In principle nothing is changed by the supposedly philosophically revolutionary critique of the 'classical law of causality' made by recent atomic physics. For in spite of all that is new, what is essential in principle, it seems to me, remains: namely, nature, which is in itself mathematical; it is given in formulae, and it can be interpreted only in terms of the formulae." (53)

253 See French 2002.

254 I am indebted to Stefania Centrone, without whose instigation, advice, and patience I would not have written this paper.

S. Centrone, *Logic and Philosophy of Mathematics in the Early Husserl* (Springer, Dordrecht, 2010)
S. Centrone, Reply to Mark von Atten. New Yearb Phenomenol Phenomenol Philos **12**, 377–384 (2012)
A. Church, The Richard's Paradox. *American Mathematical Monthly***41**, 356–361 (1934)
J. Da Silva, On the nature of mathematical knowledge, in *Brazilian Studies in the Philosophy and History of Science*, ed. by D. Krause, A. Videira (Springer, Dordrecht, 2010), pp. 151–160
J. Da Silva, *Mathematics and the Crisis in Science*, in Hill, Da Silva 2013, pp. 345–366
M. Davis (ed.), *The Undecidable* (Raven Press, New York, 1965)
J. Dauben, *Georg Cantor: His Mathematics and Philosophy of the Infinite* (Princeton University Press, Princeton, 1979)
M. Dummett, *Frege. Philosophy of Mathematics* (Harvard University Press, Harvard, 1991)
T. Elsenhans, Phänomenologie, Psychologie, Erkenntnistheorie. *Kant-Studien***20, 224**, –275 (1915)
W. Ewald, *From Kant to Hilbert. A Source Book in the Foundations of Mathematics* (Oxford University Press, Oxford, 1996)
S. French, A phenomenological solution to the measurement problem? Husserl and the foundations of quantum mechanics. Stud. Hist. Philos. Mod. Phys. **33**, 467–491 (2002)
Y. Gauthier, Husserl and the Theory of Multiplicities, in *Husserl and the Sciences*, ed. by R. Feist (University of Ottawa Press, Ottawa 2004)
C. Gethmann, Hermeneutic Phenomenology and Logical Intuitionism: On Oskar Becker's *Mathematische Existenz*. New Yearb Phenomenol Phenomenol Philos **3**, 145–160 (2003)
J. Gray, *The Hilbert Challenge* (Oxford University Press, Oxford, 2000)
M. Hartimo, Husserl's pluralistic phenomenology of mathematics. Philos. Math. **20**, 86–110 (2012)
M. Hartimo, Husserl's scientific context 1917–1938. New Yearb. Phenomenol. Phenomenol. Philos. (forthcoming)
H. Hemholtz, *Epistemological Writings* (Reidel, Dordrecht, 1977)
D. Hilbert, *Mathematical Problems* (1900a), in Gray 2000, pp. 240–282
D. Hilbert, *On the Concept of Number* (1900b), in Ewald 1996a, pp. 1092–1095
D. Hilbert, *The Foundations of Geometry* (Open Court, La Salle, 1971)
D. Hilbert, *The Theory of Algebraic Invariants* (Cambridge University Press, Cambridge, 1993)
D. Hilbert, *Lectures on the Foundations of Geometry 1891–1902*, ed. by M. Hallett, U. Majer (Springer, Heidelberg/Berlin, 2004)
D. Hilbert, *Axiomatic Thought* (1918), in Ewald 1996b, pp. 1107–1115
D. Hilbert, *The New Grounding of Mathematics* (1922), in Ewald 1996b, pp. 1117–1134
D. Hilbert, *On the Infinite* (1926), in van Heijenoort 1967, pp. 369–392
D. Hilbert, *Problems of the Grounding of Mathematics* (1928), in Mancosu 1998, pp. 227–233
D. Hilbert, W. Ackermann, *Grundzüge der theoretische Logik* (Springer, Berlin, 1928)
D. Hilbert, L. Nordheim, J. von Neumann, Über die Grundlagen der Quantenmechanik. Math. Ann. **98**, 1–30 (1928)
C. Hill, *Tackling Three of Frege's Problems* (2002), in Hill and da Silva 2013, pp. 195–218
C. Hill, J. Da Silva, *The Road Not Taken, On Husserl's Philosophy of Logic and Mathematics* (College Publications, London, 2013)
E. Husserl, *Husserliana, Gesammelte Werke*, I–XXVI: Martinus Nijhoff, Den Haag 1950+, XXVII–XXXVII: Kluwer, Dordrecht, 1989; XXXVIII: Springer, New York, 2005–
E. Husserl, *Phenomenologische Psychologie. Vorlesungen Sommersemester 1925,* Husserliana IX, 1968
E. Husserl, *Studien zur Arithmetik und Geometrie. Texte aus dem Nachlass 1886–1901,* Husserliana XXI, 1983
E. Husserl, *Aufsätze und Vorträge (1911–1921),* Husserliana XXV, 1986
E. Husserl, *Die Krisis der europäischen Wissenschaften und die transzendentale Phenomenologie: Ergänzungsbände. Texte aus dem Nachlass (1908–1921),* Husserliana XXIX, 1992
E. Husserl, *Einleitung in die Philosophie. Vorlesungen 1922/23,* Husserliana XXXV, 2002

E. Husserl, *Transzendentaler Idealismus. Texte aus dem Nachlass (1908–1921),* Husserliana XXXVI, 2003
E. Husserl, *Zur Lehre vom Wesen und zur Methode der eidetischen Variation. Texte aus dem Nachlass (1891–1935),* Husserliana XLI, 2012
E. Husserl, *Logik. Vorlesung 1896*, E. Schuhmann (ed.), *Husserliana*, Materialienbände 1 (Kluwer, Dordrecht, 2001a)
E. Husserl, *Logik. Vorlesung 1902/03*, E. Schuhmann (ed.), *Husserliana*, Materialienbände 2 (Kluwer, Dordrecht, 2001b)
E. Husserl, *Einführung in die Phenomenologie der Erkenntnis. Vorlesung 1909*, E. Schuhmann (ed.), *Husserliana*, Materialienbände 7 (Kluwer, Dordrecht, 2007)
E. Husserl, *Early Writings in the Philosophy of Logic and Mathematics*, translated by D. Willard (Kluwer, Dordrecht, 1994)
E. Husserl, [PoA] *Philosophy of Arithmetic. Psychological and Logical Investigations with Supplementary Texts from 1887–1901*, translated by D. Willard (Kluwer, Dordrecht, 2003)
E. Husserl, *Cartesian Meditations,* translated by D. Cairns (Martinus Nijhoff, The Hague, 1960)
E. Husserl, *Formal and Transcendental Logic*, translated by D. Cairns (Martinus Nijhoff, The Hague, 1969)
E. Husserl, *Shorter Works*, ed. by P. McCormick, F. Elliston (University of Notre Dame Press, Notre Dame, 1981)
E. Husserl, *Logical Investigations*, translated by J. Findlay (Humanities Press, New York, 1970)
E. Husserl, [*CoESP*] *The Crisis of European Sciences and Transcendental Phenomenology*, translated by D. Carr (Northwestern University Press, Evanston, 1970)
E. Husserl, *Phenomenology and the Foundations of the Sciences. Third Book of Ideas,* translated by T. Klein and W. Pohl (Martinus Nijhoff, The Hague 1980)
E. Husserl, *The Basic Problems of Phenomenology: From the Lectures, Winter Semester 1910–1911*, translated by I. Farin and J. Hart (Springer, Berlin, 2006)
E. Husserl, *Introduction to Logic and Theory of Knowledge. Lectures 1906/07*, translated by C.O. Hill (Springer, Dordrecht, 2008)
E. Husserl, *Ideas for a Pure Phenomenology and Phenomenological Philosophy,* translated by D. Dahlstrom (Hackett, Indianapolis/Cambridge, 2014)
C. Ierna, *Der Durchgang durch das Unmögliche*. An Unpublished Manuscript from the Husserl Archives, *Husserl Studies* 27 (2011), pp. 217–226
P. Jordan, Early years of quantum mechanics: some reminiscences, in *The Physicist's Conception of Nature*, ed. by J. Mehra (Reidel, Boston, 1973)
T. Kisiel, *The Genesis of Heidegger's 'Being and Time'* (University of California Press, Berkeley, 1993)
L. Kronecker, Über den Zahlbegriff. J. Reine Angew. Math. **101**, 337–355 (1887)
G. Leibniz, *New Essays on Human Understanding*, translated by P. Remnant and J. Bennett (Cambridge University Press, Cambridge, 1981)
H. Lipps, *Die Verbindlichkeit der Sprache* (Vittorio Klostermann, Frankfurt am Main, 1976)
D. Lohmar, *Edmund Husserl's'Formal and Transzendental Logik* (Darmstadt, 2000)
E. Mach, *Popular Scientific Lectures*, translated by T. McCormack (Open Curt, Chicago, 1895)
D. Mahnke, From Hilbert to Husserl: First Introduction to Phenomenology, especially that of Formal Mathematics. Stud. Hist. Phil. Sci. **8**, 71–84 (1977)
P. Mancosu, *From Brouwer to Hilbert* (Oxford University Press, Oxford, 1998)
P. Mancosu, T. Ryckman, Mathematics and phenomenology: the correspondence between O. Becker and H. Weyl. Philos. Math. **10**, 130–202 (2002)
I. Mehra, H. Rechenberg, *The Historical Development of Quantum Theory*, vol V (Springer, New York, 1987)
V. Pambuccian, Review of Hilbert 2004. Philosophia Mathematica **21,** 255–277 (2013)
V. Peckhaus, *Hilbertprogramm und Kritische Philosophie* (Vandenhoeck & Ruprecht, Göttingen, 1990)
V. Peckhaus, R. Kahle, Hilbert's Paradox. Hist. Math. **29**, 157–175 (2002)

O. Poggeler, Hermeneutische und mantische Phenomenologie, in *Heidegger. Perspektiven zur Deutung seines Werks*, ed. by O. Poggeler (Kiepenheuer & Witsch, Köln, Berlin, 1970)
H. PoincarÉ, *Mathematics and Science. Last Essays* (Dover, New York, 1963)
L. Pyenson, Relativity in late Wilhelmian Germany: The appeal to a pre-established harmony between mathematics and physics. Arch. Hist. Exact Sci. **27**, 137–155 (1982)
J. Richard, *Sur la Philosophie des Mathematiques* (Gauthier-Villars, Paris, 1903)
G. Rosado-Haddock, Husserl's philosophy of mathematics: Its origin and relevance. Husserl Stud. **22**, 193–222 (2006)
G. Rosado-Haddock, Platonism, phenomenology, and interderivability, in *Phenomenology and Mathematics*, ed. by M. Hortimo (Springer, Dordrecht, 2010)
G. Rota, The phenomenology of mathematical proof. Synthese **111**, 183–196 (1997)
A. Schirrmacher, Planting in his neighbor's garden: David Hilbert and early Göttingen quantum theory. Perspect. Phys. **5**, 5–20 (2003)
R. Schmit, *Husserl's Philosophie der Mathematik* (Bouvier, Bonn, 1981)
E. Scholz, Weyl's contribution to geometry, 1917–1923. Intersect. Hist. Math. Hist. Stud. **15**, 203–230 (1994)
E. Schrödinger, Über die Unanwendbarkeit der Geometrie im Kleinen. Naturwissenschaften **22**, 518–520 (1934)
K. Schulmann, *Husserl Chronik* (Martinus Nijhoff, The Hague, 1977)
S. Simpson, Ordinal numbers and the Hilbert basis theorem. J. Symb. Log. **53**, 961–974 (1988)
A. Tarski, *A Decision Method for Elementary Algebra and Geometry* (University of California Press, Berkeley, 1951)
R. Thiele, Hilbert's twenty-fourth problem. Am. Math. Mon. **110**, 1–24 (2003)
M. Toepell, 'Zur Schlüsselrolle Friedrich Schurs bei der Entstehung von David Hilberts *Grundlagen der Geometrie*', in *Mathemata. Festschrift für Helmut Gericke*, ed. by M. Folkerts, U. Lindgren (Franz Steiner, Stuttgart, 1985) pp. 637–649
T. Tonietti, Four letters of E. Husserl to H. Weyl and their context, in *Exact Sciences and Their Philisophical Foundations*, ed. by W. Deppert, 1988
M. van Atten, Construction and constitution in mathematics. New Yearb. Phenomenol. Phenomenol. Philos. **10**, 43–90 (2010)
M. van Atten, Stefania Centrone, *Logic and Philosophy of Mathematics in the Early Husserl*. New Yearb. Phenomenol. Phenomenol. Philos. **12**, 377–383 (2012)
D. van Dalen, Four letters of Edmund Husserl to Hermann Weyl. Husserl Stud. **1**, 1–12 (1984)
J. van Heijenoort, *From Frege to Gödel. A Source Book in Mathematical Logic* (Harvard University Press, Harvard, 1967)
J. von Neumann, Zur Hilbertschen Beweistheorie. Math. Z. **26**, 1–29 (1927)
H. Weyl, Über die Definitionen der mathematischen Grundbegriffe. Mathematisch-naturwissenschaftliche Blätter **7**, 93–95 (1910.) 109–113
H. Weyl, *Das Kontinuum,* Veit & Co., Leipzig 1918. English translation: *The Continuum: A Critical Examination of the Foundation of Analysis*, translated by S. Pollard and T. Bole (Thomas Jefferson University Press, Kirksville, 1987)
H. Weyl, *On the New Foundational Crisis of Mathematics* (1921), in Mancosu 1998a, pp. 86–118
H. Weyl, *The Current Epistemological Situation in Mathematics* (1925), in Mancosu 1998b, pp. 123–142
H. Weyl, *Comments on Hilbert's Second Lecture on the Foundations of Mathematics* (1927), in van Heijenoort 1967, pp. 480–484
H. Weyl, Consistency in mathematics. Rice. Inst. Pam. **16**, 245–265 (1929)
H. Weyl, David Hilbert and his mathematical work. Bull. Am. Math. Soc. **50**, 612–654 (1944)
H. Weyl, *Philosophy of Mathematics and the Natural Sciences* (Princeton University Press, Princeton, 1949)
E. Zermelo, Collected Works, vol. I, ed. by H. Ebbinghaus, A. Kanomori (Springer, Berlin, 2010)

Chapter 15
Husserl and Carnap on Regions and Formal Categories

Ansten Klev

Abstract Husserl, in his doctrine of categories, distinguishes what he calls regions from what he calls formal categories. The former are most general domains, while the latter are topic-neutral concepts that apply across all domains. Husserl's understanding of these notions of category is here discussed in detail. It is, moreover, argued that similar notions of category may be recognized in Carnap's *Der logische Aufbau der Welt*.

Keywords Categories • Husserl • Carnap

As conceived of in the *Ideas*,[1] phenomenology is what Husserl calls an eidetic science: it deals not with matters of fact, but with *Wesen* or *eidē*, what in the English literature on Husserl are usually called essences. Clarification of the nature of phenomenology therefore requires clarification of what is understood by such essences, something Husserl sets out to provide in the first section of the *Ideas*, constructively in its first chapter (§§ 1–17) and critically in its second chapter (§§ 18–26). In the course of that constructive clarification Husserl introduces the distinction between regions and formal categories. The distinction is meant to capture the different senses that the word 'category' has when one speaks on the one hand about the *categories* of the physical and the mental and on the other hand about the *categories* of individual and property: the former are most general domains, most general topics, while the latter are topic-neutral concepts that apply across all domains. In the words

[1]That is the abbreviation used here for *Ideen zu einer reinen Phänomenologie und phänomenologische Philosophie* (Husserl 1913). References to this work are of the form Ideen + paragraph number. References to the two other books of the *Ideas* (Husserl 1952a,b), not published during Husserl's lifetime, are of the form Ideen + book number + paragraph number. References to the *Logische Untersuchungen* (Husserl 1901) are of the form LU + investigation number + paragraph number. Its first volume, *Prolegomena zur reinen Logik* (Husserl 1900), is abbreviated *Prolegomena*.

A. Klev (✉)
Institute of Philosophy, Academy of the Sciences of the Czech Republic, Prague, Czech Republic
e-mail: anstenklev@gmail.com

S. Centrone (ed.), *Essays on Husserl's Logic and Philosophy of Mathematics*, Synthese Library 384, DOI 10.1007/978-94-024-1132-4_15

of Ryle, who coined the term 'topic-neutral',[2] regions may be said to provide "the fat and the lean," and formal categories the "joints and tendons," of thought. In the following pages I wish to take a closer look at this distinction, concentrating on how Husserl understood it around the time of the *Ideas* and on how Rudolf Carnap in his *Der logische Aufbau der Welt* (Carnap 1928) may have understood it.

To deal with categories in an anthology on the philosophy of logic and mathematics in Husserl is justified not only on the grounds that Husserl appears to have regarded the general theory of categories as belonging to logic,[3] but also on the grounds of the close ties that have existed between logic and the doctrine of categories throughout the history of philosophy. Although Aristotle may not have regarded the piece of writing now known as his *Categories* as belonging to logic, his ancient commentators since Andronicus of Rhodes in the first century BC did so regard it.[4] The reason given for this, for instance by the commentator Simplicius (sixth century AD), is that the *Categories* deals with the doctrine of terms, which—since syllogisms are composed of judgements, and judgements of terms—has to be taught as the first part of logic.[5] Kant's doctrine of categories falls under his so-called transcendental logic, and is intrinsically tied to the notion of judgement in general and to the so-called forms of judgement in traditional logic in particular.[6] That type theory, a cornerstone of modern logic, may be regarded as a doctrine of categories was noted already by Ryle (1938, 189).

I shall concentrate on the general theory of regions and formal categories, and hence avoid questions about the specific nature of the various regions and formal categories that come up for discussion in Husserl's work. Such questions would be proper to a different kind of study, dealing for instance with the notions of nature and spirit (*Geist*) in Husserl's work.[7] The distinction between regions and formal categories is not intrinsically tied up with phenomenology or Husserlian doctrine; indeed, I think it is of use to anyone reflecting on the notion of category. The second part of this paper will investigate how the distinction can be understood in the context of simple type theory. This is where Carnap's *Aufbau* enters the picture, since that work can be seen as suggesting one way in which regions can be mapped onto a simple type hierarchy.

Carnap attended Husserl's advanced seminar *Phänomenologische Übungungen für Fortgeschrittene* in the winter semester of 1923/1924,[8] and the influence of

---

[2]See Ryle (1954, 116).

[3]See the title of Ideen § 17 and also *Prolegomena* § 67. Husserl included a general discussion of regions and formal categories in his lectures on the theory of science; see Husserl (1996, 274–286).

[4]It was Andronicus who placed the *Categories* first in the *Organon* and the *Organon* first in the list of Aristotle's works; see e.g. Gottschalk (1990, 66).

[5]For a critical reading of Simplicius's argument, see Morrison (2005).

[6]For the background of Kant's table of judgements in traditional logic, see Tonelli (1966).

[7]A good place to begin such a study is the *Einleitung des Herasugebers* in Husserl (2001c).

[8]This is clear from Carnap's diaries of this period: on 21.11.1923 Carnap reports that Husserl has allowed him to participate in his seminar, meeting at 11.00 every Wednesday (cf. Schumann 1977, 273); after that date and until the end of February 1924 Carnap regularly mentions 'Husserl' on Wednesdays. What the topic of the seminar was, I do not know. On 13.11.1923 Carnap attended

Husserl on Carnap's dissertation, *Der Raum* (Carnap 1922), is obvious.[9] Husserl's influence on the *Aufbau* is more difficult to assess. Mayer (1991) argues for "numerous systematical and terminological parallels" between the *Aufbau* and the *Ideas*. According to Roy (2004), Carnap regarded the *Aufbau* as a realization of Husserl's idea of a *mathesis* of experience, an axiomatic counterpart to phenomenology (cf. Ideen §§ 71–75). Rosado Haddock (2008) likewise claims that Husserl's influence on the *Aufbau* was decisive. The understanding of an important aspect of Carnap's *Aufbau* in terms of a distinction from Husserl's *Ideas* suggested in the present paper can indeed be taken as an indication of the influence of the latter on the former. In the final section of this paper I shall nevertheless express some reservations against making too much of the proposed parallels between these works.

15.1 Husserl on Regions and Formal Categories

15.1.1 Region

It is a fundamental philosophical problem, according to Husserl, both to find out what regions there are (Ideen § 17) and to understand the nature of a given region and the interdependence of various regions on each other (ibid. § 152). Husserl is, as far as I have seen, never quite definite on what regions he takes there to be. It is clear that he regards physical or material nature and (pure) consciousness as two regions. An individual falling under the first of these is a physical thing and an individual falling under the second is an experience (*Erlebnis*), hence Husserl also speaks of these regions as the regions of physical thing and experience, respectively. Other regions Husserl mentions in the three books known as the *Ideas* include the region of the body (cf. Ideen III §§ 2–3), living nature (*animalische Natur*), and the region of society and culture (*die Geistige Welt*). But it is not clear from the discussion in those works whether these are all distinct, nor whether any of them encompasses several regions. Is the region of the body separate from the region of living nature? Persons, which Husserl distinguishes from living human beings, are individuals of the region of culture and society, as are institutions and nations, and also works of art; should not these be taken to belong to separate regions? It lies outside the scope of this paper to discuss these questions as well as the question of how the region of pure consciousness relates to the other regions.

Husserl's class on *Erste Philosophie*; Husserl's notes for the lecture that day (cf. Schumann 1977, 275) can be found in Husserl (1956, 44–51). In his diary Carnap writes 'nicht sehr gefallen'. Landgrebe reported in 1976 to Schumann that Carnap followed Husserl's seminars "SS 1924–SS 1925" (Schumann 1977, 281). By considering Carnap's diary entries on Wednesdays during that year—when, as in the previous years, the seminar met—one sees that this cannot be correct.

[9]For more details, see especially Stone (2010), but also Sarkar (2003) and Rosado Haddock (2008, ch. 1).

Before considering Husserl's rather technical definition of the notion of region, let us look at three important characteristics he takes regions to have. Especially important is the connection Husserl sees between regions and modes of original awareness of objects. For each region there is an original mode of awareness of objects of that region, a type of act in which items of that region are, as Husserl would say, "self-given." There is, in other words, for each region a mode of consciousness in which one has direct access to objects of that region. This connection is stated especially succinctly in *Experience and Judgement* § 4:

> every kind of object has its kind of being self-given.[10]

Similar formulations are found at many other places (e.g. Ideen §§ 1, 138). That Husserl assumes this connection is also clear from the individual discussions of the various regions, since he there typically states what is the mode of access to the region in question. We access the region of nature through ordinary perception (e.g. loc. cit.). Such perception is in fact the paradigmatic instance of a mode of original awareness, and other original modes of awareness may by analogy be called perception of objects of the kind in question. Being an original mode of awareness, perception contrasts with, for instance, remembrance and phantasy, in which the object is not present before one. The original mode of awareness of experiences, i.e., the objects of the region of pure consciousness, is phenomenological reflection (ibid. §§ 77–78). That this reflection is qualified as phenomenological means that it involves the phenomenological reduction; only when the reduction has been carried out can pure consciousness be disclosed to us (ibid. § 50). One becomes originally aware of the body in what Husserl calls bodily apprehension (Ideen II § 36; Ideen III § 2). The individuals in the region of the body are localized sensations, namely sensations localized in the body (Husserl calls such sensations "*Empfindnisse*"). In order to apprehend these sensations a special form of apprehension is required, since usually sensations are apprehended as things in the outer world.[11] I touch this table; I then speak of the sensations in my fingers on the basis of bodily sensations, but of the surface of the table on the basis of a perceptive apprehension. It is in the former sort of apprehension that I am aware of the body as a region. For the region of living beings, and that of society and culture, empathy (*Einfühlung*) plays an important role, but it is not clear to me whether Husserl regards empathy as an original mode of awareness; the question need not be settled here.

A region is, furthermore, said to prescribe a rule for how we may vary an individual of that region in imagination so that it still remains an individual, in other words, so that we still have a unitary course of experience of an individual (Ideen §§ 142, 149, 150; Ideen III § 7). Less general concepts, for instance the concept of a diamond, may also be regarded as prescribing a rule for the course of experience, for instance that the thing posited as a diamond does not bounce back in our hands in the manner of a bouncy ball when we let it fall to the ground. But we can still imagine a continuous transformation of the diamond into a bouncy ball without

[10] Husserl (1939, 12): "jede Art der Gegenständen hat ihre Art der Selbstgebung."

[11] This is the so-called content/apprehension scheme; see Klev (2013) for more details.

our experience falling apart into a series of disconnected appearances; in fact we can imagine the diamond being continuously transformed into any other physical thing, namely so long as we remain inside the region of material nature. Husserl's discussion of imaginative variation in the cited paragraphs relies on the region in question being that of nature, but he appears to have thought that the discussion would generalize to other regions.[12]

A third characteristic of regions concern their relation to sciences. A region gives rise to several "ontologies," that is, to eidetic sciences of concepts that compose the region (Ideen §§ 9, 16). In the case of nature, there is for instance geometry as the ontology of space, and there should likewise be ontologies of time (pure chronology), of movement (pure phoronomy), and of matter, since space, time, and matter are all involved in the constitution of physical objects. As a matter of fact, most ontologies have not been developed in any systematic fashion, but such developments are in principle possible. Indeed, ontologies are indispensable to the foundations of empirical sciences: any such science studies objects of some region, and so assumes the results of the ontologies associated with that region. Natural science, for instance, assumes the results of geometry and of all the other ontologies of nature (ibid. §§ 8, 9).

Let us now consider Husserl's technical definition of the notion of a region (Ideen I § 16):

> A region is nothing but the total highest genus-unity belonging to a concretum, that is, the essential unity of the highest genera to which the lowest differentiae within the concretum belongs.[13]

Several glosses are needed in order to make sense of this.[14] In particular, it must be clarified what is here meant by a highest genus, a lowest differentiae, a concretum, and by the belonging of lowest differentiae to a concretum. For the purposes of this paper we may think of Husserlian essences as objectified concepts. Essences, according to Husserl, fall into an ordering of genera and species (ibid. § 12); thus, one essence is a genus of another essence if it is more general than it; it is the species of another if it is more specific than it. That concepts, or terms, may be ordered in this way is a traditional idea, found already in Aristotle's *Topics* and often associated with the philosopher Porphyry (third century AD), who in his so-called *Isagoge* remarked that[15]

---

[12]The so-called method of eidetic variation is discussed in more detail by Husserl in lecture notes from 1925 (Husserl 1968, 69–87). Husserl notes there how the notion of genus is reached through such variation (ibid. 81–84).

[13]Husserl (1976, 36): "Region is nichts anderes als die gesamte zu einem Konkretum gehörige oberste Gattungseinheit, also die weseneinheitliche Verknüpfung der obersten Gattungen, die den niedersten Differenzen innerhalb des Konkretums zugehören."

[14]The only extended discussion of Husserl's definition of region of which I am aware is Stone (2000, 97–131). A brief discussion of Husserl's technical definition can also be found in Null (1989, 93–95).

[15]The translation is taken from Barnes (2003, 6).

> Substance is itself a genus. Under it is body, and under body animate body, under which is animal; under animal is rational animal, under which is man; and under man are Socrates and Plato and particular men.

Another, related, traditional doctrine in the logic of concepts is that of the content and extension of a concept: to the content of a concept belong all the various concepts that compose it, its marks (*Merkmale*); by the extension of a concept *C* one understood either all concepts of which *C* is a mark or all individuals falling under *C*.[16] Husserl takes over this doctrine as well. Thus, an essence is said to contain all essences that are marks of it (Ideen § 12); as we shall see in more detail below, the use of the verb 'to belong' in Husserl's definition of region refers to this notion of essence containment. Among the marks of an essence are its genera; hence an essence contains all of its genera.[17] Not all essences contained in an essence are genera of it, however; in Husserl's terminology, these are the *differentiae* of the essence.[18] Quite in line with the tradition, Husserl further distinguishes two notions of extension of an essence (Ideen § 13). The eidetical extension of an essence *g* is the collection of all the essences of which *g* is a genus. The other notion of extension—which Husserl does not give a name—is the collection of all possible instances—be they existent or not—of the essence, the collection of all the various "thisnesses" (*Diesheiten*) that instantiate the essence.[19] Thus, a particular human being instantiates the essence human being, and a particular green inhering in the cover of a particular copy of a book instantiates the essence green.

A lowest species in the ordering of genera and species is called an *eidetic singularity* by Husserl. Let *s* be an eidetic singularity falling under the essence *g*; calling *s* an eidetic *singularity* is motivated by the fact that any two instances of *s* are

---

[16]The doctrine is found, for instance, in the Port-Royal *Logique* (I.vii):

> J'appelle *comprehension* de l'idée, les attributs qu'elle enferme en soi, & qu'on ne lui peut ôter sans la détruire, comme la comprehension de l'idée du triangle enferme extension, figure, trois lignes, trois angles, & l'égalité de ces trois angles à deux droits, &c.
> J'appelle *étendue* de l'idée, les sujets à qui cette idée convenient, ce qu'on appelle les inferieurs d'un terme general, qui à leur égard est appelleé superieur, comme l'idée du triangle en general s'étend à toutes les diverses especes de triangles.

Here extension appears to be understood in the first sense. Leibniz considered extension also in the second sense; see Kauppi (1971). The doctrine seems to be have been much discussed by German logicians in the nineteenth century and was well known to Husserl; it is, for instance, taken for granted in Husserl (1891a) and it plays an important role in his 1896 logic lecture notes (Husserl 2001b).

[17]This doctrine can be found in Aristotle; see *Metaphysics* Δ 25, 1023^{b}24: "the genus is called a part of the species."

[18]In traditional terms, these are the so-called *constitutive*, rather than the *divisive*, differentiae of the essence in question. This distinction seems to have been introduced by Porphyry.

[19]Instead of 'thisness' Husserl more frequently (Ideen § 14) employs 'this-here' (*Dies-da*), which he takes as a translation of Aristotle's *tode ti*. 'Thisness' has the advantage of allowing a plural form. The distinction between thisnesses and essences is discussed in several research manuscripts from 1917/1918, in particular in Husserl (2001a, Texte Nr. 16–17, 2012, 112–154).

essentially identical g's.[20] For instance, if s is an eidetic singularity falling under the essence colour, then any two instances of s are essentially identical colours. An eidetic singularity s is either dependent or independent (LU III §§ 13, 21; Ideen § 15). It is dependent if there is another essence s', not contained in s, such that an instance of s cannot exist without an instance of s'.[21] For instance, the shape of this table cannot exist without some colour, that is without an instance of the essence colour. It is also true that this table itself cannot exist without some colour, but the essence colour is contained in the essence table, hence that fact does not make an eidetic singularity falling under the essence table dependent. An eidetic singularity is independent if it is not dependent. An independent essence is called a *concretum* and a dependent essence an *abstractum*.[22] An instance of an abstractum is thus what is often called a trope in the contemporary literature.[23] Husserl defines an *individual* to be the instance of a concretum (Ideen § 15).

An abstractum may be said to have several instances, differentiated by what they inhere in. Thus, a most specific shade of green has an instance in the cover of two books on my desk, and these instances, although they are essentially identical colours, differ in that they inhere in different books. Since an instance of a concretum does not inhere in anything else, it is unique: it cannot be differentiated from some other instance of the same concretum by reference to the individual it inheres in. Thus, to each individual there is associated a concretum that is unique to it, and which may be identified with the essence of that individual. Because of the potential infinity of individuals, there are also potentially infinite eidetic singularities. At this point Husserl is in disagreement with the tradition, since according to it the "division" of a concept into species should be dichotomous or at least finite,[24] while the potential infinity of lowest species requires that at least one division in the ordering be potentially infinite.

A differentia is itself an essence, hence it itself falls into an ordering of genera and species; whence it makes sense to speak of lowest differentiae contained in a concretum: these are differentiae that within their respective genus/species-orderings are lowest species. To each aspect of an individual there is a differentia contained in the concretum c that the individual instantiates. Under this differentia

---

[20]The assertion that there are concepts under which at most one object fall (which does not have to be an individual) can be found already in Husserl's 1896 logic lecture notes (Husserl 2001b, 124). On this topic, see also the texts cited in the previous Footnote 19.

[21]This definition is adapted from LU III §§ 13, 21. The requirement that s' not be contained in s is not stated by Husserl, but Simons (1982, 125) states something like it in his gloss on the definition of the related notion of foundation from LU III § 14. The definition given at Ideen § 15 is that s is dependent when it founds together with another essence s' "the unity of one essence"; I shall not discuss the relation of this definition to the ones found in LU III.

[22]This terminology was introduced in LU III § 17. For Husserl's use of the word 'abstract' see also LU II §§ 40–42.

[23]The use of the term 'trope' in this sense originates in Williams (1953); Stout (1923) had spoken about the same things as 'abstract particulars', in line with Husserl's use of the word 'abstract'.

[24]See the citations and references provided by Barnes (2003, 132–133).

d there falls a lowest species *s*, which also must be contained in the concretum *c*, since it is only by virtue of *c*'s containing *s* that it contains *d*. For instance, any physical thing is coloured, whence any concretum instantiated by a physical thing contains the essence colour; but since the physical thing is coloured only by virtue of having some specific colour, the concretum also contains an eidetic singularity falling under the essence colour. Thus a concretum may be thought of as a union of lowest differentiae, each such differentia corresponding to a specific characteristic of the individual that instantiates the concretum. Husserl appears to assume that if the essence *m* is a characteristic mark of the essence *s*, and the essence m' a genus of *m*, then m' is also a characteristic mark of *s*. In terms of essence containment that is to say that, if *m* is contained in *s* and if m' is a genus of *m*, then m' is contained in *s*. Hence, the highest genus of a lowest differentia contained in a concretum is itself contained in the concretum.

We are now better equipped for understanding Husserl's definition of region. It says that a region is the unification of all the highest genera contained in a concretum. A concretum contains all its characteristic marks; to each of these marks there is associated a highest genus, also contained in the concretum; the unification of all of these highest genera is a region. Husserl is unclear whether the unification of highest genera making up a region is itself an essence—that is, whether the region itself is an essence—but he appears to assume as much at least for the regions of nature and pure consciousness: the region of nature is associated with the essence of physical or material thing, while the region of pure consciousness is associated with the essence of experience (*Erlebnis*). If a region in general is itself an essence, then a region may simply be defined as the highest genus of a concretum, that is, as the highest genus of an independent eidetic singularity. In any event we can say that a region consists of several essences, each of which is the highest genus within their orderings, and each of which is a characteristic mark of that region.[25]

15.1.2 Formal Category

Let us now move to the notion of a formal category. As an intuitive description of this notion we may say that it is the form of an object. An individual differs in form from a property; a property differs in form from a state of affairs; a state of affairs differs in form from a set. Individual, property, state of affairs, and set may themselves be thought of as forms of object. According to Husserl they are formal categories (e.g. Ideen § 10). Formal categories are topic-neutral in the sense that they crosscut the regions, they apply across all regions; for instance, there are

[25]This idea is spelled out quite explicitly in a short research manuscript apparently written some time between 1924 and 1928 (Husserl 2012, 254): "Jedes Konkrete steht unter einer konkreten "Kategorie"—das ist die "Region." Jedes Abstrakte als reine Möglichkeit unter einer abstrakten Kategorie, unter einem reinen Wesensbegriff stehend, der Komponente ist der Region." From elsewhere in the manuscript it is clear that what is here called categories are highest genera.

individuals and properties in all of the regions. Husserl's official definition of the notion of formal category relies on the idea of a formal ontology. Formal ontology is the theory of topic-neutral concepts; in Husserl's words, it is the theory of the notion of an object in general (*Gegenstand überhaupt*). It studies, for instance, the part-whole relation and the formation of objects by means of categorial forms. A regional ontology is, in contrast to formal ontology, tied to a specific region. A regional ontology is the most general theory of objects falling under a region. Thus, the regional ontology of nature is the most general theory of physical, or material, objects; as already noted, it includes geometry and what Husserl calls phoronomy, the theory of movement. A formal category is defined as a primitive concept of formal ontology (*Prolegomena* § 67; Ideen § 10). A material or regional category is defined as a primitive concept of a regional ontology (LU III § 11; Ideen § 16). I mention this latter notion here only in order to exhibit the parallel between its definition and that of a formal category. In Husserl's discussions it features much less prominently than both the notion of region and the notion of formal category.

Husserl thinks of an object as a form-matter composite, its material elements being provided by the regions and its formal elements by the formal categories. The picture suggested in the *Ideas* is that regions consist of certain urelements (Husserl calls them *Urgegenständlichkeiten*), namely the individuals of that region, together with derivations of those urelements, obtained by means of the formal categories, these derivations being called syntactical objects (ibid. § 11).[26] The original mode of awareness associated with the regions, discussed in the previous subsection, are primarily modes of awareness of individuals, that is, of the urelements of the region. Thus, what we are aware of in ordinary perception are the individuals of nature. The syntactical, or categorially formed, objects of nature are not displayed in such acts. Husserl's doctrine of categorial intuition, developed in the sixth *Logical Investigation* (LU VI §§ 40–66), is meant to explain how we are originally aware of categorially formed objects.[27] There is, for instance, a way of forming, on the basis of a simple perception of a pine tree, a perception of the state of affairs that the pine tree is green (ibid. § 48). There is also a way of forming a perception of finite sets; indeed, the first part of the *Philosophy of Arithmetic* (Husserl 1891b) is mainly concerned with describing the nature of such set perception (see also LU VI § 51). Husserl also regarded so-called eidetic intuition—acts in which one perceives, or otherwise intuits, an essence—as an instance of categorial intuition (ibid. § 52); accordingly, he appears to have regarded essence as a formal category (cf. Ideen § 13).

Regions and formal categories are highly general concepts; but they are general in different ways. Husserl captures this difference with his distinction between generalization and formalization: generalization stands to regions as formalization stands to formal categories. In generalization one passes from an essence to a genus of it; in formalization one passes from an object to its formal category. Continued

---

[26] Note that the Greek word *suntaxis* simply means a putting together of certain elements, be they words or soldiers or whatnot.

[27] See Lohmar (2008) for a helpful discussion.

generalization leads to a highest genus, that is, either a region (provided a region is itself an essence), or a highest genus contained in a region. Thus, from yellow we pass to colour and from there, perhaps via some further steps, to sense quality, which according to Husserl is a highest genus (Ideen § 13). The reverse process of generalization is specialization. Continued specialization leads to an eidetic singularity, which may be a concretum or an abstractum. Thus, from colour we pass to blue, and from there, perhaps via some further steps, to a most specific shade of blue. In formalization (Ideen § 13; cf. LU III § 24), or what Husserl also calls pure categorial abstraction (LU VI § 60), we pass from an object to its formal category purely as a form or schema. In formalization we thus replace the material elements of the object by "empty forms" (*Leerformen*). Perhaps the best illustration of this process is the passage from a presentation of geometry in which the primitive terms have their intuitive, geometrical meaning to a presentation in the style of Hilbert (1899), in which the primitive terms are replaced by variables and the theory becomes "schematic."[28] (Carnap 1922, 7–8 obtains his notion of formal space from the space of traditional Euclidean geometry precisely by this process.) The reverse operation of formalization, in which the empty forms are "filled" again, Husserl calls de-formalization (*Entformalisierung*) or materialization (*Versachlichung*). In the geometrical case this corresponds to the passage from the schematic theory to a "model" of it.

Husserl also calls formal categories 'analytic categories' and material categories 'synthetic categories'. This terminology derives from his definition of analytic and synthetic laws and propositions (LU III §§ 11–12).[29] An analytic *law* is a true proposition (*Satz*) composed only of formal categories (or of concepts signifying formal categories), while a synthetic *law* is a true proposition whose only material elements are highest genera (or concepts signifying highest genera). An analytic *proposition* is one that results from an analytic law by materialization, that is, by filling the "empty forms" signifying formal categories with matter; while a synthetic *proposition* is one that results from a synthetic law by specialization of the highest genera signified in a synthetic law. Husserl gives as an example of an analytic law the proposition that the existence of a whole implies the existence of its parts; an analytic proposition derived from this law is that the existence of a particular house implies the existence of its roof. An example of a synthetic law would perhaps be that a colour cannot exist without an extension; a synthetic proposition derived from this law is that the brown of this table top cannot exist without the extension of

[28]Cf. the following remark of Hilbert from lecture notes dated 1894 (Hallett and Majer 2004, 104):

> Unsere Theorie liefert nur das Schema der Begriffe, die durch die unabänderliche Gesetze der Logik mit einander verknüpft sind. Es bleibt dem menschlichen Verstande überlassen, wie er dieses Schema auf die Erscheinung anwendet, wie er es mit Stoff anfüllt.

For more on the relation between Hilbert and Husserl, see Hartimo's contribution to this volume.

[29]Husserl made significant revisions in these paragraphs in the second edition of the *Logical Investigations*, the relevant part of which was published in the same year as the *Ideas*. For the definition of analytic and synthetic, see also Husserl (1996, 227–229).

the table top. Among the more well-known definitions of analyticity[30]—Kant's, Bolzano's, Frege's—Husserl's is perhaps closest to the latter (cf. Frege 1884, § 3), since both rest on the distinction between general logical laws and truths pertaining to a specific domain of knowledge, the notion of analyticity being related to the former and that of the synthetic to the latter. Husserl was dissatisfied with, and so tried to avoid, the terms 'a priori' and 'a posteriori' (cf. Ideen, Intro).

According to a traditional interpretation Aristotle's categories are highest genera.[31] Kant's categories, by contrast, are described as forms of thought (KrV B150, B305). Here we could thus speak of regions and formal categories respectively. It would be mistaken, however, to think that Husserl with his distinction of regions and formal categories synthesizes the doctrines of Aristotle and Kant. Firstly, none of Aristotle's categories could be regarded as Husserlian regions. In the greater scheme of things Aristotle's primary substances correspond to Husserlian individuals[32]; hence the Aristotelian category of substance splits into all the various Husserlian regions, since these are precisely the highest genera under which individuals fall. The Aristotelian category of quality would presumably be divided between the regions of nature and consciousness. Number is in the Aristotelian category of quantity (*Categories*, 4^b22), but for Husserl it is a formal category. Secondly, the role categories play in Kant's philosophy differs from the role formal categories play in Husserl's philosophy.[33] For Kant the categories are concepts of "pure synthesis" (KrV A78/B104). For Husserl, as well, the formal categories are concepts of synthesis. The synthesis corresponding to Husserl's formal categories, however, has a much narrower scope than the synthesis corresponding to Kant's categories. The latter is involved in any act; in order, for instance, to perceive objects at all the mind must synthesize a "manifold of intuition," namely by bringing it under the categories. According to Kant, the unity of an object does not lie in the object itself, to be extracted from it by perception, but is an "achievement of the understanding," by virtue of which the object has unity in the first place.[34] The synthesis corresponding to Husserl's formal categories, by contrast, is involved only in acts whose objects are higher-level, such as sets and states of affairs. The mind is in general not active in bringing about the unity of objects. The colour and the spatial form of this table, for instance, are not connected by the mind, but are given to it already connected. Likewise, the many "snapshots" I make of the table as I regard it from different sides are not synthesized in the way higher-level objects

[30] See Sundholm (2013) for a discussion of these.

[31] A critical discussion of this interpretation can be found in Klev (2014, 15–28).

[32] Stone (2000, 129) accepts this identification.

[33] On this point, see De Palma (2010).

[34] E.g. KrV B134–135: "Verbindung liegt aber nicht in den Gegenständen, und kann von ihnen nicht etwa durch Wahrnehmung entlehnt und in den Verstand dadurch allererst aufgenommen werden, sondern ist allein eine Verrichtung des Verstandes, der selbst nichts weiter ist, als das Vermögen, a priori zu verbinden, und das Mannigfaltige gegebener Vorstellungen unter Einheit der Apperzeption zu bringen." Cf. B129–130.

are synthesized from other objects. We may think of the latter as an active form of synthesis, while the former is passive (LU VI § 47; Ideen II § 9). Kant does not have this distinction between active and passive synthesis.[35] He therefore regards all experience as categorially formed; for Husserl, by contrast, only the experience of higher-level objects is categorially formed.

15.2 Regions in Type Theory: Carnap's *Aufbau*

By a simple type hierarchy I shall understand a hierarchy of types or domains of entities of the following design. At the bottom of the hierarchy there are one or more types of individuals. At the next level there is for each natural number $n > 0$ a type of n-ary relations (a 1-ary relation is a class). Then comes n-ary relations between relations of individuals, and between relations of individuals and individuals, and so on. Such a hierarchy may be recognized in the ideography of Frege's *Grundgesetze der Arithmetik* (Frege 1893), but Frege there described only its first few levels without giving a general definition. It seems to have been Carnap who first provided such a general definition in his logic textbook, *Abriss der Logistik* (Carnap 1929).[36] Following Frege as well as Russell and Whitehead, who in their *Principia Mathematica* (Russell and Whitehead 1910) had developed a more complicated type hierarchy nowadays known as a ramified type hierarchy, Carnap takes there to be only one domain of individuals, which we may denote by ι. The type of classes of individuals is denoted by (ι), that of binary relations of individuals by (ι, ι), the type of ternary relations of individuals by (ι, ι, ι), etc. The type of classes of classes of individuals is denoted by $((\iota))$, that of binary relations whose first place is a binary relation of individuals and whose second place is an individual is denoted by $((\iota, \iota), \iota)$, etc. We get the following inductive definition: ι is a type; and if $\tau_1, \ldots, \tau_n$ are types, then $(\tau_1, \ldots, \tau_n)$ is a type, namely the type of n-ary relations whose k-th place is of type τ_k.

It requires little imagination to see that a hierarchy of types offers a system of categories in the sense of general domains of entities. Indeed, from the point of view of modern logic, with its basic form of proposition $F(a)$—function F applied to argument a—simple types are the first domains of entities one sees, namely as the objectual correlates of the categories of symbols employed in that logic: individual symbols, namely individual variables and constants, perhaps of different sorts; unary functors of individual symbols; binary functors of individual symbols;

[35]Cf. the criticism of Kant found already in the *Philosophy of Arithmetic* (Husserl 1891b, 41):

> Kant übersah, daß viele inhaltliche Verbindungen uns gegeben sind, bei denen von einer synthetischen, die inhaltliche Verbundenheit schaffenden Tätigkeit nichts zu merken ist.

[36]Carnap's role in the dissemination of the simple type hierarchy is discussed in Reck (2004, 163–166).

unary functors of unary functors of individual symbols; and so on.[37] It is therefore natural to ask how the distinction between formal categories and regions can be understood against the backdrop of a simple type hierarchy. The related distinction between form and matter is not a part of the doctrine of types, just as it is not a part of the doctrine of modern logic in general. But the form/matter-distinction does suggest a useful distinction between kinds of type hierarchy. Namely, a type hierarchy may be built over the "formal" ground type of individuals—that is, the only information provided about the elements of the ground type is that they are individuals; or it may be built over a "material" ground type, namely some specific domain, like the domain of natural numbers, or the domain of real numbers, or some empirical domain, like that of living beings—in this case more specific information is provided about the nature of the elements of the ground type or types.

Distinguishing formal from material ground types is not yet to locate formal and material categories in simple type hierarchies. Husserl's conception of a region as subsuming individuals together with derivations of these by means of formal categories is in fact readily adapted to the setting of simple types. Namely, we can simply identify a region with a simple type hierarchy built over a single material ground type; thus, each region will be its own type hierarchy. The ground type of each hierarchy is the type of individuals of the region. The higher types contain precisely all the various categorial derivations of the individuals. As formal categories we should therefore recognize 'individual' and the forms by means of which higher-type objects—that is, relations—are formed. Precisely which forms these are depends on technical details, but logical operators and an abstraction form, such as lambda-abstraction, will need to be among them.

Carnap's *Der logische Aufbau der Welt* (Carnap 1928) suggests another way of mapping regions onto a simple type hierarchy. The stated goal of that work is to provide on the basis of a few primitive concepts a system of definitions of all scientific concepts (§ 1), where science is not to be equated with natural science, but must be taken to include the social sciences, including psychology, and the humanities (*Geisteswissenschaften*). Such a system of definitions is called a constitution system. Carnap sets out to construct, albeit only in outline, a constitution system by means of a simple type hierarchy with a material ground type. Two kinds of ground type are, according to Carnap, adequate for the task of yielding a constitution system, namely types whose elements are physical and types whose elements are psychological. As Carnap wishes the order of definitions in the constitution system to mirror the order of epistemic priority among concepts (§ 54), he settles on the latter kind of ground type (§ 64), and in particular on a ground type whose elements are my experiences (*Erlebnisse*) taken "in their totality and complete unity" (§ 67), what Carnap calls elementary experiences (*Elementarerlebnisse*). That these elements are experiences (in contrast to, say, electrons) and that they are *my*, the constitution-system-building subject's, experiences can, however, be

[37] The meaning categories (*Bedeutungskategorien*) of Ajdukiewicz (1935) are of course just simple types.

seen only after a significant portion of the system has been erected, namely when the domain of the physical has been been constituted, and my psyche has been distinguished from that of other subjects (§ 65).

In order to be able to define objects in this type hierarchy one or more primitive relations must be given at the outset (§ 61). According to Carnap it is enough, at least for the outlines of a constitution system he sketches, to assume one such primitive relation, namely a binary relation over the ground type, holding between elementary experiences x and y if the recollection or retention of x is in part similar to y, that is, if a part of y is similar to a part of the recollection or retention of x (§ 78). From this relation, which is asymmetric and irreflexive, it is easy to define a relation that is symmetric and reflexive and that holds between elementary experiences x and y if these are in part similar to each other, that is, if a part of x is similar to a part of y (§ 77). With these two relations in hand Carnap goes on to define classes and relations of various types that are to serve as "rational reconstructions" of scientific concepts. The definitions quickly get complicated. So called quality classes are to serve as rational reconstructions of particular sense qualities, such as the sensation of a particular shade of red at a particular point in the visual field, or a particular tactile sensation at a particular point on the body. The definition of a quality class is one of the first provided (§ 112), but it already is rather complex and requires several pages of motivation (§ 81). Carnap then defines similarity between quality classes (§ 114) and obtains what he calls sense classes as the transitive closures of quality classes under similarity (§§ 85, 115): two quality classes belong to the same sense class if there is a path of similarity leading from the one to the other; there is, for instance, for any two colour sensations such a path between them. This definition may be straightforward enough, but in order to single out the various sense classes as the visual field, the tactile field, etc., Carnap must appeal to the technically non-trivial topological notion of (inductive) dimension (§§ 86, 115). It is in fact questionable whether all of these initial definitions capture what Carnap intends them to capture, as Goodman (1951, ch. V) noted.

Our current interest in the *Aufbau* stems from the fact that Carnap there takes scientific concepts to be of different kinds, namely, he takes there to be different kinds of object (*Gegenstandsarten*) (§§ 17–25). These kinds of objects may in the greater scheme of things very well be regarded as regions more or less in the sense of Husserl. Indeed, the kinds of object that mainly concern Carnap are, in his terminology, the self-psychological (*das Eigenpsychische*), the physical, the other-psychological (*das Fremdpsychische*), and the cultural and social (*das Geistige*), and these may well be taken to correspond to Husserl's regions of consciousness, nature, living nature, and culture (*die geistige Welt*). In what follows I shall therefore also call Carnap's kinds of object 'regions'. According to Carnap an order of epistemic priority obtains between the various regions: the self-psychological is epistemically prior to the physical, the physical is epistemically prior to the other-psychological, and the other-psychological is epistemically prior to the cultural and social (§ 58). The constitution system has to reflect this order. Hence we are to think of the regions as forming strata or segments within the type hierarchy. At the bottom of the hierarchy is the region of the self-psychological; then comes a segment with the physical, then a segment with the other-psychological, and finally a segment with the cultural and social region. The formal categories of the hierarchy will be the category of individuals together with the logical forms used in constructing objects of higher types. The latter include the Sheffer stroke, universal quantification over any type, and an abstraction form (§ 107); the abstraction form takes a propositional function $\varphi(x)$ and yields, as the case may be, the proposition or propositional function $\hat{x}\varphi(x)$, where x is no longer free. These logical forms, or formal categories, Carnap calls the *Stufenformen* of his hierarchy (§ 26): "the recurring forms by which the passage from one level to the next is achieved."

Let us now consider how the conception of regions as segments in the type hierarchy is to be made precise. It seems to me that in order for the described structure to be realized, the following two requirements must be met.

(i) It need not, and usually will not, be the case that all elements of a type are used in a constitution system. That is, within any type, if it is employed at all in the system, there will, in general, be some objects that do serve and some that do not serve as rational reconstructions of concepts. Let us write $u(\tau)$, the use of τ, for those elements of type τ that do in fact serve as rational reconstructions. The first requirement says that all elements of $u(\tau)$ belong to the same region. This is a requirement of typical homogeneity: no type is to be separated by two regions. Let us, for instance, consider the type $((\iota))$ of classes of classes of elementary experiences. The so-called sense classes mentioned above belong to this type. Since sense classes serve as rational reconstructions of concepts from the self-psychological region, the requirement says that within the type $((\iota))$ there should not be another object that serves as the reconstruction of a concept belonging to some other region, for instance the region of the physical.

If $u(\tau)$ is non-empty, then let us say that the type τ is *used*. The main rationale for requirement (i) is that, provided it is met, we can define an order on used types as follows: $\sigma \leq \tau$ if and only if no element of $u(\tau)$ is epistemically prior to an element of $u(\sigma)$. Thus, if requirement (i) is met, then the relation of epistemic priority among the four regions induces a partial order on used types. For instance, if the elements of $u(\sigma)$ belong to the region of the physical and the elements of $u(\tau)$ to the region of the other-psychological, then we have $\sigma \leq \tau$. If the elements of $u(\tau)$ belong to the region of the self-psychological, however, then we have $\tau \leq \sigma$.

(ii) Let us call the trace of a type the set of all types involved in its construction from the ground type. For instance, the types involved in the construction of the type $((\iota))$ are ι and (ι); these are, as it were, the building blocks of that type. The types involved in the construction of the slightly more complicated type $\big(((\iota),\iota),\iota)\big)$ are $((\iota),\iota)$, (ι), and ι. Continuing to employ ι as a name of the ground type, the trace $\mathrm{tr}(\tau)$ of a type τ can be defined inductively as follows:

$$\mathrm{tr}(\iota) := \emptyset$$
$$\mathrm{tr}((\tau_1,\ldots,\tau_n)) := \bigcup\{\{\tau_1,\ldots,\tau_n\},\mathrm{tr}(\tau_1),\ldots,\mathrm{tr}(\tau_n)\}$$

Thus the trace of a type $(\tau_1, \ldots, \tau_n)$ is a set consisting of each of the types $\tau_1, \ldots, \tau_n$ together with the traces of each of these types. Employing the definition we find for instance that

$$\mathrm{tr}\big((((\iota), \iota), \iota)\big) = \{((\iota), \iota), (\iota), \iota\}.$$

Recall the order $\leq$, just defined, of epistemic priority among used types. The second requirement says that we should have $\sigma \leq \tau$ for all used types σ in $\mathrm{tr}(\tau)$. Spelling out the definition of $\leq$, this is to say that, if $\sigma \in \mathrm{tr}(\tau)$, then the objects in $u(\tau)$ are not to be epistemically prior to the objects in $u(\sigma)$. This requirement gives mathematical expression to the idea that in the definition of an object a that is to serve as a rational reconstruction of some concept c we shall not need to refer to an object serving as the rational reconstruction of some concept c' that, in view of its region, is epistemically posterior to c. No concept is built up from concepts epistemically posterior to it.

Carnap does not spell out these requirements in the *Aufbau*. That he entertained a picture of the construction of the world in which they are met is, however, suggested by what he says about levels of constitution (§ 41). The level of constitution of a concept is the level in the type hierarchy at which a rational reconstruction of it is defined. Carnap had defined the notion of level in a simple type hierarchy in the *Abriss*.[38] The level ℓ of the ground type is 0; the level of a higher type $(\tau_1, \ldots, \tau_n)$ is defined by

$$\ell\big((\tau_1, \ldots, \tau_n)\big) := \max\{\ell(\tau_1), \ldots, \ell(\tau_n)\} + 1$$

Those familiar with the cumulative hierarchy of sets will see the parallel to the notion of rank defined in that context. In the picture Carnap appears to entertain the regions respect levels: for used types, σ, τ, it holds that

$$\text{if } \ell(\sigma) < \ell(\tau), \text{ then } \sigma \leq \tau.$$

We thus have a situation as in Fig. 15.1. The first few levels make up the region of the self-psychological. The levels afterwards make up the region of the physical, the following levels make up the region of the other-psychological, and finally come the levels of the cultural and social. Whereas each region thus takes up only finitely many levels, the type hierarchy continues into infinity; at levels above those making up the region of culture the romantics among us can therefore imagine a region that is yet to be discovered.

While Carnap structures his sketch of a constitution system in the *Aufbau* according to what he takes to be the regions of the self-psychological, the physical, the other-psychological, and culture and society, he also maintains that the possibility of erecting a constitution system shows that there is fundamentally only one region

[38] See Carnap (1929, 32).

	Type hierarchy	Regions
	⋮	?
level $k+l+m+n$	—	
	⋮	Culture
level $k+l+m$	—	
	⋮	Other-psychological
level $k+l$	—	
	⋮	Physical
level k	—	
	⋮	Self-psychological
level 0 (ground domain)	*Elementarerlebnisse*	

Fig. 15.1 Carnap's construction of the world?

(§ 4): "objects do not fall into different, unconnected domains, rather there is just one domain of objects." Carnap seems to hold in particular that a constitution system with a self-psychological basis shows that the self-psychological region is, fundamentally, the only region. This view must be based on the assumption that if the ground type of a type hierarchy belongs to a given region, then the whole hierarchy belongs to the same region; the region of any type is inherited from the region of the types out of which it is constructed and at the base there is only the one region of the self-psychological. There seems thus to be a tension between ideas such as those expressed in Fig. 15.1 and what Carnap takes constitution theory to show, namely that there is just one region. To relieve this tension it seems to me best to distinguish between concepts before and after constitution. What Carnap calls rational reconstruction is a relation between between these, namely between ordinary scientific concepts on the one hand and relations in a certain type hierarchy on the other. The constituted concepts are ultimately to replace the ordinary scientific concepts, since only of the former do we know the precise definition in terms of elementary experiences (cf. § 179). Ultimately, therefore, it will be seen that there is only one region. For, while we can say of ordinary scientific concepts that they fall into different regions, we should not, according to Carnap, say the same about the constituted concepts. Figure 15.1 above thus does not show how regions actually live inside a constitution system; rather it shows the type hierarchy of a constitution system through the prism of a division into regions of ordinary scientific concepts; the order of epistemic priority holding between the regions is reflected by the order holding between the relations serving as rational reconstructions of the ordinary scientific concepts falling into those regions.

15.3 Carnap and Husserl

Carnap had studied the *Ideas* thoroughly. That is clear from the detailed references to different sections of this work not only in the *Aufbau* but also in Carnap's dissertation, *Der Raum* (Carnap 1922).[39] From references in the latter in particular one sees that Carnap was conversant with the distinction between regions and formal categories as well as with the related distinction between generalization and formalization, indeed with all the important notions from the first chapter of the first section of the *Ideas*.[40] It is, however, difficult to say whether Carnap thought of what he called kinds of object (*Gegenstandsarten*) as Husserlian regions. In the *Aufbau* Carnap tends to be quite generous with references to the works of others, so if he had Husserlian regions in mind with his notion of kind of object, one would have expected some reference to Husserl at the relevant places, but that is not to be found. It has, however, been noted by Mayer (1991, 301, fn. 11), and with greater emphasis by Rosado Haddock (2008), that several apparently Husserlian influences are not indicated as such by Carnap. In § 25 of the *Aufbau* Carnap lists several kinds of object apart from those already discussed: logical objects, mathematical objects, spatial forms, colours, tones, biological objects, and ethical objects. Neither of these correspond to Husserlian regions (perhaps apart from the logical and the mathematical objects, which Husserl sometimes (e.g. Ideen § 11) says belong to a formal region), a fact which suggests that Carnap's notion of kind of object is independent of Husserl's notion of region; but spatial form, colour, and tone are all essences in the region of nature, which Husserl discusses in the first chapter of the *Ideas*, so Carnap may have drawn inspiration from those discussions.

One should in any event be careful not to make too much of the parallels between the *Aufbau* and the *Ideas*. A proper assessment of the relation between these works would require more space than what I have available here, but it should be clear, I think, that the *Aufbau* in no way can be regarded as a work in phenomenology.[41] Carnap's conception of what he calls constitution is telling. It is not unlikely that he had the term 'constitution' from Husserl,[42] but his understanding of it is very

[39]Christian Damböck, who has studied Carnap's reading lists, reported in a talk at the HOPOS 2014 meeting at Ghent, 4 July, 2014 that between 1920 and 1923 Carnap worked through the *Ideas* three times.

[40]See especially Carnap (1922, 60–61), where Carnap compares the relation between the geometries related to the three kinds of space he has been studying with the relation between formal ontology, regional ontology, and factual science (*Tatsachenwissenschaft*), and where he also employs the distinction between formalization and generalization. The Husserlian notions of essence and eidetic intuition are fundamental to Carnap's treatment of what he calls intuitive space (ibid. 22–31).

[41]I take Roy (2004) and Ryckman (2007, 95–98) to argue for the same stance.

[42]In Kant it is the adjective 'constitutive' rather than the noun 'constitution' that is prominent (cf. Brockhaus 1976); 'constitution' is apparently employed by the Marburger neo-Kantians Cohen and Natorp (ibid. 1002–1003); but Carnap distances his use of 'constitution' from Marburger-school doctrine in § 5 of the *Aufbau*. The word features prominently in the *Ideas*, especially in its final

far from Husserl's. For Carnap 'constitution' means the definition of an object in the simple type hierarchy in terms of others (§ 38). For Husserl, however, 'constitution' indicates how an object presents itself to consciousness[43]; to describe the constitution of material individuals, for instance, means to describe the various components, or layers, that make up our experience of such objects (cf. Ideen II §§ 12–17). An object in Carnap's constitution system is to serve as a rational reconstruction of a concept. Carnap emphasizes that the constitution system need not reflect "the syntheses and formations of knowledge as they actually occur in the process of knowledge" (§ 54); rational reconstructions are to preserve only the "logical value" of the original concepts, they need not preserve their "cognitive value" (*Erkenntniswert*), not their sense (§§ 50–51). One of the more important methods of the *Aufbau* is called *quasi*-analysis and the result of an instance of quasi-analysis is said to be a *formaler Ersatz* for the components that a proper (*eigentliche*) analysis would yield (§§ 69–71). Such ideas are of course quite foreign to phenomenology, with its emphasis on giving a true description of experience and what is experienced, on merely explicating what originally gives itself in eidetic intuition.

References

K. Ajdukiewicz, Die syntaktische Konnexität. Studia Philosophica **1**, 1–27 (1935)

A. Arnauld, P. Nicole, *La Logique, ou l'art de penser* (Iaen Guignart/Guillame Desprez, Paris, 1662/1683). Critical edition: Vrin, Paris (1981)

J. Barnes (ed.), *The Complete Works of Aristotle. Revised Oxford Translations* (Princeton University Press, Princeton, 1984)

J. Barnes, *Porphyry. Introduction. Translated with an Introduction and Commentary*. Clarendon Later Ancient Philosophers (Oxford University Press, Oxford, 2003)

K. Brockhaus, Konstitution. III. Neuzeit, in *Historisches Wörterbuch der Philosophie*, ed. by K. Gründer, et al., volume 4 (Schwabe, Basel, 1976), pp. 997–1004

R. Carnap, *Der Raum* (Reuther & Reichard, Berlin, 1922)

R. Carnap, *Der logische Aufbau der Welt* (Weltkreis Verlag, Berlin, 1928)

R. Carnap, *Abriss der Logistik* (Springer, Vienna, 1929)

V. De Palma, Die Kategorien des Sinnlichen. Zu Husserls Kategorienlehre. Phänomenologische Forschungen **15**(NF), 23–39 (2010)

G. Frege, *Grundlagen der Arithmetik* (Verlag von Wilhelm Koebner, Breslau, 1884)

G. Frege, *Grundgesetze der Arithmetik I* (Hermann Pohle, Jena, 1893)

N. Goodman, *The Structure of Appearance* (Harvard University Press, Cambridge, 1951)

H.B. Gottschalk, The earliest Aristotelian commentators, in *Aristotle Transformed. The Ancient Commentators and Their Influence*, ed. by R. Sorabji (Duckworth, London, 1990), pp. 55–82

sections, and it may very well have come up in the Husserl seminar Carnap attended; Carnap may also have been familiar, through Ludwig Landgrebe, with the contents of the second book of the *Ideas* (cf. Stone 2010, fn. 48).

[43]Constitution is, as far as I know, not a technical term for Husserl and his employment of it is not so easy to circumscribe.

M. Hallett, U. Majer (eds.), *David Hilbert's Lectures on the Foundations of Geometry* (Springer, Heidelberg, 2004)
D. Hilbert, *Grundlagen der Geometrie* (Teubner, Leipzig, 1899)
E. Husserl, Der Folgerungskalkül und die Inhaltslogik. Vierteljahrsschrift für wissenschaftliche Philosophie **15**, 168–189 (1891a). Reprinted in Husserl (1979)
E. Husserl, *Philosophie der Arithmetik* (C.E.M. Pfeffer, Halle, 1891b). Cited from Husserl (1970)
E. Husserl, *Logische Untersuchungen. Erster Theil: Prolegomena zur reinen Logik* (Max Niemeyer, Halle, 1900)
E. Husserl, *Logische Untersuchungen. Zweiter Theil: Untersuchungen zur Phänomenologie und Theorie der Erkenntnis* (Max Niemeyer, Halle, 1901)
E. Husserl, *Ideen zu einer reinen Phänomenologie und phänomenologischen Philosophie. Erstes Buch* (Max Niemeyer, Halle, 1913). Cited from Husserl (1976)
E. Husserl, *Erfahrung und Urteil* (Academia Verlagsbuchhandlung, Prague, 1939)
E. Husserl, *Ideen zu einer reinen Phänomenologie und phänomenologischen Philosophie. Zweites Buch*, edited by M. Biemel. Volume IV of Husserliana (Martinus Nijhoff, The Hague, 1952a)
E. Husserl, *Ideen zu einer reinen Phänomenologie und phänomenologischen Philosophie. Drittes Buch*, edited by M. Biemel. Volume V of Husserliana (Martinus Nijhoff, The Hague, 1952b)
E. Husserl, *Erste Philosophie (1923/24). Erster Teil*, edited by R. Boehm. Volume VII of Husserliana (Martinus Nijhoff, The Hague, 1956)
E. Husserl, *Phänomenologische Psychologie*, edited by W. Biemel. Volume IX of Husserliana (Martinus Nijhoff, The Hague, 1968)
E. Husserl, *Philosophie der Arithmetik. Mit ergänzenden Texten (1890–1901)*, edited by L. Eley. Volume XII of Husserliana (Martinus Nijhoff, The Hague, 1970)
E. Husserl, *Ideen zu einer reinen Phänomenologie und phänomenologischen Philosophie I*, edited by K. Schumann. Volume III/1 of Husserliana (Martinus Nijhoff, The Hague, 1976)
E. Husserl, *Aufsätze und Rezensionen (1890–1910)*, edited by B. Rang. Volume XXII of Husserliana (Martinus Nijhoff, The Hague, 1979)
E. Husserl, *Logik und Allgemeine Wissenschaftstheorie. Vorlesungen 1917/18, mit ergänzenden Texten aus der ersten Fassung von 1910/11*, edited by U. Panzer. Volume XXX of Husserliana (Kluwer, Dordrecht, 1996)
E. Husserl, *Die Bernauer Manuskripte über das Zeitbewusstsein*, edited by R. Bernet, D. Lohmar. Volume XXXIII of Husserliana (Kluwer, Dordrecht, 2001a)
E. Husserl, *Logik. Vorlesung 1896*, edited by E. Schumann. Husserliana Materialen (Kluwer, Dordrecht, 2001b)
E. Husserl, *Natur und Geist. Vorlesungen Sommersemeseter 1927*, edited by M. Weiler. Volume XXXII of Husserliana (Kluwer, Dordrecht, 2001c)
E. Husserl, *Zur Lehre vom Wesen und zur Methode der eidetischen Variation*, edited by D. Fonfara. Volume XLI of Husserliana (Springer, Dordrecht, 2012)
I. Kant, *Kritik der Reinen Vernunft* (Johann Friedrich Hartknoch, Riga, 1781/1787)
R. Kauppi, Begriffsinhalt/Begriffsumfang, in *Historisches Wörterbuch der Philosophie*, ed. by J. Ritter et al., vol. 1 (Schwabe, Basel, 1971), pp. 808–809
A. Klev, Form of apprehension and the content-apprehension model in Husserl's *Logical Investigations*. Log. Anal. Hist. Philos. **16**, 49–69 (2013)
A. Klev, Categories and logical syntax. PhD thesis, Leiden University (2014)
D. Lohmar, Kategoriale Anschauung (VI. Logische Untersuchung, §§ 40–66), in *Edmumd Husserl. Logische Untersuchungen*, ed. by V. Mayer. Klassiker Auslegen (Akademie Verlag, Berlin, 2008), pp. 209–237
V. Mayer, Die Konstruktion der Erfahrungswelt: Carnap und Husserl. Erkenntnis **35**, 287–303 (1991)
B. Morrison, Les *Cátegories* d'Aristote comme introduction à la logique, in *Les Catégories et leur histoire*, ed. by O. Brun, L. Corti (Vrin, Paris, 2005), pp. 103–120
G.T. Null, Husserl's doctrine of essence, in *Husserl's Phenomenology: A Textbook*, ed. by J.N. Mohanty, W.R. McKenna (University Press of America, Washington, D.C., 1989), pp. 69–105

E.H. Reck, From Frege and Russell to Carnap: logic and logicism in the 1920s, in *Carnap Brought Home: The View from Jena*, ed. by S. Awodey, C. Klein (Open Court, Chicago, 2004), pp. 151–180
G. Rosado Haddock, *The Young Carnap's Unknown Master* (Ashgate, Aldershot, 2008)
J.-M. Roy, Carnap's Husserlian reading of the *Aufbau*, in *Carnap Brought Home: The View from Jena*, ed. by S. Awodey, C. Klein (Open Court, Chicago, 2004), pp. 41–62
B. Russell, A.N. Whitehead, *Principia Mathematica*, vol. 1 (Cambridge University Press, Cambridge, 1910)
T. Ryckman, Carnap and Husserl, in *The Cambridge Companion to Carnap*, ed. by M. Friedman, R. Creath (Cambridge University Press, Cambridge, 2007), pp. 81–105
G. Ryle, Categories. Proc. Aristot. Soc. **38**, 189–206 (1938)
G. Ryle, *Dilemmas* (Cambridge University Press, Cambridge, 1954)
S. Sarkar, Husserl's role in Carnap's *Der Raum*, in *Language, Truth and Knoweldge. Contributions to the Philosophy of Rudolf Carnap*, ed. by T. Bonk (Kluwer, Dordrecht, 2003), pp. 179–190
K. Schumann, *Husserl-Chronik* (Martinus Nijhoff, The Hague, 1977)
P. Simons, The formalisation of Husserl's theory of wholes and parts, in *Parts and Moments*, ed. by B. Smith (Philosophia Verlag, Munich, 1982), pp. 113–159
A. Stone, On Husserl and Cavellian Skepticism, with reference to the thomistic theory of creation. PhD thesis, Harvard University (2000). Available at http://www.abocalypse.com/papers/skep.pdf
A. Stone, On the sources and implications of Carnap's *Der Raum*. Stud. Hist. Philos. Sci. **41**, 65–74 (2010)
G.F. Stout, Are the characteristics of particular things universal or particular? Proc. Aristot. Soc. Suppl. Vol. **3**, 114–122 (1923)
B.G. Sundholm, Containment and variation; two strands in the development of analyticity from Aristotle to Martin-Löf, in *Judgement and the Epistemic Foundation of Logic*, ed. by van der M. Schaar (Springer, Heidelberg, 2013)
G. Tonelli, Die Voraussetzungen zur Kantischen Urteilstafel in der Logik des 18. Jahrhunderts, in *Kritik und Metaphysik. Heinz Heimsoeth zum achtzigsten Geburtstag*, ed. by F. Kaulbach, J. Ritter (Walter de Gruyter, Berlin, 1966), pp. 134–158
D.C. Williams, On the elements of being. Rev. Metaphys. **7**, 3–18; 171–192 (1953)

Chapter 16
Husserl and Gödel

Richard Tieszen

Abstract Kurt Gödel (1906–1978) began to study the philosophy of Edmund Husserl (1859–1938) in 1959. In this paper I present an overview of central themes in Gödel's study of Husserl's phenomenology. Since many of Gödel's ideas concerning Husserl were never put into a systematic form by Gödel himself, I quote fairly extensively in the paper from several sources in order to inform the reader of the nature of Gödel's interest in Husserl. Gödel prepared one manuscript specifically on Husserl, as we will see below, and many of Gödel's comments on Husserl are included in the books of Hao Wang. I will also quote some relevant texts from the Gödel *Nachlass*. In accordance with these various sources, I provide a brief overview in a later section of the paper of Gödel's interest in eidetic transcendental phenomenology as a new type of monadology. The relationship of Gödel's incompleteness theorems to Husserl's notion of 'definite' axiom systems is also discussed briefly.

Keywords Hilbert's program • Leibniz • Transcendental eidetic onadology • Constituted platonism • Incompleteness theorems

16.1 Introduction

Kurt Gödel (1906–1978) began to study the philosophy of Edmund Husserl (1859–1938) in 1959. In this paper I present an overview of central themes in Gödel's study of Husserl's phenomenology.[1] Since many of Gödel's ideas concerning Husserl were never put into a systematic form by Gödel himself, I will quote fairly

---

Professor Richard Tieszen passed away shortly after completing his contribution to this volume. He will be sorely missed.

[1] See also Tieszen 1992, 2005, Part II, 2011, 2012. Perhaps the earliest discussion of Gödel in terms of Husserl's phenomenology is in Tragesser 1977. Other papers on Husserl and Gödel include Føllesdal 1995, van Atten and Kennedy 2003, da Silva 2005, Hauser 2006, van Atten 2006, and Liu 2010.

R. Tieszen (deceased)

S. Centrone (ed.), *Essays on Husserl's Logic and Philosophy of Mathematics*, Synthese Library 384, DOI 10.1007/978-94-024-1132-4_16

extensively in the paper from several sources in order to inform the reader of the nature of Gödel's interest in Husserl. Gödel prepared one manuscript specifically on Husserl, as we will see below, and many of Gödel's comments on Husserl are included in the books of Hao Wang.[2] I will also quote some relevant texts from the Gödel *Nachlass*.[3] The content of my discussions and correspondence from the mid-1980s through early 1990s with Hao Wang will also be in the background of the paper.[4]

Gödel is known to have studied philosophy seriously from the early 1940s until the end of his life. He was exposed to the work of Kant fairly early in his studies, at around the age of 16, according to the questionnaire designed by Burke Grandjean that Gödel answered in 1974.[5] We know that he continued to think about Kant's work off and on over many years. Karl Menger indicates that Gödel had a serious interest in Leibniz already in the early 1930s,[6] and Gödel indicates in the Grandjean questionnaire that he studied Leibniz intensively from 1943 to 1946. Gödel's work on Leibniz thus antedated his study of Husserl. Gödel was perhaps first acquainted with the philosophy of Plato through the lectures of Heinrich Gomperz on the history of philosophy at the University of Vienna, but it is not clear that he studied Plato in detail. Wang tells us, notably, that Gödel wrote in 1975 that "I was [have been] a conceptual and mathematical realist [= platonist] since about 1925."[7] Gödel turned to Husserl's philosophy in 1959 and he continued to study Husserl's work through the 1970s. Gödel's library includes all of Husserl's major writings, many marked with underlinings and marginal comments and accompanied by inserted pages written mostly in Gabelsberger shorthand. "The Modern Development of the Foundations of Mathematics in the Light of Philosophy"[8] is the only text we have in which Gödel explicitly discusses Husserl's philosophy at any length. It is a very interesting text for the manner in which it connects certain ideas in Husserl's transcendental phenomenology to various central theses in Gödel's philosophical views on logic and mathematics.

---

[2]Wang 1974, 1987, 1996.

[3]The items from the *Nachlass* that I cite are not widely known. I am happy to acknowledge that I have benefitted in a number of ways from being part of the multi-year project *Kurt Gödel: Philosopher-Scientist* funded by the Agence Nationale de la Recherche (ANR) of France, headed by Professor Gabriella Crocco at Université Aix-Marseille. The project was devoted to transcribing from Gabelsberger shorthand (the shorthand used also by Husserl) Gödel's philosophical notebooks, known as the *Max-Phil* notebooks, which are also in the *Nachlass*. The entries in the notebooks evidently antedate Gödel's study of Husserl but they should help to shed light on his views on Leibniz, Kant, and other philosophers and on his own philosophical development.

[4]I met with and then later corresponded with Wang during this period about Gödel and the philosophy of mathematics. Wang read and commented on Tieszen 1989 and 1992. In our discussions he was especially interested in Gödel's attraction to Husserl's philosophy.

[5]Reproduced in Wang 1987, 16–21.

[6]Menger 1994.

[7]Wang 1996, 76.

[8]Gödel *1961/?.

I will start with a presentation of the ideas in Gödel's text *1961/? since it shows us, in Gödel's own words, how he sees Husserl's work in relation to the incompleteness theorems and other basic issues in the foundations of mathematics. In the third section of the paper I fill out the picture of Gödel's interest in Husserl by considering his comments to Wang, some of Wang's own reflections on these comments, and several relevant items from the Gödel *Nachlass*. In Sect. 16.4 I present some of the background texts of Husserl that can help to shed light on Husserl's influence on Gödel, especially texts on phenomenology as transcendental eidetic monadology.

Section 16.5 provides a brief overview of some elements of phenomenological monadology that are suggested by the previous sections of the paper. In Sect. 16.6 I comment on the issue discussed in some of the literature on Husserl of whether Husserl's ideas about what he calls "definite" axiom systems and "definite" manifolds are refuted by Gödel's incompleteness theorems. Section 16.7 contains some final observations on Husserl and Gödel.

16.2 The Modern Development of the Foundations of Mathematics in the Light of Philosophy (*1961/?)

Gödel wrote this manuscript in Gabelsberger shorthand around 1962.[9] It was found in his papers with a letter from the American Philosophical Society dated December 13, 1961, and was marked *Vortrag*, indicating that it was probably intended as a talk that Gödel would give to the Society as a newly elected member. Gödel never delivered the lecture.

In the manuscript Gödel sets up a general schema of possible philosophical worldviews according to their affinity to metaphysics. We obtain a division into two groups, with skepticism, materialism and positivism on one side and spiritualism, idealism, and theology on the other. If one thinks of philosophical doctrines as arranged along a line from left to right in this manner then empiricism is on the left side and a priorism belongs on the right. Pessimism belongs on the left side and optimism in principle toward the right, for empiricist skepticism is a kind of pessimism with regard to knowledge. Materialism, Gödel says, is inclined to regard the world as an unordered and therefore meaningless heap of atoms. On the other hand, idealism and theology see meaning, purpose, and reason in everything.

Gödel says that the development of philosophy since the Renaissance has, on the whole gone from right to left. This development also made itself felt in mathematics. Mathematics, as an a priori science, always has an inclination toward the right and has long withstood the *Zeitgeist* that has ruled since the Renaissance. The empiricist conception of mathematics, such as that set forth by John Stuart Mill in the nineteenth century, did not find much support. Indeed, mathematics evolved into ever higher abstractions, away from matter and into ever greater clarity in its foundations. The foundations of the infinitesimal calculus and the complex numbers, for example,

[9]Wang 1996, 155.

were improved. Mathematics thus moved away from skepticism. Around the turn of the century, however, the set-theoretic antinomies appeared in mathematics. Gödel says that the significance of the antinomies was exaggerated by skeptics and empiricists and that the antinomies were employed as a pretext for a leftward upheaval. He says, in response, that the contradictions did not appear in the heart of mathematics but rather near its outer boundary toward philosophy. Moreover, the antinomies have been resolved in a manner that is satisfactory to those who understand set theory.[10] These kinds of points are of no use, however, against the prevailing *Zeitgeist*. Many mathematicians came to deny that mathematics as it had developed in earlier times represented a system of truths. They acknowledged this for only a part of mathematics and retained the rest in at best a hypothetical sense.

Gödel says that the "nihilistic" consequences of the spirit of the times also led to a reaction in mathematics itself. Thus came into being "that curious hermaphroditic thing that Hilbert's formalism represents." It sought to do justice both to the leftward *Zeitgeist* and to the nature of mathematics. In conformity with the ideas prevailing in the *Zeitgeist*, it acknowledges that the truth of the axioms from which mathematics starts cannot be justified or recognized in any way and therefore that the drawing of consequences from them has meaning only in a hypothetical sense. The drawing of consequences itself, to further satisfy the spirit of the time, is construed as a mere game with symbols according to rules, where this is likewise not thought of as supported by insight or intuition.[11] In accordance with the earlier "rightward" philosophy of mathematics and the mathematicians' instinct, however, it is held that a proof of a proposition must provide a secure grounding for the proposition and that every precisely formulated yes-or-no question in mathematics must have a clear-cut answer. One aims to prove, that is, for the inherently unfounded rules of the game with symbols that of two sentences A and $\neg$A, exactly one can always be derived. Such a system is consistent if not both can be derived, and if one can be derived then the mathematical question expressed by A can be unambiguously answered. In order to justify the assertions of consistency and completeness, a certain part of mathematics must be acknowledged to be true in the sense of the old rightward philosophy. The part in question, however, is much less opposed to the spirit of the time than the high abstractions of set theory. It is the part that refers only to concrete and finite objects in space, namely the combinations of symbols. This

[10]In some places in his writings Gödel likens the antinomies to illusions of the senses. They are cases where we have not seen concepts clearly enough, and they lose their grip on us once we have achieved more clarity in our perception of concepts. See, for example, the passages in Wang 1974 on perceiving concepts clearly, 81–86. Also, Gödel *1953/9, 321; 1964, 268, and Tieszen 1992, 1998, 2002.

[11]In various places in his writings, going back to the 1930s, Gödel distinguishes the purely formal and relative concept of proof from the 'abstract' concept of proof as 'that which provides evidence' (see, e.g., 193?, 164, vol. III). To Wang (Wang 1996, 168) he said: "[A proposition or a proof is] a net of symbols associated with a net of concepts. To understand something requires introspection; for instance, the abstract idea of a proof must be seen [the idea 'behind' a proof can only be understood] by introspection." See also the comments in § 2 below on introspection.

is Hilbert's finitistic formalism. In Hilbert's program we thus see an interesting mixture of rationalist and empiricist elements.

The next step in the development comes with Gödel's incompleteness theorems: it turns out that it is impossible to rescue the older rightward aspects of mathematics in such a way as to be in accord with the spirit of the time. Even for elementary arithmetic it is impossible to find a consistent system of axioms and formal rules from which, for every number-theoretic proposition A, either A or ¬A would always be derivable. Moreover, for reasonably comprehensive axioms of mathematics it is impossible to provide a proof of consistency merely by reflecting on finite concrete combinations of symbols, without introducing more abstract elements.[12] Hilbert's combination of materialism and aspects of classical mathematics thus proves to be impossible. This means that the combination of rationalist and empiricist elements involved in Hilbert's program is unworkable. It is not possible to be a finitistic formalist and to hold that every clearly stated mathematical proposition is decidable. Nor is it possible to hold on these grounds that proofs provide a secure grounding for mathematical propositions, in the sense that they provide *evidence* for the propositions.

Gödel says that only two possibilities remain. Either we must give up the older rightward aspects of mathematics or attempt to uphold them in contradiction to the spirit of the time. The first alternative suits the *Zeitgeist* and is therefore usually the one adopted. One has to thereby give up on rightward features of mathematics that would otherwise be very desirable, namely, to (i) safeguard for mathematics the certainty of its knowledge by thinking of proof as that which provides evidence but also to (ii) uphold the optimistic belief that for clear questions posed by reason it is possible for reason to find clear answers. One would give up on these features not because any mathematical results compel us to do so but because this is the only way to remain in agreement with the prevailing leftward philosophy. Gödel grants that great advances have been made on the basis of the leftward spirit in philosophy and he thinks there have been excesses and wrong directions taken in the preceding rightward philosophies. The correct attitude, he says, is that the truth lies in the middle of these philosophies or consists in a combination of the leftward and rightward views, but not in the manner of Hilbert's conception. Hilbert's conception, like Carnap's in *The Logical Syntax of Language*, was too primitive and tended too strongly in one direction.[13] We must look elsewhere for a workable combination. If we want to preserve elements of the earlier rightward view of mathematics, then we must suppose that the certainty of mathematics is not to be secured by proving certain properties by a projection onto material systems (i.e., the mechanical manipulation of physical symbols) but rather by cultivating and deepening our knowledge of the abstract concepts that lead to setting up these mechanical systems in the first place. Furthermore, it is to be secured by seeking, according to the

[12] See also Gödel's remark in Gödel 1972, vol. II, 271–273.

[13] See Gödel, *1953/59.

same procedures, to gain insights into the solvability of all meaningful mathematical problems.

How is it possible to extend our knowledge of these abstract concepts? As Gödel commented to Wang, "One must not expect much from science [in doing philosophy]; for instance, it will not help [in learning] how to perceive concepts."[14] So Gödel asks: how can we make these concepts precise and gain a comprehensive and secure insight into the fundamental relations that hold among them, that is, into the axioms that hold for them? We cannot do this by trying to give explicit definitions for concepts and proofs for axioms, since in that case one needs other undefinable abstract concepts and axioms holding for them. Otherwise one would have nothing from which one could define or prove. Therefore, the procedure must consist to a large extent in a clarification of meaning that does not consist in giving definitions. We thus see here, as in other places in his writing, that Gödel is speaking about the need to reflect on meaning. What is required is a reflection on meaning or on concepts that is of a 'higher level' than reflection on the combinatorial properties of concrete symbols. This is a kind of ascent that is a function of reason.

In looking for a workable combination of the two directions Gödel turns to the philosophy of Husserl. He says that there exists today the beginning of a science that claims to possess a systematic method for such a clarification of meaning and that is the phenomenology founded by Husserl. Clarification of meaning, Gödel says, "consists in focusing more sharply on the concepts concerned by directing our attention in a certain way, namely onto our own acts in the use of these concepts, onto our powers in carrying out our acts, and so on." Phenomenology is not supposed to be a science in the same sense as other sciences. Rather, it is supposed to be a procedure that should produce in us a new state of consciousness in which we describe in detail the basic concepts we use in our thought, or grasp other basic concepts hitherto unknown to us.[15] Gödel says that he sees no reason to reject such a procedure at the outset as hopeless. Empiricists in particular have no reason to do so since that would mean that their empiricism is a kind of dogmatic apriorism.

Gödel says that one can in fact present reasons in favor of the phenomenological approach. His example of this in the text *1961/? is that if one considers the development of a child one sees that it proceeds in two directions. On the one hand

---

[14]Wang 1996, 166.

[15]There are references elsewhere in Gödel's thinking to meaning clarification and phenomenology. In Wang 1974, 189, for example, Wang says "With regard to setting up the axioms of set theory (including the search for new axioms), we can distinguish two questions, viz., (1) what, roughly speaking, the principles are by which we introduce the axioms, (2) what their precise meaning is and why we accept such principles. The second question is incomparably more difficult. It is my impression that Gödel proposes to answer it by phenomenological investigations." It should be noted that there is virtually no work on higher set theory in Husserl's writings. The many philosophical issues raised in connection with actual completed infinite totalities, non-denumerable sets, large cardinal axioms, impredicativity, determinateness of power set, reflection principles, forcing, etc., are thus not addressed by Husserl. The philosophy of mathematics literature contains relevant materials but a lot of the work in this direction remains to be pursued.

the child experiments with objects in the external world and with its own sensory and motor organs. On the other hand, it comes to a better and better understanding of language and of the concepts on which language rests. Concerning this second direction we can say that the child passes through states of consciousness of various heights. A higher state is attained, for example, when the child first learns the use of words and, similarly, when for the first time it understands a logical inference. We can view the whole development of empirical science as a systematic and conscious extension of what the child does when it develops in the first direction. The success of this procedure is astonishing and far greater than one might expect a priori. After all it leads to the remarkable technological development of recent times. Gödel reasons that a systematic and conscious advance in the second, rationalistic direction might also far exceed the expectations one might have a priori.

There are examples that show how considerable further development in the second direction occurs even without the application of a systematic and conscious procedure, a development that transcends 'common sense.' Gödel's example here is that in the establishment of axioms of mathematics new axioms that do not follow by formal logic alone from those previously established again and again become *evident*. His own incompleteness theorems could be used to show this, in the sense that we can augment a given formal system with its Gödel sentence, and then repeat this process indefinitely. He also has in mind the addition of more and more axioms of infinity in set theory. Gödel says that the incompleteness theorems, which are often viewed as negative results, do not exclude the possibility that every clearly posed mathematical yes-or-no question is solvable in this way, for it is just this becoming evident of more and more new axioms on the basis of the meaning of the primitive concepts that a machine cannot emulate (see Sect. 16.5 below).

At the end of his 1944 paper on Bertrand Russell Gödel had spoken favorably of Leibniz's efforts to facilitate the clarification of mathematical concepts and the systematic solution of mathematical problems on the basis of the development of the *Characteristica universalis*.[16] By the time of the text *1961/?, however, there is no mention of Leibniz. The Gödel *Nachlass* even contains the following remarks: "The universal characteristic claimed by Leibniz (1677) does not exist. Any systematic procedure for solving problems of all kinds would have to be nonmechanical."[17] Gödel amended the first sentence of this note to read: "The universal characteristic claimed by Leibniz (1677) if interpreted as a formal system does not exist." I conjecture that Gödel shifted the basis of his rationalistic optimism about solving mathematical problems from Leibniz to Husserl. The rationalistic optimism in his later writings is not based on a mechanist conception of reason but rather on a conception of the capacity of the human 'monad' (see below) for a form of reason that allows for the possibility of finding systematic and finite but non-mechanical methods for the decision of mathematical questions on the basis of clarification

---

[16]Gödel 1944, 140–141.

[17]van Atten and Kennedy 2003, 433.

of the intuition of the abstract concepts (meanings of the terms) involved in the problems.[18]

This conception of the rational capacities of human beings figures into Gödel's formulation of a disjunction that is stated at a number of places in his writing. The disjunction is that "[e]ither there exist infinitely many number-theoretic questions which the human mind is unable to answer or the human mind contains an element totally different from a finite combinatorial mechanism."[19]

In a note in the *Nachlass* Gödel says "I conjecture that the second alternative is true and perhaps can be verified by a phenomenological investigation of the processes of reasoning."[20] The human mind ('monad') could not be merely a finite combinatorial mechanism if phenomenological investigation shows that consciousness exhibits intentionality and that we can use systematic, finite and yet non-mechanical methods for the decision of open problems in number theory based on rational intuition of meanings. It should be noted that Gödel also argues against the first disjunct, in the sense that he does not think there are clearly stated but absolutely undecidable propositions in number theory. We already noted above that Gödel says the incompleteness theorems do not exclude the possibility that every clearly posed mathematical yes-or-no question is solvable. In Sect. 16.5 below I will also link this directly to his platonic rationalism.

Thus, in the text *1961/? Gödel says that the incompleteness theorems show that we have an intuitive grasp of ever newer axioms that are logically independent of the earlier ones and that this is necessary for the solvability of mathematical problems. He says that this agrees in spirit if not in letter with the Kantian conception of mathematics in the following sense: Kant asserted that in the derivation of geometrical theorems we always need new geometrical intuitions and that a purely logical derivation from a finite number of axioms is therefore impossible. We now know from work by Tarski that the decision problem for elementary geometry is solvable, so that Kant's assertion is demonstrably false. But if we replace the term "geometrical" by "mathematical" or "set-theoretical" then it becomes a demonstrably true proposition. Elsewhere Gödel says that in this case we need a conception of intuition, such as categorial intuition, that goes beyond Kant's (two forms of) sensory intuition. For Kant, Euclidean geometry was the form of outer sensory intuition.

Gödel says that many of Kant's assertions are false if understood literally but in a broader sense contain deep truths. The whole phenomenological method, according to Gödel, goes back in its central idea to Kant. What Husserl did was to formulate it more precisely, made it fully conscious, and actually carried it out for particular domains. It is because in the last analysis Kantian philosophy rests on the idea of phenomenology, albeit not in an entirely clear way, that Kant has had such an enormous influence over the entire development of philosophy. Quite divergent

[18] See also Tieszen 1992, 1998, 2002, 2011.

[19] See, e.g., Gödel *1951.

[20] Cited in van Atten 2006, 257.

directions have developed out of Kant's thought, however, due to the lack of clarity and literal incorrectness of many of his formulations. None of these have really done justice to the core of Kant's thought. Gödel says that it is Husserl's phenomenology that for the first time meets this requirement. It avoids both the death defying leap of idealism into a new metaphysics as well as the positivistic rejection of all metaphysics. The text *1961/? concludes with a question: if the misunderstood Kant has already led to so much that is interesting in philosophy and also indirectly in science, how much more can we expect from Kant correctly understood by way of Husserl?

Gödel's claim in this text that the development of philosophy since the Renaissance has, on the whole, gone from right to left should be compared with Husserl's critique of the 'positive' sciences in *The Crisis of the European Sciences and Transcendental Phenomenology*. In the *Crisis* Husserl wants to display the "one-sidedness" (see quotation W1 below) of the positive sciences, acknowledging that they reveal many remarkable things about the natural world but also indicating what they conceal about the capacities of human reason and experience. Husserl's comments about the "superficialization of reason" due to the influence of reductionist forms of empiricism and naturalism[21] would no doubt have been very appealing to Gödel. In the text *1961/? Gödel asserts that it is in Husserl's work we might find a workable combination of leftward and rightward elements. Hilbert's finitistic formalism and the early Carnap are both on the left side of Gödel's schema and are to be left behind. Kant is also too far to the left (see quotation H7 below). In the text *1961/? Gödel suggests some modifications of Kant's view. Kant is not a platonist about mathematical objects or facts, and he mounts a critique of classical rationalism (including Leibniz). For Kant, knowledge is restricted to sensory intuition and the two forms of sensory intuition, space and time. Kant distinguishes phenomena from noumena (what Wang refers to as Kant's "dualism" in quote W1 below) and is able to develop the transcendental method far enough to show how *empirical realism* is compatible with transcendental idealism, but Gödel wants to extend the transcendental method to accommodate categorial intuition and the reality of abstract concepts. Categorial intuition or *Wesensanschauung* is supposed to make it possible to see the primitive concepts clearly and distinctly (see quotes W2, W4, G3 and G4 below).

The rightward elements that Gödel is concerned to preserve include optimism about the decidability of mathematical problems by human reason and the secure grounding of mathematical propositions that is provided by the 'abstract' concept of proof, i.e., the concept of proof as that which provides evidence. On the left we have Hilbert's finitistic formalism and Carnap's conventionalism/nominalism. Both of these are to be rejected, but on the right we also want to avoid a "death-defying leap of idealism" into a dubious metaphysics. How can we avoid dubious metaphysical schemes and yet skirt Kant's objections to rationalism? Gödel evidently hopes that we can avoid the one-sidedness or prejudices of the leftward *Zeitgeist* and yet

[21] See the Vienna Lecture, Appendix I in Husserl 1954, 299.

steer clear of questionable metaphysical views by developing a scientific philosophy that employs the phenomenological *epoché* and yet still requires that knowledge in mathematics and logic depends not only on mere conception but also on intuition, only now it is categorial intuition or *Wesensanschauung* that we must cultivate.

16.3 Further Remarks of Gödel and Wang on Husserl

Much of what we know about Gödel's philosophical outlook is due to the discussions that Hao Wang conducted with Gödel. Now let us see how the picture in the text *1961/? is filled out by considering Gödel's comments to Wang, along with some of Wang's own reflections on the comments. We know from these comments that Gödel's favorite philosophers were Plato, Leibniz, and Husserl. Let me quote some passages from Wang that are, I think, important for indicating, if only very generally, how Husserl fits into Gödel's thinking about the philosophy and foundations of mathematics and logic (quotations are numbered for reference elsewhere in the paper):

> (W1) Before 1959 Gödel had studied Plato, Leibniz, and Kant with care: his sympathies were with Plato and Leibniz. Yet he felt he needed to take Kant's critique of Leibniz seriously and find a way to meet Kant's objections to rationalism. He was not satisfied with Kant's dualism or with his restriction of intuition to sense intuition, which ruled out the possibility of intellectual or categorial intuition. It seems likely that, in the process of working on his Carnap paper in the 1950s, Gödel had realized that his realism about the conceptual world called for a more solid foundation than he then possessed. At this juncture it was not surprising for him to turn to Husserl's phenomenology, which promises a general framework for justifying certain fundamental beliefs that Gödel shared: realism about the conceptual world, the analogy of concepts and mathematical objects to physical objects, the possibility and importance of categorial intuition or immediate conceptual knowledge, and the one-sidedness of what Husserl calls "the naive or natural standpoint."[22]
>
> (W2) In his philosophy Gödel tried to combine and go beyond the main contributions of his three heroes: Plato, Leibniz and Husserl. Leibniz had defined the ideal by giving a preliminary formulation of monadology. Husserl had supplied the method for attaining this ideal. Plato had proposed, in his rudimentary objectivism in mathematics, an approach that could serve as foundation for Husserl's method and, at the same time, make plausible for Gödel the crucial belief that we are indeed capable of perceiving the primitive concepts of metaphysics clearly enough to set up the axioms.[23]
>
> (W3) Gödel uses Plato, Leibniz and Husserl in a positive way, Kant and Hegel in a mixed way, and positivism and Wittgenstein negatively.[24]

Note the reservations in W1 and W3 about Kant. Moreover, the claim in W2 is that Leibniz gave a "preliminary" formulation of monadology. The method

[22] Wang 1996, 164–165.

[23] *Op.cit.*, 288–289.

[24] *Op.cit.*, 327.

that Husserl supplied, as we will see in G1, G7, G8, is the method of the phenomenological reduction or *epoché*, which would also involve what Husserl call the "eidetic reduction." Without the eidetic reduction, which is supposed to help us to make clear the *essence* of the phenomena to be studied, the entire project would be indistinguishable from the forms of naturalism, empiricism and positivism that Gödel rejects (see G4 below). Plato's "objectivism" is described in Wang's comment W2 as "rudimentary" but as providing a foundation for Husserl's method of the eidetic phenomenological reduction, especially insofar as it makes plausible for Gödel the idea that we are capable of perceiving the primitive concepts of metaphysics clearly enough to set up axioms of metaphysics. For the later Husserl, it is eidetic or 'categorial' intuition that makes it possible to intuit and clarify our grasp of abstract concepts.

Husserl is of course the most recent philosopher on Gödel's list of favorites. Husserl was thus in a position to assess, appropriate, and qualify ideas in the work of Plato, Leibniz, and Kant, and I will note a number of the developments in this direction below. I have argued elsewhere that many of Husserl's ideas lead to more defensible versions of Gödel's claims since they are informed both by philosophical arguments and by results in mathematics and logic that appeared after Kant.[25]

Some additional comments of Wang on Gödel's interest in Husserl are as follows:

(W4) For Gödel, the appeal of Husserlian phenomenology was, I think, that it developed the transcendental method in a way that accommodated his own beliefs in intellectual intuition and the reality of concepts.[26]

(W5) In the 1960s he recommended to some logicians that they should study the sixth investigation in the *Logical Investigations* for its treatment of *categorial intuition*. In his discussions with me in the 1970s he repeatedly urged me to study Husserl's later work.[27]

(W6) Gödel told me that the most important of Husserl's published works are *Ideas* and *Cartesian Meditations*: "The latter is closest to real phenomenology—investigating how we arrive at the idea of the self." According to Gödel, Husserl just provides a program to be carried out; his *Logical Investigations* is a better example of the execution of this program than is his later work, but it has no correct technique because it still adopts the "natural" attitude.[28]

(W7) Gödel's own main aim in philosophy was to develop metaphysics—specifically, something like the monadology of Leibniz transformed into exact theory—with the help of phenomenology.[29]

---

[25] Tieszen 2011.

[26] Wang 1996, 165.

[27] *Op.cit.*, 164.

[28] *Op.cit.*, 164.

[29] *Op.cit.*, 166.

Wang also recorded in his notes certain direct comments of Gödel on Husserl. I reproduce a few of these here, in order to refer to them in the analysis that follows:

(G1) Husserl's is a very important method as an entrance into philosophy, so as to finally arrive at some metaphysics. Transcendental phenomenology with *epoché* as its methodology is the investigation (without knowledge of scientific facts) of the cognitive process, so as to find out what really appears to be—to find the objective concepts.[30]

(G2) Husserl used Kant's terminology to reach, for now, the foundations and, afterwards, used Leibniz to get the world picture. Husserl reached the end, arrived at the science of metaphysics. [Wang notes that this last sentence is different from what Gödel said on other occasions.][31]

(G3) Some reductionism is right: reduce to concepts and truths, but not to sense perceptions. Really it should be the other way around: Platonic ideas [Wang includes: what Husserl calls "essences" and Gödel calls "concepts"] are what things are to be reduced to. Phenomenology makes them [the ideas] clear.[32]

(G4) Leibniz believed in the ideal of seeing the primitive concepts clearly and distinctly. When Husserl affirmed our ability to 'intuit essences' he had in mind something like what Leibniz believed.[33]

(G5) Phenomenological investigation of the constitution of mathematical objects is of fundamental importance for the foundations of mathematics.[34]

(G6) Positivists decline to acknowledge any a priori knowledge. They wish to reduce everything to sense perceptions. Generally, they contradict themselves in that they deny introspection as experience, referring to higher mental phenomena as "judgments." They use too narrow a notion of experience and introduce an arbitrary bound on what experience is, excluding phenomenological experience.[35]

Among other things, these comments of Gödel and Wang indicate that it is Husserl's *transcendental* phenomenology, with its *epoché* (= phenomenological reduction), that is of most interest to Gödel. He is interested in Husserl's later view of phenomenology as transcendental eidetic monadology. G6 contrasts positivism and phenomenology and can be seen as complementing Gödel's critique of Carnap's early view of 'mathematics as syntax of language.'[36] It is known that Gödel attended meetings of the Vienna Circle. He is said to have been rather quiet at these meetings. Many of his later comments (for example W3) show that he had a rather negative view of logical positivism.

Gödel mentions the *epoché* explicitly in G1. Here is another comment that Gödel makes about the *epoché*:

(G7) Introspection is an important component of thinking; today it has a bad reputation. Introspective psychology is completely overlooked today. *Epoché* concerns how intro-

[30] *Op.cit.*, 166.

[31] *Op.cit.*, 166.

[32] *Op.cit.*, 167.

[33] *Op.cit.*, 168.

[34] Wang 2011, 100.

[35] Wang 1996, 173.

[36] Gödel *1953/59.

spection should be used, for example, to detach oneself from influences of external stimuli (such as fashions of the day).[37]

In a further comment in which he runs together the notion of introspection with the *epoché*, Gödel says

(G8) One fundamental discovery of introspection marks the true beginning of psychology. This discovery is that the basic form of consciousness distinguishes between an intentional object and our being pointed (directed) toward it in some way (willing, feeling, cognizing). There are various kinds of intentional object. There is nothing analogous in physics. This discovery marks the first division of phenomena between the psychological and the physical. Introspection calls for learning how to direct attention in an unnatural way.[38]

Without going into details about the *epoché* I will only note for now that with the *epoché* we are supposed to suspend or 'bracket' the 'natural attitude', that is, the ordinary assumption of the existence of the world around us.[39] The suspension applies also to the sciences, including psychology, that assume the existence of the objects they study. The point of such a suspension is to shift attention away from the objects and facts in any domain to *consciousness* of the objects and facts. This shift from focusing on objects to the consciousness of objects seems, at least loosely, like introspection, but it is necessary to be very careful about this. Although Gödel uses the term 'introspection' in G6, G7, G8 above it is important to distinguish what Husserl has in mind from empirical introspectionist psychology. Husserl wants to engage in the *Wesenanalyse* of consciousness. Phenomenology, as an eidetic science, is supposed to be a priori in nature.[40] It would issue in an a priori, material or regional ontology. This would all be quite distinct from introspection, at least on standard conceptions of introspection. Essence analysis is not about what is individual, private or subjective. It does not, by its own nature, seek what is particular, what makes one human subject different from another. Essence analysis involves a kind of *abstraction*. It is supposed to be concerned with universal features of consciousness, and these features should be deliverances of *reason*. For example, the claim that beliefs exhibit intentionality is supposed to be an eidetic claim. How could there be a belief that is not about something or other? If I say I have a belief but it is not about anything then I am guilty of a material (not formal) a priori contradiction or 'countersense' involving the concept of belief. I have reduced the notion of belief to absurdity and I am therefore not in a position to analyze it philosophically.

In the *Logical Investigations* and other writings Husserl says that there will no doubt be difficulties in phenomenological analysis due to the seemingly unnatural direction of intuition and thought required by phenomenology (see the "Introduction" to the six *Logical Investigations* and Gödel's comments in G8 above).

---

[37]Wang 1996, 169.

[38]*Loc. cit.*

[39]See, e.g., Husserl 1913, 1923–24.

[40]See, e.g., Husserl 1908.

Instead of becoming lost in the performance of acts built intrinsically on one another and instead of naively positing the existence of objects, we must practice phenomenological reflection. We must, that is, make these acts themselves and their meaning-content our objects. This is a direction of thought that runs counter to deeply ingrained habits. Among other things, the *epoché* involves a shift to analysis of the *meanings* by virtue of which we are directed toward objects in our experience. This is supposed to allow us to focus on our experience itself, on the constitution of the meaning of being, without the prejudices or presuppositions that may be built into the natural attitude or the existing sciences. As Gödel says (G7), it should allow us to detach ourselves from external influences, including fashions of the day.

16.4 Some Relevant Texts of Husserl on Phenomenology as Transcendental Eidetic Monadology

In this section I present some texts of Husserl that are quite close in content to several of the sentiments of Gödel that we have seen above. It seems to be this kind of material that had the most significant impact on Gödel. In the passages that follow we can also see how several themes concerning the work of Plato, Leibniz, and Kant mentioned above come together in Husserl's own writings, especially in works such as the *Basic Problems of Phenomenology*,[41] the "London Lectures",[42] *Erste Philosophie*,[43] the drafts of the *Encyclopedia Britannica* article,[44] *Cartesian Meditations*,[45] and others.

Husserl was beginning to connect his phenomenology with ideas in Leibniz's philosophy already around 1910, and this continued in his writings throughout the 1920s and early 1930s. In the *Cartesian Meditations*, for example, Husserl says that

> (H1) The [transcendental] ego, taken in full concreteness [vs. as mere identical pole, as substrate of habitualities], we propose to call by the Leibnizian name: monad. Since the monadically concrete ego includes also the whole of actual and potential conscious life, it is clear that the problem of *explicating this monadic ego phenomenologically* (the problem of his constitution for himself) must include *all constitutional problems without exception*. Consequently the phenomenology of *self-constitution* coincides with *phenomenology as a whole*.[46]

It should be noted that Husserl speaks only about the transcendental ego in its full concreteness as a monad. We know that Leibniz has a range of different kinds of

[41] Husserl 1910–11.

[42] Husserl 1922.

[43] Husserl 1923–24.

[44] Husserl 1927–28.

[45] Husserl 1931.

[46] *Op.cit.*, 67.

monads but Husserl's focus is much narrower. It is on the kinds of 'monads' that we are. Elsewhere in Husserl's *Cartesian Meditations* we find this:

(H2) Phenomenological transcendental idealism has presented itself as a *monadology*, which, despite all our deliberate suggestions of Leibniz's metaphysics, draws its content purely from phenomenological explication of the transcendental experience laid open by transcendental reduction, accordingly from the most originary evidence, wherein all conceivable evidences must be grounded... Actually, therefore, phenomenological explication is nothing like 'metaphysical construction'....[47]

In claiming in this passage that phenomenological explication is nothing like "metaphysical construction" Husserl is saying, among other things, that phenomenology is not engaged in the naive metaphysics of earlier philosophical projects. With the *epoché* in effect we cannot speak naively, as Leibniz does, about monads as simple, active *non-material substances*. As we saw above, in his 1961 text Gödel says that phenomenology seeks to avoid "the death-defying leap into a new metaphysics" that would only amount to another dubious metaphysical scheme. Husserl wishes to distinguish naive metaphysics from phenomenological ontology.

In a long interesting passage from a draft of the *Encyclopedia Britannica* entry that contains language quite similar to some of Gödel's remarks on Husserl we are told that

(H3) Remarkable consequences arise when one weighs the significance of transcendental phenomenology. In its systematic development, it brings to realization the Leibnizian idea of a universal ontology as the systematic unity of all conceivable a priori sciences, but on a new foundation which overcomes 'dogmatism' through the use of the transcendental phenomenological method. Phenomenology as the science of all conceivable transcendental phenomena and especially the synthetic total structures in which alone they are concretely possible—those of the transcendental single subjects [monads] bound to communities of subjects [monads] is *eo ipso* the apriori science of all conceivable beings [*Seienden*]. But [it is the science], then, not merely of the totality of objectively existing beings taken in an attitude of natural positivity, but rather of the being as such in full concretion, which produces its sense of being and its validity through the correlative intentional constitution. It also deals with the being of transcendental subjectivity itself, whose nature is to be demonstrably constituted transcendentally in and for itself. Accordingly, a phenomenology properly carried through is the truly universal ontology, as over against the only illusorily all-embracing ontology in positivity—and precisely for this reason it overcomes the dogmatic one-sidedness and hence the unintelligibility of the latter, while at the same time it comprises within itself the truly legitimate content [of an ontology of positivity] as grounded originally in intentional constitution.[48]

From the notes for the "London Lectures" we have

(H4) Transcendental phenomenological subjectivity or monadologism as [is a] necessary consequence of the transcendental phenomenological attitude. The knowledge that any objectivity is only what it is through intentional meaning or significance shows that there is only one possibility for an absolute and concrete being: the being of a concretely full

[47] *Op.cit.*, 150.

[48] Husserl 1927–28, 175.

> transcendental subjectivity. *It* is the only genuine 'Substance.' The *ego* is what is it from its own fundamental meaning. The *ego* is in so far as it constitutes itself for itself as being. All other being is merely relative to the ego and is encompassed within the regulated intentionality of subjectivity.[49]

Apart from the references to Leibniz, universal science (ontology), and transcendental phenomenological method in these passages, it is important to note the language about how the monad produces the meaning (sense) of being and of validity through "intentional constitution." The idea that monads constitute the meaning of being of the objects toward which they are (intentionally) directed by their mental acts plays a very important role in my view of how Gödel's ideas can be developed and defended.

Finally, in a formulation that brings together ideas in Leibniz, Plato, and transcendental philosophy, Husserl says

> (H5) Thus, as Phenomenology is developed, the Leibnizian foreshadowing of a Universal Ontology, the unification of all conceivable a priori sciences, is improved, and realized upon the new and non-dogmatic basis of phenomenological method. For Phenomenology as the science of all concrete Phenomena proper to Subjectivity and Intersubjectivity, is *eo ipso* an a priori science of all possible existence and existences. Phenomenology is universal in its scope, because there is no a priori which does not depend upon its intentional constitution, and derive from this its power of engendering habits in the consciousness that knows it, so that the establishment of any a priori must reveal the subjective process by which it is established.
>
> ...
>
> Once the a priori disciplines, such as the mathematical sciences, are incorporated within Phenomenology, they cannot thereafter be beset by 'paradoxes' or disputes concerning principles: and those sciences which have become a priori independently of Phenomenology, can only hope to set their methods and premises beyond criticism by founding themselves upon it. For their very claim to be positive, dogmatic sciences, bears witness to their dependency, as branches merely, of that universal, eidetic ontology which is Phenomenology.
>
> ...
>
> The endless task, this exposition of the Universum of the a priori, by referring all objectives to their transcendental 'origin', may be considered as one function in the construction of a universal science of Fact, where every department, including the positive, will be settled on its a priori.
>
> ...
>
> Thus the antique conception of Philosophy as the Universal Science, Philosophy in the Platonic, Philosophy in the Cartesian, sense, that shall embrace all knowledge, is once more justly restored.[50]

Hence,

> (H6) The ideal of the future is essentially that of phenomenologically based ('philosophical') sciences, in unitary relation to an absolute theory of monads.

[49] Husserl 1922, 72.

[50] Husserl 1927–28, 191–194.

One might easily mistake this latter statement for an assertion by Gödel, but in fact it is a passage from a draft of Edmund Husserl's *Encyclopedia Britannica* article "Phenomenology", written in 1928.[51]

We know that there are notes on Leibniz in Gabelsberger shorthand in the Gödel *Nachlass* but what we do not know is exactly which parts of Leibniz's monadology Gödel would or would not have accepted.[52] Was he prepared, for example, to accept Leibniz's view that there are many different kinds of monads? From comments he made to Wang, it seems that Gödel was ready to accept some elements of Leibniz's theological monadology, such as the existence of a central monad, God, whereas Husserl's monadology is basically secular in nature. In *Ideas I*, for example, Husserl says we must apply the phenomenological *epoché* to the existence of God,[53] and it is not clear how he could support a theological monadology given his views on evidence, intuition and science. It is worth noting that Gödel read and evidently appreciated the essay *Eine neue Monadologie*[54] by one of Husserl's students, Dietrich Mahnke.[55] Mahnke obtained his doctoral degree with Husserl in 1922, writing a thesis entitled *Leibnizens Synthese von Universalmathematik und Individualmetaphysik*. This thesis was published in Husserl's *Jahrbuch für Philosophie und phänomenologische Forschung* in 1925.

After these comments on Leibniz and phenomenology I would also like to take note of some of Husserl's critical comments about Kant that would have resonated with Gödel. We should note, first of all, that the idea that human consciousness exhibits intentionality is, at best, only implicit in Kant's philosophy, while it is front and center in Husserl's work. This means that the ideas about intentionality, meaning and constitution that are so central to Husserl's philosophy are not present in Kant's thinking at all. Husserl does, however, argue for a general Kantian kind of distinction between thinking and intuiting, or signification and intuition. Departing substantially from Kant again, he argues that if we take the intentionality of human consciousness seriously then we must recognize both sensory and categorial intuition. There can be mere thinking or signification concerning sensory objects and there can also be intuition of sensory objects. Analogously, there can be mere thinking or signification concerning categorial objects and there can also be intuition of categorial objects. We saw in W5 that Gödel recommended the Sixth Investigation of Husserl's *Logical Investigations* to some logicians in the 1970s for its account of categorial intuition. Viewed in terms of genetic epistemological analysis, the thinking and intuition in the case of categorial objects, which are objects such as natural numbers, sets, propositions, and the like, is not the most basic kind of *founding* thinking or intuition but is a *founded* kind of thinking and intuition that involves acts of reflection, abstraction, idealization, formalization, and

[51] *Op.cit.*, 191–194.

[52] The ANR project mentioned in footnote 1 might shed further light on this.

[53] Husserl 1913, 133–134.

[54] Mahnke 1917.

[55] van Atten and Kennedy 2003, 457.

so on. It is a thinking about and, where possible, an intuiting of ideal objects. Husserl sometimes calls the intuition of ideal objects, especially in connection with intuition of essences, "ideation." In the *Logical Investigations* and later work he distinguishes "real" from "ideal" objects. "Real" objects are objects that are either temporal in nature (such as 'inner' mental processes), or temporal and spatial (such as 'outer' physical objects), while ideal objects such as numbers, sets, and propositions are neither temporal nor spatial. Regarding intuition, Husserl holds that there can be adequate and inadequate intuitions and, in fact, that there are degrees of (in)adequacy. We also need to recognize a difference between individual and universal intuition. Husserl says that individual intuition is usually conceived exclusively in a narrow way as sensory intuition and that this is baseless. On his alternative view the distinction between individual and universal intuition also has an application with respect to ideal objects.

Gödel's complaints about Kant, complaints which reflect his platonism (objectivism) and rationalism about mathematics and logic, are recurring themes in his philosophical remarks. Now note, by way of contrast, what Gödel says about Husserl's transcendental idealism in a draft letter of 1972 (in the *Nachlass*) to Gian-Carlo Rota:

> (G9) I believe that his [Husserl's] transcendental phenomenology, carried through, would be nothing less than Kant's critique of pure reason transformed into an exact science, except for the fact that [in footnote: Kant's subjectivism and negativism for the most part would be eliminated] the result (of the 'critique') would be far more favorable for human reason.[56]

The Kantian critique of reason was clearly too restrictive by Gödel's sights. Husserl agrees that it is too restrictive, and he portrays the phenomenological method as a way to develop and defend a new kind of rationalism that avoids the excesses of older forms of rationalism but also avoids any kind of mysticism.

Let me bring this section to a close by noting how Husserl formulates the problem of the relation of the subjectivity of consciousness to the objectivity and ideality of logic (and mathematics) in his later work. I think this was probably quite important for Gödel, given his interest in transcendental phenomenology and his remark in G5 above about the importance of the phenomenological investigation of the 'constitution of mathematical objects.' Husserl says in §100 of *Formal and Transcendental Logic* that

> (H7) Pure logic has as its thematic sphere ideal formations. But they would have had to be clearly seen, and definitely apprehended, as such ideal objectivities, before transcendental questions about them and about pure logic could have been asked. The eighteenth century and the age that followed were so strongly actuated by empiricism (or better, by anti-platonism) that nothing was remoter from them than recognition of ideal formations as being objectivities—in the manner of the good and never-relinquishable sense whose legitimacy we have established in detail.

[56] *Op.cit.*, 446.

> Nothing else hindered a clear insight into the sense, into the proper questions and methods of genuine transcendental philosophy so much as did this anti-platonism, which was so influential that it actuated all parties, and even the thinking of a Kant, struggling to free himself from empiricism.
>
> For the succeeding age this meant, however, that those investigations in the psychology of cognition, or rather those transcendental phenomenological investigations, that are the thing actually needed for a full and, therefore, two-sided logic were never seriously undertaken. But that was because no one ventured, or had the courage to venture, to take the ideality of the formations with which logic is concerned as the characteristic of a separate, self-contained, 'world' of ideal Objects and, in so doing, to come face to face with the painful question of how subjectivity can in itself bring forth, purely from sources appertaining to its own spontaneity, formations that can be rightly accounted as ideal Objects in an ideal 'world.'
>
> For only then was one faced with the unintelligibility of *how ideal objectivities* that originate purely in our own subjectivities of judgment and cognition, that are there originaliter in our field of consciousness purely as formations produced by our own spontaneity, *acquire the being-sense of 'Objects'*, existing in themselves over against the adventitiousness of the acts and the subjects. How does this sense 'come about', how does it originate in us ourselves? And where else could we get it, if not from our own sense-constituting performance?

Husserl asks how human subjectivity can bring forth formations from sources of its own spontaneity that can be considered as ideal objects in an ideal world. He asks how the objects of cognition in logic can acquire their sense or meaning as ideal and existing in themselves, over against the subjective acts in which they are known. How does this sense or meaning *originate* in us? This is the key question that leads to what I have called 'constituted platonism.'

16.5 A Brief Overview of Phenomenological Monadology and Mathematics

With the material described above in mind, the general picture of Husserl's transcendental eidetic monadology that we obtain is roughly as follows. The transcendental ego in its full concreteness is a "monad" ("substance"). It constitutes the meaning of being of the world through its intentionality. In the case of mathematics, logic, and the other a priori sciences, including phenomenology itself, it constitutes the meaning of the being of its objects (essences, categorial objects) in a rationally motivated way as ideal or abstract and non-mental. Evidence in these domains is acquired on the basis of categorial or eidetic intuition. This suggests a kind of platonism with its emphasis on non-mental and mind-independent ideal objects (in the sense of "mind-independence$_2$" discussed in Tieszen 2010 and 2011), with its rationalism, and its robust sense of objectivity. I call this kind of platonism *constituted platonism*. Constituted platonism is unlike traditional mathematical platonism since traditional platonists have not been transcendental (phenomenological) idealists. Plato certainly did not speak of the constitution of the meaning of being by "monads", and he is engaged, by Husserl's sights, in naive metaphysics. This is also true of other traditional mathematical platonists.

Leibniz is a kind of theological platonist about mathematical objects or facts. In various places in his writings he describes the region of ideas, which includes eternal objects such as essences, numbers, truths and the like, as an intelligible world in the mind of God, and he says that there are propositions that are true even though nobody is actually thinking them. Leibniz also asserts in many places that we humans (unlike God) cannot think without the aid of signs or characters (and hence a language) of some sort. There is supposed to be some non-arbitrary relation between characters (signs) and things, according to Leibniz, but what he says about this is quite murky. In any case, the universal characteristic would enter the scene at this point. The signs we use in areas such as mathematics would operate as surrogates for the abstract 'eternal' ideas. Signs are concrete, real entities that are given to us in ordinary perception and we can use them to calculate, whereas abstract ideas are harder for the mind to handle. Leibniz says that although we are able to reach some primitive and very simple ideas by *intuition*, for the most part we have only *symbolic* thought of composites. Intuition will not take us very far in the case of abstract ideas. This is quite interesting because it is very much like what Husserl says in his earliest work on the philosophy of arithmetic (Husserl 1891). Arithmetic, Husserl says early in his career, is almost entirely symbolic in nature. He was no doubt influenced by Leibniz in making these kinds of claims, but with his new-found platonism around 1900 (in the *Logical Investigations*) Husserl introduces his notion of categorial intuition to account for the possibility of knowledge of ideal or abstract objects, and it is this notion that is of great interest to Gödel. Embedded in transcendental eidetic monadology, this notion offers an alternative to a mechanist view of reason, to logical positivism, to empiricism or naturalism about mathematics and logic, and so on. Husserl also thinks that language has an important part to play in the philosophy of mathematics and logic. In later work (see, e.g., "The Origins of Geometry") he takes written language in particular to be a condition for the possibility of the constitution of mathematical and logical objects as ideal because it makes possible the sense of the persisting existence of such objects during periods when particular human subjects are not thinking about the objects or are even no longer alive. The meanings of mathematical statements become sedimented in linguistic signs and can therefore be reactivated again and again for an individual subject (monad) or for multiple subjects in communication. Husserl says that with written language the meaning of ideal objects is thus constituted as "continuing-to-be" even when no one has consciously realized the objects in self-evidence.

Leibniz is of course also a rationalist who is interested in philosophy as a rational universal science. He is interested in deciding mathematical and other problems by human reason, through the analysis of concepts, although in his writings on logic he tends to think of decidability in a mechanical way. For Leibniz, as for other classical rationalists, concepts of reason, including those of logic and mathematics, are exact and our grasp of such concepts either is or can be made clear and distinct, whereas empirical knowledge lacks, in various degrees, just these features. Leibniz holds that the science of possibilities and necessities precedes sciences of actualities. Leibniz is a monadologist but his monadology is not brought into line with the methods of transcendental phenomenology and in this respect it remains, by Husserl's sights, naive. In the quotation from the *Encyclopedia Britannica* draft above, however, Husserl says that the systematic development of transcendental phenomenology brings to realization the Leibnizian idea of a universal ontology as the systematic unity of all conceivable a priori science on a foundation that overcomes dogmatism and one-sidedness through the use of the transcendental method. Phenomenology is the science of all conceivable beings, taken not in the attitude of naive positivity but rather as understood though correlative intentional constitution.

Kant is not a monadologist, although his idea of the transcendental unity of apperception foreshadows Husserl's transcendental ego. Husserl, as we saw, refers to the transcendental ego in its full concreteness as a monad. Kant, Leibniz and Plato do not put the intentionality of human consciousness at the center of their philosophical views. Kant is also not a platonist about mathematical objects or facts, and he mounts a critique of classical rationalism (including Leibniz). For Kant, knowledge is restricted to sensory intuition and the two forms of sensory intuition, space and time. Kant, unlike Husserl, distinguishes phenomena from noumena and is able to develop the transcendental method far enough to show how *empirical realism* is compatible with transcendental idealism, but in his work there is no question of showing how a kind of platonism or mathematical objectivism is compatible with transcendental idealism.

Thus, in transcendental phenomenology the transcendental ego in its full concreteness as a monad can now be combined with a kind of constituted secular platonism about logic and mathematics (unlike in Leibniz and Kant), and with the idea of universal science (as in Leibniz and Plato) in a way that keeps Kantian transcendental method or idealism in broad outlines but extends it to mathematics, logic, and philosophy itself, avoiding Kant's dualism, his restrictions on intuition, his critique of rationalism and his skepticism about ideal or abstract objects (concepts). Elements in the work of Plato, Leibniz, Kant, and Husserl come together in one picture in which the monad (as a concrete transcendental ego), in a community of monads, constitutes the meaning of being of its objects in mathematics and logic as ideal or abstract and non-mental and acquires evidence in these domains on the basis of categorial intuition or *Wesensanalyse*.

One can in principle substitute for "monad" in the singular in this picture the plural "monads", or transcendental egos. The constitution of the meaning of being of one objective world, Husserl says, requires the community of monads, a single universe of compossible monads. Harmony among the monads is required. Each monad (transcendental ego), to extend the analogy, would presumably be "windowless" but would mirror all of the others, if there is to be constitution of one objective world. There is no objectivity about a mathematical proof, for example, if there is no harmony among the monads, each with its own perspective, regarding the proof. We see in mathematical practice how a proof offered by a particular mathematician may be corrected, clarified, simplified, extended, etc., once it is subjected to the different perspectives of various human monads, i.e., once it is made available to the mathematical community. The meaning of 'objectivity' is now not merely intrasubjective. It is not merely invariance through the multiplicity of mental states of an individual subject but is rather invariance through a multiplicity of subjects over time, often with an extensive history. It is made much more robust with the addition of this intersubjective and historical dimension. Much more can and should be said about the role of intersubjectivity in constituted platonism. Issues about intersubjectivity and the layers of constitution involved in the meaning of being of the objective world have of course been analyzed in great detail by Husserl and others in the phenomenological movement.

16.6 The Incompleteness Theorems and Husserl's Notion of "Definite" Axiom Systems

An issue that has sometimes been raised about Gödel and Husserl is whether Gödel's incompleteness theorems undermine Husserl's claims about what he calls "definite" axiom systems and their ontological correlates, "definite manifolds." Husserl's views in foundations were not informed by Gödel's incompleteness theorems. His ideas about definite axiom systems and definite manifolds were formulated early in his career, in the late 1800s through the first few years of the 1900s.[57] Gödel's results on incompleteness were published in 1931 when Husserl was in his seventies and was not focused on logic and foundations of mathematics. It should be noted that no direct comments on definite axiom systems and manifolds have been found in Gödel's writings on Husserl. It appears that Gödel was looking elsewhere in Husserl's later philosophy for help with his philosophical worries. One might conjecture that this absence of comment on definite axiom systems and manifolds is due to the fact that by Gödel's time, and in part due to Gödel himself (compare, for example Gödel 1934 with any of Husserl's writings on logic) work in mathematical logic had progressed to a point at which Husserl's somewhat vague notion of definiteness had been supplanted by precise and distinct concepts such as syntactical completeness, semantic completeness, and categoricity.[58]

Jean Cavaillès was perhaps the first person to claim that Husserl's views on "definite" axiomatic formal systems and definite manifolds, are incompatible with Gödel's incompleteness theorems.[59] In my view the issue is clouded by a number of factors: (i) it is not clear, especially in light of later developments in mathematical logic, what "definiteness" is supposed to be, and there has been disagreement about this in the secondary literature on Husserl.[60] It is apparent in any case that there

---

[57] See Tieszen 2004.

[58] Suzanne Bachelard already distinguishes these properties in her study of Husserl's *FTL* (Bachelard 1968). See her discussion on pp. 49–63.

[59] Cavaillès 1947, 72.

[60] For some of the main sources on these issues see, e.g., Lohmar 1989, Ortiz Hill 1995, 2000, Majer 1997, da Silva 2000, Hartimo 2007, Centrone 2010.

is not a sharp distinction between syntax and semantics, as we would now think of it, in Husserl's work. As just mentioned, does he have in mind something like syntactical completeness, or semantic completeness, or perhaps even categoricity? These are quite different properties and the relationships between them are nuanced in ways that are not reflected in Husserl's work. If what Husserl meant by "definite" was semantic completeness or categoricity, then Gödel's incompleteness theorems do not (directly) apply. Stefania Centrone has pointed out that in his Göttingen *Doppelvortrag* of 1901/1902 Husserl does distinguish "absolute" from "relative" definiteness.[61] On her account the former is analogous to Hilbert's axiom of completeness and the latter amounts to the idea that every proposition written in the language of a system can be decided on the basis of the axioms of the system; (ii) Did Husserl think that all axiomatic formal systems and manifolds had to be definite? He sometimes says, for example, that it is manifolds in the "pregnant sense" that are definite. Centrone notes that for Husserl relative definiteness is a property that not all formal systems possess; (iii) Is finite axiomatizability part of the notion of definiteness? (iv) in connection with (i), the very sharp, circumscribed conception of a formal system as syntax that is required for Gödel's incompleteness theorems is not clearly present in Husserl's work. Think of all of the purely syntactical properties involved in the incompleteness theorems. Where in Husserl's work do we see the notion of a formal system as pure syntax so sharply circumscribed that, for example, we could assign Gödel numbers to formulas, proofs, etc., that is, that we could arithmetize syntax? (v) nuances related to how logic should be distinguished from mathematics, e.g., is definiteness supposed to be possible in all axiomatic systems of logic *and* mathematics?

In reading Husserl's work on logic and mathematics one might have the impression that he thinks that an axiomatic grounding can be carried out once and for all. In a section on Gödel's incompleteness theorems in her study of Husserl's *FTL*, Suzanne Bachelard concludes that "the Husserlian ideal of axiomatic grounding carried out once and for all is hence transcended, for it reveals a ponderousness that has become foreign to contemporary mathematical technique."[62] "What is more", she says, "the rigidity of the Husserlian ideal of deductive science is the price paid for an *enthousiasme de la raison* toward this power of systematization."[63] How should this "rational enthusiasm", however, be interpreted? There are indications in some of Husserl's work, especially the later work, that one should think of systematization, idealization, abstraction, axiomatization and formalization as founded activities that can in principle be improved continuously through the use of human reason, where we regard the perfection of these capacities as an infinitely distant goal. In *Ideas III*, for example, Husserl speaks of "the infinite *ideal of a system of all ontologies* and eidetic disciplines in general."[64]

[61] Centrone 2010.

[62] Bachelard 1968, 54.

[63] *Op.cit.*, 55.

[64] Husserl 1980, 90.

If we focus on the main concepts involved in Gödel's incompleteness theorems—syntactic completeness or decidability and syntactic consistency—then we might think of these as something like Kantian regulative ideals for axiomatic formal theories. According to the first incompleteness theorem, for example, we can add new axioms non-arbitrarily, in the form of Gödel sentences, to existing formal systems, in order to obtain more powerful formal systems that will prove more mathematical propositions. This can be carried on indefinitely. One can think of this process as regulated by the not fully realizable ideal of achieving syntactic completeness, and therefore of the *ideal* of achieving "definiteness" *if* definiteness = syntactic completeness. Gödel applies this kind of model to various areas of mathematics, including higher set theory. He argues, for example, that we need to find a new axiom or axioms for Zermelo-Fraenkel set theory if we are to decide the Continuum Hypothesis.[65] In this case the new axioms do not just pop out of the first incompleteness theorem or out of the existing analyses of the concept of set. But this is why Gödel says we need a deeper analysis of the abstract concept of set, which is something that he evidently thinks will be possible for human monads who employ the phenomenological method with the suitable background knowledge in order to further clarify the meaning of this concept. He made suggestions for how to extend set theory with the use of axioms of infinity, but other paths might also be possible. In any case, progress in this matter will be a function of our capacity for reason and Gödel, looking at the other mostly leftward philosophical positions that were on offer at the time, evidently found them seriously wanting in this regard.

Gödel's *enthousiasme de la raison*, or what has been called his "rationalistic optimism", can thus perhaps be fruitfully compared with Husserl's. Gödel certainly knows that we cannot have a universal syntactically consistent and complete axiomatic formal system for all of mathematics, and yet he speaks of continuously clarifying our (rational) intuition of mathematical and logical concepts in order to decide open problems. This is what one would expect from his platonic rationalism about mathematics and logic: (i) *platonism*—the presumably infinite and mind-independent domain of mathematical objects and concepts transcends the grasp of finite monads at any given point in time, and yet (ii) *rationalism*—with the capacity for categorial intuition in this domain, as a form of rational intuition, there is the possibility of coming to decisions about more and more undecided mathematical propositions. The incompleteness theorems, Gödel says, do not imply that there are absolutely undecidable mathematical problems. It is possible that every clearly posed mathematical yes-or-no question is solvable because more and more new axioms might become evident to us on the basis of the meaning of the primitive concepts. Gödel says that this is something that a machine cannot emulate. He does not, that is, identify what is decidable by human reason in mathematics with what is mechanically decidable in formal systems.[66] For a platonic rationalist, abstract or ideal objects and concepts transcend concrete, mechanical finitary formal systems

[65] Gödel 1964, 1972a.

[66] See also Gödel 1934, 1972a.

and there need be no end to the process of uncovering and clarifying the mind-independent abstract concepts of mathematics and logic, where we think of this as a regulative ideal.

Let us consider a passage from Husserl in order to illustrate this kind of platonic rationalism. In the *Logical Investigations* Husserl says at one point (Investigation I, § 35) that

> (H8) There is . . . no intrinsic connection between the ideal unities which in fact operate as meanings, and the signs to which they are tied, i.e., through which they become real in human mental life. We cannot therefore say that all ideal unities of this sort are expressed meanings. Whenever a new concept is formed, we see how a meaning becomes realized that was previously unrealized. As numbers – in the ideal sense that arithmetic presupposes – neither spring forth nor vanish with the act of enumeration, and as the endless number series thus represents an objectively fixed set of general objects, sharply delimited by an ideal law, which no one can either add to or take away from, so it is with ideal unities of pure logic, with its concepts, propositions, truths, or in other words, its meanings. They are an ideally closed set of general objects, to which being thought or being expressed are alike contingent. There are therefore countless meanings which . . . are merely possible ones, since they are never expressed, and since they can, owing to the limits of man's cognitive powers, never be expressed.

With categorial intuition, also introduced in the *LI*, we should be able to realize and then express more and more, but not all, of these meanings. We should in principle be able to decide certain mathematical propositions we cannot now decide. So we do not suppose that we will reach a point at which we can *actually* decide all mathematical yes-or no questions, but we also do not suppose that there are absolutely undecidable propositions. We cannot collapse the distinction between the ideal and the real. The enthusiasm of reason, according to this platonic rationalism, says there is more to do. Keep going. In mathematics, as Hilbert said, there is no *ignorabimus*.

We should, however, recall from W6 that Gödel says that the "*Logical Investigations* . . . has no correct technique because it still adopts the 'natural' attitude." So suppose we embed our comments in Husserl's later transcendental eidetic monadology. Compare the remarks in H8 with Husserl's various comments on transcendental idealism.[67] § 94 of *FTL*, for example, has the heading "every existent is constituted in the subjectivity of consciousness." Does H8 contradict the claim that every existent is constituted in the subjectivity of consciousness? I would claim that the threat of contradiction vanishes if everything that is said in H8 is qualified with the claim that the remarks are now about how the meaning of being of ideal objects is constituted non-arbitrarily by human monads who are engaged in the practice of classical pure mathematics and logic. This is 'constituted platonism.' Much more could be said about all of this but I cannot go further into these matters here.[68]

---

[67] See, e.g., Husserl 1931, § 41 or Husserl 1929, § 94.

[68] See also Tieszen 2010, 2011.

16.7 Some Final Observations on Husserl and Gödel

On some basic issues concerning the ontology and epistemology of what he calls "ideal" objects I would separate Husserl's writings on logic and mathematics into three main stages.[69] To put it briefly, (1) in his early writings on arithmetic and formal systems, prior to 1900, there is basically no ontology or epistemology of ideal objects and there is little if anything by way of platonism; (2) with the *Logical Investigations* and subsequent writings we find a sharp distinction between the "real" and the "ideal" and fairly robust expressions of a kind of platonism about ideal objects. This underwrites his refutation of psychologism; (3) with his 'transcendental turn' around 1906 and thereafter ideal objects and the platonism are situated in the context of transcendental phenomenology. At stage 2 we do not yet have this. It is stage 3 of Husserl's work, with its introduction of the *epoché*, that is of primary interest to Gödel. I would argue, whether this would be Gödel's final position or not, that this stage issues in constituted platonism.

There are a number of themes in Husserl's work, which overlap to some extent with ideas in either Plato, Leibniz, or Kant, that stand out in Gödel's comments. He was quite interested in Husserl's idea that philosophy can be a rigorous, universal apriori science. Wang told me that he and Gödel studied Husserl's essay "Philosophy as Rigorous Science",[70] and Gödel would have known about this central aspect of Husserl's philosophy from his other readings. In light of the philosophical uses to which Gödel puts the incompleteness theorems, we might view the incompleteness theorems themselves as examples of philosophy become rigorous science. Transcendental idealism, by way of phenomenology, is important for Gödel. The idea would be to use the phenomenological method (*epoché*) to develop a new kind of monadology, but this transcendental eidetic monadology would be combined with a type of platonism or realism that recognizes the objectivity of ideal or abstract objects and concepts of mathematics, logic, and philosophy. Here we would investigate the constitution of mathematical and logical objects in the founded rational activities of human monads. Transcendental eidetic monadology, furthermore, acknowledges and seeks to cultivate categorial or eidetic intuition of ideal objects. For Gödel this holds out the possibility of clarifying the meaning of primitive concepts of logic and mathematics and could even contribute to solving open problems in the foundations of mathematics and logic, including higher set theory. Furthermore, it is in this way that one might provide a foundation for the sciences and for philosophy itself.

Gödel opposes logical positivism, naturalism, conventionalism, nominalism, and empiricism about logic, mathematics, and philosophy. He argues at length against Carnap's view of mathematics as syntax of language, and against the leftward aspects of Hilbert's formalism about the foundations of mathematics. He argues

---

[69] See Tieszen 2004.

[70] See also Wang 1996, 146; Tieszen 1998.

against the mechanist conception of the human mind as a Turing machine. He also argues against subjectivism, psychologism, and Aristotelian realism about the concepts and objects of logic and mathematics. In his thinking after 1959 it appears that Gödel was beginning to connect many of these points with his study of Husserl's transcendental eidetic phenomenology.

References

In references to Husserl's works I have used original dates to indicate historical development even though English translations, where available, are cited.

S. Bachelard, *A Study of Husserl's Formal and Transcendental Logic* (Northwestern University Press, Evanston, 1968)

J. Cavaillès, *Sur la logique et la théorie de la science* (Vrin, Paris, 1947)

S. Centrone, *Logic and Philosophy of Mathematics in the Early Husserl* (Springer, Dordrecht, 2010)

J. Da Silva, Husserl's two notions of completeness: Husserl and Hilbert on completeness and imaginary elements in mathematics. Synthese **125**, 417–438 (2000)

J. Da Silva, Gödel and transcendental phenomenology. Rev. Int. Philos. **234**, 553–574 (2005)

S. Feferman et al. (eds.), *Kurt Gödel: Collected Works*, vol I (Oxford University Press, Oxford, 1986)

S. Feferman et al. (eds.), *Kurt Gödel: Collected Works*, vol II (Oxford University Press, Oxford, 1990)

S. Feferman et al. (eds.), *Kurt Gödel: Collected Works*, vol III (Oxford University Press, Oxford, 1995)

D. Føllesdal, Gödel and Husserl, in *From Dedekind to Gödel: Essays on the Development of the Foundations of Mathematics*, ed. by J. Hintikka (Kluwer, Dordrecht, 1995), pp. 427–446

K. Gödel, On undecidable propositions of formal mathematical systems (1934), in Feferman et al. (1986), pp. 346–371

K. Gödel, Undecidable diophantine propositions (*193?), in Feferman et al. 1995a, pp. 164–175

K. Gödel, Russell's mathematical logic (1944), in Feferman et al. (1990), pp. 119–143

K. Gödel, Some basic theorems on the foundations of mathematics and their implications (*1951), in Feferman et al. (1995b), pp. 304–323

K. Gödel, Is mathematics a syntax of language? (*1953/9, III and V), in Feferman et al. (1995c), pp. 334–363

K. Gödel, The modern development of the foundations of mathematics in the light of philosophy (*1961/?), in Feferman et al. (1995d), pp. 374–387

K. Gödel, What is Cantor's continuum problem? (1964), in Feferman et al. (1990), pp. 254–270. Revised version of Gödel (1947)

K. Gödel, On an extension of finitary mathematics which has not yet been used (1972a), in Feferman et al. (1990), pp. 271–280. Revised version of Gödel (1958)

K. Gödel, Some remarks on the undecidability results (1972b), in Feferman et al. (1990), pp. 305–306

M. Hartimo, Towards completeness: Husserl on theories of manifolds 1890–1901. Synthese **156**, 281–310 (2007)

K. Hauser, Gödel's program revisited part I: The turn to phenomenology. Bull. Symb. Log. **12**, 529–588 (2006)

C. O. Hill, Husserl and Hilbert on completeness, in *From Dedekind to Gödel*, ed. by J. Hintikka (Ed), (Kluwer, Dordrecht, 1995), pp. 143–163

C. O. Hill, Husserl's *Mannigfaltigkeitslehre*, in *Husserl or Frege? Meaning, Objectivity, and Mathematics*, ed. by C.O. Hill, G.E. Rosado Haddock (Open Court, La Salle, 2000) pp. 161–178

E. Husserl, *Philosophy of Arithmetic: Psychological and Logical Investigations* (1891), translated by D. Willard (Kluwer, Dordrecht, 2003)

E. Husserl, *Logical Investigations*, vols. I, II. (1900–01), Translation of the 2nd edition by J. N. Findlay (Routledge and Kegan Paul, London, 1973)

E. Husserl, Philosophy as rigorous science (1908), translated by Q. Lauer, in *Phenomenology and the Crisis of Philosophy* (Harper, New York, 1965), pp. 71–147

E. Husserl, Ideas pertaining to a pure phenomenology and to a phenomenological philosophy. in *First Book* (1913), translated by F. Kersten (Kluwer, Dordrecht, 1982)

E. Husserl, The London lectures (Syllabus of a course of four lectures) (1922), in P. McCormick, F. Elliston (eds.), *Husserl: Shorter Works* (University of Notre Dame Press, Notre Dame, 1981), pp. 68–74

E. Husserl, Erste Philosophie. Erster Teil: Kritische Ideengeschichte (1923–24), *Husserliana*, vol. XII (Nijhoff, The Hague, 1956a)

E. Husserl, Erste Philosophie. Zweiter Teil: Theorie der phänomenologischen Reduktion (1923–24), *Husserliana*, vol. XIII (Nijhoff, The Hague, 1956b)

E. Husserl, 'Phenomenology' (drafts of the Encyclopedia Britannica article) (1927–28), in *Psychological and Transcendental Phenomenology and the Confrontation with Heidegger* (1927–1931) (Kluwer, Dordrecht, 1997), pp. 83–194

E. Husserl, *Formal and Transcendental Logic* (1929), translated by D. Cairns (Nijhoff, The Hague, 1969)

E. Husserl, *Cartesian Meditations. An Introduction to Phenomenology* (1931), translated by D. Cairns (Nijhoff, The Hague, 1960)

E. Husserl, *The Crisis of the European Sciences and Transcendental Phenomenology* (1954), translated by D. Carr (Northwestern University Press, Evanston, 1970). Original German manuscripts written 1935–38

E. Husserl, *Ideas Pertaining to a Pure Phenomenology and to a Phenomenological Philosophy. Third Book: Phenomenology and the Foundations of the Sciences*, translated by T. Klein, W. Pohl (Nijhoff, The Hague, 1980). Original manuscript dating from 1912

I. Kant, *Critique of Pure Reason*, translated by N.K. Smith (Macmillan, London 1973). First edition originally published in 1781; second edition originally published in 1787

X. Liu, Gödel's Philosophical Program and Husserl's Phenomenology. Synthese **175**, 33–45 (2010)

D. Lohmar, *Phänomenologie der Mathematik*, Elemente einer phänomenologischen Aufklärung der mathematischen Erkenntnis nach Husserl (Kluwer, Dordrecht, 1989)

D. Mahnke, *Eine neue Monadologie*. Kantstudien Ergänzungsheft 39 (Reuther and Reichard, Berlin, 1917)

D. Mahnke, Leibnizens Synthese von Universalmathematik und Individualmetaphysik. Erster Teil. Jahrbuch für Philosophie und phänomenologische Forschung **VII**, 304–611 (1925)

U. Majer, Husserl and Hilbert on completeness: A neglected chapter in early twentieth century foundations of mathematics. Synthese **110**, 37–56 (1997)

K. Menger, *Reminiscences of the Vienna Circle and the Mathematical Colloquium*, in L. Golland et al. (eds.) (Kluwer, Dordrecht, 1994)

R. Tieszen, *Mathematical Intuition: Phenomenology and Mathematical Knowledge* (Kluwer, Synthese Library series, Dordrecht, 1989)

R. Tieszen, Kurt Gödel and phenomenology. Philos. Sci. **59**(2), 176–194 (1992). Reprinted in Tieszen (2005)

R. Tieszen, Gödel's path from the incompleteness theorems (1931) to phenomenology (1961). Bull. Symb. Log. **4**(2), 181–203 (1998). Reprinted in Tieszen (2005)

R. Tieszen, Gödel and the intuition of concepts. Synthese **133**(3), 363–391 (2002). Reprinted in Tieszen (2005)

R. Tieszen, Husserl's logic, in *Handbook of the History of Logic*, ed. by D. Gabbay, J. Woods, vol. III, (Elsevier, Amsterdam, 2004), pp. 207–321
R. Tieszen, *Phenomenology, Logic, and the Philosophy of Mathematics* (Cambridge University Press, Cambridge, 2005)
R. Tieszen, Mathematical Realism and Transcendental Phenomenological Idealism, in *Phenomenology and Mathematics*, ed. by Hartimo, (Springer, Dordrecht, 2010), pp. 1–22
R. Tieszen, *After Gödel: Platonism and Rationalism in Mathematics and Logic* (Oxford University Press, Oxford, 2011)
R. Tieszen, Monads and Mathematics: Gödel and Husserl. Axiomathes **22**, 31–52 (2012)
R. Tragesser, *Phenomenology and Logic* (Cornell University Press, Ithaca, 1977)
M. van Atten, Two Draft Letters from Gödel on Self-Knowledge of Reason. Philos. Math. **14**(2), 255–261 (2006)
M. van Atten, J. Kennedy, On the philosophical development of Kurt Gödel. Bull. Symb. Log. **9**, 425–476 (2003)
H. Wang, *From Mathematics to Philosophy* (Humanities Press, New York, 1974)
H. Wang, *Reflections on Kurt Gödel* (MIT Press, Cambridge, MA, 1987)
H. Wang, *A Logical Journey: From Gödel to Philosophy* (MIT Press, Cambridge, MA, 1996)

Chapter 17
Husserl and Jacob Klein

Burt C. Hopkins

Abstract The article explores the relationship between the philosopher and historian of mathematics Jacob Klein's account of the transformation of the concept of number coincident with the invention of algebra and Husserl's early investigations of the origin of the concept of number and his late account of the Galilean impulse to mathematize nature. Klein's research is shown to present the historical context for Husserl's twin failures in the *Philosophy of Arithmetic*, (1) to provide a psychological foundation for the proper concept of number and (2) to show how this concept of number functions as the mathematical foundation of universal (symbolic) arithmetic. The argument is advanced that one significant result of bringing together Klein's and Husserl's thought on these issues is the need to fine-tune Husserl's *Crisis* project of desedimenting the mathematization of nature. Specifically, Klein's research shows that "a 'sedimented' understanding of numbers" "is superposed upon the first stratum of 'sedimented' geometrical 'evidences'" uncovered by Husserl's fragmentary analyses of geometry in the *Crisis*. In addition, then, to the task of "the intentional-historical reactivation of the origin of geometry" recognized by Husserl as intrinsic to the reactivation of the origin of mathematical physics, Klein discloses a second task, that of "the reactivation" of the "complicated network of sedimented significances" that "underlies the 'arithmetical' understanding of geometry."

Keywords Edmund Husserl • Jacob Klein • François Vieta • Arithmetic • Concept of Number • Symbolic Mathematics • Unity • Multiplicity

17.1 Introduction

The first "Klein" typically associated with Edmund Husserl is the mathematician Felix Klein (1849–1925), whose final years at Göttingen overlapped with all but two of the years Husserl spent there (1901–1915). No reference to Felix Klein's

---

B.C. Hopkins (✉)
Université de Lille, UMR-CNRS 8163 STL, Lille, France
e-mail: burthopkins1@gmail.com

S. Centrone (ed.), *Essays on Husserl's Logic and Philosophy of Mathematics*, Synthese Library 384, DOI 10.1007/978-94-024-1132-4_17

namesake, Jacob Klein (1899–1978)[1] exists in Husserl's published and unpublished work.[2] The first public connection between Husserl and Jacob Klein occurs in 1940, with the publication of Klein's article "Phenomenology and the History of Science" in *Philosophical Essays in Memory of Edmund Husserl.*[3] This article is noteworthy above all for two reasons. One, it is the first discussion in the literature of Husserl's posthumously published essays "Die Frage nach dem Ursprung der Geometrie als intentional-historisches Problem"[4] and "Die Krisis der europäischen

[1]Jacob Klein was born (1899) in Russia (in Libau, which was then in Courland and a part of the Russian Empire and which is now a part of Latvia), educated there, Belgium, and Germany (Ph.D. 1922, Marburg University). He attended Heidegger's lectures in Marburg (1924–28) and studied (1928–29) with Max Planck and Erwin Schrödinger at the Institute for Theoretical Physics in Berlin before immigrating to the United States in 1938 to escape the Nazis. He was a personal friend of Edmund Husserl's family. He taught at St. John's College Annapolis from 1938 until his death in 1978.

[2]A letter from Husserl's wife Malvine to her daughter Elisabeth (March 26, 1937) mentions a "Klein" whom the editor of Husserl's letters, Karl Schuhmann, identifies as "[d]er Altphilologe Jacob Klein (geb. 1899)." Husserl 1994a, 487. (The reference concerns Klein's written communication to Malvine expressing his positive assessment of a publication by Jakob Rosenberg, husband of Elisabeth.) According to Klein's wife, Klein "visited old Husserl in 1919 in Freiburg—he wanted to study with Husserl. He went to Freiburg and visited Husserl.... But he couldn't study with Husserl because he couldn't get a room there, because it was 1919. All the boys came back from the war, and they had preference, so he went to Marburg. Old Husserl said, 'Well, you study with my old friend Natorp'." E. Klein, 14.

[3]"Phenomenology and the History of Science," Klein 1940/1985. Hereafter cited as *PHS*. All citations from this text reflect reprinted pagination. Klein's contribution was a late addition to the volume. In a letter to Klein dated November 10, 1939, Marvin Farber, the editor of *Philosophical Essays in Memory of Edmund Husserl*, invited him to submit a paper to the volume. He wrote that Husserl's son Gerhart "has written to me about your ability to have a paper ready for the E. H. memorial volume within a week, or very soon thereafter," and that "unusual circumstances ... make it possible at this late date to consider another paper." In a letter to Farber dated November 12, 1939, Klein wrote: "Although the time is very short I can get the article written before the deadline. I shall be grateful to you, if you can extend the time limit to the end of November." Farber eventually extended the deadline to December 5, in response to Klein's November 27, 1939 telegram to Farber requesting an extension to that date. In his letter to Farber of November 12, Klein described his proposed paper as follows:

> The subject of my paper would be something like Phenomenology and History with special reference to the History of science. I have in mind the *Philosophica* essay which you mention in your letter and, in addition, Husserl's article "Die Frage nach dem Ursprung der Geometrie als intentional-historisches Problem" published in the *Revue internationale de philosophie* (Janvier 1939a). (It goes without saying that I should have to refer to other publications of Husserl as well.)
>
> I should like to add that my intention is not to give simply a commentary on those texts but also to examine the notion of History of science as such.

All of the correspondence referred to and cited above may be found among Klein's papers, which are housed in the St. John's College Library in Annapolis, Maryland. I wish to express my thanks to Mr. Elliot Zuckerman, the literary executor of Klein's estate, for permission to cite from Klein's correspondence.

[4]Husserl 1939a, b/1954/1970.

Wissenschaften und die transzendentale Phänomenologie. Eine Einleitung in die phänomenologische Philosophie."[5] Two, despite Klein's sympathetic presentation of Husserl's phenomenology and appreciation of the consistency of Husserl's late turn to historical reflection with his earlier thought, Klein critically departs from "Husserl's 'intentional-historical' analysis of the origin of mathematical physics."[6] In the place of what he characterizes as Husserl's "amazing piece of historical 'empathy'" in the *Crisis*, Klein purports "to give a general outline of that actual historical development" behind the origin of mathematical physics and with that of the origin of "modern consciousness."

Klein situates the "actual" historical development in question within the context of Husserl's statements about Galilean science in the *Crisis*, having first extracted from the OG's analysis of the concepts of 'history' and 'tradition' what he refers to as the phenomenological problem of "intentional history."[7] Klein's account of this development presents it in terms of "a 'sedimented' understanding of numbers"[8] that he maintains "is superposed upon the first stratum of 'sedimented' geometrical 'evidences'" uncovered by Husserl's fragmentary analyses of geometry in the *Crisis*. In addition, then, to the task of "the intentional-historical reactivation of the origin of geometry"[9] recognized by Husserl as intrinsic to the reactivation of the origin of mathematical physics, Klein recognizes a second task, that of "the reactivation"[10] of the "complicated network of sedimented significances" that "underlies the 'arithmetical' understanding of geometry." According to Klein, Husserl's analyses in the *Crisis* noted this network[11] but did not pursue the task of its reactivation, a task that Klein also argues is crucial for the reactivation of "the 'sedimented history' of the 'exact' nature" constructed by mathematical physics.

Three scholarly curiosities are connected with Klein's presentation of the actual historical development of the exact science of mathematics involved in the origin of modern physics. Each of these are crucial for understanding not only the relation of his thought to Husserl's but also that thought's heretofore unrecognized importance for Husserlian phenomenology's foundational aspirations in the philosophy of mathematics.

The first concerns the fact that Klein presents the reactivation of the sedimented arithmetical evidences as a "task," whereas it had in fact already been accomplished

---

[5]Husserl 1936/1954. Cited hereafter as *Crisis*.

[6]*PHS* 79.

[7]Klein's article makes repeated references to "Husserl's notion of 'intentional history'" (*PHS* 70; cf. 72–74, 76, 78, 82). However, Klein's consistent use of quotation marks when referring to the expression "intentional history" is misleading, since he and not Husserl is its originator.

[8]*PHS* 84.

[9]*PHS* 83.

[10]*PHS* 84.

[11]*PHS* 84. Klein refers to *Crisis* 44, where Husserl discusses the "arithmetization of geometry" and the consequent automatic "emptying of its meaning" as "the geometric signification recedes into the background as a matter of course, indeed drops out altogether."

by Klein himself (in two long articles published in 1934 and 1936[12]) precisely along the lines of the "actual" development leading to the origin of mathematical physics that he sketched in *PHS*.

Klein's neglect (in *PHS*) in mentioning his own earlier work on this topic in relation to Husserl's is mirrored in that earlier work, as it neglects any mention of Husserl. Thus the second scholarly curiosity in the relation of Klein's thought to Husserl's is that of his own work's silence about its relation to Husserl's phenomenology. As we shall see, given the topic of Klein's work—the transformation of the pre-modern concept of number into its modern 'symbolic' concept—reference to Husserl's own work on the topic of the concept of number in the *Philosophy of Arithmetic* and that of symbolic cognition in both that work and his *Logical Investigations* would seem to have been natural. Indeed, this has been pointed out by two among the very few scholars aware of the phenomenological horizon of Klein's work.[13]

The third scholarly curiosity concerns the speculation that the *Crisis*'s "Galileo section might have resulted from a reported visit during this period [sometime in 1934] by Husserl's friend and former student Alexandre Koyré, who published his monumental *Etudes Galiléenes* in 1940."[14] This speculation is fueled by "[t]he striking similarity between Husserl's and Koyré's interpretation of the significance of Renaissance science." However, until recently it has remained unknown that the basic ideas behind Koyré's Galileo research most likely had their origin in Klein's thought and research. Karl Schuhmann publicly called attention in 1997 to a penchant of Koyré's for appropriating without attributing the ideas of other's. Schuhmann notes that Koyré's book on Plato[15] neglected to mention that the source of many of its ideas was Adolf Reinach's lecture course on Plato that Koyré had

---

[12]See Jacob Klein 1934/1936/ 1969. Hereafter cited as *GMT*.

[13]Caton 1971 and Miller 1982. In his review of the English translation of Klein's articles, Caton (1971, 225) remarked upon Klein's "failure to cite Husserl as the source of his Husserlian terminology", that is, the terminology of the "theory of symbolic thinking" and the "concept of intentionality." It is Caton's contention that precedence for both of these should go to Husserl. In the case of the former, he appeals to Husserl's "remarkably similar theory in the *Logische Untersuchungen* (Vol. II/1, par. 20)." In the case of the latter, he points to how, "by citing the scholastic Eustachias as illustrating the sources of the thinking of Vieta and Descartes," Klein "ingeniously capitalizes on … [the] genealogy" of intentionality, which Husserl took "from Brentano, who in turn took it from medieval logic." Miller (1982, 132) writes: "Although Husserl's own analyses [i.e., in *Philosophy of Arithmetic*] move on the level of a priori possibility, Klein's work shows how fruitful these analyses can be when the categories they generate are used in studying the actual history of mathematical thought." As we shall see below, however, the relationship between Klein's analyses of natural and symbolic numbers and Husserl's is more complex than either Caton or Miller is aware. One consequence of this is that the common assumption behind Caton's and Miller's remarks here—that Husserl and Klein understand *exactly* the same thing when it comes to these kinds of numbers and their relationship—cannot withstand critical scrutiny.

[14]Carr 1970, xix n.7. This publication date of Koyré's book is incorrect; it was published in Paris in 1939.

[15]Koyré 1945.

attended.[16] And a recently discovered interview with Klein's wife mentions that her husband, together with Leo Strauss and Koyré were together in Paris in the early 1930s,[17] and that ideas Klein explained to Koyré ended up being published by Koyré without acknowledgment. According to Klein's wife,

> Strauss was furious and didn't want to have anything to do with Koyré. But instead of telling Koyré, "Why did you do that? I was present," he just didn't answer and didn't talk to him—simply mistreated him.... Jasha [Klein's nickname] simply laughed, and said, "Well, I'm very glad that he got it."[18]

Klein's ideas most likely concerned[19] "the elaboration of a study of Galileo's physics and its relation to Plato, Aristotle, and Archimedes,"[20,21] which Klein

---

[16]Schuhmann 1997, 391.

[17]She mentions the dates as "'31, or '32// (E. Klein 14).

[18]*Loc. cit.*Edmund Husserl's daughter Elisabeth (Ellie) Rosenberg, one of Klein's students in a 1933 Plato seminar he taught, invited him to visit her brother Gerhart in Kiel. Klein accepted the invitation, and soon became friends with the extended Husserl family and Gerhart's wife, Else (Dodo) (E. Klein 17). (Gerhart Husserl divorced Else in 1948; she and Klein were married in 1950 [E. Klein 9].)

[19]Klein's wife's memory that the ideas concerned "something from one of the [Plato's] dialogues" is clearly confused, since Koyré's Plato book, based on lectures he gave in Cairo in 1940, was published in 1945. However, two articles containing parts of *Etudes Galiléenes* had already appeared in 1937: Koyré 1937a, b, which makes it much more than likely that it is they that contain the unacknowledged ideas borrowed from Klein reported by his wife.

[20]Wilson 1976 ii. Wilson, whose source for this information was most likely Klein himself, reports that Klein was engaged in this study from 1935 to 1937 while a fellow of the Moses Mendelssohn Stiftung zur Förderung der Geisteswissenschaften. Klein's status as a Jew led to his exile from Germany in 1937 and the impossibility of continuing his Galileo studies during those turbulent times.

[21]Recently Moran 2012 and Parker (forthcoming) following him have suggested, contrary to Carr's suggestion that Husserl's Galileo section may be the result of a visit by Koyré in 1934, that the evidence points rather to Husserl being the source of Koyré's interest in Galileo. On the one hand, Moran points out, "Reinhold Smid has shown (Hua XXIX il n.2) that Koyré's last visit with Husserl was in July 1932, prior to the appearance of Koyré's studies on Galileo that began to appear in artilce form from 1935 on" (Moran 2012, 72). Smid, moreover, also quotes Ludwig Landgrebe, who reported that he met Koyré in Paris in 1937 and Koyré told him he was "very much in agreement with the Galileo interpretation in the *Crisis*" (Smid 1993, n.2). On the other hand, Moran speculates that "Husserl's interest in Galileo's use of geometry was most probably influenced by Jacob Klein, who had published a number of works on the origins of Greek geometry between 1934 and 1936" (Moran 2012, 72–73). Parker, in addition, cites Aaron Gurwitsch, who "recalls that Koyré once remarked that, 'even though Husserl was not a historian by training, by temperament, or by direction of interest, his analysis provides the key for a profound and radical understanding of Galileo's work. He submits [Galilean] physics to a *critique*, not (once again be it said) a criticism'" (Parker forthcoming, 3). Parker also relates that "Gurwitsch points out that some of the preparatory studies for the *Crisis* date from the late 1920s, perhaps referring to texts dealing with the 'Mathematisierung der Natur' written in 1926, and notes also that 'some of the relevant ideas can be found, at least in germinal form, as early as 1913'" (Parker forthcoming, 24). This evidence, however, is not only inconclusive but also in one instance flawed. Regarding the chronology, we've already seen that Klein's wife reports 1931 or 1932 as the dates in Paris that Koyré absorbed Klein's ideas. These dates, then, are consistent with the date Smid (following Karl

planned to develop after he completed (in the spring of 1934) the two already mentioned long articles titled "Die griechische Logistik und die Entstehung der Algebra."[22]

17.2 Foundational Problems in the Philosophy of Arithmetic

Husserl's first philosophical problem was that of the foundation of the concept of number, where number is understood to be the answer to the question: how many? His formulation of this problem, in turn, had two dimensions. The first concerned the problem of accounting for the objective unity of the most basic numbers, i.e., those with which we count. The second concerned accounting for the manner in which these numbers provide the logical foundation for the higher, symbolic numbers, including those employed in universal analysis. The issue behind the first foundational problem is the following. Each number with which we count is composed as the unity of a multiplicity, beginning with the unity of the least multiplicity, the number two. Given that the unity of each number cannot be accounted for on the basis of the individual qualities proper to the items in the multitude composing the multiplicity of which the number is the unity, the problem that presents itself is how to account for the unity of number in question. The issue

Schuhmann) reports Koyré last visited Husserl, July 1932. Moran's suggestion that Klein's articles on the origins of Greek geometry (which are dated 1934 and 1936 but actually were published together in 1936 in a single volume) most probably influenced Husserl's understanding of Galileo is very problematic. This is the case because the focus of the articles in question is *not* geometry but the transformation of the ancient Greek concept of number that occurred with the invention of modern algebra. Neither Greek geometry nor Galileo are thematically treated in Klein's articles. (See note 10 above, for bibliographic information on the German originals of the articles and their English translation by Eva Brann.) In addition, Gurwitsch's claim that the preparatory studies for the *Crisis* date from the 1920s and before, and Parker's singling out in particular Husserl's texts on the mathematization of nature in 1926, do not establish that Husserl's appreciation of Galileo's role in the establishment of modern mathematical physics and the mathematization of nature in these texts is sufficient to account for his account of Galileo's role in the mathematization of the life-world in the *Crisis*, together with Husserl's presentation of the Greek mathematical context of Galileo's mathematization in this account. In fact, close study of these texts discloses the basis for the opposite conclusion, namely, the *Crisis*'s account of the reinterpretation of Euclidean geometry that is sedimented in Galileo's mathematization of nature is unprecedented in Husserl's pre-*Crisis* discussions of Galileo and the mathematization of nature. Finally, neither Koyré's expression of appreciation for (to Gurwitsch) or agreement with (to Landgrebe) Husserl's critique of Galileo in 1937 rules out the possibility that Koyré's appropriation of Klein's ideas about the relation of Galileo's physics to ancient Greek mathematics influenced Husserl in their 1932 meeting. It's clear that Koyré's appreciation and agreement relate to the aspect of Husserl's analysis that goes beyond their historical presentation of Galileo, that is, to their phenomenological dimension, regarding which he of course could not have influenced Husserl.

[22]O. Becker 1934/1992, whose article "The Theory of Odd and Even in the Ninth Book of Euclid's *Elements*" appeared in the same journal that Klein's first article appeared in, refers therein to Klein's article as "a very important work" (Becker 1992, 100).

behind the second foundational problem is the following. Granted the mathematical-foundational role of the most basic numbers, the general concept of which is captured by the Euclidean definition of number as "a multitude composed of units," the problem that presents itself is how to account for the 'numerical' significance of the following 'kinds' of numbers: those that are either too large to permit the direct apprehension—all at once—of the multitude of discrete units that are proper it, or those that are functionally irreducible to the concept of number as a multitude of units.

Husserl's investigation of the problem of number in his first major work, *Philosophy of Arithmetic*, operates within a terminological universe that moves between the "presentation" (*Vorstellung*) and "concept" of its subject matter, and does so in a manner that distinguishes between "authentic" or "proper" (*eigentlich*) instances of each. Thus, and most crucially, Husserl distinguishes the "proper concept" of number as a "the unity of a multitude of units" from the "proper presentation" of number as the amount of units a single act of consciousness can apprehend discretely all at once. Moreover, number itself in Husserl's investigation has neither the status of a concept nor of a presentation, but rather precisely that of the unity of a multiplicity of units, of discrete items either perceived or thought. More precisely, number in general is the unity of a multiplicity of units, while its species are the natural numbers beginning with the number two, which as mentioned comprises the least multiplicity.

Had Husserl's first work not lacked philosophical maturity, it would have consistently maintained the distinction between the conscious presentation of a multitude of units as a unity and the general concept of the unity of a multitude of units, that is, the distinction between the proper presentation of number and its proper concept. In addition, the distinction between a number concept, say, the concept of five—i.e., the general concept of five (fiveness)—and the number five—i.e., the unity of any one of the unlimited many arbitrary one and one and one and one and one units that falls under this general concept—would have been maintained. Clearly, however, it was Husserl's intention to maintain these distinctions, beginning with the most important one behind which lies his first work's major discovery, namely that that which falls under the proper concept of number, to wit, the unity of a multitude of units, is *not* a concept but precisely a multitude of units. Thus, $5 + 7 = 12$ does not signify that the concept 'five' plus the concept 'seven' is identical or otherwise equivalent to the concept 'twelve' but rather it signifies that the 'five' units composed by the *number* five's unity and the 'seven' units composed by the *number* seven's unity, when combined, make up the 'twelve' units composed by the *number* twelve's unity. Moreover, Husserl clearly intended to maintain the philosophical distinction between the proper concept of number in general as the unity of a multitude of units and the proper presentation of the units signified by a number in an act of consciousness. For instance, the empirical limitation of human consciousness to be barely able to entertain in a single act more than three discrete units should in no way impinge upon the validity of the general concept of number as the unity of a multitude of units. Therefore, the inability of consciousness to apprehend the totality of five hundred units in a single act has absolutely no *mathematical* significance in relation to the proper content of the concept of the number five hundred being exactly the unity of a multitude composed of five hundred units.

It was also Husserl's intention, however, to account for the objective status of the unity of number's multiplicity on the basis of the conscious act of combining the units in the multitude composed by the number. Owing to this unity's irreducibility to the properties of the units so combined, its collective status clearly cannot be accounted for on the basis of the appeal to qualities possessed by the units. Hence, Husserl's attempt in the *Philosophy of Arithmetic*[23] to account for its origin in an act of inner reflexion that abstracts the collective unity of the assemblage (*Inbegriff*) of units united in a multiplicity from the act of collective combination in which the assemblage is initially composed. This attempt, however, was misguided, and recognized (after the *PA*'s publication) by Husserl himself as such. As he himself would put it, "from the reflexion on acts" of collecting "the concept of collecting . . . is all that can result"—not the concept of the unity of the collection.[24] The problem Husserl was trying to resolve, that of the origin of the collective unity proper to number, however, was not misguided. Moreover, it's important to note that what was misguided in Husserl's appeal to "psychological reflexion in Brentano's sense" to account for the unity of a collection, the concept of unity, and, finally, cardinal number, was not the logical psychologism Husserl's criticizes in the *Logical Investigations*. That is, Husserl's position in the *PA* wasn't that collective unity is *really* psychological because that unity makes its appearance in psychological acts. As we've seen, Husserl's intention was to keep distinct both the presentation and concept of number from number itself as the unity of a multitude of units. Rather than reduce the objective unity of number to the psychological status of a presentation, Husserl's intention was clearly to account for the origin of that objective unity in the full recognition that its collective property renders it irreducible to the properties of the individual units that it unifies. Husserl's realization of failure of the appeal to an abstraction from the act of combining units into a collection to account for the origin of collective unity's objectivity therefore leaves intact the problem of that origin.

Another problem that the *PA's* method of psychological abstraction leaves unresolved is that of the objectivity of the unity of the units that make up each number. To number, as the determinate amount of a multitude of any arbitrary items whatever, e.g., a feeling, an angel, the moon, Italy, there belongs according to Husserl the formal-logical category of "the anything" (*Etwas*), in the exact sense that any arbitrary thing whatever, no matter what it is, can be combined into an arithmetic unity. Thus the units of number, in the precise sense of the combination of "'anything' and 'anything' and . . ." fall for Husserl under a formal logical category.

[23]Husserl 1970/2003. Hereafter cited as *PA*. German page numbers, which are reproduced in the English translation, will be cited.

[24]Husserl 1939b/1975, 127. German page number, which is reproduced in the English translation.

As to the origin of this category, Husserl again appeals to psychological abstraction, this time from the act of presentation itself, which—qua act—he maintains is without individual content and thus suitable to account for the unrestricted scope of the category in question. Because the abstraction in question does not concern what is presented in the act of presentation but only the act itself, the 'anything' has for Husserl the status of a "negative" determination. What it determines is therefore not the individual content presented in a presentation but the very act presupposed by the presentation of any possible individual content. But, of course, the psychological appeal to the objective origin of this categorial determination in an act suffers in this case from the same psychologistic illegitimacy as in the similar appeal in the case of the objectivity of collective unity. Specifically, the abstractive reflexion directed to the manifold presentative acts in which determinate individual contents are given does not yield the materially indeterminate formal-logical category of the "*Etwas*" but only the materially empty concept of a determinate act of presentation.

The psychologism behind the *PA*'s response to the first dimension of the problem of the foundation of number is also related to that book's response to the second dimension of the problem of number's foundation. Or, more precisely, it is related to Husserl's initial approach to the problem of the origin of the numerical significance proper to the higher symbolic numbers, because Husserl eventually abandons that approach in the *PA*'s analyses. Husserl's initial approach to this second problem, as mentioned, is driven by the conviction that the proper concept of number, number as the unity of a multitude of units, provides the *mathematical* foundation for symbolic numbers. However, in the course of his analyses in *PA* he came to realize that the rules that govern the symbolic calculus in universal analysis assume a complete autonomy from arithmetical *concepts*, as they function instead to stipulate the "rules of the game" governing the combination of sense perceptible signs whose 'symbolic' status, being intrinsically signitive, is therefore exclusively non-conceptual. With this realization, the *PA*'s overarching project of accounting for the foundation of the higher symbolic numbers in the proper numbers, is effectively abandoned by Husserl.

Husserl's initial account of the foundational role provided by proper numbers for symbolic numbers is based on the assumption that the general concept of each is the unity of a multiplicity of units. He characterizes the identity of concepts involved here in terms of the "logical equivalence" of the objects of the concepts of each kind of number, that is, of the proper and symbolic numbers themselves, namely the concrete presentation of the unity of a multiplicity of units. On Husserl's (initial) view it is this logical equivalence that makes it possible for symbolic numbers to function as "surrogates" for number's proper presentation and it is the task of the philosophy of arithmetic to account for this possibility. The surrogate function of symbolic numbers for Husserl is bound up with the peculiar status of the signitive functioning of the sense perceptible sign in which the symbolic number is presented. Rather than simply represent an absent object, in the case at hand, the unity of a determinate amount of units, the sign that presents a symbolic number signifies the *idealized* possibility of presenting the unity of a multiplicity of units whose multitude exceeds the limits of the capacity of consciousness to intuit the totality of units that compose the multiplicity's collective unity. Husserl's account of the idealization in question presents it as an extension of the perceptual acts that collectively combine sensible multitudes. He characterizes these acts in terms of the spontaneous emergence of a "figural moment" that presents sensibly the multitude's unity as a multiplicity that encompasses the items of the multitude beyond their partial intuition. Words like "file," "row," "flock," "gaggle," "covey," "heap," etc., express for Husserl "the existence of quasi-qualitative moments"[25] that can be grasped at once "as unified intuitions analogous to sense qualities."[26] According to Husserl, these 'figural moments' naturally function as *signs* "for the full process intended,"[27] that is, for actually carrying out the psychical activity of collectively combining all the members in the multitude.

On this view, then, a symbolic number presents the idealized extension of the naturally occurring signs in the experience of collective combination, an extension that works equally well for the symbolic presentation of finite and infinite multitudes. In the latter case, the symbolic number functions to present imaginary arithmetic objects, as the impossibility in principle of a finite consciousness unifying an infinite multitude of units signals that the mathematical status of the symbolic presentation has entered the domain of the non-actual.

Husserl's account of symbolic number rehearsed here, of course, exemplifies the earlier mentioned philosophically immature inability to maintain crucial distinctions in the course of his analyses. The most obvious case in point being the distinction between the concepts of proper and symbolic numbers and the presentation of such numbers. One obvious result of this is the claim that the *arithmetical* function of symbolic numbers owes its origination to the *psychological* limits of arithmetical cognition of large multitudes. Hence, following this line of analysis, the sense perceptible mathematical sign does not symbolize a kind of number but the extension of the mind's cognitive powers beyond its natural limits to combine collectively the content of the concept of proper number, viz., multitudes of units. The further course of Husserl's analysis, as already mentioned, however, comes to reject this view of the function of symbolic number, albeit on grounds other than its peculiar psychologism. These grounds concern rather what Husserl comes to understand as the actual meaning of symbolic numbers in the algorithmic operations of universal arithmetic. Rather than function as "surrogates" for the presentation of unities of multitudes of units, which is to say, rather than refer indirectly to what falls under the proper concept of number, as employed in the symbolic calculus number has a symbolic function that is purely "signitive." According to Husserl, this means that the arithmetical meaning of the manipulation of the sense perceptible signs composing the symbolical calculus is not determined by arithmetical concepts but by the "rules of the game" stipulated for whatever operations are in question.

[25] *PA* 203.
[26] *PA* 209.
[27] *PA* 213.

These rules, in turn, are presented in the *PA* as being generated on the basis of their algorithmic "parallelism" with the proper domain of numbers and the actual operations on those numbers. Husserl, however, does not unpack this idea in terms of either that work's psychological or logical investigations. The result is what one astute commenter[28] has referred to as a "gap" in that work's account of the "step" from the original domain of proper numbers to the domain of their symbolic representation and, beyond that, to the extension of the number domain itself beyond what is admitted in the original domain. This extension includes the natural integers (thus includes 0 and 1) and imaginary numbers (fractions, negative numbers, etc.). If that original domain of numbers is identified with the cardinal numbers,[29] then the results of Husserl's investigation of the second foundational problem in the *PA* may summed up as follows: initially guided by the conviction that the cardinal numbers provide the foundation for the algorithm constitutive of the rules of calculation determinative of universal arithmetic, Husserl's discovery of the autonomy of the symbols employed by the latter from both direct references to concepts and indirect references to multitudes of units compelled the abandonment of his original conviction. Husserl's letter to Carl Stumpf written in 1890,[30] before the *PA* was complete, confirms this summary. In it Husserl relates that "[t]he view by which I was still guided in the elaboration of my *Habilitationsschrift*, to the effect that the concept of cardinal number forms the foundation of universal arithmetic, soon proved to be false."[31] This means, among other things, that "[b]y no clever devices, by no 'inauthentic presenting', can one derive negative, rational, irrational, and the various sorts of complex numbers from the concept of cardinal number."[32] Universal arithmetic, rather, "finds application to the cardinal numbers (in 'number theory'), as well as to ordinals, to continuous quantities, and to *n*-dimensional manifolds (time, space, color, force, continua, etc.)."[33]

Husserl goes on to relate how, with the realization that "no common concept underlies these various applications of arithmetic,"[34] he had to confront the question about its content, about its "conceptual objects,"[35] and that he had to reject his original supposition that this content is intrinsically conceptual. That is, he had to reject his initial view that to all of the signs of universal arithmetic there correspond, "at least potentially,"[36] "designated concepts."[37] The view that a "system of

[28] Majer 1997, 41–44.

[29] Insofar as for Husserl proper numbers begin with the least multiplicity ('two') and cardinal numbers begin with '1', this identification is not without its problems. See Majer 1997, 42.

[30] Husserl 1994b, c.

[31] Husserl 1890, 158/13.

[32] *Ibid.*

[33] *Ibid.*

[34] *Ibid.*

[35] *Ibid.*

[36] *Op. cit.*, 159/15.

[37] *Op. cit.*, 160/15.

signs and operations with signs can replace a system of concepts and operations with judgments, where the two systems run rigorously parallel,"[38] only holds (within the context of mathematics) in the case of "ordinary arithmetic [*gemeinen Rechenkunst*]."[39] Thus, for instance, if arithmetic "deals with discrete magnitudes, then 'fractions', 'irrational numbers', imaginary numbers and, in the case of cardinal numbers for example, also the negative numbers, lose all sense."[40] This is the case because the signs for these numbers, in contrast to the signs that refer to the concepts proper to discrete magnitudes, are "representatives of 'impossible' concepts." But rather than have to get clear about "how operations of thought with contradictory concepts could lead to correct theorems," which is what Husserl relates he originally tried to do, he came to realize that "through the calculation itself and its rules (as defined for those fictive numbers), the impossible falls away, and a genuine equation remains." Therefore, the system of *arithmetica universalis* "is not a matter of the 'possibility' or 'impossibility' of concepts," but "an accomplishment of the signs and their rules."[41] Universal arithmetic is, therefore, "no science, but a part of *formal* logic,"[42] which Husserl defines here "as a technique of signs (etc., etc.)" and designates "as a special—and one of the most important—chapters of logic as the technology of knowledge." Husserl also adds, very significantly, that "these investigations appear to push toward important reforms in logic, and that he knows "of no logic that would even do justice to the possibility of an ordinary arithmetic [*gemeinen Rechenkunst*]."

17.3 Historical Context of the Foundational Problems in Philosophy of Arithmetic

Husserl's discovery of the systematic discontinuity between (1) proper numbers and conceptual objects and (2) the algorithmic system of operations on signs is therefore tantamount to the realization that neither the proper concept of number—the unity of a multitude of units—nor the proper numbers themselves—any of the unlimited unities of "one and one and ..."—are capable of forming the foundation for 'universal arithmetic' (symbolic mathematics). And this is where Klein's account of the actual development of the mathematics inseparable from the origin of mathematical physics comes in, as it shows that the pre-modern concept of number as *Anzahl*—the positive natural numbers beginning with two—undergoes a radical transformation in Vieta's establishment of the algebra that lies at the basis of modern mathematics and modern science. This transformation involves the

[38] *Op. cit.*, 159/14.

[39] *Loc. cit.*

[40] *Op. cit.*, 160/15.

[41] *Op. cit.*, 160/16.

[42] *Op. cit.*, 161/17.

'formalization' of number that substitutes for the ideal numerical entities of Greek arithmetic (the unity of determinate amounts of multitudinous units) their symbolic expressions. Because the resulting universal and symbolic mathematics rest on the formalized understanding of number, number as *Anzahl* is in principle incapable of forming the foundation of the new mathematics. Thus Klein's account of the conceptual independence of modern universal and symbolic mathematics from number understood as *Anzahl* tracks Husserl's realization that universal arithmetic does not have its foundation in the proper concept of number and that which falls under that concept. Indeed, the similarity noted here is what is behind Caton and Miller characterizing Klein's research in terms of its employing the *a priori* possibilities uncovered in Husserl's systematic investigations for the study of the actual history of mathematics.

Despite, however, the similarity noted here, there are two fundamental differences between Husserl's and Klein's accounts of theses issues, differences that Klein was no doubt aware of. The first concerns that which is responsible for the universality of the units of non-formalized numbers (proper numbers in Husserl's sense and the pre-modern *Anzahlen* in Klein's sense). The second concerns the character of the initial indirect relation of symbolic numbers to the unity of the multitude of units characteristic of non-formalized numbers.

Regarding this first difference, we have seen that for Husserl the arithmetic universality of proper numbers has its basis in the formal-logical category of the 'anything' (*Etwas*). Husserl accounts for the capacity of the units of proper numbers to function as variables, such that any arbitrary object whatever may enter in to a relation with each unit and be combined in a number, on the grounds that each of these units is itself an extension of and therefore a concept that falls under the formal concept of the 'anything'. For Klein, however, the units of *Anzahlen* are not formal-logical but ontological, in the precise sense of their conception as sensibly independent units of thought. For Plato as for Aristotle, arithmetical universality has its ground in the conviction they share that what is numbered in theoretical arithmetic are units that do not have any sensible qualities and therefore are only accessible to thought. These thought-objects—*noêta*—are the true objects of counting. Rather than present logical concepts under which individual objects fall, *noêta* themselves are the true objects responsible for arithmetic's formality, that is, of number's unrestricted capacity to count any kind of object. Aristotle's dispute with Plato, according to Klein, was not over the pure status of the true objects of arithmetic treated by theoretical arithmetic, but rather over the origin of these objects. Indeed, for Klein the philosophical significance of Aristotle's account of their origin in the *abstraction* (literally, *subtraction*) of all sensible qualities from things until all that remains is their *quality* of being one, only comes to the fore when its status as a critical alternative to the origin of *noêta* found in Plato is considered. This is the case, because for Klein Plato's account of the exclusively *intelligible* origin of *noêta* brings with it the presupposition that the foundational unit of theoretical arithmetic as well as the unity of arithmetical and eidetic multitudes cannot be derived from sensible beings.

The second difference between Husserl and Klein, that regarding the relation of symbolic number to number as a multitude of units, has its basis in their radically different accounts of the symbolic number's origin. We have seen that for Husserl, prior to abandoning the conviction of the foundational role played by cardinal numbers for universal arithmetic, the property responsible for a number being symbolic was manifest in its peculiar surrogate function: rather than simply represent the totality of the units composing a multitude that is too large to be apprehended by a single conscious act, what made a number symbolic for Husserl was its idealizing extension of the natural sign (the figural moment) made manifest in the perceptual experience of collective combination. The signitive function of that sign, both in the original perceptual experience and in its idealizing extension, involves its association with the possibility in principle of extending any partial intuition of multitudes too large to be apprehended in a single act of collective combination, to other units composing the partial multitude in question.

For Klein, however, what makes a number symbolic is *not* its peculiar surrogate function in relation to the perceptual experience of collective combination or any other experience originating in perception. As shall be discussed in detail below, what makes a number symbolic is its presentation in a completely determinate intuitable mark of the indeterminate *concept* of a determinate multitude or of multitude in general. Thus, for instance, the non-symbolic number five, i.e., any multitude of units whose exact amount is five, is symbolically expressed as the general concept of 5, that is, fiveness, which as a concept is manifestly *not* something multitudinous—as it is intuited as the cipher 5—and therefore only indirectly related to the multitude of units composing any arbitrary non-symbolic number five. For Klein, as again we shall see in detail below, the step from this symbolic expression of five to the symbolic expression of any number whatever, the variable 'a', is but a short one. For the determination of the sense perceptible mark 'a' as a symbol arises with its expression of the general concept of any number, that is, the general concept of a multiplicity as such. Finally, we shall also see in detail below that the shift in number represented by either their symbolic expression as *Zahlen* or as number in general, originates for Klein in the transformation of the basic unit employed by the arithmetic art of calculation. This transformation moves from the unity of determinate amounts of the units composing *actual* multitudes to the *concepts* as such of *possible* multitudes.

These differences between Husserl's and Klein's understanding of the units that compose non-symbolic numbers and the origin and meaning of symbolic numbers, are, I submit, behind the first two scholarly curiosities mentioned above. For Klein to have called attention to the priority of his reactivation of the sedimented understanding of numbers constitutive of scientific modernity in *GMT* over Husserl's research in this area would have immediately invited questions about Klein's silence in that work about Husserl's systematic investigations of very similar phenomena. There can be no doubt that Klein was aware of Husserl's investigations when he wrote *GMT*. In *PHS* Klein writes that "Husserl's logical researches amount in fact to a reproduction and precise understanding of the 'formalization' which took place in mathematics (and philosophy) ever since Vieta and Descartes paved the way for modern science."[43] Of course, the key qualifier here is "amount in fact." In the remainder of this discussion of Husserl and Klein, I will endeavor to show two things. One, that Klein's researches into the sedimentation of number effectively establish that despite Husserl's monumental "rediscovery" of the non-conceptual and non-presentational *being* of *Anzahlen*, his account of the foundation of their universality remains formal-logical and thus cut off from the true philosophico-mathematico problem of this universality's foundation. Two, that Klein's account of the origin of symbolic numbers in the transformation of the art of calculation effectively establishes that there is a lingering psychologism in Husserl's mature attempts to account for the constitution of the formal-logical unity presupposed by modern mathematics, a transformation that for Husserl supposedly occurs on the basis of "modifications" traceable back to perceptual experience.

On Klein's view, then, the modern understanding of the basic element of arithmetic—number in the sense of *Zahl*—is inseparable from the historical origin of François Viète (Latin: Vieta) of Fonenay's invention of the "Analytic Art (*Artem Analyticen*)"[44] for Princess Mélusine (Catherine of Parthenay, 1554–1631) in 1591. According to Klein, to this day this art functions as the *sine qua non* for the formalization that makes modern mathematics possible and therefore composes its *foundation*.[45]

Klein's research shows that Vieta presented his analytical art as "the new algebra" and took its name from the ancient mathematical method of analysis, which he understood to have been first discovered by Plato and so named by Theon of Smyrna. Ancient analysis is the general half of a method of discovering the unknown in geometry, the other half, synthesis, being particular in character. According to Vieta analysis "was defined by [Theon] as 'taking of the thing sought as granted and proceeding by means of what follows to a truth that is uncontested'."[46] Synthesis, in turn, is for Theon according to Vieta "'taking the thing that is granted and proceeding by means of what follows to the conclusion and comprehension of the thing sought'."[47] The transition from analysis to synthesis was called "conversion," and depending on whether the discovery of the truth of a geometrical theorem or the solution (construction) to a geometrical problem was being demonstrated (ἀπόδειξις), the analysis was called respectively "theoretical" or "problematical."

Klein presents Vieta's innovation to involve the understanding of a novel form of arithmetical analysis found in the recently rediscovered third century AD text (titled simply *Arithmetic*) of Diophantus of Alexandria as a procedure that is completely parallel to geometrical analysis. This permitted Vieta to treat the sought after and therefore unknown numbers—understood as unities of multitudes of units—as already granted in their *species*. By the *species* of numbers he followed Diophantus'

---

[43] *PHS* 70.

[44] Vietae 1591/ 1969. Hereafter cited as *Analtyic Art*.

[45] See *GMT* and B. Hopkins 2011. Hereafter cited as *Origin*.

[46] *Analytic Art* 320.

[47] *Loc. cit.*

designations in his *Arithmetic*, i.e., square, cube, square-times-cube, and cube-times-cube. To the species of each of these unknown and therefore indeterminate quantities as well as to the species of every known quantity, Vieta assigned what he called an "everlasting and very clear symbol"[48] taken from the alphabet (vowels to the known and consonants to the unknown). This allowed both the *possibility* of there being given a determinate amount of units (that is, a number in the pre-modern and therefore non-formalized sense of the unity of a multitude) to be apprehended in a manner that functioned as if it were *actually* given and it also allowed known numbers to be expressed by their species. Following this, Klein maintains that the arithmetical need for an analogue to the second part of the geometrical method of analysis, the theoretical or problematic conversion of the synthesis that proved a *particular* theorem or solved a *particular* problem, was dispensed with by Vieta. This made possible for the first time the analytic—that is, indeterminate and therefore general—solution to arithmetical problems. Klein presents three significant results as following from Vieta's innovation: One, the geometrical distinction between the kind of object presented in a theorem and in a problem falls away, such that in the analytic art theorems are equated with problems and with this the synthetic distinction between the theoretical and problematical dissolves. Two, the exclusive calculation with the *species* of known and unknown numbers made possible by Vieta's analytic art, what he terms "*logistice speciosa*," is employed by him in the service of "pure" algebra, and therefore applied indifferently to finding unknown numbers and unknown geometrical magnitudes. And, three, because the *logistice speciosa* has but a small interest in the determinate results of the solutions to its calculations—what Vieta terms the "*logistice numerosa*"—the artful procedure of Vieta's analytic method is conceived as a general auxiliary method whose purpose is not to solve problems singly but to solve the problem of the general ability to solve problems. Characterized by Vieta as "the art of finding,"[49] the general analytic is an *instrument* in the realm of mathematics analogous to the sense in which Aristotle's *Prior and Posterior Analytics* are presented as an *organon* in the realm of all possible knowledge. In this regard, Vieta's conclusion to his *Analytic Art* is telling: "the analytic art . . . appropriates to itself by right the proud problem of problems, which is: TO LEAVE NO PROBLEM UNSOLVED."[50]

Husserl recognized Vieta's method as marking the appearance of "that 'formalization' or algebraization" which "distinguishes subsequent formal 'analysis' from all material mathematical disciplines."[51] This is consistent with Klein's view that Vieta's formalization is coincident with the invention of the mathematical formula and the first modern axiom system, whereby the syntactical rules of mathematical

[48] *Op. cit.*, 340.

[49] *Op. cit.*, 346.

[50] *Op. cit.*, 353 (bold letters in original).

[51] Husserl 1974/1969, 48. Page references refer to the German original, which are included in the English translation.

analysis defines the object to which they apply. But for Klein it is also coincident with the transformation of both the mode of being of the foundational concept of arithmetic—number—and with this, the transformation of the mode of being of the objects of mathematics in general, together with the transformation of the process of abstraction that generates the formal concepts operative in the system of knowledge in general.

Klein therefore presents Vieta's innovation as containing three interrelated and interdependent aspects. One, there is its *methodical* innovation of making calculation possible with both known and unknown indeterminate (and therefore general) numbers. Two, there is its *cognitive* innovation of resolving mathematical problems in this general mode, such that its indeterminate solution allows arbitrarily many determinate solutions based on numbers assumed at will. And, three, there is its *analytic* innovation of being applicable indifferently to the numbers of traditional arithmetic and the magnitudes of traditional geometry.

For Klein, the significance of this first innovation is the *formalization* of number and thus of its concept, such that number no longer signifies what it did in Greek arithmetic and in mathematics generally prior to Vieta's innovation, a "multitude composed of units." But, rather, number now signifies the *concept* of such a multitude in the case of known numbers and the *concept* of a multitude as such (or in general) in the case of unknown numbers. Klein maintains that this formalization of number and of its concept can be grasped neither by Aristotelian abstraction nor by Platonic dialectic.[52] This is because, as *formalized*, number is: (1) neither the product of the abstraction that yields the unit that functions to measure a multitude of items, as it is for Aristotle; nor (2) the Ideal unity of such a multitude that is grasped by dialectical thought as being irreducible to the items it unifies once the sensible suppositions of the mathematicians are left behind, as it is for Plato. Rather, for Klein, number in Vieta is the result of the conceptual process of ascending from the mind's unmediated and therefore direct relation to multitudes of items to its relation to its own apprehension of this unmediated and direct relation, while simultaneously identifying these two modes of relation. This simultaneous identification of heterogeneous relations, namely of (1) the real relation to a multitude of concrete things and (2) the cognitive relation to the concept of this multitude, is exhibited by the meaning assigned by Vieta both to ordinary number signs and to his algebraic letters. And it was exhibited and therefore manifest for him as it is (according to Klein) for us every time a sense perceptible letter is intuited *as*—and not simply as *signifying*—the general concept in question—whether that concept be of this or that number, for instance, *the concept of any two in general*, or the *concept of any number in general*. It is Klein's original claim that what

[52]On Klein's view, the prevalent attempt to capture the difference between the ancient and modern concepts of number in terms of the latter's greater abstractness falls short of the mark of the difference in question, which, as we have seen, cannot be measured in terms of degrees of abstraction but only captured in terms of the transformation of the basic unit of arithmetic from a determinate multitude to the *concept* of such a multitude.

is manifest in this intuition of *at once* a sensible mark and a general concept is precisely Vieta's invention of the mathematical *symbol*.

According to Klein, a mathematically foundational problem follows from the analytic innovation of Vieta's method that concerns the derivation of the syntactical rules that govern the axiom system and establish the systematic context that defines the indeterminate objects to which they apply. Vieta established these rules on the basis of the "*logistice numerosa*" and thus in calculations with determinate amounts of monads, which is to say, in calculations with the natural and therefore non-symbolic numbers dealt with by ancient Greek arithmetic. This is what allows letter signs with no numerical properties to nevertheless have a numerical significance in the *logistice speciosa* and in the new algebra for which it is the foundation. Vieta, however, conceptualizes these multitudes composed of units at the *same time* from the perspective of their symbolic presentation, giving rise to the symbolic expression, barely visible today, of the intrinsically non-symbolic *Anzahlen*.[53]

As we have seen, Husserl's attempt to ground the logical objectivity of the concept of number employed in the symbolic calculus of universal analysis in the psychological phenomenon of collective combination ended in self-acknowledged failure. However, as already hinted above, Husserl's later works never managed to resolve either of the mathematical problems that the *PA* failed to resolve psychologistically, namely, that of the foundation of the unity of the collective unity constitutive of proper numbers along with the constitution of the formal-logical unity constitutive of symbolic numbers.[54] The common view, that the pure logic developed in Husserl's *Logical Investigations*, and more specifically, the descriptive phenomenological foundation of this logic in categorial intuition, is capable of providing logical foundations for both of these mathematical unities, is misguided on my view.

---

[53]Vieta's conceptualization of numbers grasped as *Anzahlen*, that is, determinate amounts of units, at the same time from the conceptual level of their symbolic formulation, is the historical precedent behind Husserl's conviction that in the case of ordinary arithmetic the system of signs and operation with signs runs "rigorously parallel" to the "system of concepts and operation with judgments" (*Stumpf Letter*, 159/14). As we have seen, the symbolic level of conceptualization initiated by Vieta treats the *concepts* of determinate multitude of units (e.g., two units, three units, etc.) as numerically equivalent with their non-conceptual multitudes. Thus, the number two is conceptualized as the general concept of two, which is to say, twoness, while at the same time the numeral 2 is identified with the (non-conceptual) number itself, viz., the determinate multitude of two units. This formulation of *Anzahlen* from the conceptual level of their symbolic formulation is what, according to Klein, is responsible for what is now the matter of fact identification of ordinary (cardinal) numbers with their signs (numerals). Thus the systematic parallelism between symbolically and conceptually conceived numbers appealed to by Husserl *presupposes* rather than accounts for the symbolic expression of *Anzahlen*; this is the case, because what falls under the concepts that are expressed by the system of symbolically employed signs on Husserl's view are not determinate amounts of units (*Anzahlen*) but the self-identical and therefore manifestly non-multitudinous general concepts (the individuated species) of the cardinal numbers or the general concept of being a cardinal number as such.

[54]Hopkins 2006.

In the case of the foundation of the collective unity of pre-formalized numbers, the categorial distinction between logical species and their instances merely substitutes one philosophically unsuitable solution—psychologism—for another one—logicism. This is the case because so long as the collective unity of a multitude is held to have its foundation in the *concept* of the species of that multitude, the problem of the unity of a non-conceptual manifold is no way resolved but simply shifted to a higher level. That is, it's shifted to the problem of providing the foundation for the conceptual unity of the manifold presupposed in the extension of the concept of species (or the logically equivalent concept of class) supposedly providing the foundation of the unity of the original manifold. Thus with respect to the problem at hand, to say as, for instance, Jan Patočka says in following Husserl's *Logical Investigations* and in thinking thus that he is resolving it, that "the number five is not my counting to five, nor someone else's counting, nor is it my or some else's conceptualization of five; it is a *species*, a generality, an *ideality* which is realized or given in the individual *instance* of a class of five members,"[55] doesn't resolve the issue. And, again, it doesn't do so because the foundation of the unity of the manifold that instantiates the "class of five members" is in no way provided but simply presupposed as unproblematically given by the unity of the *class* in this account.[56]

17.4 Symbolic Abstraction

In the case of the formal-logical concept presupposed by the symbolic concept of number, the "anything whatever" (*Etwas überhaupt*) that is the presupposition behind the indeterminate generality presupposed by the various domains of non-proper numbers, cannot be generated by an "abstractive" emptying of the "material" content from categories. Because the 'material' in question is the unity of a multitude of units, the metaphor of "emptying" is clearly not pertinent to the transformation of the concept of number that occurs in Vieta's analytic art with the substitution of the 'concept' of multiplicity for the direct reference to its concrete givenness. Indeed, Klein appeals to the mediaeval distinction between first and second intentions to clarify what he terms the "symbolic abstraction" responsible for the generation of the formal-logical concept of indeterminate generality, that is, for Husserl's "anything whatever," at stake in Vieta's symbolic expression of number. Klein's clarification unpacks Descartes' philosophical attempt to understand the origin of the novel mode of being that belongs to the symbolic number concept, because on Klein's view Descartes' attempt was the first, as well as the last such, in the philosophical tradition. On Klein's view, Descartes' attempt appealed to the power of the imagination to assist the pure intellect in making visible to it (the pure

[55]Patočka 1996, 35.

[56]See *Origin* ch. 32.

intellect), as a "symbol," the indeterminate object that it has already abstracted from its own power of knowing determinate numbers. As already mentioned, abstraction in Aristotle presupposes definite beings that are intelligible in terms of common qualities, the latter being "lifted off" the former in accordance with a process that is more logical than psychological; abstraction in Descartes presupposes definite beings but not their intelligibility, in the case at hand their "intelligibility" as so many beings. Rather, Descartes' abstraction works upon the mind's act of knowing a multitude of units, separating out the mind's own conceiving of that multitude, which it immediately makes objective. The mind turns and reflects on its own knowing when it is directed to the idea of number as a multitude of units, and, in so doing, it no longer apprehends the multitude of units directly, in the "performed act" (*actus exercitus*) and thus as object of its first intention, but rather indirectly, in the "signified act" (*actus signatus*), as object of its second intention. Thus, notwithstanding the fact that what is being conceived by the intellect is a multitude of units, the intellect's immediate apprehension of its own conceiving as something, as one and therefore as a being, has the effect of transforming the multitude belonging to the number into a seemingly independent being, albeit a being that is only a "rational being" (*ens rationis*). To repeat: this rational being is the result of the intellect—and not, as in Husserl's analysis, the result of the idealization of a tendency intrinsic to the perceptual experience of collective combination—, which, secondarily (in reflection) intends a thing already conceived before, and intends it insofar as it has been conceived. When the rational being is then "grasped *with the aid of the imagination* in such a way that the intellect can, in turn, take it up as an object in the mode of a 'first intention', we are dealing with a *symbol*."[57]

Abstraction for Descartes is therefore characterized by Klein as symbolic, because the concept (*Begriff*) that it yields is manifestly not something that is lifted off the intelligible qualities of things, but rather, is something whose very mode of being is inseparable from the following: (1) the intellect's pure—by pure is meant completely separate from the things it apprehends—grasping of its own power to apprehend these qualities themselves, and (2) this power itself being apprehended as an object whose mode of being is nevertheless akin to the very things that its mode of being separates itself from. Klein stresses that the "kinship" between the power of apprehension proper to the pure intellect and that which is effectively foreign to it (i.e., the things possessing the intelligible qualities that are apprehended by the "pure" intellect's power) is established by making this power "visible." The algebraic letter signs of Vieta or the geometric figures of Descartes are what accomplish this. They are what—in the language of the Schools—allow the object of a second intention to be apprehended as the object of a first intention, and are therefore symbols. The indeterminate or general object yielded in "symbolic abstraction" is neither purely a concept nor purely a sign, but precisely the unimaginable and unintelligible identification of the object of a second intention with the object of a first. This identification is unimaginable because

[57] *GMT* 208.

images properly—both for the ancient Greeks and for Descartes—refer to either particular objects of first intentions or to their particular "common qualities."[58] The identification between second and first intentional objects is unintelligible because for natural predication, to say that a concept is both general and particular at the same time is nonsensical.

The analytical philosopher of science Ernst Nagel's critical comments on Klein's *PHS* in his 1941 review of *Philosophical Essays in Memory of Edmund Husserl*[59] provide an instructive occasion to conclude our considerations of Husserl and Klein. They do so by allowing us to address the unintelligibility to natural predication of the symbolic expression of modern mathematics within the context of the third scholarly curiosity characteristic of Klein's relation to Husserl noted above. That curiosity, recall, concerned Klein's indirect role as the source of Husserl's account of Galileo's geometrically mathematizing impulse. After granting that Klein's paper "raises by implication the important question to what extent and in what sense genetic analysis is relevant to the understanding of scientific concepts,"[60] Nagel concludes "it is not clear how the psychological (or phenomenological) analysis which Dr. Klein suggests does bear upon the nature and history of geometry or how it contributes to the solution of the concrete problems connected with the use of geometry in the natural sciences."[61] Nagel's lack of clarity has its source in what he sees as Klein's agreement with Husserl "that to be intelligible an object must be understood in terms of its 'constitutive origins' and its 'intentional genesis'."[62] The problem with this for Nagel is that it places the impossible demand of having to "be at least the demi-urge in order to know anything," because as he understands "Dr. Klein's position, it is only in so far as we grasp the method by which we are supposed to construct or 'constitute' objects that we can really be at home with them intellectually."[63]

Nagel's criticism is informed by the understandable yet misguided presupposition that the aim of phenomenology's "genetic analysis" of scientific concepts is to enhance our understanding of their meaning for contemporary science on the

[58]Indeed, it is for this reason that Descartes, on Klein's view, stresses the "power" of imagination, and not the imagination's images, to assist the pure intellect in grasping the completely indeterminate concepts that it has separated from the ideas that the imagination offers it, because these ideas are precisely "determinate images"—and therefore, intrinsically unsuitable for representing to the intellect its indeterminate concepts. The imagination's power, however, being indeterminate insofar as it is not limited to any particular one of its images, is able to use is own indeterminateness to enter into the "service" of the pure intellect and make visible a "symbolic representation" of what is otherwise invisible to it, by facilitating, as it were, the identification of the objects of first and second intentions in the symbol's peculiar mode of being. The imagination's facilitation involving, as it were, its according its "power" of visibility to the concept's invisibility.

[59]Nagel 1941.

[60]*Op. cit.*, 303.

[61]*Ibid.*

[62]*Ibid.*

[63]*Ibid.*

basis of an investigation of their meaning in the science antedating ours. The latter meaning, while out of date, is nevertheless enlisted in the service of contributing to the clarification of our contemporary concepts. Nagel cites "the writings of such men as Mach and Duhem" as exemplifying "[t]hat a historical-genetic account of science can be illuminating and clarifying."[64] This presupposition is of course understandable, since the prevailing philosophical understanding of natural science in the 1930s as well as today takes it for granted that its method and results provide knowledge of the "true" nature, or what amounts to the same thing, the "true" knowledge of nature. It is misguided, however, because phenomenology's genetic analysis of historical meaning cannot be characterized without further ado as the straightforward investigation of historically datable 'concepts' against the backdrop of the a priori assumption that those with the most recent date are necessarily the ontologically superior. Husserl's key methodical term in this context, "reactivation," is crucial for understanding what is at stake in Husserl's *Crisis*-texts' appeal to 'history'. The history at issue in those texts, as mentioned, for Husserl is "intentional," in the precise sense of the layers of meaning whose methodically reflective exposure provides foundational evidence for the constitution of the very meaning of the most recently dated concepts, for instance, of their meaning proper to an exact science such as geometry. As presented by Husserl in the *Crisis*-texts, the method behind these reflections aspires not to the grandeur fitting only the demiurge but rather to the recovery of the integrity of knowledge in the face of the crisis precipitated by knowledge's peculiar "naturalization," and, indeed, its naturalization made possible on the foundation of the symbolic formalization of the mathematics employed by modern "mathematical physics."

Given Klein's indirect albeit seminal role in the Galilean focus of Husserl's *Crisis* writings, it's therefore no accident that he not only was the first, but also, that he remains to this day the only thinker to have grasped the peculiar phenomenological significance of Husserl's late turn to historical reflection; in the case at hand, to historical reflection on the origin of an exact science like geometry. The significance at stake here may be succinctly characterized in terms of the following insight. That to the reactivation of the sedimentation of an historically prior "conceptuality" (*Begrifflichkeit*) belonging to a science in the constitution of its historically later conceptuality, there corresponds, in the case of the exact sciences of analytic geometry and universal analysis, two things: 1) the rediscovery of that prior conceptuality and 2) the disclosure of the latter conceptuality's ontological status as "a symbolic disguise concealing the original 'evidence' and the original experience of things."[65]

[64] *Ibid.*

[65] *PHS* 84.

References

O. Becker, Die Lehre vom Geraden und Ungeraden im Neunten Buch der Euklidischen Elemente, Quellen und Studien zur Geschichte der Mathematik, Astronomie und Physik, Abteilung B: Studien. **3**, 533–553 (1934). English translation: The theory of odd and even in the Ninth Book of Euclid's elements, trans. by O. and P. Adler, Grad Fac Philos. J. **16**, 87–110, (1992)

D. Carr, Translator's introduction, in Husserl (1970)

H. Caton, Review of Jacob Klein's *Greek Mathematical Thought and the Origin of Algebra*. Studi Internazionali di Filosofia **3**, 222–226 (1971)

B. Hopkins, Husserl's psychologism, and critique of psychologism, revisited. Husserl Stud. **22**, 91–119 (2006)

B. Hopkins, *The Origin of the Logic of Symbolic Mathematics. Edmund Husserl and Jacob Klein* (Indiana University Press, Bloomington, 2011)

E. Husserl, Die Krisis der europäischen Wissenschaften und die transzendentale Phänomenologie. Eine Einleitung in die phänomenologische Philosophie. Philosophia **I**, 77–176 (1936). The text of this article is reprinted as §§ 1–27 of Husserliana VI

E. Husserl, Die Frage nach dem Ursprung der Geometrie als intentional-historisches Problem. Revue internationale de Philosophie **I** (1939a), ed. by E. Fink, 203–225. Fink's typescript of Husserl's original, and significantly different, 1936 text (which is the text translated in English) was published as Beilage III, in W. Biemel (ed.), *Die Krisis der europäischen Wissenschaften und die transzendentale Phänomenologie. Eine Einleitung in die phänomenologische Philosophie*, Husserliana VI (Nijhoff, The Hague, 1954, 1976). English translation: 'The Origin of Geometry', trans. by D. Carr, in *The Crisis of European Sciences and Transcendental Phenomenology* (Evanston, Northwestern University Press, 1970a)

E. Husserl, *Philosophie der Arithmetik* (Nijhoff, The Hague, 1970b). English translation: *The Philosophy of Arithmetic*, trans. by D. Willard (Kluwer, Dordrecht, 2003)

E. Husserl, Entwurf einer 'Vorrede' zu den *Logischen Untersuchungen* (1913). *Tijdschrift voor Philosophie***1**, 106–133 (1939b). English translation: *Introduction to the Logical Investigations*, trans. by P.J. Bossert, C.H. Peters (Nijhoff, The Hague, 1975)

E. Husserl, *Formale und transzendentale Logik* (Nijhoff, The Hague, 1974). English translation: *Formal and Transcendental Logic*, trans. by D. Cairns (Nijhoff, The Hague, 1969)

E. Husserl, in *Briefwechsel*, vol. IX, ed. by K. Schuhmann. (in cooperation with E. Schumann) (Martinus Nijhoff, The Hague, 1994a)

E. Husserl, Husserl an Stumpf, ca. Februar 1890, in *Briefwechsel*, ed. by K. Schuhmann. vol. I (Kluwer, Dordrecht, 1994b), 157–164. English translation: 'Letter from Edmund Husserl to Carl Stumpf', trans. by D. W. Kluwer, in *Early Writings, Early Writings in the Philosophy of Logic and Mathematics* (Dordrecht 1994c), pp. 12–19

E. Klein, (Wife of J. Klein), 14-page transcript of a tape recording [the original tape recording is apparently lost] among J. Klein's papers, which are housed in St. John's College Library, Annapolis, MD

E. Klein, Die griechische Logistik und die Entstehung der Algebra, *Quellen und Studien zur Geschichte der Mathematik, Astronomie und Physik,* Abteilung B: Studien **3** no. 1 (1934), 18–105 (Part I), and no. 2 (1936), 122–235 (Part II). English translation: *Greek Mathematical Thought and the Origin of Algebra*, trans. by E. Brann (MIT Press, Cambridge, 1969), reprint: Dover, New York, 1992

J. Klein, Phenomenology and the history of science, in M. Farber (ed.), *Philosophical Essays in Memory of Edmund Husserl* (Harvard University Press, Cambridge, 1940) 143–163, reprinted in: R. Williamson, E. Zuckerman (eds.), *Jacob Klein, Lectures and Essays* (St. John's Press, Annapolis, 1985) 65–84

A. Koyre, Galilée et l'expérience de Pise. Annales de l'Université de Paris **12**, 441–453, (1937a)

A. Koyre, Galilée et Descartes. Travaux du IXe Congrés international de Philosophie **2**, 41–47 (1937b)

A. Koyré, *Introduction à la lecture de Platon* (Gallimard, Paris, 1945)

U. Majer, Husserl and Hilbert on completeness: A neglected chapter in early twentieth century foundations of mathematics. Synthese **110**, 37–56 (1997)

J. Miller, *Numbers in Presence and Absence: A Study of Husserl's Philosophy of Mathematics* (Nijhoff, The Hague, 1982)

D. Moran, *Husserl's Crisis of the European Sciences and Transcendental Philosophy: An Introduction* (Cambridge University Press, Cambridge, 2012)

E. Nagel, Review of *Philosophical Essays in Memory of Edmund Husserl*', ed. by M. Farber. J. Philos. **38**, 301–306, (1941)

R. Parker, The history between Koyré and Husserl, forthcoming, in Pisano, Agassi, Drozdova (eds.), *Hypotheses and Perspectives within History and Philosophy of Science. Homage to Alexandre Koyré 1964–2014* (Springer, Berlin, forthcoming)

J. Patocka, The philosophy of arithmetic, in J. Dodd (ed.), *An Introduction to Husserl's Phenomenology*, trans. by E. Kohák (Open Court, Chicago, 1996)

K. Schuhmann, Alexandre Koyré, in *Encyclopedia of Phenomenology*, (Kluwer, Dordrecht, 1997), pp. 391–393

R. Smid, Einleitung des Herausgebers, in *Die Krisis der Europäischen Wissenschaften und die Transzendentale Phänomenologie, Ergänzungsband Texte aus dem Nachlass 1924–1937* (Kluwer, Dordrecht, 1993)

F. Vietae, *In Artem Analyticem* (sic) *Isagoge*, Seorsim excussa ab opere restituate Mathematicae Analyseo, seu, Algebra Nova (*Introduction to the Analytical Art*, excerpted as a separate piece from the *opus* of the restored Mathematical Analysis, or *The New Algebra*, Tours 1591. English translation: *Introduction to the Analytic Art*, trans. by J. W. Smith, appendix to J. Klein (1969/1992)

C. Wilson, *'Preface' to Essays in Honor of Jacob Klein* (St John's College Press, Annapolis, 1976)

References

E. Adickes, *Kant als Naturforscher*, 1924

K. Ajdukiewicz, Die syntaktische Konnexität. Studia Philosophica **1**, 1–27 (1935)

L. Albertazzi, M. Libardi, R. Poli (eds.), *The School of Franz Brentano* (Kluwer, Dordrecht, 1996)

R. Anderson, It adds up after all: Kant's philosophy of arithmetic in light of the traditional logic. Philos. Phenomenol. Res. **69**, 501–540 (2004)

Aristotle, *The Complete Works*, ed. by J. Barnes, vol. I-II (Princeton University Press, Princeton, 1984)

R. Argawal, Lambert series and Ramanujan. Indian Acad. Sci. (Math. Sci.) **103**, 269–293 (1993)

A. Arnauld and P. Nicole, *La Logique, ou l'art de penser*, Iean Guignart/Guillame Desprez, Paris 1662/1683. Critical edn. (Vrin, Paris, 1981)

S. Bachelard, *A Study of Husserl's Formal and Transcendental Logic* (Northwestern University Press, Evanston, 1968)

R. Baldus, Zur Axiomatik der Geometrie I. Über Hilberts Vollständigkeitsaxiom. Math. Ann. **100**, 321–333 (1928)

J. Barnes, *Porphyry. Introduction* (Oxford University Press, Oxford, 2003)

J. Barwise (ed.), *Handbook of Mathematical Logic* (North-Holland, Amsterdam, 1977)

F.L. Bauer, Lamberts Kettenbruch. Informatik-Spektrum **28**, 303–309 (2005)

O. Becker, *Mathematische Existenz*, Jahrbuch für Philosophie und phänomenologische Forschung 8, 439–809. 2nd unrevised edn. (Max Niemeyer, Tübingen 1927)

O. Becker, Das Symbolische in der Mathematik. Blätter für deutsche Philosophie **1**, 329–348 (1928)

O. Becker, The Philosophy of Edmund Husserl (1930), in R.O. Elveton, *The Phenomenology of Husserl, Selected Critical Readings* (Quadrangle Books, Chicago, 1970), pp. 40–72

O. Becker, Die Lehre vom Geraden und Ungeraden im Neunten Buch der Euklidischen Elemente, in *Quellen und Studien zur Geschichte der Mathematik, Astronomie und Physik,* Abteilung B: Studien **3**, 533–553 (1934). English translation: The Theory of Odd and Even in the Ninth Book of Euclid's *Elements*, trans. by O. and P. Adler, Grad. Fac. Philos. J. **16**, 87–110 (1992)

O. Becker *Grundlagen der Mathematik in geschichtlicher Entwicklung* (Alber, Freiburg/München, 1954; Suhrkamp, Frankfurt, 1975)

O. Becker, D. Mahnke, Briefwechsel mit Dietrich Mahnke, With an introduction by B. Aust and J. Sattler, in Peckhaus 2005, pp. 245–278

Y. Belaval, *Leibniz Critique de Descartes* (Gallimard, Paris, 1960)

P. Benacerraf, H. Putnam (eds.), *Philosophy of Mathematics: Selected Readings* (Cambridge University Press, Cambridge, [1]1964, [2]1983)

S. Centrone (ed.), *Essays on Husserl's Logic and Philosophy of Mathematics*, Synthese Library 384, DOI 10.1007/978-94-024-1132-4

G. Berkeley, The analist; or a discourse addressed to an infidel mathematician (1734), in A.A. Luce, T.E. Jessop (eds.), *The Works of George Berkeley Bishop of Cloyne*, vol. 4. (Nelson, London/Edinburgh, 1948–57), pp. 53–102

P. Bernays, Probleme der theoretischen Logik (1927), in Bernays 1976a, pp. 1–16.

P. Bernays, *Abhandlungen zur Philosophie der Mathematik* (Wissenschaftliche Buchgesellschaft, Darmstadt, 1976b)

R. Bernet, I. Kern, E. Marbach, *Edmund Husserl: Darstellung seines Denkens* (Meiner, Hamburg, 1989)

E. Beth, H. Pos, J. Hollak (eds.), *Proceedings of the 10th International Congress of Philosophy, Amsterdam 1948,* vol. 2, bk. 1, 1949

E.W. Beth, *The Foundations of Mathematics* (North Holland, Amsterdam, 1965)

C. Beyer, *Von Bolzano zu Husserl. Eine Untersuchung über den Ursprung der phänomenologische Bedeutungslehre* (Kluwer, Dordrecht, 1996)

C. Beyer, *Von Bolzano zu Husserl. Eine sprachanalytische Studie zu Husserls transzendentaler Phänomenologie* (Mentis, Paderborn, 2000)

C. Beyer, *Subjektivität, Intersubjektivität, Personalität. Ein Beitrag zur Philosophie der Person* (De Gruyter, Berlin/New York, 2006)

C. Beyer, M. Weichold, Philosophy of language, in *The Routledge Companion to Phenomenology*, ed. by S. Luft, S. Overgaard (Routledge, London/New York, 2011), pp. 406–416

C. Beyer, Husserl's transcendental phenomenology considered in the light of (recent) epistemology, in *Constructive Engagement of Analytic and Continental Approaches in Philosophy. From the Vantage Point of Comparative Philosophy*, ed. by Bo Mou, Richard Tieszen (Brill Academic Publishing, Leiden, 2013)

C. Beyer, Self-identification, intersubjectivity, and the background of intentionality. Reply to Anita Pacholik-*Żuromska*, in *OpenMIND*, ed. by T. Metzinger, J. Windt (MIND Group, Frankfurt/M, 2015)

G. Birkhoff, *Lattice Theory*, 3rd edn. (AMS Colloquium Publications, Providence R.I, 1940)

E. Bishop, *Foundations of Constructive Analysis* (McGraw-Hill, New York, 1967)

L. Boi, Questions regarding Husserlian geometry and phenomenology. A study of the concept of manifold and spatial perception. Husserl Stud. **20**, 207–267 (2004)

L. Boi, P. Kerszberg, F. Patras (eds.), *Rediscovering Phenomenology* (Springer, Dordrecht, 2007)

B. Bolzano, *Betrachtungen über einige Gegenstände der Elementargeometrie*, Barth, Prague 1804. Reprint: L. Novy (ed.), *Acta historiae rerum naturalium necnon technicarum* 12, Prague 1981. Engl. transl. in: S. Russ (ed.), *The mathematical works of Bernard Bolzano*, Oxford University Press, Oxford, 2004, pp. 25–81

B. Bolzano, *Beyträge zu einer begründeteren Darstellung der Mathematik. Erste Lieferung*, Caspar Widtmann, Prague 1810. Reprint: Wissenschaftliche Buchgesellschaft, Darmstadt 1974. English translation in S. Russ (ed.), *The Mathematical Works of Bernard Bolzano,* Oxford University Press, Oxford, 2004, pp. 82–137. B.

B. Bolzano, *Dr. Bolzanos Wissenschaftslehre. Versuch einer ausführlichen und größtentheils neuen Darstellung der Logik mir steter Rücksicht auf deren bisherige Bearbeiter.* Her. von mehren seiner Freunde, Sulzbach 1837. 2nd ed. by A. Hofler, W. Schultz, Meiner, Leipzig 1928. Engl. trans. by R. George and P. Rusnock, *Theory of Science, Vols. 1–4*, Oxford University Press, Oxford, 2014

B. Bolzano, *Bolzano's Wissenschaftslehre und Religionswissenschaft in einer beurtheilenden Uebersicht. Eine Schrift für Alle, die dessen wichtigsten Ansichten kennen zu lernen wünschen*, Sulzbach, Seidel 1841.

B. Bolzano, *Paradoxien des Unendlichen*, her. aus dem schriftichen Nachlasse des Verfassers von Fr. Prihonsky, Reclam, Leipzig 1851. Reprint: A. Hofler, with Notes by H. Hahn, Reclam, Leipzig 1921. Reprint with Notes by B. van Rootselaar, Meiner, Hamburg 1975. English translation: *Paradoxes of the Infinite*, in Ewald 1996, vol. 1, pp. 249–292.

B. Bolzano, in *Gesamtausgabe*, ed. by E. Winter, J. Berg, F. Kambartel, J. Louzil, E. Morscher, B van Rootselaar (Stuttgart-Bad Cannstatt, Frommann, 1969)ff.

G. Boole, *The Mathematical Analysis of Logic* (Macmillan & Co., Cambridge, 1847)

G. Boole, *An Investigation of the Laws of Thought* (Macmillan & Co., Cambridge, 1854)

G. Boolos, The iterative concept of set. J. Philos. **68**, 215–231 (1971)

P. Bossert (ed.), *Phenomenological Perspectives* (Martinus Nijhoff, Den Haag, 1975)

C. Brady, *From Peirce to Schroeder. A neglected chapter in the history of logic* (North Holland, Amsterdam, 2000)

F. Brentano, *Psychologie vom empirischen Standpunkte* (Duncker & Humblot, Leipzig, 1874)

F. Brentano, Die elementare Logik und die in ihr nötigen Reformen I. Manuscript Y 2, Vienna 1884/85

D. Bridges, Constructive mathematics, in Zalta 1997–, Summer 2009., http://plato.stanford.edu/archives/sum2009/entries/mathematicsconstructive/

K. Brockhaus, Konstitution. III. Neuzeit, in K. Gründer et al. (eds.), *Historisches Wörterbuch der Philosophie*, vol. 4 (Schwabe, Basel, 1976), pp. 997–1004

L.E.J. Brouwer, *Die mögliche Mächtigkeiten,* Lecture (1908), published in Castelnuovo 1909, pp. 569–571. Facsimile reprint in Brouwer 1975, pp. 102–104

L.E.J. Brouwer, Over de grondslagen der wiskunde, PhD diss., Universiteit van Amsterdam 1907. English translation in Brouwer 1975, pp. 11–101

L.E.J. Brouwer, De onbetrouwbaarheid der logische principes, Tijdschrift voor Wijsbegeerte **2**, 152–158 (1908). English translation in Brouwer 1975, pp. 107–111

L.E.J. Brouwer, *Intuïtionisme en formalisme* (Clausen, Amsterdam, 1912). English translation in Benacerraf and Putnam 1983, pp. 77–89

L.E.J. Brouwer, Intuitionism and formalism, Bull. Am. Math. Soc. **20**, 81–96 (1913), Facsimile reprint in Brouwer 1975, pp. 123–138

L.E.J. Brouwer, Begründung der Mengenlehre unabhängig vom logischen Satz vom ausgeschlossenen Dritten: Erster Teil: Allgemeine Mengenlehre. KNAW Verhandelingen **5**, 1–43 (1918). Facsimile reprint in Brouwer 1975, pp. 150–190

L.E.J. Brouwer, Über Definitionsbereiche von Funktionen, *Mathematische Annalen* 97 (1927), pp. 60–75. Facsimile reprint in Brouwer 1975, pp. 390–405. English translation of Sects. 1–3 in van Heijenoort 1967, pp. 457–463

L.E.J. Brouwer, Intuitionistische Betrachtungen über den Formalismus, KNAW Proc. **31**, 374–379 (1928), . Facsimile reprint in Brouwer 1975, pp. 409–414. English translation in Mancosu 1998, pp. 40–44

L.E.J. Brouwer, Mathematik, Wissenschaft und Sprache, Monatshefte für Mathematik und Physik. **36**, 153–164 (1929), . Facsimile reprint in Brouwer 1975, pp. 417–428. English translation in Mancosu 1998, pp. 45–53

L.E.J. Brouwer, *Die Struktur des Kontinuums,* Komitee zur Veranstaltung von Gastvorträgen ausländischer Gelehrter der exakten (Wissenschaften, Wien, 1930). Facsimile reprint in Brouwer 1975, pp. 429–440. English translation in Mancosu 1998, pp. 54–63

L.E.J. Brouwer, Willen, weten, spreken, In Brouwer, Clay, de Hartog, Mannoury, Pos, Révész, Tinbergen and van der Waals 1933, 45–63. English translation of Sect. 3 in Brouwer 1975, pp. 443–446. Full English translation in van Stigt 1990, pp. 418–431

L.E.J. Brouwer, Essentieel negatieve eigenschappen. Indagationes Mathematicae **10**, 322–323 (1948). English translation in Brouwer 1975, pp. 478–479

L.E.J. Brouwer, Consciousness, philosophy and mathematics, in Beth, Pos and Hollak 1949, pp. 1235–1249. Facsimile reprint in Brouwer 1975, pp. 480–494

L.E.J. Brouwer, Historical background, principles and methods of intuitionism, S. Afr. J. Sci. **49**, 139–146 (1952). Facsimile reprint in Brouwer 1975, pp. 508–515

L.E.J. Brouwer, Points and spaces, Can. J. Math. **6**, 1–17 (1954). Facsimile reprint in Brouwer 1975, pp. 522–538

L.E.J. Brouwer, *Philosophy and Foundations of Mathematics,* vol. 1 of *Collected Works,* ed. by A. Heyting (North-Holland, Amsterdam, 1975)

L.E.J. Brouwer, *Brouwer's Cambridge Lectures on Intuitionism, edited by Dirk van Dalen* (Cambridge University Press, Cambridge, 1981)

L. E. J. Brouwer, J. Clay, A. de Hartog, G. Mannoury, H. Pos, G. Révész, J. Tinbergen, J. van der Waals Jr (eds.), *De uitdrukkingswijze der wetenschap: Kennistheoretische openbare voordrachten gehouden aan de Universiteit van Amsterdam gedurende de kursus 1932–1933* (Noordhoff, Groningen, 1933)

L.E.J. Brouwer, F. van Eeden, J. van Ginneken, G. Mannoury, Signifische dialogen, Synthese **2**, 168–174, 261–268, 316–324 (1937)

L.E.J. Brouwer, F. van Eeden, J. van Ginneken, G. Mannoury, *Signifische dialogen* (Erven J. Bijleveld, Utrecht, 1939). English translation in Brouwer 1975, pp. 447–452

P. Bucci, *Husserl e Bolzano. Alle origini della fenomenologia* (Edizioni Unicopli, Milano, *2000*)

M. Bullynck, Decimal periods and their tables: A German research topic (1765–1801). Hist. Math. **36**, 137–160 (2009)

J. Burgess, Frege and arbitrary functions, in *Frege's Philosophy of Mathematics*, ed. by W. Demopoulous (Harvard University Press, Cambridge, 1995)

H. Busemann, *The Geometry of Geodesics* (Academic, New York, 1955)

R. Butts, J. Hintikka (eds.), *Logic, Foundations of Mathematics and Computability Theory* (Reidel, Dordrecht, 1977)

D. Byers, *Intentionality and Transcendence* (University of Wisconsin Press, Wisconsin, 2002)

A. Cantini, *I fondamenti della matematica da Dedekind a Tarski* (Loescher, Torino, 1979)

G. Cantor, Grundlagen einer allgemeinen Mannigfaltigkeitslehre. Ein mathematisch-philosophischer Versuch in der Lehre des Unendlichen (1883), (Teubner, Leipzig. in Cantor, 1932), pp. 165–246

G. Cantor, Rezension von Freges *Grundlagen der Arithmetik.* Deutsche Literaturzeitung **VI**(20) 728–729 (1885)

G. Cantor, Mitteilungen zur Lehre vom Transfiniten. Zeitschrift für Philosophie und philosophische Kritik **91**, 81–125 (1887); **92**, 240–265 (1888). in Cantor 1932, pp. 378–439

G. Cantor, *Briefbücher I (1884–1888), II (1890–1895), III (1895–1896)* at the Niedersächsische Staats-und Universitätsbibliothek Göttingen, Abteilung Handschriften und Seltene Drucke (Cod. Ms. 18)

G. Cantor, in *Gesammelte Abhandlungen* ed. by E. Zermelo (Springer, Berlin 1932)

G. Cantor, in *Briefe* ed. by H. Meschkowski, W. Nilson (Springer, Berlin 1991)

G. Cardano, *The Rules of Algebra (Ars Magna)* (Dover, Mineola, 2007)

W. Carl, *Frege's Theory of Sense and Reference. Its Origins and Scope* (Cambridge University Press, Cambridge, 1994)

R. Carnap, *Der Raum* (Reuther & Reichard, Berlin, 1922)

R. Carnap, *Der logische Aufbau der Welt* (Weltkreis Verlag, Berlin, 1928)

R. Carnap, *Abriss der Logistik* (Springer, Vienna, 1929)

R. Carnap, *Die logische Syntax der Sprache* 1934. Expanded English edition, *The Logical Syntax of Language* (Routledge, London, 1937); reprint, Chicago et al. 2003

D. Carr, Translator's introduction, in Husserl 1970

E. Casari, *La filosofia della matematica del 900* (Sansoni, Firenze, 1973)

E. Casari, Anche la Filosofia dell'Aritmetica compie cent'anni, in *Atti del Convegno: Peano e i fondamenti della matematica*, (Mucchi, Modena, 1991), pp. 35–49

E. Casari, *Sull'origine dell'«oggettivo» in Bolzano*, in *Logica e teologia. Studi in onore di Vittorio Sainati*, ed. by A. Fabris, G. Fioravanti, E. Moriconi (ETS, Pisa, 1997), pp. 93–115

E. Casari, Logic and the laws of possible being, in *Logic and Metaphysics*, ed. by M. Marsonet, M. Benzi (Name, Genova. 2004), pp. 16–43

E. Casari, *Bolzano's Logical System* (Oxford University Press, Oxford, 2016)

G. Castelnuovo (ed.), *Atti del IV Congresso internazionale dei matematici, Roma, 6–11 aprile 1908: Comunicazioni delle sezioni III-A, III-B e IV* (Tipografia della Reale Accademia dei Lincei, Roma, 1909)

H. Caton, Review of Jacob Klein's *Greek Mathematical Thought and the Origin of Algebra.* Studi Internazionali di Filosofia **3**, 222–226 (1971)

J. Cavaillès, *Sur la logique et la théorie de la science* (Vrin, Paris, 1947)

J. Cavaillès, On Logic and the Theory of Science, in *Phenomenology and the Natural Sciences*, ed. by J.J. Kockelmans, T.J. Kisiel (Northwestern University Press, Evanston, 1970), pp. 353–409
S. Centrone, Husserl on the Totality of all conceivable arithmetical operations. Hist. Philos. Log. **27**(3), 211–228 (2006)
S. Centrone, *Logic and Philosophy of Mathematics in the Early Husserl* (Springer, Berlin, 2010)
S. Centrone, Bolzano und Leibniz über Klarheit und Deutlichkeit. Archiv für Geschichte der Philosophie **92**(3), 256–289 S (2010)
S. Centrone, Functions in Frege, Bolzano and Husserl. Hist. Philos. Log **31**, 315–336 (2010)
S. Centrone, Reply to Mark von Atten. New Yearb. Phenomenol. Phenomenol. Philos. **12**, 377–384 (2012)
S. Centrone, Aspekte des Psychologismus-Streits. Husserl und Frege über Anzahlen und logische Gesetze, in *Versuche über Husserl*, ed. by S. Centrone (Meiner, Hamburg, 2013)
S. Centrone, *Begründungen bei Bolzano and beim frühen Husserl, in: Studien zu Bolzano* (Academia, Sankt Augustin, 2015), pp. 117–143
A. Church, The Richard's Paradox. Am. Math. Mon. **41**, 356–361 (1934)
L. Couturat, *La Logique de Leibniz: d'après des documents inédits, F* (Alcan, Paris, 1901)
L. Couturat, *Opuscules et fragments inédits de Leibniz, F* (Alcan, Paris, 1903)
G. Crapulli, *Mathesis universalis. Genesi di un'idea nel XVI secolo* (Edizioni dell'Ateneo, Roma, 1969 G)
J.J. da Silva, Husserl's phenomenology and Weyl's predicativism. Synthese **110**, 277–296 (1997). Reprinted in Hill & da Silva 2013, pp. 265–284
J.J. da Silva, Husserl's conception of logic. Manuscrito **22**(2), 367–397 (1999)
J.J. da Silva, Husserl's two notions of completeness: Husserl and Hilbert on completeness and imaginary elements in mathematics. Synthese **125**, 417–438 (2000). Reprinted as Husserl and Hilbert on Completeness and Imaginary Elements in Mathematics in Hill & da Silva 2013, pp. 115–136
J.J. da Silva, The many senses of completeness. Manuscrito **23**(2), 41–60 (2000). Reprinted in Hill & da Silva 2013, pp. 137–150
J.J. da Silva, Gödel and transcendental phenomenology. Rev. Int. Philos. **234**, 553–574 (2005)
J. da Silva, On the nature of mathematical knowledge, in *Brazilian Studies in the Philosophy and History of Science*, ed. by D. Krause, A. Videira (Springer, Dordrecht, 2010), pp. 151–160
J.J. da Silva, Away from the facts, symbolic knowledge in Husserl's philosophy of mathematics, in *Symbolic Knowledge from Leibniz to Husserl*, Studies in Logic, vol. 41, ed. by A.L. Casanave (College Publications, London, 2012), pp. 115–136
J.J. da Silva, Husserl on geometry and spatial representation, Axiomathes **22** (2012), 5–30. Reprinted in Hill & da Silva 2013, pp. 31–60
J.J. da Silva, Beyond Leibniz, Husserl's vindication of symbolic mathematics, in *The Road Not Taken, On Husserl's Philosophy of Logic and Mathematics*, ed. by C.O. Hill, J.J. da Silva (College Publications, London, 2013), pp. 61–80
J.J. da Silva, Husserl and the principle of bivalence, in Hill & da Silva 2013, pp. 285–298
J.J. da Silva, Mathematics and the crisis of science, in: Hill & da Silva 2013, pp. 345–367
J.J. da Silva, Husserl and Hilbert on completeness, still, Synthese (2015), 10.1007/s11229-015-0821-2.
C. Dasypodius, *Euclidis quindecim elementorum geometriae primum* (C. Mylius, Strasburg, 1564a)
C. Dasypodius, *Euclidis quindecim elementorum geometriae secondum* (C. Mylius, Strasburg, 1564b)
C. Dasypodius, *Euclidis elementorum liber primus* (C. Mylius, Strasburg, 1571)
C. Dasypodius, *Protheoria Mathematica* (Argentorati, Strasburg, 1593)
J. Dauben, *Georg Cantor: His Mathematics and Philosophy of the Infinite* (Princeton University Press, Princeton, 1979)
J. Dauben, *Georg Cantor: His Mathematics and Philosophy of the Infinite* (Princeton University Press, Princeton, 1979)

M. Davis (ed.), *The Undecidable: Basic Papers on Undecidable Propositions, Unsolvable Problems and Computable Functions* (Raven Press, New York, 1965)

R. Dedekind, *Was sind und was sollen die Zahlen?* (Vieweg & Sohn, Braunschweig, 1888)

V. De Palma, Die Kategorien des Sinnlichen. Zu Husserls Kategorienlehre. Phänomenologische Forschungen NF **15**, 23–39 (2010)

Descartes, *Oeuvres*, Vol. X, Ch. E., ed. by Adam, P. Tannery (Léopold Cerf, Paris, 1908)

R. Descartes, in *The Geometry of René Descartes*, ed. by D.E. Smith, M.L. Lathan (Dover, Mineola, 1954)

J. Dodd, Husserl between formalism and intuitionism, in Boi, Kerszberg and Patras 2007, 261–265

K. Donnellan, Reference and definite descriptions. Philos. Rev. **75**(3), 281–304 (1966)

B. Dreben, A. Kanamori, Hilbert and set theory. Synthese **110**, 77–125 (1997)

H. Dreyfus (ed.), *Husserl, Intentionality and Cognitive Science* (MIT Press, Cambridge, 1982)

M. Dummett, *Frege. Philosophy of Language* (Duckworth, London, 1973)

M. Dummett, *Frege. Philosophy of Mathematics* (Harvard University Press, Harvard, 1991)

M. Dummett, *Ursprünge der analytischen Philosophie* (Suhrkamp, Frankfurt/M, 1992)

M. Dummett, Is time a continuum of instants? Philosophy **75**, 497–515 (2000)

M. Dummett, *Elements of Intuitionism.* 2nd, rev. edn. (Clarendon Press, Oxford, 2000)

M. Du Sautoy, *The Music of Primes. Searching to Solve the Greatest Mysteries of Mathematics* (Harper Perennial, New York, 2003)

P. Dybjer, S. Lindström, E. Palmgren, G. Sundholm (eds.) 2012. *Epistemology versus Ontology: Essays on the Philosophy and Foundations of Mathematics in Honour of Per Martin-Löf* (Springer, Dordrecht, 2012)

H. D. Ebbinghaus in cooperation with V. Peckhaus, *Ernst Zermelo, An Approach to His Life and Work* (Springer, Berlin/Heidelberg/New York, 2007)

L. Eley, *Metakritik der Formalen Logik* (Martinus Nijhoff, Den Haag, 1969)

T. Elsenhans, Phänomenologie, Psychologie, Erkenntnistheorie. Kant-Studien **20**, 224–275 (1915)

G. Evans, *The Varieties of Reference* (Clarendon, Oxford, 1982)

W. Ewald, *From Kant to Hilbert: A Source Book in the Foundations of Mathematics*, vol II (Clarendon Press, Oxford, 1996)

L. Falkenstein, Kant, Mendelssohn, Lambert, and the subjectivity of time. J. Hist. Philos. **29**, 227–251 (1991)

E. Fink, The Phenomenological Philosophy of Edmund Husserl and Contemporary Criticism, *The Phenomenology of Husserl,* edited and translated by R. O. Elveton, Quadrangle books, Chicago. 1970

D. Føllesdal, *Husserl und Frege*, Aschehoug, Oslo 1958. English translation in L. Haaparanta (ed.), *Mind, Meaning and Mathematics*, Kluwer, Dordrecht 1994, pp. 3–47

D. Føllesdal, Husserl's notion of noema. J. Philos. **66**, 680–688 (1969)

D. Føllesdal, Husserl's conversion from psychologism and the Vorstellung-meaning-reference distinction: Two separate issues, in Dreyfus 1982, pp. 52–56

D. Føllesdal, Gödel and Husserl, in *From Dedekind to Gödel: Essays on the Development of the Foundations of Mathematics*, ed. by J. Hintikka (Kluwer, Dordrecht, 1995), pp. 427–446

D. Føllesdal, Indikatoren und der Geist, in *Phänomenologie und Sprachanalyse*, ed. by G. Keil, U. Tietz (Mentis, Paderborn, 2006)

G. Frege, *Begriffsschrift. Eine der arithmetischen nachgebildete Formelsprache des reinen Denkens*, Verlag von Louis Nebert, Halle 1879. Reprinted in Frege 1964. English translation in Van Heijenoort 1967

G. Frege, *Die Grundlagen der Arithmetik. Eine logisch mathematische Untersuchung über den Begriff der Zahl*, Koebner, Breslau 1884. Centenary Edition, ed. by C. Thiel (Meiner, Hamburg, 1986). English translation: *The Foundations of Arithmetic* (1884), Blackwell, Oxford, 1986

G. Frege, Über formale Theorien der Arithmetik (1885), in *Kleine Schriften*, pp. 103–111

G. Frege, Funktion und Begriff, Pohle, Jena 1891. Reprinted in *Kleine Schriften*, pp. 125–142

G. Frege, Über Begriff und Gegenstand. Vierteljahresschr. für wissenschaftliche Philoso. **16**, 192–205 (1892)

G. Frege, Über Sinn und Bedeutung. Zeitschrift für Philosophie und philosophische Kritik NF **100**, 25–50 (1892). Reprinted in *Kleine Schriften*, 143–162

G. Frege, Ausführungen über Sinn und Bedeutung (1892–95), in *Nachgelassene Schriften*, pp. 128–136

G. Frege, Rezension von E. G. Husserl: *Philosophie der Arithmetik I*. Zeitschrift für Philosophie und philosophische Kritik NF **103**, 313–332 (1894). Reprinted in *Kleine Schriften*, 179–192. English translation in *Collected Papers on Mathematics, Logic und Philosophy*, pp. 195–209

G. Frege, Kritische Beleuchtung einiger Punkte in E. Schröders Vorlesungen über die Algebra der Logik. Archiv für systematische Philosophie **1**, 433–456 (1895). English translation in *Collected Papers on Mathematics, Logic and Philosophy*, pp. 210–228

G. Frege, Über Euklidische Geometrie (1899?), in *Nachgelassene Schriften*, pp. 182–184

G. Frege, *Grundgesetze der Arithmetik, begriffschriftlich abgeleitet*, Band 1, Pohle, Jena 1893; Band 2, Pohle, Jena, 1903

G. Frege, Der Gedanke, *Beiträge zur Philosophie des deutschen Idealismus* I. Heft 2 (1918), pp. 58–77. Reprinted in *Kleine Schriften*, pp. 342–362

G. Frege, *Begriffsschrift und andere Aufsätze. Mit E. Husserls und H. Scholz' Anmerkungen*, I. Angelelli (ed.), Wissenschaftliche Buchgesellschaft, Darmstadt 1964.

G. Frege, *Kleine Schriften*, I. Angelelli (ed.), Olms, Hildesheim, 1967. Revised edition, Olms, Hildesheim, 1990

G. Frege, in *Nachgelassene Schriften*, ed. by H. Hermes et al., 1969. Revised edition, G. Gabriel, W. Rödding (eds.), Meiner, Hamburg, 1983

G. Frege, *Wissenschaftlicher Briefwechsel*. G. Gabriel et al. (eds.), Meiner, Hamburg, 1976. Abridged English edition: *Philosophical and Mathematical Correspondence*, Basil Blackwell, Oxford, 1980

G. Frege, in *Posthumous Writings*, ed. by H. Hermes, et al. (Blackwell, Oxford, 1979)

G. Frege, in *Philosophical and Mathematical Correspondence*, ed. by B. McGuinness, G. Gabriel, et al. (Blackwell, Oxford, 1980)

G. Frege, in *Collected Papers on Mathematics, Logic und Philosophy*, ed. by B. McGuinness (Blackwell, Oxford, 1984)

S. French, A phenomenological solution to the measurement problem? Husserl and the foundations of quantum mechanics. Stud. Hist. Philos. Mod. Phys. **33**, 467–491 (2002)

M. Friedman, *Kant and the Exact Sciences* (Harvard University Press, Cambridge, MA, 1992)

Y. Gauthier, Husserl and the Theory of Multiplicities, in *Husserl and the Sciences*, ed. by R. Feist (University of Ottawa Press, Ottawa, 2004)

H.M. Gerlach, H. R. Sepp (eds.), Es ist keine Seligkeit 13 Jahre lang Privadocent und Tit. 'prof'. zu sein. Husserls hallesche Jahre 1887 bis 1901, *Husserl in Halle* (Peter Lang, Bern, 1994), pp. 15–39

C. Gethmann, Hermeneutic Phenomenology and Logical Intuitionism: On Oskar Becker's *Mathematische Existenz*. New Yearbook Phenomenol. Phenomenol. Philos. **3**, 145–160 (2003)

K. Gödel, On Undecidable Propositions of Formal Mathematical Systems (1934), in Gödel 1986, pp. 346–371

K. Gödel, Russell's mathematical logic (1944), in Schilpp 1944, 123–153. Reprinted, with original page numbers in the margin, in Gödel 1990, pp. 119–141

K. Gödel, *Remarks before the Princeton bicentennial conference on problems in mathematics*, Lecture (1946), first published in Davis 1965, pp. 84–88

K. Gödel, What is Cantor's continuum problem?, Am Math Mon **54**, 515–525 (1947). Reprinted, with original page numbers in the margin, in Gödel 1990, pp. 176–187

K. Gödel, *The modern development of the foundations of mathematics in the light of philosophy*, Lecture draft in German (1961?), published, with an English translation, in Gödel 1995, pp. 374–387

K. Gödel, What is Cantor's continuum problem?, in Benacerraf and Putnam 1964, 258–273. Revised and expanded version of Gödel 1947. Reprinted, with original page numbers in the margin, in Gödel 1990, pp. 254–270

K. Gödel, On an Extension of Finitary Mathematics Which Has Not Yet Been Used (1972), in Gödel 1990, pp. 271–280. Revised version of Gödel 1958
K. Gödel, *Publications 1929–1936.* Vol. 1 of *Collected Works,* ed. by S. Feferman et al. (Oxford University Press, Oxford, 1986)
K. Gödel, *Publications 1938–1974.* Vol. 2 of *Collected Works,* ed. by S. Feferman et al. (Oxford University Press, Oxford, 1990)
K. Gödel, Some Remarks on the Undecidability Results (1972a), in Gödel 1990, 305–306
K. Gödel, *Unpublished Essays and Lectures.* Vol. 3 of *Collected Works,* ed. by S. Feferman et al. (Oxford University Press, Oxford, 1995)
K. Gödel, Undecidable Diophantine Propositions (*193?), in Gödel 1995, pp. 164–175
K. Gödel, Some Basic Theorems on the Foundations of Mathematics and Their Implications (*1951), in Gödel 1995, pp. 304–323
K. Gödel, Is Mathematics a Syntax of Language? (*1953/9, III and V), in Gödel 1995, pp. 334–363
K. Gödel, *Correspondence A–G.* Vol. 4 of *Collected Works,* ed. by S. Feferman et al. (Oxford University Press, Oxford, 2003)
K. Gödel, *Correspondence H–Z.* Vol. 5 of *Collected Works,* ed. by S. Feferman et al. (Oxford University Press, Oxford, 2003)
N. Goodman, *The Structure of Appearance* (Harvard University Press, Cambridge, 1951)
H.B. Gottschalk, The earliest Aristotelian commentators, in *Aristotle Transformed. The ancient commentators and their influence*, R. Sorabji (Duckworth, London, 1990), pp. 55–82
H. Grassmann, *Die Wissenschaft der extensiven Grösse oder die Ausdehnungslehre* (Wigand, Leipzig, 1844)
H. Grassmann, *Lehrbuch der Arithmetik für höhere Lehranstalten* (Enslin, Berlin, 1861)
I. Grattan-Guinness, Towards a Biography of Georg Cantor. Ann. Sci. **27**(4), 345–391 (1971)
I. Grattan-Guinness, How Russell Discovered His Paradox. Hist. Math. **5**, 127–137 (1978)
I. Grattan-Guinness, Georg Cantor's Influence on Bertrand Russell. Hist. Philos. Log. **1**, 61–93 (1980)
I. Grattan-Guinness, *The Search for Mathematical Roots, Logics, Set Theories and the Foundations of Mathematics from Cantor through Gödel* (Princeton University Press, Princeton, 2000)
J. Gray, *The Hilbert Challenge* (Oxford University Press, Oxford, 2000)
L. Haaparanta, Analysis as the method of logical discovery. Some remarks on Frege and Husserl. Synthese **77**, 73–97 (1988)
M. Hallett, *Cantorian Set Theory and Limitation of Size* (Clarendon, Oxford, 1984)
M. Hallet, Hilbert and logic, in *Quebec Studies in the Philosophy of Science,* 1995
M. Hallett, U. Majer (eds.), *David Hilbert's Lectures on the Foundations of Geometry* (Springer, Heidelberg, 2004)
H. Hankel, *Vorlesungen über die complexen Zahlen und ihre Functionen I* (Leopold Voss, Leipzig, 1867)
G. Hardy, L. Littlewood, On a Tauberian Theorem for Lambert Series, and some Fundamental Theories in Analytic Number Theory. Proc. Lond. Math. Soc. **19**, 21–29 (1921)
M. Hartimo, Mathematical roots of phenomenology: Husserl and the concept of number. Hist. Philoso. Log. **27**, 319–337 (2006)
M. Hartimo, From geometry to phenomenology. Synthese **162**, 225–233 (2008)
M. Hartimo, Towards completeness: Husserl on theories of manifolds 1890–1901. Synthese **156**, 281–310 (2007)
M. Hartimo (ed.), *Phenomenology and Mathematics* (Springer, Dordrecht, 2010)
M. Hartimo, Husserl's Pluralistic phenomenology of mathematics. Philos. Math. **20**, 86–110 (2012)
M. Hartimo, Husserl's scientific context 1917–1938. New Yearb. Phenomenol Phenomenol Philos. forthcoming (2016)
M. Hartimo, M. Okada, Syntactic reduction in Husserl's early phenomenology of arithmetic. Synthese (2015). doi:10.1007/s11229-015-0779-0
K. Hauser, Gödel's program revisited, Part I: The turn to phenomenology. Bull. Symb. Log. **12**, 529–588 (2006)

M. Heidegger, *Prolegomena zur Geschichte des Zeitbegriffs*, in P. Jaeger (ed.), *Heidegger Gesamtausgabe*, vol. 20 (Klostermann, Frankfurt am Main, 2009)
E. Hellinger, Logische Prinzipien des Mathematischen Denkens. Vorlesungen von Professor Dr. Hilbert im Sommer-Semester 1905, unpublished lecture notes 1905
D. Hesseling, *Gnomes in the Fog: The Reception of Brouwer's Intuitionism in the 1920s* (Birkhäuser, Basel, 2003)
A. Heyting, Die formalen Regeln der intuitionistischen Logik, I, *Sitzungsberichte der Preussischen Akademie der Wissenschaften* (1930), 42–56. English translation in Mancosu 1998, pp. 311–327
A. Heyting, *Intuitionism: An Introduction* (North-Holland, Amsterdam, 1956)
A. Heyting, L.E.J. Brouwer, in Klibansky 1968, pp. 308–315
D. Hilbert, On the Concept of Number (1900), in Ewald (1996), 1092–1095
D. Hilbert, Mathematical Problems (1900), in Ewald (1996), 1096–1105
D. Hilbert, Über die Grundlagen der Geometrie (1902), *Mathematische Annalen* 56, 381–422. English translation in Hilbert (1990), pp. 150–190
D. Hilbert, Axiomatic Thought (1918), in Ewald 1996, 1107–1115
D. Hilbert, Neubegründung der Mathematik (Erste Mitteilung), *Abhandlungen aus dem Mathematischen Seminar der Hamburgischen Universität* 1 (1922), 157–177. English translation: The New Grounding of Mathematics, in Ewald 1996, pp. 1117–1134
D. Hilbert, On the INFINITE (1926), in van Heijenoort 1967, pp. 369–392
D. Hilbert, The foundations of mathematics (1927), in van Heijenoort 1967), pp. 464–479
D. Hilbert, Problems of the Grounding of Mathematics (1928), in Mancosu 1998, pp. 227–233
D. Hilbert, in *Lectures on the Foundations of Geometry 1891–1902*, ed. by M. Hallett, U. Majer (Springer, Heidelberg, 2004)
D. Hilbert, W. Ackermann, *Grundzüge der theoretische Logik* (Springer, Berlin, 1928)
D. Hilbert, *The Foundations of Geometry*. Authorized translation by E. J. Townsend. Open Court, La Salle, Illinois 1950
D. Hilbert, *The Theory of Algebraic Invariants* (Cambridge University Press, Cambridge, 1993)
D. Hilbert, P. Bernays, *Grundlagen der Mathematik I* (Springer, Berlin, 1934)
D. Hilbert, L. Nordheim, J. von Neumann, Über die Grundlagen der Quantenmechanik. Math. Ann. **98**, 1–30 (1928)
C.O. Hill, *Word and Object in Husserl, Frege and Russell* (Ohio University Press, Athens, 1991)
C.O. Hill, Frege Attacks Husserl and Cantor. The Monist **77**(3), 347–357 (1994). Also in: Hill & Rosado Haddock 2000
C.O. Hill, Husserl and Hilbert on completeness, in Hintikka (ed.), *From Dedekind to Gödel*, Synthese Library v. 251, Kluwer, Dordrecht 1995. Reprinted in C. O. Hill & G. Rosado Haddock, *Husserl and Frege. Meaning, objectivity, and mathematics* (Open Court, Chicago/La Salle, 2000, pp. 179–198
C.O. Hill, Did Georg Cantor Influence Edmund Husserl?, Synthese **113**, 145–170 (1997a). Also in: Hill & Rosado Haddock 2000
C.O. Hill, *Rethinking Identity and Metaphysics. On the Foundations of Analytic Philosophy* (Yale University Press, New Haven, 1997b)
C.O. Hill, The varied sorrows of logical abstraction, Axiomathes **1–3**, 53–82 (1997c). Also in: Hill & Rosado Haddock 2000
C.O. Hill, From Empirical Psychology to Phenomenology: Husserl on the Brentano Puzzle in *The Brentano Puzzle*, ed. by R. Poli (Ashgate, Aldershot, 1998), pp. 151–168
C.O. Hill, Abstraction and Idealization in Georg Cantor and Edmund Husserl', in F. Coniglione et al. (eds.), *Abstraction and Idealization. Historical and Systematic Studies, Poznan studies in the philosophy of the sciences and the humanities* (Rodopi, Amsterdam, 1999). Also in: Hill & Rosado Haddock 2000
C.O. Hill, Husserl's *Mannigfaltigkeitslehre*, in: Hill & Rosado Haddock 2000, pp. 161–177
C.O. Hill, On Husserl's Mathematical Apprenticeship and Philosophy of Mathematics, in: Anna-Teresa Tymieniecka (ed.), *Phenomenology World Wide* (Kluwer, Dordrecht, 2002a), pp. 76–92. Also in: Hill & da Silva 2013

C.O. Hill, Tackling Three of Frege's Problems: Edmund Husserl on Sets and Manifolds. Axiomathes **13**, 79–104 (2002b). Also in Hill & da Silva 2013

C.O. Hill, Phenomenology from the Metaphysical Standpoint, Diálogos **XLIII**, 91, January (2008), 19–35

C.O. Hill, Husserl and phenomenology, experience and essence, in *Phenomenology and Existentialism*, (Springer, Dordrecht, 2009), pp. 9–22

C.O. Hill, Husserl on axiomatization and arithmetic, in Hartimo 2010, pp. 47–71

C.O. Hill, Cantor's Paradise, Metaphysics and Husserlian Logic, in: H. Koskinen (ed.), *Categories of Being, Essays on Metaphysics and Logic* (Oxford University Press, Oxford, 2012), PP. 217–240. Also in Hill & da Silva 2013

C.O. Hill, J.J. da Silva, *The Road Not Taken, on Husserl's Philosophy of Logic and Mathematics* (College Publications, London, 2013)

C.O. Hill, G.E. Rosado Haddock, *Husserl or Frege? Meaning, Objectivity, and Mathematics* (Open Court, La Salle, 2000)

J. Hintikka (ed.), *From Dedekind to Gödel: Essays on the Development of the Foundations of Mathematics* (Kluwer, Dordrecht, 1995)

A. Höfler, Besprechung von Kerry, Über Anschauung und ihre psychische Verarbeitung, Husserl, Philosophie der Arithmetik, von Ehrenfels, Zur Philosophie der Mathematik, Zeitschrift für Psychologie und Physiologie der Sinnesorgane VI (1894)

B. Hopkins, Husserl's Psychologism, and Critique of Psychologism, Revisited, Husserl Stud. **22**, 91–119 (2006)

B. Hopkins, *The Origin of the Logic of Symbolic Mathematics. Edmund Husserl and Jacob Klein* (Indiana University Press, Bloomington, 2011)

Husserl Archive, Mitteilungsblatt 34. Leuven 2011., http://hiw.kuleuven.be/hua/mitteilungsblatt/mitteilungsblatt34.pdf. Accessed 19 December 2013

Husserl Archive, Mitteilungsblatt 36, Leuven 2013., http://hiw.kuleuven.be/hua/mitteilungsblatt/mitteilungsblatt36.pdf Accessed 19 December 2013

E. Husserl, [HGW] *Husserliana, Gesammelte Werke*, I–XXVI: Martinus Nijhoff, Den Haag 1950–, XXVII–XXXVII: Kluwer, Dordrecht 1989– ; XXXVIII–: Springer, New York 2005

E. Husserl, [HD] *Husserliana, Dokumente*, U. Melle (Series ed.), I: Martinus Nijhoff, Den Haag 1981; II–IV: Kluwer, Den Haag 1989

E. Husserl, [HCW] *Husserliana, Collected Works*, U. Melle (Series ed.), I-XIII: Martinus Nijhoff, Den Haag/Springer, New York 1981

E. Husserl, Recollections of Franz Brentano, in *Husserl: Shorter Works*, ed. by P. McCormick, F. Elliston (University of Notre Dame Press, Notre Dame, 1981), pp. 342–49

E. Husserl, *Beiträge zur Variationsrechnung*, Vienna 1883 (unpublished dissertation)

E. Husserl, *Studien zur Arithmetik und Geometrie (1886–1901)*, I. Strohmeyer (ed.), HGW XXI, 1983

E. Husserl, *Über den Begriff der Zahl: Psychologische Analysen*, Heynemann'sche Buchdruckerei (F. Beyer), Halle 1887

E. Husserl, *Lecture on the Concept of Number* (WS 1889/90). C. Ierna (ed. & tr.), The new yearbook for phenomenology and phenomenological philosophy V (2005)

E. Husserl, Begriff der Allgemeinen Arithmetik (1890), in HGW XII, 1970, 375–79. English translation: The Concept of General Arithmetic, in HCW V, 1994, 1–6

E. Husserl, Zur Logik der Zeichen (1890), in HGW XII, 1970, 340–73. English translation: On the Logic of Signs (Semiotic), in HCW V, 1994, 20–51

E. Husserl, Husserl an Stumpf, ca. Februar 1890, in *Briefwechsel* I, 1994, 157–164. English translation: Letter from Edmund Husserl to Carl Stumpf, in HCW V, 1994, 12–19

E. Husserl, Besprechung von: E. Schröder, *Vorlesungen über die Algebra der Logik*, I. Band, Leipzig 1890, *Göttingsche gelehrte Anzeigen* (1891), 243–270. English translation in HCW V, 1994, pp. 52–91

E. Husserl, Das Folgerungskalkül und die Inhaltslogik. Vierteljahrsschrift für wissenschaftliche Philosophie **15**, 168–189 (1891). English translation in HCW V, 1994, 92–114

E. Husserl, *Philosophie der Arithmetik. Logische und psychologische Untersuchungen*, Bd. I, Pfeffer, Halle a.d.S. 1891. Critical Edition: [*PdA*] *Philosophie der Arithmetik. Mit ergänzenden Texten (1890–1901)*, ed. by L. Eley. HGW XII, 1970. English translation: [*PoA*] *Philosophy of Arithmetic, Psychological and Logical Investigations with Supplementary texts from 1887 – 1901*, ed. by D. Willard (Kluwer, Dordrecht, 2003)

E. Husserl, Brief an Stumpf (1891), in HGW XXI, 1983, 244–51. English translation: Letter from Edmund Husserl to Carl Stumpf, in HCW V, 1994, pp. 12–19

E. Husserl, *Briefwechsel*, in K. Schuhmann (ed.), HD III/1-10, 1994

E. Husserl, [HM] *Husserliana, Materialien*, I-VII: Kluwer, Dordrecht 2001–; VIII–IX: Springer, New York 2006

E. Husserl, [1893] A. Voigts "Elementare Logik", in HGW XXII, 73–82. English translation: A. Voigt's "Elemental Logic", in HCW 1994, pp. 121–130

E. Husserl, [1894] Psychologische Studien zur elementaren Logik, in HGW XXII, 92–123. English translation: Psychological Studies in the Elements of Logic, in HCW 1994, pp. 139–170

E. Husserl, Phenomenology (drafts of the *Encyclopedia Britannica* Article) (1927–28), in HCW VI, 1997, pp. 83–194

E. Husserl, *Cartesian Meditations. An Introduction to Phenomenology* (1931), translated by D. Cairns (Nijhoff, The Hague, 1960)

E. Husserl, [*FTL*] *Formale und transzendentale Logik. Versuch einer Kritik der logischen Vernunft*, Jahrbuch für Philosophie und phänomenologische Forschung X (1929). Critical Edition: *Formale und transzendentale Logik. Versuch einer Kritik der logischen Vernunft*, ed. by P. Jannsen, HGW XVII, 1975. English translation: *Formal and Transcendental Logic*, ed. by D. Cairns (Martinus Nijhoff, Den Haag, 1969)

E. Husserl, *Logische Untersuchungen*, Bd I, *Prolegomena zur reinen Logik*, Max Niemeyer, Halle, 1900 (^{2}1913); Bd II, *Untersuchungen zur Phänomenologie und Theorie der Erkenntnis*. Max Niemeyer, Halle a.d.S., 1901 (^{2}1913–1921, ^{7}1993). Critical edition, E. Holenstein/U. Panzer (ed.), HGW XVIII/XIX, 1984. English translation of HGW XVIII-XIX: *Logical Investigations*, J. N. Findlay (ed.), Routledge, New York 1970

E. Husserl, Notiz einer mündlichen Mitteilung Zermelos an Husserl (1902), in HGW XXII, 1979, 399. English translation: Memorandum of a Verbal Communication from Zermelo to Husserl, HCW V, 1994, 442

E. Husserl, Bericht über deutsche Schriften zur Logik in den Jahren 1895–99. Dritter Artikel (1903), in HGW XXII, 1979, 201–15. English translation: A Report on German Writings in Logic from the Years 1895–1899, Third Article, in HCW V, 1994, pp. 246–59

E. Husserl, [1903b] Rezension von Palaygi, in HGW XXII, 152–61. English translation: Review of Melchior *Palágyi's Der Streit der Psychologisten und Formalisten in der Modernen Logik*, in HCW 1994, pp. 197–206

E. Husserl, Husserl an Brentano, 27. III. 1905, in *Briefwechsel* I

E. Husserl, Persönliche Aufzeichnungen (1906), in HGW XXIV, 1985, 442–449. English translation: 'Personal Notes', in HCW V, 1994, 490–500.

E. Husserl, Philosophie als strenge Wissenschaft. Logos **1**, 289–341 (1910/11). English translation: Philosophy as Rigorous Science, in *Phenomenology and the Crisis of Philosophy*, Harper, New York 1965, pp. 71–147

E. Husserl, *Ideen zu einer reinen Phänomenologie und phänomenologischen Philosophie. Erstes Buch. Allgemeine Einführung in die reine Phänomenologie*, Jahrbuch für Philosophie und phänomenologische Forschung, I, 1913, 1–323. As a separate volume: Niemeyer, Halle a.d.S. 1913; ^{2}1922; ^{3}1928. Critical edition: K. Schuhmann (ed.), HGW III/1-2, 1950. English translation: *Ideas Pertaining to a Pure Phenomenology and to a Phenomenological Philosophy. First Book: General Introduction to a Pure Phenomenology*, HCW II, 1982

E. Husserl, Die Krisis der europäischen Wissenschaften und die transzendentale Phänomenologie. Eine Einleitung in die phänomenologische Philosophie, *Philosophia* I (1936), 77–176. In W. Biemel (ed.), HGW VI, 1954. English translation: *The Crisis of European Sciences and Transcendental Phenomenology*, Northwestern University Press, Evanston, Ill. 1970

E. Husserl, *Erfahrung und Urteil*, Academia Verlagsbuchhandlung, Prag 1939. Reprint: Meiner, Hamburg 1972. English translation: *Experience and Judgement*, Northwestern University Press, Evanston 1973.
E. Husserl, Die Frage nach dem Ursprung der Geometrie als intentional-historisches Problem, E. Fink (ed.), *Revue internationale de Philosophie* I (1939), pp. 203–225. English translation: The Origin of Geometry, in: *The Crisis of European Sciences and Transcendental Phenomenology*, Evanston, Northwestern University Press 1970
E. Husserl, *Cartesianische Meditationen und Pariser Vorträge*, S. Strasser (ed.), HGW I, 1950
E. Husserl, Persönliche Aufzeichnungen. Philos. Phenomenol. Res. **XVI**, 293–302 (1956)
E. Husserl, *Erste Philosophie. Erster Teil: Kritische Ideengeschichte* (1923–24), R. Boehm (ed.), HGW VII, 1956
E. Husserl, *Erste Philosophie. Zweiter Teil: Theorie der phänomenologischen Reduktion* (1923–24), R. Boehm (ed.), HGW VII, 1956.
E. Husserl, 1966. *Analysen zur passiven Synthesis (1918–1926),* M. Fleischer (ed.), HGW XI, 1966. English translation, *Analyses Concerning Passive and Active Synthesis: Lectures on Transcendental Logic*, translated by A. Steinbock, HCW IX, 2001
E. Husserl, *The Paris Lectures,* translated by P. Koestenbaum (Martinus Nijhoff, The Hague, 1967)
E. Husserl, *Phänomenologische Psychologie*, W. Biemel (ed.), HGW IX, 1968
E. Husserl, *Phenomenologische Psychologie. Vorlesungen Sommersemester 1925*, HGW IX, 1968
E. Husserl, *Ideen zu einer reinen Phänomenologie und phänomenologischen Philosophie, Drittes Buch*, M. Biemel (ed.), HGW V, 1971. English translation: *Ideas Pertaining to a Pure Phenomenology and to a Phenomenological Philosophy. Third Book*: *Phenomenology and the Foundations of the Sciences*, translated by T. Klein & W. Pohl, HCW I, 1980
E. Husserl, Kant and Idea of Transcendental Philosophy", *Southwestern Journal of Philosophy,* 9 (1974), pp. 9–56, translated from Husserl 1956 by T. Klein and W. Pohl
E. Husserl, *Aufsätze und Rezensionen (1890–1910)*, B. Rang (ed.), HGW XXII, 1979.
E. Husserl, *Shorter Works*, ed. by P. McCormick, F. Elliston (University of Notre Dame Press, Notre Dame, 1981)
M. Husserl, Skizze eines Lebensbildes von E. Husserl. Husserl Stud. **5**, 105–125 (1988)
E. Husserl, *On the Phenomenology of the Consciousness of Internal Time*, translated by J. Brough, (Kluwer, 1991)
E. Husserl, The London Lectures (Syllabus of a Course of Four Lectures) (1922), in *Husserl: Shorter Works*, ed. by P. McCormick, F. Elliston (University of Notre Dame Press, Notre Dame, 1981), pp. 68–74
E. Husserl, *Logische Untersuchungen. Ergänzungsband. Erster Teil. Entwürfe zur Umarbeitung der VI. Untersuchung und zur Vorrede für die Neuauflage der Logischen Untersuchungen* (Sommer 1913), *Tijdschrift voor Philosophie* 1 (1939), 106–133. In U. Melle (ed.), HGW XX/1, 2002. English translation: *Introduction to the Logical Investigations. A Draft of a Preface to the Logical Investigations* (1913), E. Fink (ed.), Tijdschrift voor Philosophie 1939. English translation with Introductions by P. J. Bossert and C. H. Peters, M. Nijhoff, The Hague, 1975)
E. Husserl, in *[LV'06–07] Einleitung in die Logik und Erkenntnistheorie Vorlesungen 1906/07*, ed. by U. Melle, HGW XXIV, 1985. English translation: *Introduction to Logic and Theory of Knowledge* (Lectures 1906/1907), ed. by C.O. Hill (Springer, Dordrecht, 2008)
E. Husserl, *Aufsätze und Vorträge 1911–21*, Th. Nenon and H. R. Sepp (ed.), HGW XXV, 1986
E. Husserl, *Aufsätze und Vorträge 1922–37*, Th. Nenon and H. R. Sepp (ed.), HGW XXVII, 1988
E. Husserl, *Early Writings in the Philosophy of Logic and Mathematics*, HCW V, 1994
E. Husserl, *Logik und Allgemeine Wissenschaftstheorie. Vorlesungen 1917/18*, mit ergänzenden Texten aus der ersten Fassung von 1910/11, U. Panzer (ed.), HGW XXX, 1996
E. Husserl, in *Ding und Raum. Vorlesungen 1907*, ed. by U. Glaesges, HGW XVI, 1973. English translation: *Thing and Space: Lectures of 1907*, HCW VII, 1997
E. Husserl, *Logik Vorlesung 1896*, in ed. by E. Schuhmann, HM I, 2001
E. Husserl, Aus der Einleitung der Vorlesung Erkenntnistheorie und Hauptpunkte der Metaphysik 1898/99, in E. Schuhmann (ed.), HM III, 2001

E. Husserl, in *Logik Vorlesung 1902/03*, ed. by E. Schuhmann, HM II, 2001

E. Husserl, *Natur und Geist. Vorlesungen Sommersemeseter 1927*, M. Weiler (ed.), HGW XXXII, 2001

E. Husserl, *Die Bernauer Manuskripte üuber das Zeitbewusstsein*, R. Bernet and D. Lohmar (eds.), HGW XXXIII, 2001.

E. Husserl, *Einleitung in die Philosophie. Vorlesungen 1922/23*, HGW XXXV, 2002

E. Husserl, *Zur phänomenologischen Reduktion: Texte aus dem Nachlass (1926–1935)*, S. Luft (ed.), HGW XXXIV, 2002

E. Husserl, *Logische Untersuchungen* (Ergänzungsband: Erster Teil), U. Melle (ed.), HGW XX/1, 2002

E. Husserl, *Transzendentaler Idealismus. Texte aus dem Nachlass (1908–1921)*, HGW XXXVI, 2003

E. Husserl, *The Basic Problems of Phenomenology: From the Lectures, Winter Semester 1910–1911*, translated by I. Farin and J. Hart, Springer, Berlin, 2006

E. Husserl, *Einfuhrung in die Phenomenologie der Erkenntnis. Vorlesung 1909*, E. Schuhmann (ed.), HM VII, 2007

E. Husserl, *Untersuchungen zur Urteilstheorie: Texte aus dem Nachlass (1893–1918)*, R. Rollinger (ed.), HGW XL, (2009)

E. Husserl, *Zur Lehre vom Wesen und zur Methode der eidetischen Variation. Texte aus dem Nachlass (1891–1935)*, HGW XLI, 2012

E. Husserl, *Zur Lehre vom Wesen und zur Methode der eidetischen Variation*, D. Fonfara (ed.), HGW XLI, 2012

E. Husserl, *Ideas for a Pure Phenomenology and Phenomenological Philosophy*, translated by D. Dahlstrom/Hackett, Indianapolis/Cambridge, 2014

E. Husserl, *Ms A 1 35*, Manuscript on set theory available in the Husserl Archives in Leuven, Cologne and Paris

R. Iemhoff, Intuitionism in the philosophy of mathematics, in Zalta 1997–, Winter 2009., http://plato.stanford.edu/archives/win2009/entries/intuitionism/

C. Ierna, The Beginnings of Husserl's Philosophy (Part 1: From *Über den Begriff der Zahl* to *Philosophie der Arithmetik*). New Yearb. Phenomenol. Phenomenol. Philos. **5**, 1–56 (2005)

C. Ierna, The Beginnings of Husserl's Philosophy (Part 2: Philosophical and Mathematical Background). New Yearb. Phenomenol. Phenomenol. Philoso. **6**, 23–71 (2006)

C. Ierna, Husserl et Stumpf sur la Gestalt et la fusion. Philosophiques **36**(2), 489–510 (2009)

C. Ierna, Brentano and Mathematics. Revue Roum. Philosophie **55**(1), 149–167 (2011a)

C. Ierna, Der Durchgang durch das Unmögliche. An Unpublished Manuscript from the Husserl-Archives. Husserl -Stud. **27**(3), 217–226 (2011b)

C. Ierna, La notion husserlienne de multiplicité : au-delà de Cantor et Riemann, *Methodos* 12 (2012)

C. Ierna, Husserl's Psychology of Arithmetic. Bull. d'Analyse Phénoménologique **8**, 97–120 (2012)

C. Ierna, La notion husserlienne de multiplicité: au-delà de Cantor et Riemann, Methodos, Savoirs et Textes 12, n. 23 (2012), http://methodos.revues.org/2943?lang=en#bibliography

C. Ierna, Husserl and the Infinite', Studia Phaenomenologica III No. 1–2 (2003), 179-194C. Ierna, Carl Stumpf's Philosophy of Mathematics, in *Philosophy from an Empirical Standpoint: Essays on Carl Stumpf*, ed. by D. Fisette, R. Martinelli (Rodopi, Amsterdam, 2015)

C. Ierna, On Christian von Ehrenfels' Dissertation, Meinong Studies 201*, forthcoming

H. Jervell, Constructing ordinals, in *Cahier spécial 6: Constructivism: Mathematics, Logic, Philosophy and Linguistics*", G. Heinzmann and G Ronzitti (eds.), *Philosophia Scientiae* (2006), pp. 5–20

P. Jordan, Early Years of Quantum Mechanics: Some Reminiscences, in*The Physicist's Conception of Nature*, ed. by J. Mehra (Reidel, Boston, 1973)

I. Kant, *Universal Natural History and Theory of the Heavens*, translated with an introduction and notes by S. L. Jaki (Scottish Academic Press, Edinburgh, 1981)

I. Kant, *Kant's Gesammelte Schriften*, edited by the Koniglich Preussischen Akademie der *Wissenschaften*, subsequently the Deutsche Akademie der Wissenschaften, 29 volumes, Reimer, Berlin 1900–1919, De Gruyter, Berlin 1920–

I. Kant, *The Cambridge Edition of the Works of Immanuel Kant in English Translation,* P. Guyer and A. Wood, General Co-editors, 13 volumes (Cambridge University Press, 1992–)

I. Kant, *Theoretical Philosophy 1755–1770*, translated and ed. by D. Walford and R. Meerbote (Cambridge University Press, Cambridge 1992a)

I. Kant, *Lectures on Logic*, translated and ed. by J. Young (Cambridge University Press, Cambridge 1992b)

I. Kant, *Opus Postumum*, translated by E. Förster and M. Rosen (Cambridge University Press, 1993)

I. Kant, *Critique of Pure Reason*, translated and ed. by P. Guyer and A. Wood (Cambridge University Press, Cambridge, 1997)

I. Kant, *Correspondence,* translated by A. Zweig (Cambridge University Press, Cambridge, 1999)

I. Kant, *Critique of Aesthetic Judgment,* translated by P. Guyer and E. Matthews (Cambridge University Press, Cambridge, 2000)

I. Kant, in *Theoretical Philosophy after 1781,* ed. by H. Allison, P. Health, translated by G. Hatfield, M. Friedman, H. Allison, and P. Health (Cambridge University Press, Cambridge, 2002)

I. Kant, *Notes and Fragments*, translated by P. Guyer, C. Bowman, F. Rauscher (Cambridge University Press, Cambridge, 2005)

I. Kant, *Anthropology from a Pragmatic Point of View*, translated and ed. by R. Louden with an introduction by M. Kuehn (Cambridge University Press, Cambridge, 2006)

R. Kauppi, Begriffsinhalt/Begriffsumfang, in *Historisches Wöorterbuch der Philosophie*, vol. 1, ed. by J. Ritter et al. (Schwabe, Basel, 1971), pp. 808–809

I. Kern, *Husserl und Kant. Eine Untersuchung über Husserls verhältnis zu Kant und zum Neukantianismus* (Martinus Nijoff, Den Haag, 1964)

I. Kern, The Three Ways to the Transcendental Phenomenological Reduction in the Philosophy of Edmund Husserl, in *Husserl. Expositions and Appraisals*, ed. by F. Elliston, P. McCormick (University of Notre Dame Press, Notre Dame, 1977)

S. Kierkegaard, *Fear and Trembling* (1843), S. Evans and S. Walsh (eds.), Cambridge University Press, Cambridge, 2006

T. Kisiel, *The Genesis of Heidegger's Being and Time'* (University of California Press, Berkeley, 1993)

E. Klein (Wife of J. Klein), 14-page transcript of a tape recording [the original tape recording is apparently lost] among J. Klein's papers, which are housed in St. John's College Library, Annapolis

J. Klein, Die griechische Logistik und die Entstehung der Algebra, *Quellen und Studien zur Geschichte der Mathematik, Astronomie und Physik,* Abteilung B: Studien 3, no. 1 (1934), pp. 18–105 (Part I), and no. 2 (1936), pp. 122–235 (Part II). English translation: *Greek Mathematical Thought and the Origin of Algebra*, trans. by E. Brann, MIT Press, Cambridge 1969, reprint: Dover, New York 1992

J. Klein, Phenomenology and the history of science, in M. Farber (ed.), *Philosophical Essays in Memory of Edmund Husserl* (Harvard University Press, Cambridge, 1940) 143–163, reprinted in: R. Williamson, E. Zuckerman (eds.), *Jacob Klein, Lectures and Essays* (St. John's Press, Annapolis, 1985) 65–84

A. Klev, Form of apprehension and the content-apprehension model in Husserl's Logical Investigations. Log. Anal. Hist. Philos. **16**, 49–69 (2013)

A. Klev, *Categories and Logical Syntax*, PhD thesis, Leiden University (2014)

R. Klibansky (ed.), *Logic and Foundations of Mathematics.* Vol. 1 of *Contemporary Philosophy. A Survey* (La Nuova Italia editrice, Firenze, 1968)

M. Kline, *Mathematical Thought*, vol. 1 (Oxford University Press, New York, Oxford, 1972) M

L. Königsberger, *Vorlesungen über die Theorie der elliptischen Functionen nebst einer Einleitung in die allgemeine Functionenlehre*, vol 1 (Teubner, Leipzig, 1874)

A. Koyré, Galilée et l'expérience de Pise. Annales de l'Université de Paris **12**, 441–453 (1937a)
A. Koyré, Galilée et Descartes. Travaux du IX Congrés international de Philosophie **2**, 41–47 (1937b)
A. Koyré, *Introduction à la lecture de Platon* (Gallimard, Paris, 1945)
O. Kraus, *Franz Brentano. Zur Kenntnis seines Lebens und seiner Lehre* (Beck, München, 1919) (includes: E. Husserl, *Erinnerungen an Franz Brentano*)
G. Kreisel, Mathematical logic, in Saaty (1965), pp. 95–195
L. Kronecker, Über den Zahlbegriff. Crelle's Journal für die reine und angewandte Mathematik **101**, 337–355 (1887)
G. Küng, Das Noema als reelles Moment, in Bossert (1975), pp. 151–153
W. Künne, Indexikalität, Sinn und propositionaler Gehalt. Grazer Philosophische Studien **18**, 41–74 (1982)
W. Künne, *Analyticity and logical truth. From Bolzano to Quine, in Textor* (2006), pp. 184–249
W. Künne, *Abstrakte Gegenstande. Semantik und Ontologie* (Klostermann, Frankfurt, 2007)
W. Künne, *Die Philosophische Logik Gottlob Freges. Ein Kommentar* (Klostermann, Frankfurt, 2010)
S. Lapointe, *Bolzano e Husserl*. Discipline filosofiche. **XXI**(2), 228–244 (2011)
J. Lambert, *Neues Organon oder Gedanken über die Erforschung und Bezeichnung des Wahren und des Unterscheidung vom Irrthum und Schein* (Leipzig, 1764)
J. Lambert, Memoire sur quelque propiétés remarquables des quantités transcendentes circularies et logarithmeques. Akademie de Berlin **17**, 265–322 (1768)
J. Lambert, Solution general et absolut du problem de Trois Corps moyen des suites infinites. Memoires de Academie **22**, 353–364 (1769)
J. Lambert, Vorlaufige Kenntnisse fur die, so die Quadratur und Rectification des Circuls suchen, Beitrage zum Gebrauche der Mathematik und der Anwendung, Bd. 2 (1770), 140–169
J. Lambert, *Anlage zur Architectonic,oder Theorie des Einfachen und des Ersten in der philosophischen und mathematischen Erkenntniss,* Riga, 1771
J. Lambert, Theorie der Parallelinien, *Die Theorie der Parallellinien van Euklid bis auf Gauss*, edited by P. Stäckel and F. Engel, Tuebner, Leipzig, 1895
G.W. Leibniz, *Dissertatio de Arte Combinatoria* (1666), in *Sämtliche Schriften und Briefe* VI.1.A, pp. 163–230
G.W. Leibniz, *Confessio Philosophi* (1672/73), in *Sämtliche Schriften und Briefe* VI.3, pp. 115–149
G.W. Leibniz, *Quid sit Idea?* (1677), in *Sämtliche Schriften und Briefe* VI.4.B, pp. 1369–1371
G.W. Leibniz, *Dialogus* (1677), in *Sämtliche Schriften und Briefe* IV.4.A, pp. 20–25
G.W. Leibniz, *La vraie méthode* (1677), in *Sämtliche Schriften und Briefe* VI, 4/A, pp. 3–7
G.W. Leibniz, *Quid sit Idea* (1677), in *Sämtliche Schriften und Briefe* VI.4.B, pp. 1369–1371
G.W. Leibniz, *Lingua generalis* (1678), in *Sämtliche Schriften und Briefe* VI, 4/A, pp. 65–67
G.W. Leibniz, *Lingua universalis* (1678), in *Sämtliche Schriften und Briefe* VI,4/A, pp. 67–68
G.W. Leibniz, *Leibniz an die Pfalzgräfin Elizabeth* (?) (1678), in *Sämtliche Schriften und Briefe* II.1.A, 659–666. English translation: *Letter to Countess Elizabeth. On God and Formal Logic*, R. Ariew & D. Garber (ed.), Hackett Publ. Co., Indianapolis & Cambridge 1989
G.W. Leibniz, *Elementa Nova Matheseos Universalis* (1683?), in *Sämtliche Schriften und Briefe* IV.4.A, pp. 513–524
G.W. Leibniz, *Meditationes de cognitione, veritate et ideis* (1684), in *Sämtliche Schriften und Briefe* IV.4.A, pp. 585–592
G.W. Leibniz, in *Nova Methodus* (1684), ed. by D.J. Struik, *A Source Book in Mathematics, 1200–1800* (Princeton University Press, Princeton, 1969)
G.W. Leibniz, [1686] in *Generales inquisitiones de analysi notionum et veritatum,* ed. by F. Schupp (Meiner, Hamburg, 1982)
G.W. Leibniz, *Definitiones: ens, possibile, existens* (1687–96?), in *Sämtliche Schriften und Briefe* VI.4.A, pp. 867–870

G.W. Leibniz, *Non inelegans specimen demonstrandi in abstractis* (1687), in *Sämtliche Schriften und Briefe* IV.4.A, pp. 845–855

G.W. Leibniz, *De Characteribus et de Arte Characteristica* (1688?), in *Sämtliche Schriften und Briefe* IV.4.A, 916

G.W. Leibniz, *Definitiones: terminus vel aliquid, nihil* (1688–89), in *Sämtliche Schriften und Briefe* VI.4.A, pp. 934–936

G.W. Leibniz, *Nouveaux Essais sur l'entendement humain* (1704), in *Sämtliche Schriften und Briefe* IV.6. English translation: *New Essays on Human Understanding*, translated by P. Remnant and J. Bennett (Cambridge University Press, Cambridge, 1981)

G.W. Leibniz, *Sämtliche Schriften und Briefe* (Akademie-Ausgabe) (Akademie Verlag, Leipzig/Darmstadt/Berlin, 1923+)

G.W. Leibniz, in *Logical Papers*, ed. by G.H.R. Parkinson (Clarendon Press, Oxford, 1966)

G.W. Leibniz, in *Die Grundlagen des logischen Kalkuls*, ed. by F. Schupp, S. Weber (Meiner, Hamburg, 2000)

O. Linnebo, The context principle in Frege's *Grundgesetze*, in *A Companion to Frege's Grundgesetze*, ed. by P. Ebert, M. Rossberg (Oxford University Press, Oxford, forthcoming)

H. Lipps, *Die Verbindlichkeit der Sprache* (Vittorio Klostermann, Frankfurt am Main, 1976)

X. Liu, Gödel's Philosophical Program and Husserl's Phenomenology. Synthese **175**, 33–45 (2010)

D. Lohmar, *Phänomenologie der Mathematik. Elemente einer phänomenologischen Aufklärung der mathematischen Erkenntnis nach Husserl* (Kluwer, Dordrecht, 1989)

D. Lohmar, On the relation of mathematical objects to time: Are mathematical objects timeless, overtemporal or omnitemporal? J. Indian Counc. Philos. Res. **10**, 73–87 (1993)

D. Lohmar, *Edmund Husserl's 'Formal and Transzendental Logik'* (Darmstadt, 2000)

D. Lohmar, The transition of the principle of excluded middle from a principle of logic to an axiom: Husserl's hesitant revisionism in the field of logic. New Yearb. Phenomenol. Phenomenol. Philos. **4**, 53–68 (2004)

D. Lohmar, Husserl's Concept of Categorial Intuition, in D. Zahavi and F. Stjernfelt (eds.), *One Hundred Years of Phenomenology*, (2002), pp. 125–145

D. Lohmar, Kategoriale Anschauung (VI. Logische Untersuchung, §§ 40–66), in *Edmumd Husserl. Logische Untersuchungen, Klassiker Auslegen*, ed. by V Mayer (Akademie Verlag, Berlin, 2008), pp. 209–237

D. Lohmar, Intuition in Mathematics: On the Function of Eidetic Variation in Mathematical Proofs, in M. Hartimo (ed.), *Phenomenology and Mathematics*, Springer (2010)

D. Lohmar, Ego and Arch-Ego in Husserlian Phenomenology, in *Life, Subjectively, and Art: Essays in Honor of Rudolph Bernet* (Springer, 2012), pp. 277–302.

R.H. Lotze, in *Logic, in three Books: of Thought, of Investigation, and of Knowledge* (1874), ed. by B. Bosanquet (Clarendon Press, Oxford, 1884, 2nd edn., 1887)

E. Mach, *Beiträge zur Analyse der Empfindungen* (Gustav Fischer, Jena, 1886)

E. Mach, *Popular Scientific Lectures*, translated by T. McCormack (Open Curt, Chicago, 1895)

E. Mach, in *Space and Geometry in the Light of Physiological, Psychological, and Physical Inquiry,* translated by T. McCormak (Open Court, LaSalle, 1960)

D. Mahnke, Leibniz auf der Suche nach einer allgemeinen Primzahlgleichung, *Biblioteca Mathematica*, 13 (1912/1913), 3 Folge, pp. 29–61

D. Mahnke, *Eine neue Monadologie*. Kantstudien Ergänzungsheft 39 (Reuther and Reichard, Berlin, 1917)

D. Mahnke, From Hilbert to Husserl: First Introduction to Phenomenology, especially that of Formal Mathematics (1923), Stud. Hist. Philos. Sci. **8**, 71–84 (1977)

D. Mahnke, Leibnizens Synthese von Universalmathematik und Individualmetaphysik. Erster Teil. Jahrbuch für Philosophie und phänomenologische Forschung **VII**, 304–611 (1925)

U. Majer, Husserl and Hilbert on Completeness: A Neglected Chapter in Early Twentieth century Foundations of Mathematics. Synthese **110**, 37–56 (1997)

P. Mancosu, *Philosophy of Mathematics and Mathematical Practice in the Seventeenth Century* (Oxford University Press, Oxford, 1996)

P. Mancosu, *From Brouwer to Hilbert* (Oxford University Press, Oxford, 1998)

P. Mancosu (ed.), *The Debate on the Foundations of Mathematics in the 1920s*. In *From Brouwer to Hilbert* (Oxford University Press, Oxford, 1998)

P. Mancosu, T. Ryckman, Mathematics and phenomenology: The correspondence between O. Becker and H. Weyl. Philos. Math. **10**, 130–202 (2002)

P. Mancosu, Das Abenteuer der Vernunft: O. Becker and D. Mahnke on the phenomenological foundations of the exact sciences, in Peckhaus 2005, pp. 229–243

P. Mancosu, *The Adventure of Reason. Interplay between Philosophy of Mathematics and Mathematical Logic 1900–1940* (Oxford University Press, Oxford, 2010)

S. Marcucci, Kant, Gustav von Stark, und die Pythagorischen Ternen, in *Vernunftkritik und Aufklarung*, 2001, ed. by M. Oberhausen, pp. 251–263

G. Martin, *Arithmetic and Combinatorics, Kant and his Contemporaries, translated by J* (Southern Illinois University Press, Wubnig, 1985)

V. Mayer, Die Konstruktion der Erfahrungswelt: Carnap und Husserl. Erkenntnis **35**, 287–303 (1991)

R. McIntyre, Husserl and Frege. J. Philos. **84**, 528–535 (1987)

I. Mehra, H. Rechenberg, *The Historical Development of Quantum Theory*, vol V (Springer, New York, 1987)

K. Menger, in *Reminiscences of the Vienna Circle and the Mathematical Colloquium*, ed. by L. Golland et al. (Kluwer, Dordrecht, 1994)

J. Mensch, After Modernity: Husserlian Reflections on a Philosophical Tradition, Chap. Intersubjectivity and the constitution of time, pp. 57–66 (State University of New York Press, Albany, 1996)

J.S. Mill, *A System of Logic. Ratiocinative and Inductive*, 2nd edn. (Harper & Brothers, New York, 1846)

J.P. Miller, *Numbers in Presence and Absence. A Study of Husserl's Philosophy of Mathematics* (Martinus Nijhoff, The Hague, 1982)

J.N. Mohanty, *Husserl & Frege* (Indiana University Press, Bloomington, 1982)

D. Moran, *Introduction to Phenomenology* (Routledge, London, 2000)

D. Moran, *Husserl's Crisis of the European Sciences and Transcendental Philosophy: An Introduction* (Cambridge University Press, Cambridge, 2012)

B. Morrison, Les Cátegories d'Aristote comme introduction a la logique, in O Brun, L. Corti, (eds.), *Les Cátegories et leur histoire*, Paris, Vrin, 2005, 103–120

J. Moschovakis, Intuitionistic logic. In Zalta 1997–, Fall 2008., http://plato.stanford.edu/archives/fall2008/entries/logic-intuitionistic/

E. Nagel, Review of *Philosophical Essays in Memory of Edmund Husserl*, ed. M. Farber. J. Philos **38**, 301–306 (1941)

P.J. Nahin, *An Imaginary Tale. The History of* $\sqrt{-1}$ (Princeton University Press, Princeton, 1998)

G. T. Null, Husserl's doctrine of essence, in J. N. Mohanty and W. R. McKenna (eds.), *Husserl's Phenomenology: A textbook* (University Press of America, Washington D.C. 1989), pp. 69–105

G.T. Null, Entities without identities vs. temporal modalities of choice: Review of Mark van Atten, *Brouwer Meets Husserl*. Husserl Stud. **24**, 119–130 (2008)

M. Okada, Husserl and Hilbert on Completeness and Husserl's Term Rewrite-based Theory of Multiplicity, in 24th International Conference on Rewriting Techniques and Applications (RTA'13), ed. by F. van Raamsdok (LIPIcs, Eindhoven, 2013), pp. 4–19

E. Orth, Can 'Phenomenology' in Kant and Lambert be connected with Husserlian Phenomenology, *Kant and Phenomenology*, ed. by T. Seebohm, J. Kockelmans (Center for Advanced Research in Phenomenology and University Press of America, Washington, DC 1984)

V. Pambuccian, Review of Hilbert 2004. Philos. Math. **21**, 255–277 (2013)

R. Parker, The History Between Koyré and Husserl, forthcoming in: Pisano, Agassi and Drozdova (ed.), *Hypotheses and Perspectives within History and Philosophy of Science. Homage to Alexandre Koyré 1964–2014* (Springer, Berlin, forthcoming)

C. Parsons, Mathematical Intuition. Proc. Aristot. Soc. **80**, 145–169 (1980)

C. Parsons, What is the iterative conception of set?, in Parsons (1983), pp. 268–297. Originally in Butts and Hintikka 1977, pp. 335–367
C. Parsons, *Mathematics in Philosophy: Selected Essays* (Cornell University Press, Ithaca, 1983)
C. Parsons, Platonism and mathematical intuition in Kurt Gödel's thought. Bull. Symb. Log. **1**, 44–74 (1995)
C. Parsons, Finitism and intuitive knowledge, in Schirn (1998), pp. 249–270
J. Patočka, The Philosophy of Arithmetic, in *An Introduction to Husserl's Phenomenology*, ed. by J. Dodd, trans. by E. Kohák (Open Court, Chicago, 1996)
G. Peano, *Arithmetices Principia, Nova Methodo Exposita* (Bona, Torino, 1889). Partial English translation in Van Heijenoort 1967
N.R. Pearcey, C.B. Thaxton, *The Soul of Science* (Crossway Books, Wheathon, 1994)
V. Peckhaus, *Hilbertprogramm und Kritische Philosophie. Das Göttinger Modell interdisziplinärer Zusammenarbeit zwischen Mathematik und Philosophie* (Vandenhoeck & Ruprecht, Göttingen, 1990)
V. Peckhaus, Wozu Algebra der Logik? Ernst Schröders Suche nach einer universalen Theorie der Verknüpfungen. Mod. Log. **4**, 357–381 (1994)
V. Peckhaus (ed.), *Oskar Becker und die Philosophie der Mathematik* (Fink, München, 2005)
V. Peckhaus, R. Kahle, Hilbert's Paradox, Report No. 38 (2000/2001), Institut Mittag-Leffler, The Royal Swedish Academy of Sciences, https://www.mittag-leffler.se/preprints/files/IML-0001-38.pdf
C.S. Peirce, On the algebra of logic. Am. J. Math. **3**, 15–57 (1880)
C. Peirce, in *Collected Papers of Charies Sanders Peirce,* vol. 7, ed. by A. W. Burks (The Belknap Press of Harvard University, Cambridge, MA, 1966)
J. Perry, A problem about continued belief. Pac. Philos. Q. **61**(4), 317–322 (1980)
A. Pfänder, in *Philosophie auf phänomenologischer Grundlage: Einleitung in die Philosophie und Phänomenologie*, ed. H. Spiegelberg (Fink, München, 1973)
A. Piccolomini, *Commentarius de certitudine mathematicarum* (Antonio Baldum Asulanum, Roma, 1547)
T. Placek, *Mathematical Intuitionism and Intersubjectivity: A Critical Exposition of Arguments for Intuitionism* (Kluwer, Dordrecht, 1999)
O. Poggeler, Hermeneutische und mantische Phenomenologie, in *Heidegger. Perspektiven zur Deutung seines Werks*, ed. by O. Poggeler (Kiepenheuer & Witsch, Köln, Berlin, 1970)
H. Poincare, *Mathematics and Science. Last Essays* (Dover, New York, 1963)
G. Primiero, S. Rahman (eds.), *Judgement and Knowledge: Papers in honour of B.G. Sundholm* (College Publications, London, 2009)
L. Pyenson, Relativity in Late Wilhelmian Germany: The Appeal to a Pre-established Harmony between mathematics and physics. Arch. Hist. Exact Sci. **27**, 137–155 (1982)
W.V.O. Quine, *Word and Object* (MIT Press, Cambridge, MA, 1960)
B. Rang, W. Thomas, Zermelo's Discovery of Russell's Paradox, *Historia Mathematica***8**, 16–22 B (1981)
E.H. Reck, From Frege and Russell to Carnap: Logic and logicism in the 1920's, in *Carnap Brought Home: The View from Jena*, ed. by S. Awodey, C. Klein (Open Court, Chicago, 2004), pp. 151–180
C. Reid, *Hilbert* (Springer, Berlin, 1970)
J. Richard, *Sur la Philosophie des Mathematiques* (Gauthier-Villars, Paris, 1903)
B. Riemann, *Über die Hypothesen, welche der Geometrie zugrunde liegen, 1867*. 3rd edn. Berlin 1923; reprint (Chelsea, New York, 1973)
R. Rollinger, *Husserl's Position in the School of Brentano*, Phaenomenologica 150 (Kluwer, Dordrecht, 1999)
R. Rollinger, Brentano's Logic and Marty's Early Philosophy of Language. Brentano Studien **XII**, 77–98 (2009)
G.E. Rosado Haddock, Husserl's epistemology of mathematics and the foundations of Platonism in mathematics. Husserl Stud. **4**, 81–102 (1987)

G. Rosado Haddock, Review of *Mathematical Intuition*, by Richard Tieszen. J. Symb. Log. **56**, 356–360 (1991)

G.E. Rosado Haddock, Husserl's Relevance for the Philosophy and Foundations of Mathematics. Axiomathes **1–3**, 125–142 (1997)

G.E. Rosado Haddock, Husserl's Philosophy of Mathematics: Its Origin and Relevance, Husserl Stud. **22**, (2006). Reprinted as Chapter 5 of *Against the Current*, pp. 145–181

G.E. Rosado Haddock, *The Young Carnap's Unknown Master* (Ashgate, Aldershot, 2008)

G. Rosado-Haddock, Platonism, phenomenology, and interderivability, in *Phenomenology and Mathematics*, ed. by M. Hortimo (Springer, Dordrecht, 2010)

G.E. Rosado Haddock, Husserl's Conception of Physical Theories and Physical Geometry in the Time of the Prolegomena: a Comparison with Duhem's and Poincaré's Views, *Axiomathes 22*, (2012). Reprinted as Chapter 6 of *Against the Current*, pp. 183–214

G.E. Rosado Haddock, *Against the Current* (Ontos Verlag, Frankfurt, 2012)

G. Rota, The phenomenology of mathematical proof. Synthese **111**, 183–196 (1997)

D.E. Rowe, Klein, Hilbert, and the Göttingen Mathematical Tradition, Osiris, 2nd Series, vol. V, Science in Germany: The Intersection of Institutional and Intellectual Issues (1989), pp. 186–213

J.-M. Roy, Carnap's Husserlian reading of the Aufbau, in *Carnap Brought Home: The View from Jena*, ed. by S. Awodey, C. Klein (Open Court, Chicago, 2004), pp. 41–62

B. Russell, *My Philosophical Development (1959)* (Allen and Unwin, London, 1975)

B. Russell, *Principles of Mathematics* (Norton, New York, 1903)

B. Russell, A.N. Whitehead, *Principia Mathematica*, vol 1 (Cambridge University Press, Cambridge, 1910)

T. Ryckman, Carnap and Husserl, in *The Cambridge Companion to Carnap*, M. Friedman and R. Creath (Cambridge University Press, Cambridge, 2007), pp. 81–105

G. Ryle, Categories. Proc. Aristot. Soc. **38**, 189–206 (1938)

T. Saaty (ed.), *Lectures on Modern Mathematics*, vol 3 (Wiley, New York, 1965)

S. Sarkar, Husserl's role in Carnap's *Der Raum*, in *Language, Truth and Knoweldge. Contributions to the Philosophy of Rudolf Carnap*, ed. by T. Bonk (Kluwer, Dordrecht, 2003), pp. 179–190

P. A. Schilpp (ed.), *The Philosophy of Bertrand Russell* (Nortwestern University Press, Evanston, 1944)

M. Schirn (ed.), *The Philosophy of Mathematics Today* (Oxford University Press, Oxford, 1998)

A. Schirrmacher, Planting in his Neighbor's Garden: David Hilbert and Early Göttingen Quantum Theory. Perspect. Phys. **5**, 5–20 (2003)

R. Schmit, *Husserl's Philosophie der Mathematik* (Bouvier, Bonn, 1981)

E. Scholz, Weyl's Contribution to Geometry, 1917–1923. Intersect. Hist. Math. Hist. Stud. **15**, 203–230 (1994)

E. Schröder, *Lehrbuch der Arithmetik und Algebra für Lehrer und Studirende. Erster Band. Die Sieben algebraischen Operationen* (Teubner, Leipzig, 1873)

E. Schröder, *Vorlesungen über die Algebra der Logik (Exakte Logik)*, vol I (Teubner, Leipzig, 1890)

E. Schröder, *Vorlesungen über die Algebra der Logik (Exakte Logik)*, vol. II, part 1 (Teubner, Leipzig, 1891)

E. Schröder, *Vorlesungen über die Algebra der Logik (Exakte Logik)*, vol. III, part 1: *Algebra und Logik der Relative* (Teubner, Leipzig, 1895)

E. Schröder, *Vorlesungen über die Algebra der Logik (Exakte Logik)*, vol. II, part 2, (Teubner, Leipzig 1905)

E. Schrödinger, Über die Unanwendbarkeit der Geometrie im Kleinen. Naturwissenschaften **22**, 518–520 (1934)

K. Schuhmann, *Reine Phänomenologie und phänomenologische Philosophie* (Martinus Nijhoff, Den Haag, 1973)

K. Schuhmann, *Husserl-Chronik (Denk- und Lebensweg Edmund Husserls)*, HD I, 1977

K. Schuhmann, Malvine Husserls "Skizze eines Lebensbildes von E. Husserl". Husserl-Stud. **5**, 105–125 (1988)

K. Schuhmann, Alexandre Koyré, in *Encyclopedia of Phenomenology*, (Kluwer, Dordrecht, 1997), pp. 391–393
K. Schuhmann, Brentano's impact on twentieth-century philosophy, in *The Cambridge Companion to Brentano*, ed. by D. Jacquette (Cambridge University Press, Cambridge, 2004), pp. 277–297
E. Schuhmann, K. Schuhmann, Husserls Manuskripte zu seinem Göttinger Doppelvortrag von 1901. Husserl Stud. **17**(2), 87–123 (2001)
J. Schultz, *Prüfung der Kantischen Critik der reinen Vernunft. Part I* (Hartung, Königsberg, 1789)
J. Schultz, *Anfangsgründe der reinen Mathesis* (Hartung, Königsberg, 1790)
J. Schultz, *Prüfung der Kantischen Critik der reinen Vernunft. Part II* (Hartung, Königsberg, 1792)
J. Schultz, *Sehrleichte und kurze Entwicklung einiger wichtigsten mathematischen Theorien* (F. Nicolovius, Königsberg, 1803)
F. Schur, Johann Heinrich Lambert als Geometer. Jahresberichte der deutsche Mathematiker Vereinigung **14**, 186–198 (1905)
J. Sebestik, Husserl Reader of Bolzano, in *Husserl's Logical Investigations Reconsidered*, ed. by D. Fisette (Kluwer, Dordrecht, 2003), pp. 59–81
H.R. Sepp (ed.), *Edmund Husserl und die Phänomenologische Bewegung. Zeugnisse in Text und Bild* (Alber, Freiburg/München, 1988 H. R).
J. Shoenfield, Axioms of set theory, in Barwise (1977), pp. 321–344
W. Sieg, Hilbert's Programs: 1917–1922. Bull. Symb. Log. **5**, 1–44 (1999)
W. Sieg, Beyond Hilbert's Reach?, in S. Lindström et al. (eds.), *Logicism, Intuitionism, and Formalism* (Springer, 2009), pp. 449–483
W. Sieg, *Hilbert's Programs and Beyond* (Oxford University Press, Oxford, 2013)
P. Simons, The formalisation of Husserl's theory of wholes and parts, in *Parts and Moments*, ed. by B. Smith (Philosophia Verlag, Munich, 1982), pp. 113–159
P. Simons, *Austrian philosophers on truth*, in M. Textor 2006, pp. 159–83
S. Simpson, Ordinal Numbers and the Hilbert Basis Theorem. J. Symb. Log. **53**, 961–974 (1988)
T. Skolem, Begründung der elementaren Arithmetik durch die rekurrierende Denkweise ohne Anwendung scheinbarer Veränderliche mit unendlichem Ausdehnungsbereich (1923), reprinted in Skolem 1970, 153–188. English translation in Van Heijenoort (1967)
T. Skolem, in *Selected Works in Logic*, ed. by J.E. Fenstad (Universitetsforlaget, Oslo, 1970)
R. Smid, Einleitung des Herausgebers, in *Die Krisis der Europäischen Wissenschaften und die Transzendentale Phänomenologie, Ergänzungsband Texte aus dem Nachlass 1924–1937* (Kluwer, Dordrecht, 1993)
D.W. Smith, R. McIntyre, *Husserl and Intentionality. A Study of Mind, Meaning and Language* (Reidel, Dordrecht, 1982)
B. Smith, D. Smith (eds.), *The Cambridge Companion to Husserl* (Cambridge University Press, Cambridge, 1995)
J. Sommer, Review of Hilbert's *Festschrift*. Bull. Am. Math. Soc. **6**, 289–295 (1900)
D. Spalt, Die mathematischen und philosophischen Grundlagen des Weierstraßschen Zahlbegriffs zwischen Bolzano und Cantor. Arch. Hist. Exact Sci. **41**(4), 311–362 (1991)
A. Steve & E. H. Reck, Completeness and Categoricity: 19th Century Axiomatics to 21st Century Semantics, Technical Report no. CMU-PHIL-118, Carnegie Mellon, Pittsburgh, Pennsylvania (2001), http://www.hss.cmu.edu/philosophy/techreports/118_Awodey.pdf, accessed January 9, 2014
O. Stolz, B. Bolzanos Bedeutung in der Geschichte der Infinitesimalrechnung. Math. Ann. **18**, 255–279 (1881)
O. Stolz, *Vorlesungen über allgemeine Arithmetik* (Teubner, Leipzig, 1885)
A. Stone, *On Husserl and Cavellian Skepticism, With Reference to the Thomistic Theory of Creation*, PhD thesis, Harvard University (2000). Available at http://www.abocalypse.com/papers/skep.pdf
A. Stone, On the sources and implications of Carnap's *Der Raum*. Stud. Hist. Philos. Sci. **41**, 65–74 (2010)

G.F. Stout, Are the characteristics of particular things universal or particular?, Proc. Aristotelian Soc. Supplementary Volumes, **3**, 114–122 (1923)
P.F. Strawson, *On referring*. Mind. New Ser. **59**(235), 320–344 (1950)
C. Stumpf, *Vorlesungen über Psychologie*, manuscript Q 11/I, Halle WS 1886/87 (unpublished lecture notes)
C. Stumpf, *Logik und Enzyklopädie der Philosophie*, manuscript Q 14, Halle SS 1887 (unpublished lecture notes)
C. Stumpf, *Tonpsychologie*, vol 1 (Hirzel, Leipzig, 1883)
C. Stumpf, *Tonpsychologie*, vol 2 (Hirzel, Leipzig, 1890)
C. Stumpf, *Erkenntnislehre* (Barth, Leipzig, 1939)
C. Stumpf, in *Über die Grundsätze der Mathematik*, ed. by W. Ewen (Königshausen & Neumann, Würzburg, 2008)
G. Sundholm, Containment and variation; two strands in the development of analyticity from Aristotle to Martin-Löf, in *Judgement and the Epistemic Foundation of Logic*, ed. by M. van der Schaar (Springer, Heidelberg, 2013)
G. Sundholm, M. van Atten, The proper interpretation of intuitionistic logic: On Brouwer's demonstration of the Bar Theorem, in van Atten, Boldini, Bourdeau and Heinzmann 2008, pp. 60–77
W. Tait, Finitism. J. Philos. **78**, 524–546 (1981)
A. Tarski, *A Decision Method for Elementary Algebra and Geometry* (University of California Press, Berkeley, 1951)
M. Textor, *The Austrian Contribution to Analytic Philosophy* (Routledge, London, 2006)
R. Thiele, Hilbert's twenty-fourth problem. Am. Math. Mon. **110**, 1–24 (2003)
R. Tieszen, *Mathematical Intuition: Phenomenology and Mathematical Knowledge* (Kluwer, Synthese Library series, Dordrecht, 1989)
R. Tieszen, Kurt Gödel and phenomenology. Philos. Sci. **59**(2), 176–194 (1992). Reprinted in Tieszen 2005
R. Tieszen, The Philosophy of Arithmetic: Frege and Husserl, in *Mind, Meaning and Mathematics. Essays on the philosophical views of Husserl and Frege*, ed. by L. Haaparanta (Kluwer, Dordrecht, 1994), pp. 85–112
R. Tieszen, Mathematics, in Smith and Smith 1995, pp. 438–462
R. Tieszen, *Logicism, Impredicativity, Formalism. Some remarks on Poincaré and Husserl* (1996), in Tieszen 2005, pp. 294–313
R. Tieszen, Gödel's Path from the Incompleteness Theorems (1931) to Phenomenology (1961), Bull. Symbolic Log. **4**(2), 181–203 (1998). Reprinted in Tieszen 2005
R. Tieszen, Gödel and the Intuition of Concepts, Synthese **133**(3), 363–391 (2002). Reprinted in Tieszen 2005
R. Tieszen, Husserl's logic, in *Handbook of the History of Logic*, ed. by D. Gabbay, J. Woods, vol. III, (Elsevier, Amsterdam, 2004), pp. 207–321
R. Tieszen, *Phenomenology, Logic and the Philosophy of Mathematics* (Cambridge University Press, Cambridge, NY, 2005)
R. Tieszen, Mathematical Realism and Transcendental Phenomenological Idealism, in *Phenomenology and Mathematics*, ed. by Hartimo, (Springer, Dordrecht, 2010), pp. 1–22
R. Tieszen, *After Gödel: Platonism and Rationalism in Mathematics and Logic* (Oxford University Press, Oxford, 2011)
R. Tieszen, Monads and Mathematics: Gödel and Husserl. Axiomathes **22**, 31–52 (2012)
M. Toepell, Zur Schlüsselrolle Friedrich Schurs bei der Entstehung von David Hilberts *Grundlagen der Geometrie*, in *Mathemata. Festschrift für Helmut Gericke*, ed. by M. Folkerts, U. Lindgren (Franz Steiner, Stuttgart, 1985) pp. 637–649
G. Tonelli, Die Voraussetzungen zur Kantischen Urteilstafel in der Logik des 18. Jahrhunderts, in *Kritik und Metaphysik. Heinz Heimsoeth zum achtzigsten Geburtstag*, ed. by F. Kaulbach, J. Ritter (Walter de Gruyter, Berlin, 1996), pp. 134–158

T. Tonietti, Four letters of E. Husserl to H. Weyl and their context, in *Exact Sciences and Their Philosophical Foundations*, ed. by W. Deppert (1988)

R. Tragesser, *Phenomenology and Logic* (Cornell University Press, Ithaca, 1977)

F.A. Trendelenburg, Über Leibnizens Entwurf einer allgemeinen Charakteristik, in *Philosophische Abhandlungen der Königl* (Akadademie der Wissenschaften zu Berlin, 1856) pp. 37–69

A. Troelstra, *Choice Sequences: A Chapter of Intuitionistic Mathematics* (Oxford University Press, Oxford, 1977)

A. Troelstra, On the origin and development of Brouwer's concept of choice sequence, in Troelstra and van Dalen 1982, pp. 465–486

A. Troelstra, D. van Dalen (eds.), *The L.E.J. Brouwer Centenary Symposium* (North-Holland, Amsterdam, 1982)

A. Troelstra, D. van Dalen, *Constructivism in Mathematics: An Introduction*, vol 2 (North-Holland, Amsterdam, 1988)

M. van Atten, Why Husserl should have been a strong Revisionist in mathematics. Husserl Stud. **18**, 1–18 (2002)

M. van Atten, Two Draft letters from Gödel on self-knowledge of reason. Philos. Math. **14**(2), 255–261 (2006)

M. van Atten, Construction and Constitution in Mathematics. New Yearb. Phenomenol. Phenomenol. Philos. **10**, 43–90 (2010)

M. van Atten, Stefania Centrone, *Logic and Philosophy of Mathematics in the Early Husserl*. New Yearb. Phenomenol. Phenomenol. Philos. **12**, 377–383 (2012)

M. van Atten, Intuitionistic remarks on Husserl's analysis of finite number in the *Philosophy of Arithmetic*. Grad. Fac. Philos. J. **25**, 205–225 (2004)

M. van Atten, Review of *Gnomes in the Fog. The Reception of Brouwer's Intuitionism in the 1920s*, by Dennis Hesseling. Bull. Symbolic Log. **10**, 423–427 (2004)

M. van Atten, *Brouwer Meets Husserl: On the Phenomenology of Choice Sequences* (Springer, Dordrecht, 2007)

M. van Atten, Luitzen Egbertus Jan Brouwer, in Zalta 1997–, Winter (2008), http://plato.stanford.edu/archives/win2008/entries/brouwer

M. van Atten, Monads and sets: On Gödel, Leibniz, and the reflection principle, in Primiero and Rahman (2009), pp. 3–33

M. van Atten, The development of intuitionistic logic, in Zalta 1997–, Summer (2009), http://plato.stanford.edu/archives/sum2009/entries/intuitionistic-logicdevelopment

M. van Atten, Kant and real numbers, in Dybjer, Lindström, Palmgren and Sundholm (2012), pp. 3–23

M. van Atten, *Essays on Gödel's Reception of Leibniz, Husserl, and Brouwer* (Springer, Dordrecht, 2013)

M. van Atten, On the fulfillment of certain categorial intentions. New Yearb. Phenomenol. Phenomenol. Philos. **13**, 173–185 (2015)

M. van Atten, P. Boldini, M. Bourdeau, G. Heinzmann (eds.), *One Hundred Years of Intuitionism (1907–2007): The Cerisy Conference* (Birkhäuser, Basel, 2008)

D. van Dalen, Four letters of Edmund Husserl to Hermann Weyl. Husserl Stud. **1**, 1–12 (1984)

D. van Dalen, Four Letters of Edmund Husserl to Hermann Weyl. Husserl Stud. **1**, 1–12 (1984)

D. van Dalen, *L.E.J. Brouwer en de grondslagen van de wiskunde* (Epsilon, Utrecht, 2001)

D. van Dalen, A bibliography of L.E.J. Brouwer, in van Atten, Boldini, Bourdeau and Heinzmann (2008), pp. 343–390

J. van Heijenoort, *From Frege to Gödel. A Source Book in Mathematical Logic, 1979–1931* (Harvard University Press, Cambridge, MA/London, 1967)

A. Van Romen, *Apologia pro Archimede* (Fleischmann, Würzburg, 1597)

W. van Stigt, *Brouwer's Intuitionism* (North-Holland, Amsterdam, 1990)

F. Vietae, *In Artem Analyticem* (sic) *Isagoge*, Seorsim excussa ab opere restituate Mathematicae Analyseo, seu, Algebra Nova (*Introduction to the Analytical Art*, excerpted as a separate piece

from the *opus* of the restored Mathematical Analysis, or *The New Algebra*, Tours 1591). English translation: *Introduction to the Analytic Art*, trans. by J. W. Smith, appendix to J. Klein 1969/1992
C. von Ehrenfels, Zur Philosophie der Mathematik. Vierteljahrsschrift für wissenschaftliche Philosophie **15**, 285–347 (1891)
H. von Helmholtz, Űber Zählen und Messen, erkenntnistheoretisch betrachtet, in *Philosophische Aufsätze, Eduard Zeller zu seinem fünfzigjährigen Doctorjubiläum gewidmet*, (Fues's Verlag, Leipzig, 1887), pp. 17–52
H. von Helmholtz, *Epistemological Writings* (Reidel, Dordrecht, 1977)
J. von Neumann, Zur Hilbertschen Beweistheorie. Math. Z. **26**, 1–29 (1927)
J. von Plato, In search of the roots of formal computation, in F. Gadducci and M. Tavosanis (eds.), *Proceedings of the 2015 HaPoC Conference*, Springer, to appear.
J. Vuillemin, Sur la demonstration de l'irrationalité de π chez Leibniz, Lambert et Kant, Revue philosophique. **43**, 417–431 (1961)
H. Wang, *From Mathematics to Philosophy* (Humanities Press, New York, 1974)
H. Wang, *Reflections on Kurt Gödel* (Massachusetts Institute of Technology Press, Cambridge, MA, 1987)
H. Wang, *A Logical Journey: From Gödel to Philosophy* (Massachusetts Institute of Technology Press, Cambridge, MA, 1996)
J. Webb, *Mechanism Mentalism and Metamathematics* (Reidel, Dordrecht, 1980)
H. Weyl, Über die Definitionen der mathematischen Grundbegriffe. Mathematisch-naturwissenschaftliche Blätter **7**, 93–95 (1910) 109–113
H. Weyl, On the New Foundational Crisis of Mathematics (1921), in Mancosu 1998, pp. 86–118
H. Weyl, *The Current Epistemological Situation in Mathematics* (1925), in Mancosu 1998b, pp. 123–142
H. Weyl, *Comments on Hilbert's Second Lecture on the Foundations of Mathematics* (1927), in van Heijenoort 1967, pp. 480–484
H. Weyl, Consistency in mathematics. Rice. Inst. Pam. **16**, 245–265 (1929)
H. Weyl, Levels of Infinity (1930), in Weyl 2012, pp. 17–31
H. Weyl, The Open World: Three Lectures on the Metaphysical Implications of Science (1932), in Weyl 2009, pp. 34–82
H. Weyl, Mind and Nature (1934), in Weyl 2009, pp. 83–150
H. Weyl, The Mathematical Way of Thinking (1940), in Weyl 2012, pp. 67–84
H. Weyl, David Hilbert and his mathematical work. Bull. Am. Math. Soc. **50**, 612–654 (1944)
H. Weyl, Address at the Princeton Bicentennial Conference (1946), in Weyl 2009, pp. 162–174
H. Weyl, Mathematics and Logic (1946), in Weyl 2012, pp. 133–146
H. Weyl, Man and the Foundations of Science (ca. 1949), in Weyl 2009, pp. 175–193
H. Weyl, *Raum, Zeit, Materie. Vorlesungen über allgemeine Relativitätstheorie*, Springer, Berlin, 1918. English translation: *Space, Time, Matter*, Dover, New York 1952
H. Weyl, Axiomatic Versus Constructive Procedures in Mathematics (1953), in Weyl 2012, pp. 191–202
H. Weyl, The Unity of Knowledge (1954), in Weyl 2009, pp. 194–2003
H. Weyl, Insight and Reflection (1955a), in Weyl 2009, pp. 204–221
H. Weyl, Why is the World For-Dimensional? (1955b), in Weyl 2012, pp. 203–216
H. Weyl, *Philosophy of Mathematics and Natural Science*, Princeton University Press, Princeton 1949. Revised and augmented English edition based on a translation by Olaf Helmer, Atheneum, New York (1963)
H. Weyl, *The Continuum, A Critical Examination of the Foundation of Analysis* (Dover, New York, 1994), translated by S. Pollard, Th. Bole, originally published as *Das Kontinuum. Kritische Untersuchungen über die Grundlagen der Analysis* (Veit, Leipzig, 1918)
H. Weyl, *Mind and Nature. Selected Writings on Philosophy, Mathematics, and Physics*, ed. by P. Pesic (Princeton University Press, Princeton, 2009)

H. Weyl, in *Levels of Infinity. Selected Writings on Mathematics ad Philosophy*, ed. by P. Pesic (Dover, New York, 2012)

D. Willard, *Logic and the Objectivity of Knowledge: A Study in Husserl's Philosophy* (Ohio University Press, Athens, 1984)

D.C. Williams, On the elements of being. Rev. Metaphys. **7**, 3–18 (1953) 171–192

C. Wilson, *Preface' to Essays in Honor of Jacob Klein* (St John's College Press, Annapolis, 1976)

L. Wittgenstein, *Philosophical Investigations* (Blackwell, Oxford, 2001)

L. Wittgenstein, in *Tractatus logico-philosophicus* (1921), ed. by J. Schulte (Suhrkamp, Frankfurt am Main, 2013)

C. Wolff, in *Mathematisches Lexicon*, ed. by J. E. Hofmann (Georg Olms, Hildesheim, 1965)

J. Yoshimi, Mathematizing phenomenology. Phenomenol. Cogn. Sci. **6**, 271–291 (2007)

E. Zermelo, Collected Works, vol. I, ed. by H. Ebbinghaus, A. Kanomori (Springer, Berlin, 2010)

W. Zage, The geometry of binocular visual space. Math. Mag. **53**, 157–175 (2002)

E. Zalta (ed.), *The Stanford Encyclopedia of Philosophy,* The Metaphysics Research Lab, CSLI, Stanford University (1997)–. http://plato.stanford.edu

K. Zindler, Beiträge zur Theorie der mathematischen Erkenntnis. Sitzungsberichte der Wiener Akademie. *Philos-Histor. Abth.*, Bd. 118 (1889)

Author Index

S. Centrone (ed.), *Essays on Husserl's Logic and Philosophy of Mathematics*, Synthese Library 384, DOI 10.1007/978-94-024-1132-4

Subject Index

S. Centrone (ed.), *Essays on Husserl's Logic and Philosophy of Mathematics*, Synthese Library 384, DOI 10.1007/978-94-024-1132-4

D

L

M

FSC
www.fsc.org
MIX
Papier | Fördert
gute Waldnutzung
FSC® C083411

Kolibri 360 GmbH
Ferdinand-Jühlke-Straße 7
99095 Erfurt, Deutschland
produktsicherheit@kolibri360.de